Beyond the Meme

Minnesota Studies in the Philosophy of Science 22

BEYOND THE MEME

Development and Structure in Cultural Evolution

ALAN C. LOVE AND WILLIAM C. WIMSATT, EDITORS

Minnesota Studies in the Philosophy of Science 22

University of Minnesota Press

Minneapolis

London

The University of Minnesota Press gratefully acknowledges the
financial assistance provided for the publication of this book by
the Winton Chair in the Liberal Arts at the University of Minnesota
and the John Templeton Foundation.

Published by the University of Minnesota Press
111 Third Avenue South, Suite 290
Minneapolis, MN 55401-2520
http://www.upress.umn.edu

Printed in the United States of America on acid-free paper

The University of Minnesota is an equal-opportunity educator and employer.

Library of Congress Cataloging-in-Publication Data
Names: Love, Alan C., and William C. Wimsatt, editors.
Title: Beyond the meme : development and structure in cultural evolution /
    Alan C. Love and William C. Wimsatt, editors.
Description: Minneapolis : University of Minnesota Press, [2019] | Series: Minnesota
    studies in the philosophy of science ; 22 | Includes bibliographical references and index.
Identifiers: LCCN 2018055274 (print) | ISBN 978-1-5179-0689-4 (hc) |
    ISBN 978-1-5179-0690-0 (pb)
Subjects: LCSH: Social evolution. | Cultural diffusion.
Classification: LCC HM626 .B476 2019 (print) | DDC 303.4—dc23
LC record available at https://lccn.loc.gov/2018055274

UMP BmB 2019

CONTENTS

EXPLAINING CULTURAL EVOLUTION
An Interdisciplinary Endeavor

ALAN C. LOVE AND WILLIAM C. WIMSATT

A CONTESTED EXPLANANDUM

Scholars are prone to dispute. However, two words in particular can provoke a full-scale brawl across the humanities and various sciences: *cultural evolution*. For some, the phrase conjures up shadows of an older anthropological tradition in which diverse peoples, groups, and societies were fitted—in procrustean fashion—onto a template of evolutionary progress with Western categories and social organization presumed as the apotheosis of cultural development. For others, it sounds like a route designed to eliminate thick, narrative-focused studies of variation in human groups at different times and places with abstract modeling from the natural sciences, which reduces cultural heterogeneity to a small set of idealized biological factors. Perhaps the most infamous poster child for this latter strategy is the meme, a purportedly basic unit of culture claimed to be analogous to the gene. Problems with meme-based approaches include their lack of structural detail to account for why specific entities might acquire particular memes, how they do so (e.g., acquisition order), and how memetic dynamics in aggregate illuminate the complex architecture of culture and its changes through time (Wimsatt 2010; see also Lewens 2015). However, apart from these particular deficiencies (discussed further below), there are still genuine concerns about the very idea of cultural evolution; it is, at least, a contested explanandum.

   Cultural evolution's status as a contested explanandum has two dimensions. The first is whether it is even a single thing in need of explanation. Worries about whether "culture" is being reified in problematic ways in order to be explained might encourage a different strategy altogether: giving up trying to explain cultural evolution because there is no such process. Here, the concern about colonial impulses to conceptualize culture or its essential traits is palpable. To advance a model of cultural evolution seemingly requires smuggling in a host of assumptions about the nature of culture that we should be suspicious of based on a checkered history of past attempts. A less strident version of this dimension simply emphasizes that there is not sufficient commonality among items frequently referred to under the rubric of culture and therefore little rationale to offer a more unified account of its supposed dynamics. One might try to individually explain the origin and proliferation of gasoline-powered engines, the manual skill involved in crafting stone tools, or specific variations in cooking, dialect, or marriage practices, but nothing is gained, on this complaint, by shoving them underneath a common theoretical blanket labeled "cultural evolution."

The second dimension of cultural evolution's contested explanandum status emerges from a less skeptical posture. Assuming that cultural evolution is something in need of explanation, how should we proceed? What kinds of disciplinary approaches are needed or should be emphasized in offering explanations (Lewens 2015)? Here, what is contested are the criteria of adequacy because the standards for what counts as a genuine explanatory account differ across disciplines. Is there some reason to privilege a perspective that focuses on biological factors rather than social factors? If so, in what contexts, and for what kinds of factors? If not, how do we build integrated models that articulate both biological and social factors? Should we even look to biology for analogical inspiration for modes of cultural change? And if so, is an evolutionary approach to culture mere analogy, or can it provide something more? Are there interpretive issues involved in deciphering cultural formats that outstrip the analytical capacities found in abstract modeling? Does an evolutionary approach to culture necessarily exclude social science disciplines, as many investigators have assumed? To some degree, answers to these questions of methodological or disciplinary appropriateness and priority depend on answers to another question: What *is* culture? Needless to say, opinions differ. However, it is instructive to pause for a moment over the diversity of answers available.

In their magisterial book, *Culture: A Critical Review of Concepts and Definitions* (1952), A. L. Kroeber and Clyde Kluckhohn documented six thematic groupings of definitions of culture: (1) descriptive, (2) historical, (3) normative, (4) psychological, (5) structural, and (6) genetic. Descriptive definitions emphasize the enumeration of specific content types, such as social customs, property systems, and artistic expression, including the norms that govern each of these, which comprise the "whole" or "sum" of culture. Historical definitions focus on social heritage, such as items, forms, or institutions inherited from earlier generations, either through explicit teaching or implicit exemplars. Normative definitions isolate rules or sanctioned ways of living that are typical of different communities and constitute membership identity. Psychological definitions revolve around capacities of problem solving or adjustment to environments of different kinds—how culture is a means to different ends. These capacities can be learned through formal or informal education, be instilled by habit, or be present in common attitudinal orientations. Structural definitions concentrate on predominant patterns of organization in a society that play particular functional roles. Genetic definitions conceptualize culture as a created product or artifact of recurrent human activities, including central ideas, sacred rituals, or ubiquitous symbols. Treated more abstractly, culture (from this vantage point) can be seen as a type of information (for discussion, see Lewens 2015). For each of these definitions, change over time will be conceptualized differently, both in terms of the relevant units (e.g., rules, material artifacts, or problem-solving strategies) and their dynamics, which will range over their origination, diversification, and (sometimes) extinction. At a minimum, talk of the "nature" of culture and its evolution is strained in light of this definitional diversity.

A more recent and focused discussion subdivides senses of culture that have been relevant to evolutionary analyses (Driscoll 2017; cf. Lewens 2015). Driscoll identifies five different groupings: (1) behaviors or artifacts used by individuals that are typically acquired by social learning in a particular community environment or population; (2) behaviors or artifacts used by nonhuman individuals, especially primates, that require trial-and-error learning on the part of an individual in emulation of achieving a particular goal and shed light on homologous capacities or traits that might underlie corresponding or allied human traits; (3) the environmental features that are modified to transform selective forces transgenerationally (e.g., in terms of niche construction); (4) properties of groups (e.g., traditions) that yield differential survival and reproduction of these higher level units over time; and, (5) the origin of social learning mechanisms that yield cumulative effects on particular features in a society, such as a socially learned trait, or the processes from which these mechanisms, and hence capacities for culture, originated. Driscoll then resists two divergent interpretations of these groupings. First, she argues that these are not five separate approaches or inquiries. In fact, standard theoretical perspectives (e.g., dual inheritance or multilevel selection) overlap in addressing different conceptions of culture and deal with various facets of these conceptions in their models and explanations. This is a nuanced reply to the less strident form of the first dimension of cultural evolution's contested status. Although there is not a single endeavor or project in view, there is sufficient commonality among these items to talk of accounting for the dynamics of cultural change: "The current single definitions of culture . . . seem to refer to different levels or parts of cultural evolutionary phenomena" (36).

The second interpretation Driscoll resists is that sufficient analytic fiddling will permit collecting the diversity of cultural phenomena under a single definitional umbrella. Through a survey of literature in which culture is treated as information and phenotypic traits, it is clear that what counts as either of these is variable (e.g., mental representations versus socially learned behaviors). We agree: culture is a complex beast. Although its many facets evolve, moving beyond the truism ("culture changes") requires dealing specifically with the facets of this complexity and their many interconnections. In teasing apart the individual level from the group level or environmental culture from cultural psychology, we have circled back to the insights of Kroeber and Kluckhohn: culture, in virtue of its complexity, admits of many characterizations. Although we have found at least one way to address the less strident form of the skeptical dimension for cultural evolution being a contested explanandum—there are significant, overlapping connections within the complexity of items referred to as "culture"—we still must face the second dimension of cultural evolution's contested status: how should investigation, modeling, and explanation proceed?

FROM CONTESTED EXPLANANDUM TO AN INTERDISCIPLINARY RESEARCH AGENDA

Culture is not unique as a concept in admitting of different characterizations. Philosophers of biology have wrestled with a number of concepts that fit this description: gene, species, individual, and homology (inter alia). Driscoll (2017) explicitly recognizes this relationship: "It seems the absence of a single definition of culture . . . is a feature of other important scientific concepts and exists because of the increasing understanding of cultural evolutionary processes in the cultural evolutionary sciences" (52). For concepts like gene or individual, the complexity of processes and entities involved suggests that seeking a single "correct" concept is methodologically ill-advised. In fact, a variety of relevant and related conceptions can emerge from scientific success (i.e., having learned more about features of the phenomenon), such as in the case of genes (Griffiths and Stotz 2013). The epistemic response of proliferating distinct but allied senses of culture (or genes, or individuals) can be taken as a positive signal of a different methodology. Researchers use divergent characterizations of a concept that represents a complex phenomenon because they have different explanatory aims. These characterizations are then justified by reasons related to those aims. An account of those reasons starts with the identification and rejection of an implicit premise in the search for a single correct concept: that the primary task of a concept is to categorize phenomena or provide a classification of objects or processes that fall under the concept.

Conceptual Roles: Representing Structured Research Agendas

Debates over gene concepts or species concepts can frustrate many scientists. They are often dismissed as "merely semantic." This frustration arises naturally from the implicit premise of a single-concept methodology. If the primary task of a concept is to identify a single, correct definition, then multiple characterizations are a kind of failure. They do not univocally tell you what is in the category and what falls outside of it. However, one way forward is to shift from finding the definition of a concept, where the goal is to formulate criteria for delineating the set of entities a term classifies or categorizes, to characterizing the explanatory agenda associated with a concept (Brigandt and Love 2012), where the goal is to map out a space of explanatory expectations for the study of diverse features of a complex phenomenon. Mapping this problem space promotes the construction of theory and an understanding of the processes of change. Different conceptions of culture (or genes, or species) involve different commitments to what counts as an adequate explanation for the particular features of a complex entity. For example, consider the emergence of interchangeable parts in what came to be characterized as the "American system of manufacture." A full account would need to address: (1) the motivations of the U.S. Bureau of Ordinance interested in the benefits of repairing arms in the field; (2) the development of machine tools allowing for the adjustability that contributes to different functions and high precision in reproducible manufacture; (3) the practice of using sets of templates to give precision in assessments of dimensions; (4) the change in labor practices from the production of individual muskets by craftsmen to the piecemeal production of "lock, stock, and barrel" parts that could be produced and assembled by relatively untrained labor (with consequent labor unrest); and (5) the spread of the method of manufacture as "mechanics," who were expert in the use of the new tools and methods, migrated into other manufacturing industries.

Different disciplinary approaches are relevant for addressing these distinct facets of the complex phenomenon, such as sociological aspects of changes in labor organization, technological aspects of templates and machine tools, and historical aspects of the development and spread of manufacturing practices (see Smith 1977; Hounshell 1984; Wimsatt 2013). These differential commitments structure the investigative and explanatory efforts of researchers and provide criteria for how to address distinct scientific questions associated with the different conceptions. As a consequence, different models and theories with distinct causal factors represented in various fashions become more or less germane to different characterizations. These models and theories derive from a range of disciplinary approaches and therefore not only speak to the fact that interdisciplinary explanations are needed but also inform how interdisciplinary contributions should be coordinated to meet the criteria of explanatory adequacy.

Several corollaries follow from adopting this perspective of concepts as markers of structured explanatory agendas (Brigandt and Love 2012; Love 2014). The first is that there are different sources for the structure manifested in these explanatory agendas. One of these is historical debate, which has carved out dominant theoretical positions in the landscape of questions related to the complex phenomenon. This debate includes controversy over how to characterize culture and the manner in which particular characterizations, such as group properties or social learning mechanisms, have crystallized over time. Another is epistemic heterogeneity: different kinds of questions are being asked about cultural change. Some of these are empirical. *Has the rate of technological change increased with the onset of more rapid diffusion of innovations? Does it depend more on the rate of transmission or the size of the cultural breeding population?* Some research questions are theoretical. *How can we characterize cultural heredity with multiple parents making contributions of different sizes at different stages in the developmental processes of enculturation?* Other questions are conceptual. *Can the dependency structure of knowledge and skills acquired in ontogeny provide structure analogous to genetic architecture in population genetics?*

A further source of structure results from relationships between research questions, sometimes as nested, component hierarchies (e.g., research questions divided into subproblems) and sometimes as functional, control hierarchies (e.g., answers to one research question are presupposed in another question). Consider again the question of how interchangeable machine parts emerged in the early nineteenth century. It subdivides into a question about the origin of motivations to do so, a question of what technological innovations were necessary to support the increased precision of manufacture, and a question of how the requisite conditions of production were put into place. The question about conditions of production can be subdivided further into the relevant changes in labor, the organization of parts manufacture and assembly, and the support tools that facilitated this production, such as lathes and templates. An example of functional hierarchies in problem structure can be seen in the research question of how to create autonomous vehicles that operate in standard (and nonstandard) traffic situations. In order to address this question, different levels of autonomy need to be distinguished. Once this is done, the requirements for creating a control system for a particular level of autonomy can be specified appropriately, such as the degree of reliance on computational analysis of multimodal sensory information versus accumulated information about road conditions. Combinations of different disciplinary approaches and methods will be required to address these structured sets of research questions in the problem agenda. The focus of one discipline on some questions rather than others creates a fruitful division of labor and organizes different lines of investigation in terms of the kinds of questions they tackle, such as different teams working on different aspects of the problem of autonomous vehicles in the same company.

Structure that derives from history, heterogeneity, or hierarchy is significant because of how criteria of explanatory adequacy are embedded within them. For example, heterogeneous questions have distinct standards for what counts as an adequate answer. An acceptable explanation of how and when the rate of technological change depends more on the rate of transmission than on the size of the cultural breeding population will differ from an acceptable explanation of how the dependency structure of skills acquired through a developmental sequence is analogous to genetic architecture in population genetics. Relevant criteria of adequacy are localized to different types of questions and different hierarchical levels, as well as conditioned on trajectories of historical debates. However, this localization facilitates systematic understanding as a consequence of the problem structure. Explanations of what changes in labor were relevant to the conditions of production for interchangeable machine parts will differ from explanations of how support tools facilitated this production. However, both explanations jointly increase our knowledge of this nineteenth-century episode of cultural evolution. At a more theoretical level, criteria of adequacy for the problem agenda of cultural evolution will include attention to what transmissible elements are relevant, what kinds of biological individuals are involved, what developmental sequences are germane for the transmission and expression of cultural traits, what types of social organizations and societal institutions are involved, what kinds of material artifacts are manifested, and what forms of scaffolding relationships obtain between these various items (see chapter 1). Once these criteria of adequacy are made explicit, they comprise a strong rationale for an interdisciplinary approach to cultural evolution. No single discipline or approach will be sufficient to fulfill these criteria. As a result, tendencies to ignore or selectively eliminate aspects of the complex phenomenon not amenable to particular disciplinary approaches are lessened (a perennial complaint about abstract modeling that reduces cultural heterogeneity to a small set of idealized factors). The criteria of adequacy embedded in the structured problem agenda not only speak to what disciplines or approaches are required but show exactly where they need to make their contributions.

Combining Thick and Thin Descriptions of Cultural Phenomena

An often-discussed locus of controversy related to the selective elimination of aspects of complex, heterogeneous cultural phenomena is the methodological distinction between "thick" and "thin" descriptions (Geertz 1973). The former, exemplified in the practices of disciplines like cultural anthropology or natural history, involves descriptions that embody details about intentions, history, context, and related cultural practices or analogous details for natural history. Thin description, exemplified in the practices of disciplines like population genetics, involves descriptions keyed to traditional mathematical modeling that adopt only a few variables in the equations used to model the dynamics of phenomena. Those disciplines that value thick description are most skeptical about evolutionary approaches to culture because they assume thin description is constitutive of these approaches. They argue that thin descriptions treat cultural phenomena too abstractly and therefore fail to engage its multilayered character. However, those disciplines that value thin description complain that the rich detail of thick descriptions sacrifices quantitative prediction, generalizability, and explanation. But the methodological distinction between thick and thin description is not a dichotomy. Because of the complexity of cultural change over time, both are often necessary at multiple levels of detail and from multiple perspectives. Which perspective and what level of detail is needed will depend on the question being addressed. Often, thick descriptions will be involved in analyzing how to operationalize the variables of thin models. Moreover, the need to combine thick and thin descriptions in various ways for different questions points us to methods of intermediate or heterogeneous *viscosity* (Wimsatt and Griesemer 2007). These methods generate characterizations that can provide explanations of intermediate grain and varying textures, which help to capture real-world detail while still utilizing general mathematical theories and quantitative predictions. Incorporating development and culturally induced population structure to study cultural evolution brings us into this domain (see chapter 1).

A lovely example that involves integrating multiple thin models with thick descriptions and methods of intermediate viscosity is William Durham's complex narrative for the maintenance of sickle cell anemia through heterozygote superiority driven by malaria resistance (Durham 1991, 103–53). Although the spread through West Africa of three different HbS mutations, which cause sickle cell anemia when in homozygous genotypes, can be modeled abstractly within population genetics, a complete explanation of the relevant phenomena must deal with spatially structured migrations, different degrees and patterns of rainfall, the cultivation of different crops that differentially favor mosquitoes (and exposure to them), and various cultures and language groups that affect interbreeding and cultural practices related to mosquito control. Templeton's (1982) analysis of the interactions between the HbA, HbS, and HbC loci, as well as the effects of inbreeding, further complements Durham's analysis by suggesting an explanation for the relationships between the distribution of the HbS and HbC alleles. Combining Templeton's abstract population genetic model of the temporal evolution of the phenomena with Durham's descriptions of environmental heterogeneity shows how the complementary application of multiple thin models of different types yields a thicker overarching narrative. The parameters of abstract mathematical theory are calibrated with the spatially and temporally variable patterns of rainfall and the complexities of migration, including population flow and linguistic group structure, affecting the degree of interbreeding. This calibration modulates our expectations for how close populations should be in relation to predicted equilibrium frequencies.

Regularities emerge from the data but curve fits are noisy at best; deviations from quantitative expectations are sometimes explained by thick descriptions of the characteristics of local, individual populations.

Despite the fact that Durham's analysis serves as an exemplar of combining thick and thin descriptions in different ways to achieve a rich explanatory tapestry that is also generalizable, it is an ongoing effort to address the variegated structure of this problem by articulating different approaches. Even with a "thin" mathematical model only containing a few explicit variables, the complexity of cultural phenomena can make it difficult to determine which specific causes are relevant to a modeling result. Although it may be experimentally tractable in principle, the specificity of complex causal relationships might resist generalization. The only reasonable response in these situations—common to the study of cultural evolution—is to practice multiple approaches simultaneously at different grains of analysis appropriate to different questions and then integrate the answers in a piecewise fashion to better comprehend such multifaceted and multidimensional phenomena.

Beyond Partitioning Theoretical Approaches

Once we have adopted the perspective that there is a structured problem agenda with diverse explanatory questions about the evolution of culture construed as a complex phenomenon, it becomes clear why partitioning the research landscape in terms of broad theoretical approaches alone might be less fruitful. For example, Lewens (2015) offers a tripartite division of approaches to cultural evolution: historical (scrutinizing facets of culture as products of historical processes), selectionist (analyzing cultural change in terms of selective dynamics operating on individual behaviors or group-level units as replicators and interactors), and kinetic (emphasizing the capacity for learning to modulate cultural change over time). These different approaches are not mutually exclusive but tend to have clusters of shared theoretical commitments. Selectionist approaches are sometimes motivated by the aim of offering a fully general account of selection; kinetic approaches are often focused on generating increased understanding of the mechanisms of learning. While Lewens is fully aware that this taxonomy does not capture everything relevant in studies of cultural evolution (e.g., cultural phylogenetics), a more important concern from our perspective is that it does not offer enough structure for answering the key methodological question: How should investigation, modeling, and explanation proceed?

Kinetic approaches might concentrate on learning models and their explanatory power, but what these models represent and whether they explain depends on what research question is being asked. And these research questions about the dynamics of cultural change require more than one disciplinary contribution. Without an explicit account of the problem structure (in terms of history, heterogeneity, and hierarchy) and its associated criteria of adequacy, the complexity of the contested explanandum of cultural evolution is elided, and the necessary articulation of diverse explanatory resources—both thick and thin—is elusive. However, embracing the need to flesh out the problem architecture and its evaluative standards yields a broad outline of answers to our earlier questions. An evolutionary approach to culture necessarily *includes* social science disciplines precisely because there are interpretive issues involved in deciphering cultural formats that outstrip the analytical capacities found in abstract modeling. Considerations of thick and thin description indicate that there are sometimes reasons to privilege a perspective that focuses on biological factors (such as the relation between number of mosquitoes at a time and the temporal and spatial distribution of rainfall) and sometimes reasons to privilege a perspective that focuses on social factors (such as cultural practices for rooting out evil spirits that involve waving firebrands at dusk near the roofs inside residences, which happens to be when mosquitoes tend to congregate there). The problem structure and criteria of adequacy govern in what contexts and for what kinds of factors privileging is warranted, while pointing toward the ongoing need to build integrated models that articulate both biological and social factors together. An evolutionary approach to culture (sensu lato) provides far more than mere analogy or inspiration.

The criticisms of how Lewens partitions the research landscape for studying cultural evolution remind us why approaches anchored in memes are woefully inadequate. It is not clear that the central role of finding a unit of heredity in biology should be paralleled in studies of culture. (In chapter 1, Wimsatt suggests not, or at least not directly.) Not only are there concerns about a nontrivial characterization of what memes are and how they can be transmitted or replicated but memetic approaches do not provide any significant structure for coordinating different explanatory resources to account for the complex phenomenon of cultural evolution. The structure in the problem anatomy described above in terms of history, heterogeneity, and hierarchy gives scaffolding to theorizing that articulates the necessary diverse perspectives. It is not enough to assume people can be infected by memes and that there are different rates of infection for different memes. What exactly are the relevant subpopulations, the cultural histories, their connections, and the population dynamics? Is cultural evolution best modeled epidemiologically? Why do some people "catch" the meme in question and others do not? Adequate answers reliant on memetics alone seem unlikely; one disciplinary approach will not be adequate even if it could determine a population dynamics for memes. Instead, we need strategies for articulating diverse perspectives in order to comprehend cultural evolution.

BEYOND THE MEME: ARTICULATING THE EXPLANANS

Although memetic approaches suffer from a variety of irremediable problems, one motivation for their introduction was venerable: start simple. It is a time-honored modeling practice to begin with simple models that involve relatively strong idealizations and appear disconnected from the phenomena of interest. (Recall the humorous jab at theoretical physicists modeling biological phenomena: "Assume a spherical cow in a vacuum . . .") The virtue of starting simple is visible in early population genetics, as well as in dual-inheritance theories of cultural evolution (e.g., Boyd and Richerson 1985). Consider the former. The assumption of panmixia, or random mating, played an important role in developments of neo-Darwinian evolutionary theory. This is marked by its prime location near the beginning of textbooks and alongside accompanying discussions of the Hardy-Weinberg principle in population genetics. Its use as a simplifying assumption nurtured the elaboration of several aspects of the mathematical theory. However, as is true of other simplifying assumptions, there also were drawbacks. For example, we have learned subsequently that population structures that violate the assumptions central to Hardy-Weinberg equilibrium are critical for engendering biological evolution. Population structure like groups, which arise either through selective breeding or localized interaction, can facilitate and elaborate adaptations that could not be supported at the individual level (reviewed in Wade 2016; for culture, see Sterelny 2012). The denial of any population structure in mating at the group level was equivalent to assuming an extremely strong form of blending inheritance, which rendered group selection and local population differentiation difficult or impossible (Wade 1978; Wimsatt 1980, 1981, 2002).

Although population structure was initially ignored in evolutionary theory, there were other systematic sources of structure within population genetics that derive from the architecture of the genome. They can be recognized by the fact that these features of genomic structure retard the rate of approach to Hardy-Weinberg equilibrium, causing deviations from random assortment or a maximally mixed distribution of elements. Representations of these sources of genomic structure have played crucial roles in the elaboration of evolutionary genetic theory and contributed new complexities in modeling the dynamics. First among these are linkage relations arising from the location of genes at different distances along the same chromosome. Other aspects of structure that originate in genetic architecture and were incorporated into population genetic models include diploidy (chromosomes in pairs, as in whole genotypes), in contrast with haploidy (single chromosome sets, as found in sperm and egg), haplodiploid mating systems as found in some of the social insects (with diploid queens and haploid sterile castes), and the role of diploid gametic organization in life cycles alternating between haploid gametes and diploid zygotes in sexual reproduction, which imposes a seldom recognized correlation in linkage models (Wimsatt 2007, 287–93). The effects of sex-linkage and age structure act as *segregation analogs* by retarding the rate of mixing and therefore attenuate the approach to a (maximally mixed) Hardy-Weinberg equilibrium of gene frequencies. Any element of population structure can act as a segregation analog with similar effects (Wimsatt 1981, 152–64; 2002, S9).

Evolution is substantially affected by *both* internal and external sources of structure. To emphasize this, Michael Wade coined the terms *endogenetics* (for what we call genetics) and *exogenetics* (for what we call population structure; Wimsatt 2002). We tend to treat genetics as crucial and population structure as a subsidiary complication, but they are equally important in determining evolutionary outcomes. This is nicely illustrated in the population genetics of the system of alleles affecting sickle cell anemia and malaria resistance (HbA, HbS, and HbC; Templeton 1982). The HbS allele causes sickle cell anemia in the homozygote (HbS/HbS) and confers malaria resistance in the heterozygote (HbS/HbA). The HbS allele arose four separate times and increased in frequency in regions of Africa where the incidence of malaria was high (Durham 1991). However, one also finds pockets of the apparently more recent HbC allele. When homozygous (i.e., HbC/HbC), this genotype confers malaria resistance without the ravages of sickle cell disease and is thus higher in fitness than any alternative. In this case, inbreeding (as would occur within small, relatively isolated groups) allows the HbC allele to grow in frequency because it occurs more frequently in the homozygote.

But under conditions of random mating (panmixia), this is not possible (Templeton 1982). In a population at equilibrium with HbA and HbS, HbC cannot invade, even with unusually high fitness, because at low frequencies it would occur primarily in heterozygotes of much lower fitness. Thus, HbC alleles would be eliminated before they could achieve the higher frequencies of homozygotes necessary to become established.

Michael Wade's distinction between endogenetic and exogenetic structure for population genetics and evolutionary biology is paralleled for cultural evolution by sources of internal and external structure. This volume explores the nature, variety, and impact of features that add structural elements that amplify evolutionary potential, as well as other characteristic features that must be accounted for in formulating an adequate theory of cultural evolution. These elements include the impact of sequential dependencies in the acquisition during development of cultural traits—a prime example of internal structure—and the roles of external structure, such as social institutions, organizations, and technological infrastructure, which scaffold segregation, learning, and cumulative culture in individuals and groups. Including these structures provides resources to deal with cultural traits that satisfy the diverse definitions surveyed by Krober and Kluckhohn (1952) or Driscoll (2017). As a consequence, this allows for more unified and compelling accounts of cultural evolution. The diversity of the kinds of structural elements yields both a more abundant range of phenomena and an increased number of evolutionary possibilities for cultural evolution than for biological evolution. However, many current models of cultural evolution are, for the most part, stuck at the earlier stages of theoretical development where the modeling assumptions do not include or recognize these diverse kinds of structure. One reason for this situation may be that researchers have not yet found good ways of incorporating this structure into their models. Another is that their assumptions hide the relevance of these factors (as with ignoring the role of technology). Regardless, the potential of these structural elements for mediating far more complex forms of adaptive evolution is therefore not usually taken into account.

A crucial structural element for cultural evolution is the fact that different aspects of culture are acquired sequentially throughout the life cycle. As a result, earlier acquisitions can act as necessary precursors that facilitate, inhibit, or transform the reception of later ones. This corresponds to the endogenetic structure provided in biology by the architecture of the genome. For instance, the language an individual learns channels all subsequent cultural additions. Humans acquire many complex skills that show strong sequential dependencies of this kind, especially in societies with robust social institutions. Consider the inculcation of mathematical skills, where arithmetic precedes algebra, which precedes geometry, which precedes calculus. A closer inspection of this standard sequence would reveal multiple intermediate dependencies, such as the pathway from elementary algebra, through intermediate algebra, and on to advanced algebra. Each of these introduces new tools, procedures, and concepts used at later stages. Similar patterns of dependencies are true for most of the sciences and for reading, as well as the modes of thought mediated by them. This is no less true for manual skills and for our social modes of interaction.

Much of culture can be seen as the construction of external structures, like our schools and learning curricula, to support the sequential acquisition of these competencies. Our culturally induced group structures—things like universities, business firms, and religious communities—interact with us and with these institutions, mediating knowledge acquisition and modes of collective action that we could not do individually. Furthermore, our technologies, while often credited with increased powers of production, have also become ubiquitous elements of scaffolding for our cognitive and cultural development. These operate both individually and collectively through an infrastructural generative reconstruction and extension of our cognitive and social niches. The interactive character of this scaffolding makes the articulation of endogenetic and exogenetic factors far more interpenetrating for culture than for biology.

The essays contained herein explore the impact of these structuring elements and their interactions in various elements of culture. These include spoken and written language, the institutional structure and interest groups of science, the evolution and descent relations of technology as reflected in patents, the role of prior theory in scaffolding the development of new theory, the cumulative effects of lithic technology, religion, irrigation practice, and the costs and adaptations required when adopting new technologies that challenge our entrenched practices. These cases begin to illustrate the interdisciplinary combinations required to address the problem agenda of cultural evolution. They document and account for interactive dynamics dependent on multiple structural dimensions and thereby encourage new directions for elaborating theory to explain the diverse possibilities for cultural evolutionary processes. Although the contributions do not always mention cultural evolution per se, they focus on relevant factors from theories in the social sciences. This is as it should be; social structures are of central import for cultural evolution and, from our perspective, should be incorporated into theoretical approaches to adequately account for the complex phenomenon. The evaluative standards for the problem agenda demand an evolutionary approach that genuinely integrates existing social, cultural, and technological theory, not one that trades social science approaches for simplistic genetic models of social structures and practices (cf. Wilson 1975). An evolutionary approach to culture necessarily *includes* social science disciplines. There is far too much of value in existing social science theory and allied analyses of phenomena. Existing attempts at theories of cultural evolution typically lack the intellectual resources to generate stable explanatory combinations incorporating the riches these accounts provide. The contributions to this volume jointly accent what is needed and point us, sometimes forcefully, in the direction of how to accomplish it.

RESEARCH AGENDA EXEMPLARS

In chapter 1, Wimsatt reviews the conceptual geography of cultural evolution and the kinds of elements required for an adequate explanatory account. This involves several additions to those factors typically considered in extant theoretical formulations. In particular, development plays a central role and includes two main interacting components: the developmental dependencies of individuals in acquiring complex skills and the social and institutional structures that scaffold this development. Interactions among these components involve an intercalation of both endogenetic and exogenetic elements. In turn, these elements interact with the development of groups (like business firms or professions) and the institutions they construct to mediate their interactions. All of these interactions are significantly scaffolded by evolving artifact structures. Some of these are general infrastructure, such as written language, exchange markets, or power and communication networks, whereas others are specialized to particular tasks and roles, such as mathematics curricula, machine tools, computer hardware and software, scientific theories, or medical training.

Wimsatt argues that an adequate theory of cultural evolution must be capable of incorporating, describing, and explaining these complex interactions. In order to do so, the roles of transmissible elements, developing biological individuals, organizations, institutions, and artifact structures must be delineated individually and articulated jointly, with special attention to understanding the scaffolding relations between them. The conceptual geography proposed is intended to provide a landscape in which the different forms of structure and scaffolding relations found in the other contributed papers can be situated in appropriate contexts of interpretation. This landscape then offers a template to guide the process of articulating the Babel of different approaches, perspectives, and subjects that are necessary to comprehend the polyphony of culture and its change through time. The inclusion of additional considerations of structure amplifies the range of phenomena that can be accounted for by theories of cultural evolution. It also encourages the synthesizing of theories for the evolution of culture, cognition, and technology. Furthermore, the use of these elements to make the criteria of adequacy within the problem agenda explicit encourages the use of categories and processes drawn from traditional social sciences, which then makes it possible to increase the explanatory power of an interdisciplinary theory of cultural evolution.

The next three articles (chapters 2–4) explore dimensions of scientific and technological change that operate at three different scales. Sabina Leonelli discusses the formation of two organizations for managing biological data and research that have become central to tens of thousands of researchers. The first involves an organization that acts to standardize ontologies in genomic and proteomics research. This standardization is an institutional creation that is crucial to communication across different databases and facilitates the conjoint utilization of the data contained therein. Although incommensurability was never a problem for communication between scientists across revolutions in the way some philosophers imagined in the 1970s, these fixed and institutionalized artifact structures turn out to be a crucial element for communication between modern computerized databases whose syntax is less tolerant of variation than the negotiated meanings of conversing scientists. Leonelli's second case involves the emergence of steering committees for model organism research in the United Kingdom, which play a central role in determining priorities, funding, and coordinating research. In both cases, these organizations emerged "spontaneously" (i.e., without central planning), and Leonelli documents the different factors and features of how they came into existence. These organizations have developed and maintain institutions that mediate communication and organize research, as well as scaffold activities, on national and international scales. Leonelli employs research from sociology on the formation of social movements to further understand the processes relevant to the origination of these organizations. Her analysis provides a paradigm for how work on cultural evolution can articulate with existing theory in sociology.

Nancy Nersessian employs her deep and multifaceted ethnographic research on the development of interdisciplinary investigations in bioengineering to look at knowledge production in the laboratory and the creation of new multidisciplinary communities. Her study illuminates how researchers from biology and bioengineering learn to build bridges between their disciplinary perspectives. This "bridge-building" activity includes the construction of laboratory systems, the integration of modeling and "soft" bioengineered experimental systems, and the training of both graduate and undergraduate students. Nersessian becomes a participant–observer in this activity through her involvement in the design of curricula to systematize such interdisciplinary training. All of this was possible because of her detailed tracking of laboratory life, the dynamics of research, and the intercalation of training and research practices in unparalleled breadth and depth. The resulting account of the coevolution of practices, experimental systems, research, training of individuals, and curricula simultaneously deals with multiple dimensions of experimental practice and culture, the interactive evolution of models and knowledge, and the role of all these factors in the generation of scientific careers. This is a remarkable exemplar for science studies generally, as well as for cultural evolution in particular, and a penetrating reflection of the necessary articulation of multiple analytical perspectives.

In an intricate and technically demanding narrative, Michel Janssen offers a groundbreaking analysis of how prior theory and mathematical methods can scaffold and structure the development of new theory. In doing so, he details how different elements facilitate this process in a productive manner. He considers five historical cases involving the transformation of classical mechanics and electromagnetic theory into quantum mechanics and relativity theory. Janssen exploits the nature of scaffolding explicitly to argue against a Kuhnian picture of scientific revolutions as destructive replacement and new reconstruction on different foundations. He paints a picture of subtle transformation and extension of theoretical structures. These often leave crucial elements of the older structure informing or supporting the newer edifice, which yields a continuity that makes the transformation intelligible and the progressive evolution of science plausible. The elements that are preserved in the transformation and the roles they play suggest problem-solving heuristics with broader import in cognitive psychology and elsewhere in science. Janssen's account is applicable to theoretical change in other domains and is an important contribution to the literature on scientific change more generally.

Chapters 5–7 provide different models of cultural processes and point toward more systematic perspectives in the study of cultural change. Jacob Foster and James Evans offer a theoretical structure that has been largely missing from theories of cultural evolution: an account of heredity in cultural and technological systems based on a general treatment of reticulate phylogenies. Although this allows for traditional tree-like phylogenies as a special case, their analysis makes it possible to treat cultural heredity in all of its complexities, including not only multiple parentage with contributions of different degrees but also skipped generations. Examples include the recovery of a buried artifact, inspiring subsequent invention, recombination of elements from different lineages, and the black boxing of sets of features that are subsequently inherited as a unit. The absence of such an account has been especially vexing because of the central role that heredity played in the development of neo-Darwinian evolutionary theory. This important theoretical contribution regarding formal characteristics of the reticulate aspects of cultural inheritance involves new concepts (e.g., transmission isolating and accelerating mechanisms) and new inferential tools that are particularly appropriate to the more plentiful "fossil" records we often find with technology in comparison to paleontology. In a striking parallel, most aspects of this structure also apply to the acquisition of information and skills in individual development, giving a further tool for the analysis of generative entrenchment throughout ontogeny.

Mark Bedau gives us a superb case study in his analysis of descent relations within the patent system. This is an unusually tractable and rich example that relates directly to technology but just as adequately represents the characteristics of descent and the modification of theories in the sciences. The patent record contains tremendous detail about inventions and their varied ancestors. Well-developed software tools are available for mining this detail. This combination of detail and methods to explore it establish a parallel with biological model systems, making the patent system an excellent model for cultural evolution ("the right organism for the job"). (These features also suggest it is an unusually appropriate case to apply Foster and Evans' analysis of cultural inheritance.) One of Bedau's striking findings is just how promiscuous technological inventions are (in the sense of the multiplicity of their parentage). The patent system model makes it possible to ask and answer new questions: Has the multiplicity of parentage increased since

WWII with increasing cross-disciplinary communication? This analysis nicely documents "door-opening" inventions where one technological invention stimulates other inventions in diverse areas.

Marshall Abrams uses agent-based modeling (ABM) to study the evolution of coordinated irrigation practices by different communities to manage limited water resources and pests in Bali (a paradigm case for anthropologists). ABM has the advantage of modeling a population of individual agents that may have different and modifiable characteristics and has become a common tool in modeling cultural evolution. These diverse characteristics could be the product of different programs for behavior, different experience (if their behavior is modifiable through learning), or both. In such populations, both individual characteristics and the spatial distribution (or population structure) of the agents they interact with matter and can be used to model the formation of complex task groups (see, e.g., chapter 12). (This kind of structural diversity was inaccessible to modeling before the advent of ABM and is an important move toward "thick description.") These interactions are studied with a large number of *Monte Carlo* simulations that have randomized values of variables other than those being scrutinized to get averaged effects of the experimental treatments. Although ABM dramatically increases the degrees of freedom one can model (parameters must be specified for each agent), it is also fraught with problems of how to interpret the results. Abrams's model succeeds in explaining the phenomena robustly, though it is relatively complex. However, he considers a simpler model that bundles interactions into a single parameter and shows that it does not work except under very limited circumstances. This demonstrates that under some circumstances model complexity is necessary to get empirically adequate results. It also corroborates earlier social science claims that the spread of religious practices favoring cultural *coherence,* which are also correlated with local success in crop yields, could have mediated the coordination of irrigation practices.

The next six chapters (8–13) explore diverse methodological problems and structural factors relating to the reproduction of skills and the transmission of knowledge. Flintknapping was a skill critical to the emergence of culturally abundant societies because of the role that cutting tools played in the production of many other artifacts that made diverse new practices possible. It required extended manual practice and tutelage from an accomplished master and is therefore a plausible prototype for the acquisition of specialized skills and knowledge. Gilbert Tostevin is a Paleolithic archaeologist who has both practiced and taught flintknapping. His chapter addresses the question of what the appropriate unit of analysis for the cultural replication process should be given that nothing is materially transmitted. Earlier generations of archaeologists focused on the finished product of the toolmaking process, but these can be produced in many different ways. Moreover, what is taught and learned is *how* to make it; the final product is the wrong target of analysis. (This is often true for the questions we want to answer for artifacts.) Tostevin applies the distinction between the intimate knowledge a member of the culture possesses (-emic or *savoir faire*) and anthropological nonnative knowledge (-etic or *connaisance*) to the teacher and the learner within the culture, respectively. This sensible observation broadens the application of the *emic-etic* distinction from one of methodological precaution in the interpretation of anthropological results to a widespread and important process in the transmission of culture. Here, the study of cultural evolution suggests an ampliative reinterpretation of anthropological theory. It also illuminates how the learning process is conceptualized. Different theories presume different processes for scaffolding the learning: simple reverse engineering, observation of the teacher making the product, gesturally and tactually assisted manipulation of the learner's hands, and language-assisted teaching. These can also be understood as one or more stages in a longer procedure. By carefully dissecting which details of the making process are visible to the learner who views it from a different perspective than the teacher, deeper insights into the complicated learning process and conditions for cultural reproduction emerge.

Linguist Salikoko Mufwene argues that spoken languages should be seen as communicative technologies that are hybrid biological–cultural products. These exist in and are conditioned by a variegated social ecology and constraints engendered by their mode of expression. According to Mufwene, the Chomskyian picture of an "innate language module" is inadequate on both biological and cultural grounds, whether in terms of the supposition of a single macromutation generating the language capacity or the independence of that capacity from a host of other cultural capacities that travel with it. In contrast, Mufwene uses evidence from phonology, morphology, and syntax to show that language displays the combination of constraints and variation one would expect for the evolution of any adaptation that is directed toward the solution of a common set of problems. For example, one cannot make multiple diverse sounds in parallel, and the resulting linear stream of language is unavoidable for spoken discourse. However, modality matters; sign language escapes this constraint because gestures can take place in three dimensions. Other technologies manifest similar patterns—how something is produced constrains the product. For language, Mufwene claims that naming comes first, followed by predication and an increase of vocabulary. Recursion increases economy and facilitates greater complexity of expression. In this sequence, there is a significant role for generative entrenchment and scaffolding. Spoken language made possible more effective cooperation and diffusion of skills and may have been required for the sophistication of many complex skills. Written language emerged slowly from numerical tallies. Advances in representing sounds increased the power and economy of language, which made cumulative culture possible. These together comprise the most general of infrastructural scaffolds in a society and essentially midwife all other skills.

Massimo Maiocchi reviews the origins of writing, which appears to have occurred independently four times (Mesopotamia, Egypt, China, and Mesoamerica), and then elaborates the case of Mesopotamia, which is better documented and researched than the other three. Written signs or counters first appeared in the eighth millennium B.C.E., but these became more elaborate between 4500 and 3500 B.C.E. with the appearance of inscribed counters and then bullae (hollow, sealed clay pockets containing counters) in Uruk. Large numbers of diverse clay tablets with cuneiform records from a slightly later time were also found there, some with an emerging syntax for the representation of numbers. Flat tablets made storage and indexing simpler and may have become more common for those reasons. Inscriptions originally served accounting purposes, and bullae probably validated legal contracts. This need and the use of clay left an entrenched legacy for subsequent written forms. Cuneiform writing grew out of signs that depicted the kinds of items represented. Lexical lists of diverse kinds proliferated, categorizing both objects and professions, accompanied by a system of weights and measures. However, it was hundreds of years before writing expanded to serve other functions, such as state administration, religious practice, and narrative history. Throughout this period writing was known and used only by a restricted class of scribes. Subsequently, phonetic languages permitted the representation of phonemes, making up words from other languages with a reduction of signs; alphabetic languages went further in reducing the number of signs required. Written language came to structure both social practices and individual cognition generally after moving beyond its more circumscribed role in governance, interpersonal interaction, and information storage.

Chapters 12 and 13 return us to more general issues. Joseph Martin focuses on the role that scaffolding plays in changes that accompany the adoption of new technologies. Scaffolding is relevant to support existing practices and those associated with new technologies, as well as the transitions between them. Martin distinguishes three ways in which newer technologies may relate to older ones: (1) displacement, such as when the internal combustion engine as a power source replaced the horse in propelling cars and trucks; (2) combination, in which a newer technology interacts with and complements an older one (e.g., the Internet can be scaffolded by cable networks); and (3) catalysis, in which a new technology interacts with an older one to generate new capabilities (e.g., how the Internet, with the computer, catalyzes a host of new activities, from electronic payment to the streaming of movies). Although new technologies can spread due to advantages manifested in any of these three ways, we must also consider trade-offs—what they may prevent or inhibit. Frozen dinners contributed to the downfall of family dinners and the interactions they facilitated; the advent of automobile-based suburbs made popular the construction of houses surrounding cul-de-sacs that protected children from through traffic, but their topology made bus or tram-based public transport impractical. What is lost in adopting new technologies leads us to focus on what changes to scaffolding are required to make the transition and what sources of resistance might be present. As Martin discusses, this can lead to better policy decisions when developing and introducing technology.

Paul Smaldino analyzes the function of social identity in facilitating cooperative group formation. Social identity is a particularly important tool in navigating affiliations in complex societies, which have large numbers of different social roles and many individuals who do not know each other personally but must interact with multiple groups in different contexts for different ends. How do such individuals assort into appropriate groups to serve their interests, develop competencies needed for professions, and find mates (inter alia)? For this we need a multidimensional social identity in which different aspects can be expressed in different contexts. (The phenomenon of register switching in language is one sign of changing behavior for these differing contexts.) As a consequence, we can participate in religions, condominium associations, professions, departments, neighborhoods, sports preferences, and team affiliations, plus have ethnicities, sexual identities, and age groups, each of which may compel us to act within that group in different ways to serve our needs and interests. As Smaldino notes, the needs of affiliation involve not only who to cooperate with for common interests but who *best* to cooperate with to serve those interests. Decisions of this kind may demand further differentiating information. In larger societies with more roles, stereotypes associated with identities may serve cognitive functions in conveying relevant information for coordination decisions, such as by signaling a high probability of common knowledge. Smaldino discusses different interactions that could be involved in group formation and how these play specific roles in societies of different size and structure. This chapter provides a crucial theoretical plank in understanding the emergence of complex societies and articulates naturally with Wimsatt's discussion of the need to deal with career trajectories that involve multiple, coordinated cultural breeding populations.

It is implicit in Wimsatt's "Articulating Babel" that an account of cultural evolution requires an unprecedented marshaling of diverse perspectives with local theories. Practitioners often have overestimated the generality, power, and completeness of their particular perspective. Combining these perspectives requires two things. First, practitioners must recognize how and where their perspectives are relevant to generate an adequate explanatory account while accepting that, as perspectives, each of their vantage points is individually incomplete in addressing the complex phenomenon of cultural evolution. The endeavor of making the problem agenda structure explicit and detailing the associated criteria of adequacy provides a rationale for both the relevance and incompleteness of individual theoretical perspectives.

Second, the structure and criteria of adequacy for the problem agenda of cultural evolution demand that the relevant but incomplete theoretical and methodological perspectives articulate with one another in a coordinated fashion to answer different research questions. This is the topic of chapter 13 by Claes Andersson, Anton Törnberg, and Petter Törnberg. They describe *wicked systems* generally, which have characteristics common to the complexity we have observed for cultural evolution. How do such systems arise? Ecological and societal systems combine bottom-up features of complex systems (e.g., path dependence, nonlinearity, chaotic dynamics, and multiple relevant overlapping boundaries) with the top-down organization of complicated systems (e.g., many components, different relaxation times of their interactions, and irregular connectedness). The fact that these diverse facets comprise wicked systems and are studied in different disciplines, which reveal different aspects of the phenomenon of interest, increases the urgency of heeding the organizational structure and criteria of adequacy inherent in the problem agenda. Only this provides an antidote to claims that one theoretical perspective derived from a particular discipline offers a uniquely systematic viewpoint on cultural evolution. The essay by Andersson et al., along with the other contributions to this volume, substantially augment the number and kind of handles available for managing the complex domain of cultural evolution and determining the biases inherent in our modeling simplifications. This will encourage us not only to go beyond the meme but also to better marshal our collective investigative efforts to interdisciplinarily explain the evolutionary dynamics of different facets of culture.

REFERENCES

Boyd, R., and P. Richerson. 1985. *Culture and the Evolutionary Process.* Chicago: University of Chicago Press.

Brigandt, I., and A. C. Love. 2012. "Conceptualizing Evolutionary Novelty: Moving beyond Definitional Debates." *Journal of Experimental Zoology Part B: Molecular and Developmental Evolution* 318B:417–27.

Driscoll, C. 2017. "The Evolutionary Culture Concepts." *Philosophy of Science* 84:35–55.

Durham, W. 1991. *Coevolution: Genes, Culture, and Human Diversity.* Stanford, Calif.: Stanford University Press.

Geertz, C. 1973. "Thick Description: Toward an Interpretive Theory of Culture." In *The Interpretation of Cultures: Selected Essays,* 3–30. New York: Basic Books.

Griffiths, P. E., and K. Stotz. 2013. *Genetics and Philosophy: An Introduction.* New York: Cambridge University Press.

Hounshell, D. A. 1984. *From the American System to Mass Production, 1800–1932.* Baltimore: Johns Hopkins University Press.

Kroeber, A. L., and C. Kluckhohn. 1952. *Culture: A Critical Review of Concepts and Definitions.* Vol. 47, no. 1. Cambridge, Mass.: Peabody Museum of American Archaeology and Ethnology, Yale University.

Lewens, T. 2015. *Cultural Evolution: Conceptual Challenges.* New York: Oxford University Press.

Love, A. C. 2014. "The Erotetic Organization of Developmental Biology." In *Towards a Theory of Development,* edited by A. Minelli and T. Pradeu, 33–55. Oxford: Oxford University Press.

Smith, M. R. 1977. *Harper's Ferry Armory and the New Technology: The Challenge of Change.* Ithaca, N.Y.: Cornell University Press.

Sterelny, K. 2012. *The Evolved Apprentice: How Evolution Made Humans Unique.* Cambridge, Mass.: MIT Press.

Templeton, A. C. 1982. "Adaptation and the Integration of Evolutionary Forces." In *Perspectives on Evolution,* edited by R. Milkman, 15–31. Stamford, Conn.: Andrew Sinauer.

Wade, M. J. 1978. "A Critical Review of the Models of Group Selection." *Quarterly Review of Biology* 53:101–14.

Wade, M. J. 2016. *Evolution in Metapopulations: How Interaction Changes Evolution.* Chicago: University of Chicago Press.

Wilson, E. O. 1975. *Sociobiology: The New Synthesis.* Cambridge, Mass.: Harvard University Press.

Wimsatt, W. C. 1980. "Reductionistic Research Strategies and Their Biases in the Units of Selection Controversy." In *Scientific Discovery (Volume 2): Case Studies,* edited by T. Nickles, 213–59. Dordrecht, The Netherlands: Reidel.

Wimsatt, W. C. 1981. "Units of Selection and the Structure of the Multi-level Genome." In Vol. 2 of *PSA-1980,* edited by P. D. Asquith and R. N. Giere, 122–83. Lansing, Mich.: Philosophy of Science Association.

Wimsatt, W. C. 1987. "False Models as Means to Truer Theories." In *Neutral Models in Biology,* edited by M. Nitecki and A. Hoffman, 23–55. London: Oxford University Press.

Wimsatt, W. C. 2002. "False Models as Means to Truer Theories: Blending Inheritance in Biological vs. Cultural Evolution." *Philosophy of Science* 69:S12–24.

Wimsatt, W. C. 2007. *Re-engineering Philosophy for Limited Beings: Piecewise Approximations to Reality.* Cambridge, Mass.: Harvard University Press.

Wimsatt, W. C. 2010. "Memetics Does Not Provide a Useful Way of Understanding Cultural Evolution: A Developmental Perspective." In *Contemporary Debates in Philosophy of Biology,* edited by F. J. Ayala and R. Arp, 273–91. Malden, Mass.: Blackwell.

Wimsatt, W. C. 2013. "Scaffolding and Entrenchment: An Architecture for a Theory of Cultural Change." In *Developing Scaffolds in Evolution, Culture, and Cognition,* edited by L. Caporael, J. Griesemer, and W. C. Wimsatt, 77–105. Cambridge, Mass.: MIT Press.

Wimsatt, W. C., and J. R. Griesemer. 2007. "Reproducing Entrenchments to Scaffold Culture: The Central Role of Development in Cultural Evolution." In *Integrating Evolution and Development: From Theory to Practice,* edited by R. Sansom and R. Brandon, 228–323. Cambridge, Mass.: MIT Press.

ARTICULATING BABEL
A Conceptual Geography for Cultural Evolution

WILLIAM C. WIMSATT

AFTER THE INITIAL CONSTRUCTION of the mathematical theory of population genetics in the first third of the twentieth century, its amplification and application took place largely through an elaboration of a mostly simplified theory based on panmixia, or random mating. This assumption was equivalent to ignoring any significant form of population structure. Much richer theories that explore dimensions of population structure and evolution in metapopulations have only taken off within the last generation—despite the foresight and theoretical directions provided by Sewall Wright. A corresponding uptake within theories of cultural evolution has not yet occurred despite the multiplicity of developmentally and culturally mediated sources of population structure. In an effort to facilitate an increasing incorporation of these factors, allowing a richer set of theories, I describe the many ways that developmental and population structure can enter into processes of cultural evolution, with special attention to the kinds of results they can produce. Corresponding to the architecture of the genome as an important source of structure in genetics (e.g., for linkage mapping), cultural evolution has as a source of structure the developmental dependencies (or what I have termed *generative entrenchment*) among cultural elements acquired over time in the learning of complex abilities. This structure is not as simply specifiable as the genetic structure of linkage relations (being more akin to the causal network structure of gene action), but it is well defined and, for culture, relatively accessible for reasons I discuss below. Corresponding to the external population structure of biology, cultural evolution has structured environments of learning and production that are scaffolded by social organizations, institutions, technological infrastructure, and specialized material artifacts and practices. The technological infrastructure and its articulation with our practices are particularly important since they are formative elements in our cognitive constructed niche (e.g., Wolf 2008) and not merely a kind of transmitted cultural content.

I begin with a survey of the multifarious disciplinary approaches to cultural evolution. Next, I characterize the structures that guide and amplify cultural change processes, as well as examine how they articulate with diverse disciplinary approaches, and then discuss their implications for what counts as relevant cultural units required for an adequate theory. Just as evolutionary theory can be seen as an organized series of heterogeneous models in relation to several general principles, so also theories of the evolution of culture should exhibit a similar structural heterogeneity, given the large diversity of hereditary and production systems that interact in rich and varied ways.

THE SCOPE OF THE PROBLEM

The field of evolutionary developmental biology (evo-devo) has emerged as a rich and multifaceted paradigm over the past three decades (Love 2003, 2013, 2015). It includes cross talk between developmental genetics, genomics, evolutionary genetics, cell biology, morphology, embryology, paleontology, systematics, and even, increasingly, behavioral biology and ecology. Each of these areas has provided important perspectives on the evolutionary role of development and, through substantial interactions, changed both the problem space and what count as acceptable solutions across these disciplines. Evo-devo is a natural paradigm for the interdisciplinary linkages one should expect to appear in the study of cultural evolution. Although it is tempting to think of "culture" simply as a complex adaptation of one species, this ignores the internal structural and dynamic detail of human cultures. Cultures are complex beasts, in many ways more analogous to evolving ecosystems, in part because of the richness and diversity of modes of horizontal transmission. This tends to break down what might otherwise be seen as species boundaries. But there are other complexities. The multiple evolving and interdependent lineages acting on different time and size scales within cultures and the recursive embedding of cultural elements and processes operate similarly to richly interacting species in an ecology that contains everything from primary producers to keystone predators, bacteria-mediating digestion, and, ultimately, the recycling of its constituents into the biosphere.[1]

Evolutionary theory is interdisciplinary; it spans the whole of biology and draws insights, concepts, and tools from many diverse disciplines. For the study of cultural evolution, almost all of these insights, concepts, and tools are relevant. However, the emerging interdisciplinarity surrounding the study of cultural evolution is in its early stages—the range of relevant theories substantially exceeds (and, in many cases, complements) those of the biological theory. Pertinent dimensions include an elaboration of the ecology of human evolution and the characterization of new hereditary channels, many of which are transmission pathways for information not present in most biological cases, such as spoken language,[2] written language, the telegraph, the telephone, and the Internet. Each of these provided new channels for information transmission that were independent of and yet complementary to the others. Although written language was probably the greatest facilitator of cumulative culture, the relative contributions of different transmission pathways in different contexts demand further study.

Inquiry into the evolution of different cultural forms requires combinations of approaches within recognized disciplines (e.g., combinations from cultural anthropology, paleoanthropology, and archaeology within anthropology, broadly construed), as well as combinations from across disciplines (e.g., from genetics, epidemiology, history of technology, and linguistics). For example, to understand a particular exemplar of culture, say, a primitive wheel, researchers might require an array of diverse techniques that includes a search for related artifacts; radioisotope dating for archaeological artifacts; phylogenetic methods, modified to handle reticulation resulting from horizontal transfer, to analyze data from comparative linguistics; and agent-based modeling to ascertain likely patterns of spatial movements for and between groups.

Our growing technologies have midwifed elaboration of our constructed niches, scaffolding differentiated roles and communities of language and practice. These in turn yield new complex ecosystems of production. As a consequence, an interdisciplinary approach to cultural evolution is necessary. This applies in discussions where the focus is the intersection of cultural evolution and biological evolution because *gene–culture coevolution* perspectives must articulate with various dimensions of biological theory as expanded to include evo-devo. It also applies when dimensions of culture have become sufficiently autonomous, and their rates of change have escalated to the point that standard genetic variation is largely irrelevant, though even there, in some cases, epigenetic variation may still be important. Even in these circumstances, when we consider culture as an evolutionary system in its own right, we need new techniques and new conceptual frameworks—not simply new massive data sets—to develop an adequate theoretical viewpoint from which to investigate and explain cultural evolution.

Currently, there is no consensus paradigm for how to approach cultural evolution, not even along the lines of the synthetic theory of evolution or the neo-Darwinian paradigm, which itself is being openly scrutinized (Laland et al. 2015). However, there are signs that different theoretical accounts from relevant disciplines could be on trajectories that lead to increased coordination in the near future. Take, for instance, the *dual-inheritance* approach spawned by Cavalli-Sforza and Feldman (1981) and subsequently elaborated by many others (Boyd and Richerson 1985; Richerson and Boyd 2005; Durham 1992). Rich developments of these ideas have appeared in economics (e.g., Nelson and Winter 1982; Mokyr 2002; Murmann 2003), linguistics (e.g., Mufwene 2008; Pagel 2009), and archaeology (e.g., O'Brien and Shennan 2010; Andersson 2011, 2013; Tostevin 2013). These and other expanding research programs are often governed by a specific paradigm that is manifested in graduate students, programs, conferences, and jobs. Of these, the work of Boyd and Richerson has had the broadest influence. Other areas, such as the history of technology (e.g., Basalla 1988; Arthur 2009) or history of scientific change (Griesemer and Wimsatt 1989; Wimsatt and Griesemer 2007; Wimsatt 2012; see chapter 4 of this book) have practitioners with an evolutionary perspective, and there is a large literature on innovation studies (e.g., in management and business) relevant to but mostly not integrated with work on cultural evolution (e.g., Lane and Maxfield 2005; Lane et al. 2009).

Given this diversity, it is not surprising that no common unifying focus has emerged for studies of cultural evolution.[3] Indeed, besides difficulties arising from trying to cover such a complex problem space, some researchers resist these kinds of studies altogether. For example, symbolic anthropologists tend to deny the relevance of any evolutionary or biological perspective on culture and sometimes ignore or dismiss the cultural changes that do occur.[4] This type of view is more widely distributed among social scientists and some biologists (e.g., Fracchia and Lewontin 1999; Gerson 2013b). In cognitive psychology from the mid-1960s up into the 1990s, psychologists cultivated a perspective that was both anti-evolutionary and asocial.[5]

Language competency was a paradigm case that spearheaded Chomsky's attack on Skinner's behaviorism. But the internalism of psychology that emerged in response to behaviorism (complete with methodological justifications; e.g., Fodor 1980) was a swing too far and especially paradoxical as applied to language because Chomsky's claims about the unity of a "language module" and his assumed account of innateness were both biologically problematic (Wimsatt 1986; Dove 2012; see chapter 9 of this book) and because language is such a richly social and cultural phenomenon.

Although we must reject some of the attitudes of the critics of an evolutionary approach to culture, that does not mean we can afford to ignore their chosen phenomena and subject matter. The realm of intention, meaning, and symbolic thought is crucial to understanding the nature of culture. Connections between thought, culture, and language are very deep. However, intentionality and the symbolic character of thought are not isolated logically from evolutionary processes. A perception of this isolation may derive from how critics of evolutionary approaches represent these aspects of human culture. Language is profoundly developmental and social; the formation of dialects (Mufwene 2008) and the differential stability of words for commonly used concepts across languages (Pagel 2009) exhibit some of the clearest evidence and crispest data for cultural evolution. The origins of language (see chapter 9) and the coevolution of linguistic capacities with human sociality, cognition (Sperber 1996, 2001), and tool use pose one of the most challenging and rewarding areas of study for archaeology (Sterelny 2012; see chapters 7 and 8 of this book). More recently, the explosive acceleration and expansion of cultures with the invention, adaptive radiation, and divergence of *written* language is a prime area of study for the evolution of technology (see chapter 9 of this book; Woods 2010) Moreover, there is a correlative coevolution of cognitive skills and reading that reflect the impact of our technology on our own evolution (Wolf 2008).

Subsequent chapters in this volume explore some of the new interdisciplinary linkages that might move us toward a more productive articulation of existing disciplines bearing on cultural evolution, including new kinds of cases to act as *model organisms* (see chapter 6) and new concepts to aid in their analysis (e.g., the temporally structured characterization of differentiated cultural breeding populations). One strategy allied with these approaches is the use of more general themes, principles, structures, and causal processes to midwife new linkages among relevant disciplines. For example, are there core problems or techniques that can provide foci for organization, as well as residua not captured by existing theoretical frameworks or methods that expose revealing exceptions and point to other relevant perspectives? Biologists have taught us that we must be opportunistic in seeking cases that are tractable and generate meaningful data. We must "seek the right organism for the job," and this volume brings together promising approaches to finding and leveraging suitable models to comprehend the evolutionary dynamics of different aspects of culture.

Commonly, the first theories to be formulated in a domain focus on dominant causal factors in a maximally simplified context to illuminate the operation of major mechanisms. Later theoretical developments amplify the explanatory power of earlier theories by adding layers of context that characterize real systems. In this the likely most productive strategy will be to place special emphasis on localizing the failures of these simpler models and on the estimated effects of including what is left out (Wimsatt 1987, 2002a). The chapters herein emphasize the roles of diverse types of structure in guiding and scaffolding cultural change in acting agents engaged in diverse relationships of interaction, which manifest more broadly as organizations and institutions that themselves facilitate and channel cultural change. The results point to much more powerful, versatile, and realistic theoretical resources.

ELEMENTS OF A PROTOTHEORY FOR CULTURAL EVOLUTION

The different disciplinary approaches to culture demonstrate that it has an impact on an observer at multiple levels and from multiple perspectives. An account of culture must show how these act and how they articulate. Thus, we can track different ideas, practices, habits, conventions, skills, individuals, artifacts, technologies, art forms, disciplines, religions, social structures, institutions, organizations, theories, and more. The list goes on almost indefinitely. These are not different species, however, for the multiplicity of their interactions—symbiotic, parasitic, and competitive—make them too interdependent, so the relevant unit for most cultural evolution is more like a richly interacting ecosystem or subsystem.

It may be too soon to construct a full-blown dynamic account of cultural evolution, but I wish to lay out here the conceptual geography of the necessary factors and elements of such an account, and why they are necessary. A fuller theory can then result by constructing different submodels of the interaction of these elements and studying their behavior. Two additions are necessary elaborations of the present population-based dual-inheritance theory as it stands, one emphasizing a perspective on developmental dependencies and another elaborating a notion of breeding populations relevant to cultural evolution. In conjunction with these additions, I delineate five crucial elements that comprise the components whose interactions produce cultural maintenance and change and a crucial relationship articulating them.

One cannot have an adequate account of cultural evolution without recognizing a central role for cognitive and social development. Different dimensions of culture are cumulatively acquired by individuals through a life cycle in which there is a rich structure of sequential and parallel dependencies mediating the sequential acquisition of skills (Wolf 2008; Hiscock 2014) and the parallel development of different facets of the individual. A rich and variegated array of social and cultural organizations and institutions support and structure these developmental processes (Wimsatt and Griesemer 2007; Sterelny 2012; Andersson 2011, 2013; Caporael, Griesemer, and Wimsatt 2013; Anderson, Tornberg, and Tornberg 2014). In light of this, what form should an evolutionary theory of culture take? Part of this task lies in characterizing the kinds of units that must be *used* in theories of cultural evolution, as well as what units must be *accounted for* since they are themselves cultural products. Then we have to look at the kinds of things that are evolving through interactions among these kinds of units and determine what heuristics are available for coming to understand their dynamics.

The Roles of Development and Population Structure for Culture

Two things seem central to account for the reticulate complexity of culture from an evolutionary point of view. The first is an insight from evo-devo applied to culture: development is even more important to the dynamics and structure of cultural transmission and change than it is to biological evolution. The second is a transformed notion of "breeding population" derived originally from population genetics but modified to reflect the nature and modes of cultural inheritance in individuals.

Aspects of Culture Are Acquired Sequentially by Individuals throughout a Life Cycle

I would argue that the primary target of analyses of cultural evolution should be skills. Some must be acquired before others can be. This sequential acquisition means that dependency relations within and among complex, sequentially acquired skills should play a central role in characterizing aspects of culture and their evolution. All skills are honed through practice, and complex skills are taught and assembled through a succession of stages where performance at later stages requires mastering and assimilating earlier ones (Greenfield, Maynard, and Childs 2000; Stout 2005; Thornton and Raihani 2008). Wolf (2008) documents this extensively for the coevolution and development of written language, reading, and cognition, both in history and in the developing child, and Hart and Risley (1999, 2003) show how different exposures to spoken language in deprived versus enriched language environments have multiple downstream impacts on cognition and socialization in young children. Many cultural elements cannot be understood or used unless appropriately prepared by prior experience and training. This provides a rationale for ordered teaching curricula that are organized and structured to facilitate the seamless assimilation of both intellectual content and practical skills (Warwick 2003). And one generally requires a deeper mastery of an element to teach it or to demonstrate its use than merely to use it.

A particularly striking case is the interweaving of mathematics and quantitative sciences through mutually supporting skills that are first acquired during primary and secondary school and further developed in college and graduate school. However, sequential dependencies are found as formative structures throughout most complex skills, including reading and reading-dependent cognition as well as argumentative skills (Wolf 2008). Thus, we must understand cognitive development, both in general and in particular kinds of cases. It is inadequate to scrutinize only the acquisition of general skills, like language use in speaking, along with the socialization processes studied by cognitive psychologists (many of which are acquired "spontaneously" in normal interaction). One must also investigate more particular forms of cognitive development, especially those that demand explicit training or the design of curricula, such as reading, writing, mathematics, and more specialized and professional skills such as car repair, animal husbandry, medical diagnosis, engineering drawing, chemistry, the solution of ordinary differential equations, programming in Java, and genetics. Humans have many such skills, which are deployed in different combinations. The notion of scaffolding is crucial in understanding and facilitating their acquisition (see below, section 4). These skills and their codeployment, such as the role they play in constructing group identities (see chapter 12) and configuring the status of such groups in a society, are the source of most differentiated complexity observed in culture. The work on the generation and nature of subcultures in the professions, or even within individual corporations, document what makes an "IBM man" or a photovoltaic engineer (Bucciarelli 1996). A major problem for memetic and most dual-inheritance theories is their inability to recognize the organizing structure provided by these dependency relations.

Although the curricula of the natural sciences and other academic disciplines reflect this kind of sequential dependency for complex skills, it is no less characteristic of complex manual skills in prehistory or in contemporary life (Hiscock 2014). Tostevin (2013, chapter 8 of this book) argues that the products of lithic technology can be produced in multiple ways and therefore do not reveal (by themselves) the culture transmitted to produce them. As a consequence, he formulates an observational and experimental methodology to uncover the sequences developed to make stone tools and lithic-dependent technologies in order to track their evolution. Mathematical development requires developing the subject sequentially (Warwick 2003), as does the experimental methodology of classical genetics. More advanced techniques require the mastery and practice of more basic ones.

> We have attempted to treat the subject . . . as a logical development in which each step depends upon the preceding ones. This book should be read from the beginning, like a textbook of mathematics or physics, rather than in an arbitrarily chosen order. (Sturtevant and Beadle 1939, 11)

Yet to learn a skill is not enough for the elements to be presented in the right order. The earlier elements and techniques must be mastered through practice so that their execution becomes habitual, quasi-automatic, and standardized. This is what makes assimilating these skills possible.

> Genetics also resembles other mathematically developed subjects, in that facility in the use and understanding of its principles comes only from using them. The problems at the end of each chapter are designed to give this practice. It is important that they actually be solved. (Sturtevant and Beadle 1939, 11)

The mastery of earlier skills, including the modulation of different steps, allows their *chunking (or articulation)* and deployment as components in still more complex skills in a semiautomatic manner. These, in turn, are mastered similarly, thereby creating a hierarchy of increasingly complex skills. The skills and competencies acquired in one discipline affect the possible reach of individuals into other disciplines without substantial supplementary training or close collaboration. They also attune individuals to the relevance of other disciplines for their research problems. This dependency structure of skills and knowledge thus modulates likely directions for forming interdisciplinary linkages.[6] Dependencies recapitulate the order of instruction and the design of curricula. Biochemistry presupposes organic chemistry, which presupposes general chemistry. And the dependencies continue: biochemistry is presupposed by cell biology, as cell biology is for developmental biology. And correlative skills are required as well: none of it can be taught without mathematics—often, increasingly sophisticated mathematics to master the details of some kinds of interactions. In instruction within a discipline, the same topic is often revisited multiple times as more sophisticated and powerful methods enable a more detailed and deeper analysis of the subject matter. Janssen (see chapter 4) demonstrates how earlier theory guides and scaffolds the creation of later theories in the transition from classical mechanics and electromagnetic theory to relativity theory and quantum mechanics, which suggests how this dependency structure plays a role not only in learning but also in complex theory development. So this structure of dependencies is reflected both in the evolution of experimental methodologies and in the construction of successor theories.

Such dependencies exist everywhere in culture, and there are broader consequences of this generative entrenchment for culture and technology. They affect what we can learn, what we must learn first, and where we can go from what we have learned so far. But these dependencies affect more than learning. They condition what changes can be made in our technologies and institutions, and in what order. A deeply entrenched trait in biology or culture is one that is difficult or impossible to change because so many other things depend upon it, and virtually any other change wreaks havoc elsewhere in the system (Wimsatt and Griesemer 2007). Thus, the dimensions of English and metric threaded fasteners are deeply entrenched, fixed within their respective mechanical technologies, and mutually incompatible (Wimsatt 2013).

Such changes are relatively rare, but analyzing them has methodological consequences. A successful change in a functional element of an adaptively integrated system requires that the main functions of the existing element and compatibilities with other parts be maintained. The more downstream dependencies that exist, the more demanding are the standards for a successful replacement.[7]

This process of replacement can span years or generations and is nicely illustrated with the development of the IBM 704, 709, and 7090 computers. In 1958, IBM released its vacuum-tube 709 scientific computer, the first to emulate an older computer, the IBM 704 of 1954. The move from the 704 to the 709 took four years. The emulation was so the 709 could continue to run older software, particularly FORTRAN, which ran for the first time on the 704 and rapidly become crucial for scientific computing. (This backward compatibility became a virtual requirement for newer computers and software packages from then on and illustrates my point.) IBM then put the 709 team to work producing the 709T, (or 7090), a logically identical computer substituting transistors for tubes. This is conservation of function in spades! Released only a year later, the 7090 was smaller, more reliable, and half the cost, with much lower (five volt) power requirements (no high-voltage filament transformers for vacuum tubes) and a much reduced need for air conditioning. It also ran six times faster. With higher reliability, it had much less downtime, and its higher speed allowed real-time control of processes that the 709 could not manage. But for all of these massive (and advantageous) changes in support structures, the 7090 was logically identical as far as running programs was concerned. This conserved function made it quick to develop, since it immediately had functioning software, and its other characteristics gave it much wider distribution and use. It fomented a revolution and guaranteed the role of the transistor and, in later descendants, the integrated circuit, as the basic construction element in future computers. The broader use of integrated circuits spawned an information technology revolution that has penetrated all aspects of our other technologies and has deeply modified our behavior, connectivity, and culture.

A successful change in a deeply entrenched element can play a major generative role in the elaboration of downstream elements, effectively producing in science or technology an adaptive radiation or a scientific or technological revolution, as exemplified above in the development of computers and also by the development of the internal combustion engine as a power source. It is now used in applications ranging all the way from chain saws and lawn mowers, through automotive engines, to truck and marine diesels. In each case it provided a lighter, more tractable, and more powerful substitute for steam power or, at the smaller end, a power source where steam would have been impracticable. (This proliferation was noted extremely early: Page

[1918] lists a three-page classification of types of internal combustion engines only thirty years after their invention.) And it reaches far beyond the target element. Thus, the Chicago yellow pages for 2001 had ninety-five pages, at five columns per page, listing thousands of businesses falling under dozens of different categories relating to "automotive," and the integrated chip spawned an information technology revolution that has penetrated all aspects of our other technologies and deeply modified our behavior.

The probability of a successful change declines with the acquisition of further dependencies. Elements with different degrees of entrenchment can be expected to evolve at different rates. Working from different evolutionary rates to degrees of entrenchment, together with looking at which things change or are conserved together, is a fundamental tool of inference in untangling developmental programs and in constructing phylogenies in biological evolution (Wimsatt 2015, see below). Similar design principles are integral to biological organisms, which can be seen as complex variations on the theme of cellular organization and conserve the entrenched features necessary to cellular function and reproduction. Our technologies are even more obviously organized; dependencies that recapitulate their histories exist in the design of our computer software and hardware, as well as in other technological systems, where sequential acquisition and hierarchical modularity is endemic (Arthur 2009; Wimsatt 2013), and early contingent commitments can leave a long shadow as they become increasingly entrenched. Such things are difficult and expensive to change, as was illustrated in the massive readjustments involving reprogramming software and the purchase of new computer hardware to address the Y2K threat posed by the two-digit representation of years widely embedded in software. The two-digit representation would in 2000 AD have become ambiguous between that date and 1900 AD and wreak havoc on financial and other time-sensitive data (Webster 1999). The necessary changes to a four-digit representation were extremely far reaching and costly, including massive reprogramming (contracting out work to programmers in India, thus creating an industry to compete with our own) and a substantial peak in the purchase of new computers with the appropriate hardware.

Individuals Participate in Multiple Sequential and Parallel Cultural Breeding Populations

Biological evolution (in sexual species) has a single breeding population in which diploid mating mediates heredity in a systematic fashion. The genetic bases for all traits are inherited together at the same time. Cultural evolution, by contrast, takes each individual through learning trajectories that traverse multiple successive and simultaneous parallel cultural breeding populations. The acquisition and transmission of diverse skills acquired over time include multiple "parents" in proportions that can vary from person to person and generation to generation (Wimsatt 1999, 2010; chapter 5 of this book). We inhabit and pass through a number of culturally defined peer groups (or *reference groups*) in our life cycle that delineate our identities and inform our skills, sometimes sequentially and sometimes contemporaneously (see chapter 12). The structure of such groups, membership criteria, migration patterns between groups, and the factors mediating these movements are proper objects of study for sociology. The identities we acquire in the process of participating in and across different breeding populations (Smaldino) inform us and provide values that shape future choices and trajectories.

Depending on the mobility and degree of role differentiation within a society, these trajectories may differ substantially from individual to individual. However, they can still exhibit strong similarities from "common education" or within "trades" and "professions" with standardized content and modes of training, often with certification exams or procedures to increase the heritability of skills and standardize knowledge and competence, generating subcultures within the society. Medical doctors display diplomas on the walls of their office but so do many auto mechanics! It is crucial to understand the production, maintenance, and articulation of these groups through individuals that participate first as students in successively more advanced training and subsequently as teachers to those earlier in the learning sequence for groups they comprise (Wimsatt 2001). Some of this knowledge and competence is transmitted to other groups that will use it but only require training up to a less advanced level; secondary math teachers, physicists, and engineers are typically less advanced users of mathematics than mathematicians. This is the analog of *age-structure* models in biology, though they must be elaborated further to capture cultural phenomena.[8] Such professions are in many respects organism-like and self-reproducing within the context of a broader supportive society. How groups articulate with one another is also crucial; societies in which reading and writing are promulgated only through a priestly class or in monasteries are very different from those in which these skills are acquired in a universal public education system. The groups themselves have an identity and characteristic content.

The sustained interruption of a profession that teaches and practices a complex of sequentially acquired skills can lead to the cultural equivalent of species extinction, or even the disappearance of a whole ecosystem. The destruction of the giant Chinese wooden junks—far more advanced than European vessels of the same period—by the Ming dynasty in the fifteenth century and the halting of their production for three generations led to the irreversible loss of the associated skill complex (Diamond 1997). U.S. leadership in public education in the nineteenth and twentieth centuries generated a workforce capable of mastering new machine technologies. This education, in combination with the GI Bill that fostered college education after World War II, midwifed our technological and economic ascendancy. Our current shortfall in elementary and secondary school education with respect to mathematics and the sciences is crippling our technological society. Our universities increasingly instruct able and enthusiastic foreign students who will return home after their education, thereby changing the geography of international economic competition.

A variety of training curricula have established trajectories through our complex culture where individuals experience differential exposure to parts of it and isolation from other parts.[9] The distribution and interrelation of social roles in society help to link these population groups, their distribution, the support that society provides them, and the migration of individuals through them. A prominent example is the concentration of innovation, education, jobs, marriage patterns, other institutions and organizations, and financial well-being in city hubs at the national level (Moretti 2012). Moretti uses well-structured economic and sociological information to reveal insights into what fosters creativity and innovation and builds the institutions that scaffold them. His account articulates naturally with economics, psychology, and education, as well as with how developing technologies radiate invention and expansion across fields. It is a paradigm of the kind of cross-disciplinary study required to comprehend the various strands of evolving cultural lineages.

These two elements—sequential acquisition through a life cycle and multiple cultural breeding populations—serve to indicate the central role that development and population structure play in the maintenance, transmission, and elaboration of culture. Population structure is a central feature of modern evolutionary theory, but its importance is far greater for culture than for biology due to the role of development in the acquisition of knowledge or skills and the elaboration of social structure.[10] The critical role of sequential acquisition indicates that we need to actively combine evolution and development. To account for cultural change over time, we must adopt an evo-devo perspective. Cultural population structure, mediated by social institutions, organizations, and technological scaffolding must be integrated with this and incorporate new dimensions of theorizing about cultural transmission and evolution. Few of the extant perspectives on cultural evolution have considered either (but see Wimsatt and Griesemer 2007; Wimsatt 2010; Sterelny 2012; Caporael, Griesemer, and Wimsatt 2013). An expanded ontology is needed for an integrated account of cultural evolution that accommodates these complexities (see below, section 4).

A Curious Theoretical Inversion in Biological versus Cultural Evolution

An intriguing and important difference between biological and cultural evolution is that the study of biological heredity has become more tractable with technological progress in classical transmission genetics, population genetics, and (subsequently) molecular genetics.[11] By comparison, the study of development or developmental genetics in biology, though getting easier, is a much more difficult topic. As a consequence, it is tempting to see genetics at the center of the theoretical structure of biological evolutionary theory, with development and even ecology being derivatively informed by the same source. For culture, by contrast, heredity is a mess. The possibility of multiparental inheritance of varying degrees, latencies of transmission (e.g., cultural influences can skip generations; Temkin and Eldredge 2007), and diverse modes of transfer that can vary irregularly makes the study of cultural transmission enormously complex (Wimsatt 1999; chapter 5 of this book). *However, the developmental acquisition of a cultural element has to be possible for learners in the relevant audience so that it can be transmitted and employed. If the appropriate subjects can learn it, then it should be easier for us to study and untangle.*[12]

Specialization reduces the technological overhead that must be mastered by any one individual, vastly expanding the complexity that can be managed by a culture. But there is another crucial element. Learning the technology becomes manageable in part through the fact that past technologies can be chunked or "black boxed" and used without understanding or transmitting all of the knowledge necessary to generate them (Wimsatt 2013); any generation need only study the outermost layer of an accreting onion. Thus, in the teaching of science, it is not necessary to engage in a complete recapitulation of theory development; a designed representation of theoretical accounts resembling earlier simpler stages suffices, as it captures simpler phenomena or acts as a scaffolding for more complex cases (see chapter 4). Technology represents an even more extreme case. Since massive amounts of detail often can be collapsed into a portable result, highly complex nested sets of technological dependencies can be transmitted.[13] Although it would take an enormous team of specialists to dissect, understand, and be capable of reproducing any piece of modern technology from scratch, we only need to master its outer "user interface" or in design, the chunked components that are assembled and articulated at that level.

Therefore, heredity and development in some respects interchange roles in the study of biology and culture. For cultural evolution, a study of developmental, learning, and teaching processes could provide essential levers in understanding cultural heredity and supply the core for understanding cultural evolutionary processes, just as for biology the study of heredity has provided a crucial tool in understanding development. Indeed, I believe this will prove to be the case and yield theoretical perspectives for cultural evolution that will look quite different in spite of many recognizable similarities with what is found in biological evolutionary theory. Will development emerge as providing the core architectural elements for cultural evolution in a way similar to the role played by genetics in biological evolution? It will be interesting to see how this develops.

Additional factors common to both are the layered complexity and generative entrenchment of elements, both in biological evolution and technological development, leading to the evolutionary conservation and cumulative architecture that makes the study of their histories an essential source of insight in both areas. Thus, I have found that my study of original sources in the history of genetics rendered intelligible otherwise mysterious aspects of modern theory, or its choice of certain problems, as crucial, and often with new handles on modern disputes.[14] So developmental generative entrenchment has had a systematic effect on evolving systems from biology and culture.

RELEVANT UNITS OF THE CULTURAL SYSTEM FOR A THEORY OF CULTURAL EVOLUTION

The roles of development and population structure for culture suggest five kinds of units that must be included in any adequate theory of cultural evolution to properly capture the dynamics of cultural change. In discussing them, I will attempt to suggest how they articulate with relevant capabilities and disciplines, sketching the causal linkages pertinent to cultural evolution that are scaffolded by culture itself and characterize the scaffolding relation. Second, I argue that cultural evolution can be seen as interrelations of evolutionary change in several different kinds of processes, which are driven or modulated by a number of correlative changes in other evolving lineages.[15] Finally, I comment on the relevance of time scale in the analysis of these processes and their interactions.

The five kinds of elements necessary for an adequate theory of cultural evolution that account for the role of scaffolding in articulating these elements can be divided into two main categories.

Category 1

1. *Transmissible or replicable elements* (TREs). Examples of TREs include artifacts, practices, and ideas that are taught, learned, constructed, or imitated. These include ideational, behavioral, and material items, which are capable of being modularly decomposed or chunked and black boxed hierarchically.[16] Thus, they can engender multiple levels of organization that may not all be accessible to inspection at a given time. Their modular structure can be circumscribed either within an individual's cognition, capabilities, and interactions with an environment or by an organization or profession that assembles a team of individuals that collectively have the necessary capabilities. There will be populations of TREs at different levels of organization that show variation and therefore can be targets of differential selection.

Conceptualizing TREs as memes has been criticized heavily. The loose characterization of memes allows almost anything to count as one. As a result, it is not possible to explicate how the resulting heterogeneity of items can be reproduced or transmitted in any unitary way. This is especially problematic for an account that focuses so strongly on heredity. This heterogeneity becomes more manageable when one sees that particular kinds of TREs are part of a complex array of elements that interact to produce cultural change and that many of these causal structures facilitate or constrain their reproduction (Griesemer 2013b). Unlike memes (Dawkins 1976), TREs are not autonomous, self-replicating elements. Their spread is conditioned by developing individuals through a life cycle, an aspect not utilized in standard dual-inheritance accounts, and their reproduction is mediated by scaffolding elements from category 2 (Wimsatt 2010, see below). We must not make the mistake of memeticists and fail to see the contextual forest for the TREs.

2. *Developing biological individuals* (DBIs). DBIs develop, are socialized, and are trained over time in multiple cultural breeding populations. The earlier training of DBIs affects their capabilities, exposure, and receptivity to subsequent TREs or to participation in or interaction with elements from category 2 (below). A developmental process of sequential acquisition and assimilation is crucial because the developmental state of an individual determines whether they are "infectible" by a TRE, as well as how they will interpret and use it. The culturally induced population structure of individuals that mediates the exchange and development of TREs is the main driver of cultural evolution and is also a major element of social structure, especially for generating identities, and this has an impact on power structure. This population structure is generated as a consequence of various lower-level units that compose the population (Wimsatt 2010, 2013). Thus, the cognitive and social characteristics of DBIs matter, and the study of cognitive heuristics is pertinent to elucidating the architecture of culturally induced population structure (Sperber 1996; Gigerenzer et al. 1999; Heintz 2013). Individuals may differ in their success or competence at specific skills and therefore be preferred targets for imitation or association (see chapter 12); they may use other heuristics, such as conformity bias, in deciding who to imitate (Richerson and Boyd 2005).[17]

DBIs are socialized through their developmental life histories and make culture through social and encultured interactions, especially in the acquisition, application, and extension of complex skills. These include both common skills (e.g., language use or socialization, largely in family dyads and family or small peer groups) and specialized skills, such as those acquired and practiced in differentiated roles attached to institutionalized task groups. A distinctive array of specialized skills can be grouped together as a *repertoire* (Leonelli and Ankeny 2015), which gives unity to scientific specialties and helps to organize their research efforts institutionally (Gerson 2013a, 2017). The structure and texture of repertoires characterize much of the complexity we find in culture. DBIs also have psychological tendencies that affect who they interact with and how and what they draw from others.

Gene–culture coevolution (e.g., Richerson and Boyd 2005) and memetic-inspired theories incorporate only some of the structure of TREs and DBIs.

Development and the order-dependent sequential acquisition of complex skills is ignored in extant theories. Population structure is not a significant element in most gene–culture coevolution accounts except for the recognition of biological kin and group selection, along with the fitness possibilities of trait-group effects. The most significant omission is culturally induced population structure and its scaffolding effect on training for complex skills, including coordinated tasks with role differentiation and group identity formation (see chapter 12).

Structures of this kind emerge naturally from interactions of DBIs with elements of category 2, providing further reason for their inclusion in any adequate theory of cultural change.[18] These culturally created structures are constructed parts of the human cognitive, normative, and affective environment that scaffold the acquisition and the performance of knowledge and skills and coordinate their acquisition. Thus, the choice of a profession (an organization with richly structured curricula and institutional norms) scaffolds subsequent learning and commits one to a trajectory of exposure to relevant knowledge and procedures, institutions, and population structures that condition their life course (B. Wimsatt 2013; Warwick 2003). These trajectories structure the sequence of the peer groups we move through and the dependency relations among skills utilized during this migration. This substantially reduces the complexity of social and cultural structure that an individual must face, making the cognitive tasks more manageable. Whether it is promotion to middle management (which may change friends and neighborhood as well as job tasks) or a group identity change associated with age-structured roles (like becoming parents or grandparents), this culturally induced population structure brings order—both in navigating and in theorizing—to an otherwise forbidding complexity of overlapping peer groups.

Category 2

1. *Institutions.* Institutions are ideational structures at a social or group level that constitute or contain explicit or implicit (and commonly internalized) normative rules or frameworks that guide the behavior of individuals: "A collective enterprise carried on in a somewhat established and expected way" (Gerson 2013b). These rules or frameworks apply to individuals either universally or as classified by society for a certain role, class, or profession (e.g., social norms of behavior, legal codes, and transition rituals like bar/bat mitzvahs and graduations). They are diverse and can be quite complex.

The complicated expectations for individuals participating simultaneously in diverse institutions indicate an important role for habit in the formation and explanation of behavior (Duhigg 2012). More broadly, larger swaths of culture can be seen as systems of institutions that are made up of conventions where each institution mediates a collective capacity to carry out a task (Gerson 2013a). Institutions also evolve under changing conditions and demands from social groups. The promulgation and elaboration of engineering standards for different kinds of interchangeable parts in manufacturing was a critical element in the explosion of technology beginning in the nineteenth century and serves an important coordinating function for the design and manufacturing of parts that must meet many constraints to function properly in diverse complex mechanisms (Wimsatt 2013).

2. *Organizations.* Self-maintaining groups of individuals that have self-organized for some purpose or set of shared purposes are organizations. These are like DBIs, but at a social/group level, and include interest groups, such as unions and political parties, firms, nations, and professions.[19] Departments at universities are an excellent example. They recruit students and faculty, produce academic products (papers, books, technology, students), teach classes, and inculcate professional values. They may undergo development as a function of their size, demography, and histories. Sometimes, they reproduce, either with characteristic members that propagate to constitute similar groups or by spinning off new organizations that reflect some of their values, aims, and structure. Although I focus here on their role in transmitting elements of culture, such groups are also commonly foci of political action and the expression of power through their common purposes.

Complex group interactions in organizations allow the production of entities, artifacts, and practices that individuals could not generate on their own (Theiner, Allen, and Goldstone 2010). Organizations develop the capability for cooperative and coordinative interaction and socialization and may also interact competitively. Group structure manifests on different size and time scales, sometimes as a hierarchical organization and sometimes in a stable manner that cuts across hierarchical relations. Organizations mediate much of the specialized role differentiation and training that make our society and others so reticulate.

Organizations can be seen as socially or culturally determined *core configurations* that are widely found in different human populations; they are naturally configured groups of individuals of different characteristic sizes adapted to different functions (Caporael 1997; 2013; Sterelny 2012). These act as cultural breeding populations to define, maintain, elaborate, and teach knowledge, procedures, and values and are central elements in identity formation. Organizations and their interactions play a formative role in generating institutions that provide further structure to their identity and interactions (Murmann 2013). Individuals follow trajectories through organizations, pursuing their ends while at the same time having them shaped by the groups they inhabit or pass through, with the institutions appropriate to those organizations coming to bear in relevant contexts along the way.

3. *Artifact structures.* Artifacts or physical structures mediate short-term activities or processes (like those found or used in a work environment, including physical tools, and reading and writing or utilizing or producing specialized language, serving multiple functions) or provide physical infrastructure that is maintained on transgenerational time scales to yield "public goods." These may be produced, interacted with, and maintained by organizations like manufacturing firms or by institutions in society at large. Both units can facilitate a range of activities or, in other circumstances, provide specific infrastructure for a delimited subgroup, such as practitioners of a specialty or users of a specialized technology. Markets mediate the development and distribution of new or transformed artifacts or procedures involved in using them. Complex technologies require and generate complex distribution networks and a host of standardized practices (Wimsatt 2013; Arthur 2009).

Many regard artifacts only as products of culture rather than as elements or producers of culture, especially if artifacts are treated as external tools for accomplishing tasks instead of integral parts of thought processes (e.g., Richerson and Boyd 2005).[20] However, embodied theories of cognition and of distributed cognition reveal that artifacts and the structured interactions and motor activities they induce play an essential part of the cognition of individuals and groups; they must be recognized as components of thought processes (Wilson and Clark 2009; Theiner, Allen, and Goldstone 2010). Artifacts not only extend and change our cognitive skills (Wolf 2008) but also facilitate the formation of new kinds of groups as cognitive units and help segment us into new skill groups and cultural population structures (see chapter 3). Although this can be seen as a friendly extension of niche construction theory (Odling-Smee, Laland, and Feldman 2003), the conceptual tools required for the "cultural niche" must encompass much more than niche theory has currently embraced (see Wimsatt and Griesemer 2007). These components of distributed cognition make possible particular cultural interactions (e.g., Internet communication for the collective solution of complex problems in data analysis) and products (e.g., open source software; Nielsen 2012). Scientific research, practice, and institutions are important examples of this collective activity that is a technological and cognitive expansion of our niche (see chapters 2 and 3).

Institutions, organizations, and artifact structures are components of a society and of many things we find in culture. Government bodies are hybrids of all three of these entities, as are most other complex cultural constructions. Organizations at one level are the primary source of formal institutions at another level; networking interest groups are the source of informal institutions.[21] An important contrast between biological and cultural evolution enters here: the single breeding population for biology is replaced by multiple overlapping reference groups of culture, each being a possible source of interaction and learning or the transmission of knowledge and practices (e.g., professional associations, places of employment, political and governmental affiliations, and religious congregations, inter alia). Each has characteristic norms of behavior and modes of interaction—a subculture—and their structure is modulated by core configurations of people of various sizes that we find natural (Caporael 1997).

Scaffolding

Scaffolding refers to structures or structure-like dynamic interactions among performing individuals that are the means through which other structures or competencies are constructed or acquired by individuals or organizations (Wimsatt and Griesemer 2007; Caporael, Griesemer, and Wimsatt 2013). Scaffolding serves a function and is thus a many-termed relation (Wimsatt 2002b); *something* scaffolds an *action* or class of actions for an *individual* or group of individuals, often *in a larger system* of interactions, in a characteristic *environment* or set of environments relative to *a goal*. Material or ideational entities that contribute to achieving this goal are *scaffolds*.

How does scaffolding emerge? Common patterns become habitual, and if widespread through conformity bias, coordination games generated by common advantage, or other means, can become standardized. This generates normal modes of behavior for all sorts of regular behavior and activities. Many cultural elements emerge in conjunction with this standardization and are specifically designed to aid in constructing or developing competencies among individuals and organizations. Thus, chaperone molecules scaffold the correct configuration for folding proteins, and the cell scaffolds gene replication and expression so profoundly that the cell is arguably the relevant reproductive unit, rather than the gene or genome.[22] A similar perspective points to the insufficiency of methodological individualism, which is the view that higher levels of social organization can be characterized exhaustively in terms of component individuals, including their internalized thoughts and actions (e.g., "*Homo economicus*" of rational decision theory and economics). For the encultured and socialized human, whose agency is richly scaffolded in multiple dimensions, this perspective is empirically and conceptually inadequate.

It is critical to distinguish agent scaffolding, artifact scaffolding, and infrastructural scaffolding (Wimsatt and Griesemer 2007) because they cross-classify the foregoing units of and for theories of cultural evolution. Scaffolding is not necessarily introduced intentionally, but its presence is part of a means-end chain of action directed toward one or more goals. *Scaffolding for individuals* includes family structure, schools, curricula, disciplines, professional societies, church, work organizations, interest groups, governmental units, and laws. Some of this scaffolding is imposed by organizations or institutions, though individuals also pursue it actively, such as embarking on a normal training trajectory to achieve competence and certification in a profession. *Scaffolding for organizations* include (for businesses) articles of incorporation, corporate law, codes of ethics, manufacturers' organizations, dealerships, chambers of commerce, and distribution networks for manufactured parts. *Infrastructural scaffolding* is so broadly applicable that it is sometimes difficult to specify the pertinent individuals and organizations or what competencies it facilitates. Language, both spoken and written, is so obvious as to be easily overlooked. Mathematics and computer languages are natural technological extensions. Janssen (chapter 4) documents how earlier theoretical structures in physical science provided crucial scaffolding for the development of newer theories; specialized experimental technologies—from microscopes to statistical techniques—can do the same. Our technological civilization has many systems of infrastructural scaffolding: highway, sea, rail, and air networks; shopping centers; containerized shipping; distribution networks for gas, water, power, telephone, and sewage; warehouses and reservoirs; public transport; Internet; and waste removal.

The census offers an especially poignant example. Its diverse uses by governments for the distribution of resources or structuring and the distribution of political power means it is a deeply entrenched feature of our society. Markets, as institutions, are also infrastructural scaffolds that elaborate and coordinate a host of businesses, products and practices, and the choices and activities of agents.

TREs, DBIs, institutions, organizations, and artifact structures are the requisite kinds of units to formulate minimally adequate accounts of cultural evolution in its current complexity. However, what we still lack are substantive analyses that show how these units articulate in more elaborate regulative and production structures, and this should be a topic of continued and elaborative research. These structural configurations would presumably differ for different levels of organization in society, for its differently articulated microcultures, and for different cultures as a whole. This would be akin to different phyla in the biological world that are elaborations of different major body plans, each of which have diversified within different ecosystems. Thus, mathematics through calculus and statistics provide a common backbone for all the mathematical sciences, which add further differentiated skills, and programming is rapidly becoming equally essential. Structural configurations of these units generate and mediate power relations and regulate the distribution of information and resources. They also indicate what kinds of disciplinary approaches, in various combinations, are required to understand the dynamics of cultural change, which is likely most fruitfully tackled by examining how the different units are articulated in a given domain of culture. This, in addition, would tend to highlight the fact that many cultural processes require inputs from more disciplinary perspectives than they now receive and point to new interdisciplinary projects. The notion of scaffolding is crucial throughout; it creates and assists processes of varying degrees of entrenchment that extend or facilitate the exercise of our capacities (Caporael, Griesemer, and Wimsatt 2013).

ARE OTHER ELEMENTS REQUIRED?

I have introduced and articulated five kinds of units and the relations between them (especially scaffolding) that provide a more structured account of cultural evolution and begin to coordinate the joint contributions of many diverse disciplines and approaches. Are these sufficient? At least two other perspectives have claimed universality (both applicability and sufficiency) over the range of human behavior: intentionality and markets. How are they relevant, and what relations do they bear to the above discussion?

The Role of Intentionality

Means-end reasoning, planning, and the construction and use of complex tools are crucial to human intentional actions. Lane et al. (2009) see scaffolding and action as so integrally linked that they advocate a conjoint agent–artifact space to delineate the basic entities of culture. Many of the intentions implicit in cultural entities that are not features of explicit conscious plans will be scaffolding relations, such as products of intentional actions by others. These intentions are realized in a complex cognitive niche that is a product of multiple institutions, organizations, artifact structures, and standardized modes of individual behavior. They are woven deeply in the fabric of artifact design and construction, patterns of convention, standards, norms, institutions, and the acquisition of skills. That scaffolding is so central to the analysis proposed here is a reflection of the importance of intentions. Intentions manifest as emergent meanings in spoken and written language through combinatorial generative systems of communication (Wimsatt 2013). Thus, we must recognize the intentionality of complex differentiated groups that cooperate to produce technological and intellectual products (Kidder 2001; Theiner, Allen, and Goldstone 2010). Meanings and intentions both derive and emerge from heterogeneous relationships among ideational and material structures and processes.[23] Adding an explicit treatment of them is desirable and will be necessary to comprehend how they articulate with and emerge from the other five elements of culture, but this is a task for another time.

Markets

Many economists behave as if the market (or markets), together with the *Homo economicus* of rational decision theory (or its satisficing successor agent from behavioral decision theory), is an adequate framework for understanding all cultural activity. Human social and behavioral practices that facilitated institutions of exchange have been a crucial element in the evolution of culture and in coordinating behavior across distant places (Seabright 2004). In many respects, it has operated as an integrating force. Why is it absent from the primary catalog of elements required for an evolutionary theory of culture? It plays such a central role in Western economies and their colonial activities throughout the world. Perhaps it is an institution that requires special attention, though I confess to some uncertainty about how to introduce it. Although it is clearly infrastructural, does it play a special or distinctive role as an institution that is more important than others? Spoken and written language are two other infrastructural elements that act in related but disparate ways with similar breadth and import (see chapters 9 and 10). What about governance and power relations, ethnic identities, and other elements that achieve a coordination of societies as a whole? Perhaps each of these demands special treatment.

The perspective of the market does capture some features that are difficult to isolate as localizable effects of the other five elements of culture. For example, Moretti (2012) returns again and again to the "spillover effects" that arise from the concentration of individuals with technological skills in companies that make a city or region an "innovation center," improving the number of jobs, salaries, the quality of education, and the standard of living for unskilled laborers in the same area. He demonstrates how these effects emerge out of an interaction of market conditions and other relationships in a way that is not exhausted by the consideration of specific organizations and institutions. Thus, there is clearly more work to do in analyzing the unique contributions of markets to our understanding of cultural evolution.

TIME SCALES AND ENTRENCHMENT

I have already discussed (section 3.1) how widespread generative entrenchment is in the organization of complex systems in biology, culture, and technology (Wimsatt 2013). Differential entrenchment and consequent differences in evolutionary rates have been powerful inferential tools for analyzing the structure of developmental dependencies and the structure of phylogenetic relations in biology (Wimsatt 2015). They should be also for technology and for culture. The more stable elements play an important role as architectural foundations for the construction and elaboration of adaptive structures of increasing complexity and are recognizable in biology (e.g., body plans), in culture (through the roles of language and socialization processes), and in technology (through the roles of mathematics, science, and the methods of mass production). These are all important handles for the analysis of complex systems. But the existence of processes acting at different rates has other important consequences for the structure of evolving systems.

The rate at which processes happen and the rates at which they can affect change are crucial elements for understanding the dynamics of evolutionary change in biological systems. The same is true for culture. It is commonly claimed that cultural evolution operates much more quickly than biological evolution. However, it would be more accurate to say that evolutionary processes in both domains operate on a wide range of time scales, some of them overlapping to a significant degree. Bacteria can evolve significantly in weeks, with the measles virus becoming more virulent as it moves from host to host within a family while adapting to their common genetic architectures (Wills 1997), and insect pesticide resistance emerges in a few years. In contrast, in cultural evolution, some things evolve very slowly: Acheulean lithic point technology persisted and evolved gradually for over a million years.

The issue of time scale is important in part to sort the relative significance of different transmission processes. It is usually assumed that processes acting faster in time tend to dominate those acting more slowly and (intentionally or unintentionally) evolve to act as control structures for slower processes. In this context of cultural evolution, it seems clear that the maximum and average rates of change have increased substantially. We can document an interesting transformation from societies that valued stability and resisted change (perhaps culturally adapted to oral transmission) to those that valued innovation and change (often dated to the Renaissance and the concomitant rise of capitalism). In addition to time scale, the magnitude of the effects of cultural evolution has also increased, primarily through our development of methods of mass production (Wimsatt 2013) and the consequent increasing mobilization of energy and reticulate complexity of our technology. Indeed, anthropogenic global warming shows how these effects can threaten our very existence.

HOW CAN WE bring order into the study of such a multifaceted entity as cultural evolution? Our characterization of the five kinds of entities, plus scaffolding, that are required for any adequate account gives more room and resources to classify phenomena and comprehend diverse mechanisms of change that can relate productively to the approaches of existing social sciences, and this should be an aim for future development of the theory. Since our cultural activities take place in a much richer and more structured context than is typically adumbrated, our analyses must be adjusted accordingly. Despite its obvious power and adaptability, the absence of a detailed developmental component is a major lacuna in the Boyd and Richerson account; without it, all sorts of dependency relations cannot be explained or utilized in the explanation of other features of culture, and we have no structure on which to hang the different breeding populations we experience through our life trajectories. Accounts of sequential acquisition are necessary to understand who is able or likely to acquire specific cultural skills and traits or be influenced by certain ideas and forces. Additionally, the absence of diverse forms of culturally induced population structure hamstrings theoretical frameworks from capturing the reticulate and interwoven character of cultural evolution. We need to recognize that organizations and institutions develop and that relations of scaffolding and entrenchment offer tools for understanding the interlocking means-end structure of social action—organizations and individuals interact with and through artifact structures as guided by institutions. Even niche construction, which includes a developmental component, lacks the necessary theoretical perspectives on diverse forms of scaffolding (Wimsatt and Griesemer 2007), and the role of technology in facilitating our cognitive capabilities lies unelaborated.

The fact that TREs, DBIs, institutions, organizations, and artifact structures relate naturally to work in sociology, history, ethnography, the history of technology, and the history of science shows both the need for implementing *interdisciplinary* approaches to cultural evolution and finding specific resources that can enrich the connections among elements in our theoretical framework. This is a welcome change from prior approaches, such as that of the reductive sociobiology of the 1970s. Then, the suggestion was that social theory should give way to a sociobiological framework through displacement—like "urban renewing" a neighborhood with a bulldozer. In this approach, our cultural evolutionary perspective should articulate with developments in the traditional social sciences in a negotiation between equals—how can the new perspective enrich traditional insights? But this suggests a new danger: Do we need to study everything in order to understand anything? How can we avoid making the investigation of cultural change an impossibly complex task? There are reasons why the dual-inheritance theory of Boyd and Richerson has been so successful in terms of the elements they chose to model and reasons why we should be careful in arguing that the further complexities discussed here must be considered.

First, I want to note that the aim of this chapter is not to develop a complete adequate theory of cultural evolution. It is, rather, to sketch and to argue for a conceptual geography of the major elements required and how they articulate. Presumably, progress will be made by developing parts of this framework. We should not aim to capture all details of cultural phenomena but rather ask what aspects of culture might be usefully systematized. Then, efforts can be directed at including the major features and aspects of culture relevant to its evolutionary change. Progress was made in studies of heredity with Mendel's systematic work on pea plants and the Morgan school's mapping of *Drosophila* chromosomes. Crucial in both cases was the right methodology, which included significant simplifications in the experimental system (Kohler 1994) and "the right organism for the job." But *Drosophila* proved intractable for questions of development until the discovery of the *Hox* gene complex and its use as both a subject and as a tool in developmental genetics. This articulates naturally with the "problem-centered" approach argued for by Brigandt and Love (2012) since "the job" is always an identified problem with its own history and structure. Such problems are elaborated and restructured through a productive research program, but their identification and operationalization is crucial. In this we must remember that some problems are tractable with the resources at hand and others are not. And this reinforces that finding the "right organisms"—the peas of cultural heredity—is equally critical; patents and scientific diagrams are just two of many promising candidates (see chapter 6 of this book; Griesemer and Wimsatt 1989; Wimsatt 2012). But we also need to expect that different methodologies will be appropriate to different problems—for example, due to limitations of data, relevant theory, or computational complexity. No one would propose a population genetics analysis of the terrestrial origin of the vertebrates (even though it surely applies in principle), but we would look for handles within developmental genetics and within functional morphology that could give insight into particular aspects or stages of the emergence. We should expect similar disciplinary handles to give leverage on different aspects of cultural evolution.

We should expect cultural change to provide these kinds of paradigmatic examples of evolutionary change while investigation can steer away from intractable complexities that would make any such account exceedingly difficult. It will often be possible to study and confirm the operation of some of the elements producing cultural change by abstracting away from or idealizing others and through the comparative analyses of cases with selective similarities. However, an adequate evolutionary account that emerges from paradigmatic examples should offer reasonable explanations for why such complex cases are so refractory to an evolutionary analysis. For example, the characterization of the properties of "evolutionary meta-ontogenies" as a complex of interacting and embedded entrenchment processes acting on different time scales (Wimsatt 2013, 91) provides an evolutionary account that explains why some cultural elements seem to resist precise characterization as either "developing" or "evolving." Thus, habits develop through repetition, and skills develop through the accumulation and coarticulation of habits. Both of these develop through the maturing capabilities of a growing and developing individual, who develops the capabilities for a given career track. That individual may then go to work for IBM back when it was known for punch-card readers and mechanical adding machines. IBM developed to become the prime provider of computing machinery, which developed from mechanical relays through vacuum tubes to transistors and integrated circuits. In the early stages, IBM also wrote software and provided integrated business solutions, but the development of minicomputers gave other firms like DEC and Data General room to grow, particularly for scientific applications. The DEC-20 provided a new paradigm of multiuser computing and the emerging "mainframe," and the emergence of the microcomputer and Microsoft as an independent software producer spawned an efflorescence of third-party hardware and software accessories. AT&T, originally a telecommunications company, produced UNIX, and the government spawned AR-PANET, which became the Internet, and the evolution and development continues. So here we have articulated developing habits, skills, individuals, firms, industries, and technologies, all on different time and size scales, with a host of emergent properties at all scales. This is clearly both development and evolution, in multiple places on different scales, depending upon the problem and the question regarding it. Given the crucial differences between evolution and development in biological theories of evolution, this has suggested to some a dangerous sloppiness that seriously compromises theories of cultural evolution (Fracchia and Lewontin 1999; Gerson 2013b), but we have the tools to address this in the dissection of cases like the preceding.

Second, following up on using abstraction and idealization, I suggest we take seriously the exploratory use of "false models" in which we construct accounts incorporating some, but not all, of the elements of which we are aware (Wimsatt 1987, 2002b), sometimes with additional false simplifying assumptions. This was characteristic of the panmixia assumption discussed in the introduction to this volume (Wimsatt 2002a). These partial accounts of the structures relevant to cultural change can be mobilized to see both what else we can relate to them and also what we *cannot* account for. The latter (especially) can suggest other structural elements or perspectives to include in a more robust theoretical framework. Agent-based models would be a particularly salient tool for this task, although here (where it is relatively easy to add a great amount of detail) it is particularly important to start with a simple orientation, to which various complexities are added, to better assess their effects (e.g., Andersson 2011, 2013). Given the diversity of cultural systems, this endeavor will surely yield a branching tree of multiple models rather than a linear sequence of increasingly "better fits" of a single model. (Schank and Koehnle [2007] consider an example of such a branching model tree.) The necessity to make central both the role of dependency in the acquisition of complex, sequentially acquired skills and the culturally induced population structures through which we proceed in acquiring and practicing them, as well as to explicitly utilize all five elements (TREs, DBIs, organizations, institutions, and artifact structures) in rich relationships of scaffolded interactions, reaches across the necessary variety of disciplines to apply in any contexts where culture or cultural change are studied. The question of whether and how structures of intentionality, economics, and power relations are integrated into this conceptual architecture remains to be answered, and other perspectives not covered here will need to be recognized. Our work is just beginning.

NOTES

I thank two anonymous reviewers for their suggestions on an earlier version of this chapter; Marshall Abrams for his commentary on a talk given at the Pacific American Philosophical Association drawing on some of these ideas; Alan Love, as always, for his multifarious and detailed insightful suggestions; and Barbara Wimsatt for her suggestions to improve intelligibility. More broadly, I thank Penny Winton and the support of the Winton Professorship since 2010, which has made possible my stay at Minnesota, my research, and the conference at Minnesota in the fall of 2014 that midwifed this book.

1. Actually, if we incorporate the complexities introduced by our microbiota, the biological and cultural ecosystems are closer (the microbiota are richly horizontally and vertically transmitted, for one), but traditional neo-Darwinism has only begun to address this and is far from incorporating its complexities.

2. It is tempting to think that these are alternative channels that merely duplicate one another, with later channels just faster, but this would be a serious mistake. Thus, the telephone not only is faster than the telegraph but

captures vocal emotive information that the telegraph does not. And written language stimulated a massive increase of a more sophisticated and detailed vocabulary in addition to leaving a persisting and potentially cumulative record (Wolf 2008).

3. Even among those who accept a "blind variation and selective retention" paradigm (Campbell 1965) or "heritable variance in fitness" (Lewontin 1970), schematic requirements for an evolutionary process leave the relevant details frustratingly underspecified, with no tools for further guidance. The diversity of possible units, complexity of hereditary processes (Jablonka and Lamb 2005), and fusion of heredity, selection, and developmental processes for various aspects of culture (Wimsatt 1999) pose challenges unique in comparison to biological evolution.

4. Part of this is due to a rejection of both earlier (largely nineteenth-century) progressivist evolutionary views in anthropology and imperialistic (and simplistic) approaches to human behavior from sociobiology in the 1970s.

5. Is evolutionary psychology, with its focus on heuristics and the search for "Machiavellian intelligence," correcting this? No, because the theoretical resources in this area are too narrow for what is required of an adequate account of cultural evolution (see below, section 3).

6. This is a double-edged sword: differences in the characteristic methodology of two different disciplines may be misleading when trying to understand how they use a common resource or tool. For example, the relative certainty characterizing mathematical inferences may lead empirical scientists to misunderstand how mathematics is used as a tentative and exploratory tool in constructing possible templates for patterns of phenomena. These templates do not give certainty to the results of the models, which often deliberately use false assumptions. Instead, these models are more instructive for the ways they fail than for how they succeed (Wimsatt 1987, 2007). Scientists who are not modelers may be improperly skeptical of the usefulness of "unrealistic" or "simplistic" mathematical modeling in their empirical area. This makes it crucial to be aware of these methodological differences.

7. Strategies for facilitating a change in a deeply entrenched element can include constructing a supportive environment to meet some of the functional requirements in other ways (common in major organ surgery), such as duplication (as in dipoidy or gene duplication) and encapsulation (so that

the bad consequences of not meeting some of the requirements are not allowed to propagate into the broader system).

8. Biological age-structure models focus on viability (what proportion of individuals survive to the next stage) and fertility (how many new organisms are produced per individual at that stage). This has cultural analogs in professional training, where administrators of programs must worry about how many students survive through a given level of training and whether enough of them begin teaching the relevant skills to maintain the profession in the numbers required. Cultural models using detail of this kind can yield useful information (e.g., Andersson, Törnberg, and Törnberg 2014), though further elaboration is necessary to answer other questions.

9. Selective isolation is no less important than selective exposure. There are limited resources for learning, and exposure to multiple diverse things may dilute and frustrate those efforts.

10. The elaboration of social structure has led some to argue that we must abandon the population structure characteristic of evolutionary biology in favor of an organizational and institutional structure to account for cultural evolution (e.g., Lane et al. 2009). I think we need both perspectives.

11. Although epigenetic processes and their interactions with developmental and ecological factors are demonstrating a greater complexity than originally thought (Jablonka and Lamb 2005).

12. Here, acquired elements (reading, writing, arithmetic, and other taught skills) are easier to investigate because we have teaching methods for them, unlike "innate" skills such as spoken language, whose scaffolding for acquisition has become internalized and must be studied experimentally and through the study of cognitive anomalies.

13. This phenomenon is visible in the evolution of automobile owners' manuals. The owner's manual for the Ford Model T (made from 1908 to 1926) dealt with topics that were quite complex. It gave detailed instructions for all but the most demanding repair operations (e.g., a paragraph lists the eleven steps necessary to remove the engine). The owner's manual for my 1962 Volvo 122S was still quite detailed, though much less so. The gory details had been moved to the "shop" manual (which I purchased)—the owner was no longer expected to play a role in the repairs, although doing so was still possible for simple to moderately complex tasks. By the time I bought my 2013 Audi A4, the diagnosis of repairs had become fully computerized, in part because integrated circuit chips had taken over multiple regulatory

and sensory roles. Diagnosis and repair have become only possible at the service department of a dealer. Repairs involve multiple specialized tools and often involve computerized notification that a module is defective, rather than needing to understand what is wrong with it. The suitably longer "shop manual" is available only on CDs, for which you need the correct computer and software in order to read it. More generally, the complexity of automobiles has grown exponentially, necessitating this increasing specialization and knowledge segregation of roles, as well as technology for scaffolding the diagnosis, maintenance, and repair.

14. For example, the importance of blending inheritance and its role in the history of population genetics (and its further application in understanding the units of selection controversy in modern times) was particularly illuminating (Wimsatt 1980, 2002a).

15. The focus here is on mature culture, not the emergence of culture in the course of evolution. An account of this, which interdigitates naturally with Wimsatt and Griesemer (2007), is Sterelny (2012). See also Tostevin (2013), Hiscock (2014), Morgan et al. (2015), and Stout et al. (2015) on the importance of the evolution of lithic technology).

16. Iterative modular decomposition, or chunking and black boxing, is a crucial feature of both the mechanical and the cognitive assembly of larger complexes of machinery and practice, going back to Miller (1956) and applied to more complex perceptual tasks by Chase and Simon (1973). For technology, see Latour 1987 (who introduced the term *black box* in this context), as well as Arthur (2009) and Wimsatt (2013) for further development. Black boxing is a crucial feature of most complex sequential skill acquisition.

17. Richerson and Boyd elaborate these heuristics of social learning but ignore the sequential dependencies in the development and practice of skills. Skills also have a structure, which is realized as individuals acquire them from experts and apprentices, with those of greater skill playing a role in the instruction of those earlier in a trajectory. This yields a hierarchy of training where top-level experts are not responsible for training early neophytes (Wimsatt 2001).

18. Boyd and Richerson (2008) analyze the properties and evolution of social institutions, but they do not address how institutions structure learning in development. These effects and the norms associated with such institutions should, for example, increase the heritability of the affected cultural traits. If changing environments are reflected in institutions, these can mobilize systematic changes in transmitted characters, such as rapidly updat-

ing the content taught in a class by requiring continuing education among teachers.

19. These similarities are not identities, and some have overextended the analogy in the U.S. legal system.

20. Richerson and Boyd (2005) characterize culture as transmissible information, which they further characterize as a mental state (conscious or not) that affects behavior (5). This rules out material artifacts, an important and problematic move. They discuss technology primarily to make the point that it evolves through piecemeal incremental improvement (51–53).

21. Informal institutions require their own treatment and should be targets for social psychology. The processes through which they are formed, as well as how and when they are formalized, are key elements in the elaboration of culture.

22. Selfish genes and selfish memes are conceptual mistakes for parallel reasons. Memetics ignores the role of organized context, internal and external, which enables or facilitates memetic transmission (Wimsatt 2010).

23. Wittgenstein's (2009) notion of a "language game" that articulates language and interactions with material artifacts is pertinent here, but the simplicity of his examples may be misleading.

REFERENCES

Andersson, C. 2011. "Paleolithic Punctuations and Equilibria: Did Retention Rather than Invention Limit Technological Evolution." *PaleoAnthropology* 2011:243–59.

Andersson, C. 2013. "Fidelity and the Emergence of Stable and Cumulative Sociotechnical Systems." *PaleoAnthropology* 2013:88–103.

Andersson, C., A. Törnberg, and P. Törnberg. 2014. "An Evolutionary Developmental Approach to Cultural Evolution." *Current Anthropology* 55 (2). http://doi.org/10.1086/675692.

Arthur, B. 2009. *The Nature of Technology.* New York: Free Press.

Basalla, G. 1988. *The Evolution of Technology.* Oxford: Oxford University Press.

Boyd, R., and P. Richerson. 1985. *Culture and the Evolutionary Process.* Chicago: University of Chicago Press.

Boyd, R., and P. Richerson. 2008. "Gene–Culture Coevolution and the Evolution of Social Institutions." In *Better Than Conscious? Decision Making, the Human Mind, and Implications for Institutions,* edited by C. Engel and W. Singer. Cambridge, Mass.: MIT Press, 305–24.

Brigandt, I., and A. C. Love. 2012. "Conceptualizing Evolutionary Novelty: Moving Beyond Definitional Debates." *Journal of Experimental Zoology Part B: Molecular and Developmental Evolution* 318B:417–27.

Bucciarelli, L. L. 1996. *Designing Engineers.* Cambridge, Mass.: MIT Press.

Campbell, D. T. 1965. "Variation and Selective Retention in Socio-cultural Evolution." In *Social Change in Developing Areas: A Reinterpretation of Evolutionary Theory,* edited by H. Barringer, G. Blanksten, and R. Mack. Cambridge, Mass.: Schenkman.

Caporael, L. 1997. "The Evolution of Truly Social Cognition: The Core Configurations Model." *Personality and Social Psychology Review* 1:276–98.

Caporael, L. 2013. "Evolution, Groups, and Scaffolded Minds." In Caporael, Griesemer, and Wimsatt 2013, 57–76.

Caporael, L., J. Griesemer, and W. Wimsatt, eds. 2013. *Developing Scaffolds in Evolution, Culture and Cognition.* Cambridge, Mass.: MIT Press.

Cavalli-Sforza, L., and M. Feldman. 1981. *Cultural Evolution and Transmission: A Quantitative Approach.* Princeton, N.J.: Princeton University Press.

Chase, W. G., and H. A. Simon. 1973. "Perception in Chess." *Cognitive Psychology* 4:55–81.

Dawkins, R. 1976. *The Selfish Gene.* Oxford: Oxford University Press.

Diamond, J. 1997. *Guns, Germs, and Steel.* New York: W. W. Norton.

Dove, G. 2012. "Grammar as a Developmental Phenomenon." *Biology and Philosophy.* doi:10.1007/s10539-012-9324-4.

Duhigg, C. 2012. *The Power of Habit: Why We Do What We Do in Life and Business.* New York: Random House.

Durham, W. 1992. *Coevolution: Genes, Culture, and Human Diversity.* Palo Alto, Calif.: Stanford University Press.

Fodor, J. 1980. "Methodological Solipsism Considered as a Research Strategy in Cognitive Psychology." *Behavioral and Brain Sciences* 3 (1): 63–109.

Foster, J., and J. Evans. 2018. "Promiscuous Inventions: Modeling Cultural Evolution with Multiple Inheritance." In Love and Wimsatt 2019.

Fracchia, A., and R. C. Lewontin. 1999. "Does Culture Evolve?" *History and. Theory* 38:52–78.

Gerson, E. M. 2013a. "Integration of Specialties: An Institutional and Organizational View." *Studies in History and Philosophy of Biological and Biomedical Sciences.* doi:10.1016/j.shpsc.2012.10.002.

Gerson, E. M. 2013b. "Some Problems of Analyzing Cultural Evolution." In Caporael, Griesemer, and Wimsatt 2013, 265–82.

Gerson, E. M. 2017. "Institutions and Repertoires." Unpublished manuscript. San Francisco: Tremont Research Institute.

Gigerenzer, G., P. M. Todd, and the ABC Research Group. 1999. *Simple Heuristics That Make Us Smart*. New York: Oxford University Press.

Greenfield, P. M., A. E. Maynard, and C. P. Childs. 2000. "History, Culture, Learning, and Development." *Cross-Cultural Research* 34:351–74.

Griesemer, J. 2013a. "Integration of Approaches in David Wake's Model-Taxon Research Platform for Evolutionary Morphology." *Studies in History and Philosophy of Biological and Biomedical Sciences*. doi:10.1016/j.shpsc.2013.03.021.

Griesemer, J. 2013b. "Reproduction and the Scaffolded Development of Hybrids." In Caporael, Griesemer, and Wimsatt 2013, 23–55.

Griesemer, J., and W. C. Wimsatt. 1989. "Picturing Weismannism: A Case Study in Conceptual Evolution." In *What Philosophy of Biology Is,* edited by M. Ruse, 75–137. Essays for David Hull. Leiden, The Netherlands: Martinus-Nijhoff.

Hart, B., and T. Risley. 1999. *The Social World of Children Learning to Talk*. Baltimore: Paul H. Brookes.

Hart, B., and T. Risley. 2003. "The Early Catastrophe: The 30 Million Word Gap by Age 3." *American Educator,* Spring, 4–9.

Heintz, C. 2013. "Scaffolding on Core Cognition." In Caporael, Griesemer, and Wimsatt 2013, 209–27.

Hiscock, P. 2014. "Learning in Lithic Landscapes: A Reconsideration of the Hominid 'Toolmaking' Niche." *Biological Theory* 9:27–41.

Jablonka, E., and M. Lamb. 2005. *Evolution in Four Dimensions*. Cambridge, Mass.: MIT Press.

Kidder, T. 2001. *The Soul of a New Machine*. New York: Little, Brown.

Kohler, Robert. 1994. *Lords of the Fly: Drosophila Genetics and the Experimental Life*. Chicago: University of Chicago Press.

Laland, K. N., T. Uller, M. W. Feldman, K. Sterelny, G. B. Müller, A. Moczek, E. Jablonka, and J. Odling-Smee. 2015. "The Extended Evolutionary Synthesis: Its Structure, Assumptions, and Predictions." *Proceedings of the Royal Society B: Biological Sciences* 282:20151019.

Lane, D., and R. Maxfield. 2005. "Ontological Uncertainty and Innovation." *Journal of Evolutionary Economics* 15:3–50.

Lane, D., R. Maxfield, D. Read, and S. Van der Leeuw. 2009. "From Population to Organization Thinking." Chapter 1 in *Complexity Perspectives on Innovation and Social Change*. Berlin: Springer.

Latour, B. 1987. *Science in Action*. Cambridge, Mass.: Harvard University Press.

Leonelli, S., and R. Ankeny. 2015. "Repertoires: How to Transform a Project into a Research Community." *BioScience*. May 20. doi:10.1093/bioscience/biv061.

Lewontin, R. C. 1970. "The Units of Selection." *Annual Review of Ecology and Systematics* 1:1–18.

Love, A. C. 2003. "Evolutionary Morphology, Innovation, and the Synthesis of Evolutionary and Developmental Biology." *Biology and Philosophy* 18:309–45.

Love, A. C. 2013. "Dimensions of Integration in Interdisciplinary Explanations of the Origin of Evolutionary Novelty." *Studies in History and Philosophy of Biological and Biomedical Sciences* 44(4): 537–50.

Love, A. C., ed. 2015. *Conceptual Change in Biology: Scientific and Philosophical Perspectives on Evolution and Development*. Boston Studies in the Philosophy of Science, no. 307. New York: Springer.

Love, A. C., and W. C. Wimsatt, eds. 2019. *Beyond the Meme: Development and Structure in Cultural Evolution*. Minneapolis: University of Minnesota Press.

Miller, G. A. 1956. "The Magical Number Seven, Plus or Minus Two: Some Limits on Our Capacity for Processing Information." *Psychological Review* 63: 81–97.

Mokyr, J. 2002. *The Gifts of Athena: The Origins of the Knowledge Economy*. Princeton, N.J.: Princeton University Press.

Moretti, E. 2012. *The New Geography of Jobs*. New York: Houghton-Mifflin.

Morgan, T. J. H., N. T. Uomini, L. E. Rendell, L. Chouinard-Thuly, S. E. Street, H. M. Lewis, C. P. Cross, C. Evans, R. Kearney, I. de la Torre, A. Whiten, and K. N. Laland. 2015. "Experimental Evidence for the Co-evolution of Hominin Tool-making Teaching and Language." *Nature Communications*. doi:10.1038/ncomms70.

Mufwene, S. 2008. *Language Evolution: Contact, Competition, and Change*. New York: Continuum.

Murmann, P. 2003. *Knowledge and Competitive Advantage: The Coevolution of Firms, Technology, and National Institutions*. New York: Cambridge University Press.

Murmann, P. 2013. "Scaffolding in Economics, Management, and the Design of Technologies." In Caporael, Griesemer, and Wimsatt 2013, 57–76.

Nelson, R., and S. Winter. 1982. *An Evolutionary Theory of Economic Change*. Cambridge, Mass.: Harvard University Press.

Nielsen, M. 2012. *Reinventing Discovery: The New Era of Networked Science*. Princeton, N.J.: Princeton University Press.

O'Brien, M., and S. Shennan, eds. 2010. *Innovation in Cultural Systems: Contributions from Evolutionary Anthropology.* Cambridge, Mass.: MIT Press.

Odling-Smee, F. J., K. Laland, and M. Feldman. 2003. *Niche Construction: The Neglected Process in Evolution.* Princeton, N.J.: Princeton University Press.

Page, V. W. 1918. *Aviation Engines: Design, Construction, Operation, Repair.* New York: Norman Henley.

Pagel, M. 2009. "Language as a Culturally Transmitted Replicator." *Nature Reviews Genetics* 10:405–15.

Richerson, P., and R. Boyd. 2005. *Not by Genes Alone.* Chicago: University of Chicago Press.

Schank, J. C., and T. J. Koehnle. 2007. "Modeling Complex Biobehavioral Systems." In *Modeling Biology: Structures, Behaviors, Evolution,* edited by M. D. Laubichler and G. B. Muller, 219–44. Cambridge, Mass.: MIT Press.

Seabright, D. 2004. *The Company of Strangers: A Natural History of Economic Life.* Princeton, N.J.: Princeton University Press.

Sperber, D. 1996. *Explaining Culture: A Naturalistic Approach.* Oxford: Blackwell.

Sperber, D. 2001. "Conceptual Tools for a Natural Science of Society and Culture." *Proceedings of the British Academy* 111:297–317.

Sterelny, K. 2012. *The Evolved Apprentice: How Evolution Made Humans Unique.* Cambridge, Mass.: MIT Press.

Stout, D. 2005. "The Social and Cultural Context of Stone Knapping Skill Acquisition." In *Stone Knapping: The Necessary Conditions for a Uniquely Hominid Behaviour,* edited by V. Roux and B. Bril, 331–40. Cambridge: McDonald Institute for Archaeological Research.

Stout, D., E. Hecht, N. Khreisheh, B. Bradley, and T. Chaminade. 2015. "Cognitive Demands of Lower Paleolithic Toolmaking." *PLOS One.* doi:10.1371/journal.pone.0121804.

Sturtevant, A., and G. Beadle. 1939. *An Introduction to Genetics.* New York: W. B. Saunders.

Temkin, Ilya, and Niles Eldredge. 2007. "Phylogenetics and Material Culture Evolution." *Current Anthropology* 48 (1): 146–53.

Theiner, G., C. Allen, and R. Goldstone. 2010. "Recognizing Group Cognition." *Cognitive Systems Research* 11:378–95.

Thornton, A., and N. Raihani. 2008. "The Evolution of Teaching." *Animal Behaviour* 75 (6): 1823–36.

Tostevin, G. 2013. *Seeing Lithics: A Middle-Range Theory for Testing for Cultural Transmission in the Pleistocene.* American School of Prehistoric Research

Monograph Series. Oxford and Oakville, CT: Peabody Museum, Harvard University, and Oxbow Books.

Warwick, A. 2003. *Masters of Theory: Cambridge and the Rise of Mathematical Physics.* Chicago: University of Chicago Press.

Webster, B. F. 1999. *The Y2K Survival Guide: Getting to Getting through, and Getting Past the Year 2000 Problem.* Upper Saddle River, N.J.: Prentice Hall.

Wills, Christopher. 1997. *Yellow Fever, Black Goddess: Coevolution of Peoples and Plagues.* Boston: Addison-Wesley.

Wilson, R., and A. Clark. 2009. "How to Situate Cognition: Letting Nature Take Its Course." In *The Cambridge Handbook of Situated Cognition,* edited by P. Robbins and M. Aydede, 55–77. Cambridge, Mass.: Cambridge University Press.

Wimsatt, B. 2013. Footholds and Handholds: Scaffolding Cognition and Careers." In Caporael, Griesemer, and Wimsatt 2013.

Wimsatt, W. C. 1980. "Reductionistic Research Strategies and Their Biases in the Units of Selection Controversy." In *Scientific Discovery—Vol. II: Case Studies,* edited by T. Nickles, 213–59. Dordrecht, The Netherlands: Reidel.

Wimsatt, W. C. 1986. "Developmental Constraints, Generative Entrenchment, and the Innate-Acquired Distinction." In *Integrating Scientific Disciplines,* edited by P. W. Bechtel, 185–208. Dordrecht, The Netherlands: Martinus-Nijhoff.

Wimsatt, W. C. 1987. "False Models as Means to Truer Theories." In *Neutral Models in Biology,* edited by M. Nitecki and A. Hoffman, 23–55. London: Oxford University Press reprinted in Wimsatt (2007).

Wimsatt, W. C. 1999. "Genes, Memes, and Cultural Heredity." *Biology and Philosophy* 14:279–310.

Wimsatt, W. C. 2001. "Generative Entrenchment and the Developmental Systems Approach to Evolutionary Processes." In *Cycles of Contingency: Developmental Systems and Evolution,* edited by S. Oyama, R. Gray, and P. Griffiths, 219–37. Cambridge, Mass.: MIT Press.

Wimsatt, W. C. 2002a. "False Models as Means to Truer Theories: Blending Inheritance in Biological vs. Cultural Evolution." *Philosophy of Science* 69: S12–24.

Wimsatt, W. C. 2002b. "Functional Organization, Functional Inference, and Functional Analogy." In *Functions: New Essays in the Philosophy of Psychology and Biology,* edited by R. Cummins, A. Ariew, and M. Perlman, 174–221. Oxford: Oxford University Press.

Wimsatt, W. C. 2007. *Re-engineering Philosophy for Limited Beings: Piecewise Approximations to Reality.* Cambridge, Mass.: Harvard University Press.

Wimsatt, W. C. 2010. "Memetics Does Not Provide a Useful Way of Understanding Cultural Evolution: A Developmental Perspective." In *Current Controversies in Philosophy of Biology,* edited by F. Ayala and R. Arp, 273–91. London: Blackwell.

Wimsatt, W. C. 2012. "The Analytic Geometry of Genetics: Part I: The Structure, Function, and Early Evolution of Punnett Squares." *Archive for the History of the Exact Sciences* 66:359–96.

Wimsatt, W. C. 2013. "Scaffolding and Entrenchment: An Architecture for a Theory of Cultural Change." In Caporael, Griesemer, and Wimsatt 2013, 77–105.

Wimsatt, W. C. 2015. "Entrenchment as a Theoretical Tool in Evolutionary Developmental Biology." Chapter 18 in *Conceptual Change in Biology: Scientific and Philosophical Perspectives on Evolution and Development,* edited by A. C. Love, 365–402. Boston Studies in Philosophy of Science, vol. 307. Berlin: Springer.

Wimsatt, W. C., and J. Griesemer. 2007. "Reproducing Entrenchments to Scaffold Culture: The Central Role of Development in Cultural Evolution." In *Integrating Evolution and Development: From Theory to Practice,* edited by R. Sansom and R. Brandon, 228–323. Cambridge, Mass.: MIT Press.

Wittgenstein, L. 2009. *Philosophical Investigations.* 4th ed. Edited by P. M. S. Hacker and J. Schulte. London: B. H. Blackwell. First published in 1953.

Wolf, M. 2008. *Proust and the Squid: The Story and Science of the Reading Brain.* New York: Harper.

Woods, C., ed. 2010. *Visible Language: Inventions of Writing in the Ancient Middle East and Beyond.* Chicago: Oriental Institute.

2

SCIENTIFIC AGENCY AND SOCIAL SCAFFOLDING IN CONTEMPORARY DATA-INTENSIVE BIOLOGY

SABINA LEONELLI

Philosophers of science are starting to pay attention to the impact of communication technologies, particularly those functioning as means to share results and resources, such as data or materials, on scientific methods and epistemology (Callebaut 2012; Leonelli 2012; O'Malley and Soyer 2012; Ratti 2015). This is especially salient in so-called big data initiatives, where high-throughput means of data production (such as sequencing machines, particle colliders, and space telescopes) are coupled with new technologies for the dissemination, integration, and visualization of the resulting masses of data (such as online databases and software for data analysis). Several commentators have described this phenomenon as an "information turn" in the practices of knowledge production (Castells 1996; Gibbons et al. 1996; Hey, Tansley, and Tolle 2009; Floridi 2013). What philosophers tend to overlook, however, is the significant role of social scaffolds in the development and implementation of these technologies toward generating new research. Social scaffolds include project teams, research networks, scientific institutions, policy bodies, learned societies, governmental committees, and other relevant forms of social engagement and governance. Here, I explore the circumstances under which specific types of social scaffolding facilitate advances in research and the reasons why some forms of sociality are effective in promoting certain kinds of scientific work. I concentrate on cases in which scientists coordinate their efforts with the goal of creating groups responsible for articulating common concerns, making these concerns visible to peers as well as funders and publishers, and developing ways to address them in everyday research practice. As I will show, these groups need to acquire resilience to endure the ever-shifting landscape of short-term funding agreements, fast-moving technologies, and multiple clusters of expertise that support research in any given field. This resilience is necessary, given the challenges and time involved in gaining enough visibility to command the attention of well-established regulatory institutions, such as governmental funders and learned societies. At the same time, these groups of scientists also need to be flexible and responsive enough to retain their usefulness vis-à-vis the shifting needs of relevant scientific communities. I argue that in their attempts to straddle these requirements, scientists tend to rely on well-entrenched social configurations and coordination strategies, some of which political theorists looking at the emergence and establishment of social movements have singled out and examined. Borrowing key ideas from social movement theory, I show how they can help us to understand the evolution of regulatory structures aimed at facilitating scientists' engagement with new technologies to enhance research outputs.

My discussion will be grounded in the examination of two types of organizations that have been heavily involved in developing practices of data dissemination through digital means within the life sciences over the last decade. These are (1) ontology *consortia,* which were created by biologists to promote online tools to classify and disseminate data and have evolved into *de facto* regulatory bodies in bioinformatics and data curation in the United States and Europe, and (2) *steering committees* for model organism communities, whose success in enhancing the cohesion, visibility, and reputation of biological research resulted in their playing significant roles in the governance of research. These are cases in which individual researchers successfully joined forces to build representation and political agency for their scientific concerns that resulted in the creation of organizations with regulatory power over research activities at the national and sometimes even the international level. They are also instances of two broader types of social structures that play a crucial role in the management of virtually every field: consortia and steering committees. Yet these have received little attention from science studies scholars, especially in comparison to "networks" and "laboratories," which have been central units of analysis for social scientific work in this area over the last twenty years.[1]

The chapter is structured as follows. In the first section, I briefly document the emergence of these groups and their successful transformation into scientific institutions with political and epistemic visibility and agency. Next, drawing on ideas from political theory, I argue that viewing these organizations as social movements is a fruitful strategy to make sense of their development from informal groups into well-recognized regulatory bodies. In the third section, I discuss how this process of institutionalization builds on highly entrenched forms of group socialization (*core configurations*) while also fitting the modular and highly dynamic nature of current research networks (Caporael 1997), which typically involve short-term collaborations around individual projects. In conclusion, I reflect on how my analysis could inform studies of the interrelation between institutional and infrastructural scaffolding involved in the evolution of scientific knowledge-making activities.

REGULATING DATA DISSEMINATION IN CONTEMPORARY BIOLOGY

Over the last three decades, scientific societies, governmental bodies, and industry have devoted increasing attention to the opportunities offered by the implementation of new technologies for the production and dissemination of biological research data (Leonelli 2016).[2] The sheer amount of organization, standardization, and infrastructure required to store and disseminate biological data—as well as the bureaucracy, institutional accountabilities, and red tape developed to that end—arguably exceeds anything previously experienced within the life sciences. In the words of prominent scientific commentators: "The introduction in 2005 of so-called next generation sequencing instruments that are capable of producing millions of DNA sequences has not only led to a huge increase in genetic information but has also placed bioinformatics, and life science research in general, at the leading edge of infrastructure development for the storage, movement, analysis, interpretation and visualisation of petabyte-scale datasets" (Southan and Cameron 2009, 119).[3]

The development of efficient data-sharing practices requires insights from the producers and users of data, whose understanding of their quality and significance as research materials is unparalleled. At the same time, individual scientists are not typically in a position to control the considerable resources and man power required to build relevant infrastructures, policies, and standards nor does scientific expertise constitute the only source of insight with regard to the value of research data. Indeed, data management on such a large scale requires a variety of skills, expertise, and insight, which include not only scientific assessment but also social, political, legal, and economic understanding of the circumstances under which data can be stored, maintained, and reused. Biologists interested in data dissemination have long struggled with the complex cluster of expertise and political visibility needed to debate—let alone decide upon—data-management and data-sharing strategies, as demonstrated by the history of data-sharing agreements like the Bermuda Rules (Harvey and McMeekin 2007; Jones, Ankeny, and Cook-Deegan 2018). Two initiatives that groups of biologists have taken in order to organize the public dissemination of research data produced within their field—the Gene Ontology and the Genomic Arabidopsis Resource Network— illustrate how scientists can and do join forces to influence the governance of their research in ways that favor their professional interests and intellectual commitments. Both types of collective action required the development of common standards and practices geared toward the resolution of scientific problems emerging in specific research contexts. At the same time, establishing such standards was intertwined with developing and implementing a regulatory system for scientific research targeted toward addressing the needs and characteristics of the groups involved.

Consortia and the Case of the Gene Ontology

The term *consortium* has recently acquired popularity within the life sciences as a way to refer to scientific collectives brought together by a common set of concerns. These span from an interest in specific phenomena (e.g., the Beta Cell Biology Consortium, devoted to pancreatic islet development and function, http://www.betacell.org/) to solving a common technical problem (e.g., the Flowers Consortium in the United Kingdom, aimed at creating a common infrastructure for synthetic biology, http://www.synbiuk.org/) or promoting a specific standard or technique (e.g., the Molecular Biology Consortium [MBC], founded to further high-throughput analysis of biomolecular and subcellular structures via a superbend X-ray beamline at the advanced light source, http://www.mbc-als.org/). The members of a consortium, which can be individuals as well as groups, labs, and institutes, do not need to be located in the same geographic site or belong to the same discipline. Indeed, the term is typically used to designate groups of scientists based in different institutions around the world and from a variety of disciplinary backgrounds. Consortia are sometimes fueled by dedicated funding, most often provided by governmental bodies interested in supporting a specific area of scientific work. In other cases, financial support is achieved by bringing together a variety of public and private resources. One example is the

Gene Ontology Consortium, which was created to develop and promote a particular tool for online data dissemination: the Gene Ontology (GO).

GO was created in 1999 as an alternative to the classification systems for genomic data proposed within medical informatics. The group of curators involved in the GO Consortium started their involvement as scientists discontented with how data were organized in databases at that time. They set out to create a resource that would do a better job of representing biologists' needs. In 1998, the group consisted of only five representatives from the yeast, mice, and fly communities, who saw themselves as fighting for a biology-driven bioinformatics. Their involvement with GO stemmed from their dissatisfaction with the ways in which medical informatics, as a field, was handling the setup of data-sharing tools in biomedicine, particularly model organism biology. They felt that the voices of biologists actually producing and working with these data were not being heard and endeavored to produce a set of tools grounded in biological know-how and geared toward the expectations and needs of biology users (for more historical detail, see Leonelli 2009, 2010). In 2000, funding for their efforts started to trickle in, and they found themselves in a position to recruit more like-minded researchers from other model organism communities. Following the explosion of data-intensive methods and related data infrastructures, these efforts came to be more widely recognized as crucial to the future development of biological research as a whole. The GO group expanded to include a head office based at the European Bioinformatics Institute (EBI) in the United Kingdom, counting up to ten researchers at any one time and at least twenty affiliated data curators spread around the world. These curators come together as a collective in regular meetings, online discussions, and funding applications. While many of the curators shift periodically, depending on project funding and local institutional arrangements, some have persisted as a long-term core group of affiliated scholars since the start of the project. GO has been increasingly institutionalized, both as part of the EBI and through strong links with the National Centre for Biomedical Ontology in the United States. Still, it continues to rely on voluntary contributions of participants, both financially and in terms of man power and data donation. For example, representatives from FlyBase, the database devoted to the dissemination of data on the fruit fly *Drosophila melanogaster,* contribute as much as they can justify under the remit of their project funding. Many others involved with organism databases do the same (e.g., the Arabidopsis Information Resource and WormBase, for the nematode *Caenorhabditis elegans*).

In previous work (Leonelli 2009, 2010), I have discussed the function of the GO Consortium as a powerful force within biology and beyond. The consortium has successfully developed procedures and technologies through which users can interact and upload, retrieve, and analyze data. It has also strongly influenced what counts as professional training for data curators in model organism databases—most notably, by helping to establish the International Society for Biocuration, which largely defined best practices for this field and strengthened its professional standing. Moreover, it has contributed to promoting values such as open access to data, intercommunity cooperation, and diversity in epistemic practices across biology, as well as fostering the pursuit of common goals, including specific kinds of cross-species integrative biology. All these activities involve networking with both the biological communities interested in the data being disseminated and the funding bodies and learned societies involved in supporting the relevant biological fields. The successes of GO signal the impressive increase in regulatory power, international visibility, and political resonance that this group has enjoyed since its origin. The GO Consortium has played an important role as an agent of change within the biological community.

Steering Committees and the Case of GARNet

A similar case study demonstrates the ways in which model organism communities have organized and coordinated themselves, resulting in an affirmation of their identity as key actors within the scientific landscape. Such organization is provided largely by steering committees: groups of representatives from the community who meet regularly to discuss future directions for the community as a whole (typically, some of the most active principal investigators [PIs], either elected by the community or sometimes self-appointed). One of these steering committees is GARNet, the Genomic Arabidopsis Resource Network. GARNet consists of plant scientists working on the model organism *Arabidopsis thaliana*. Most committee members are elected for a three-year term by UK researchers who self-identify as having an interest in *Arabidopsis* research, with efforts made at every election to ensure a fair representation in terms of research interests, gender, and geographical spread. Coordination and long-term memory is provided by two GARNet coordinator posts, one that has been in place since the committee's birth and another consisting of different individuals over the years; the committee chairs and PIs of the GARNet grant, who have shifted over the years but continue to maintain a close affiliation with the group even after the end of their mandates; and two ex officio committee members (the director of the European Arabidopsis Stock Centre, who has been part of the committee since its birth, and myself as an *Arabidopsis* historian and plant data expert since 2009). GARNet was created in 2000 as part of the gene function initiative funded by the Biotechnology and Biological Sciences Research Council (BBSRC) in the United Kingdom. While its initial remit was to ensure the availability of functional genomic technologies across UK plant science labs (Beale et al. 2002), GARNet has succeeded in obtaining two further rounds of funding from the BBSRC and has established itself as one of the most important organizations for the coordination, steering, and representation of basic plant research in the United Kingdom and internationally. This has happened through several initiatives, including (1) establishing a website and regular newsletter, which constitute unique information sources for new resources and initiatives in the field (principally concerning data but also embracing experimental techniques and instruments, as well as new funding opportunities); (2) organizing annual meetings attracting *Arabidopsis* scientists but also, increasingly, other plant scientists interested in updates on opportunities, techniques, and technologies for cross-species research; (3) coordinating dialogue among key stakeholders in the field, including learned societies like the Society of Biology, key funders such as the BBSRC, and the publishing industry responsible for the leading journals in plant science; (4) setting up surveys across the plant community, with the objective of articulating scientists' perception of what constitutes interesting new research directions and communicating it to funders (e.g., a survey commissioned by the BBSRC on the status of system biology in plant research); and (5) monitoring the number of resources funding bodies allocate to plant science vis-à-vis other parts of biology and lobbying for more resources and attention to be allocated to plant scientists.

As a result of these activities, GARNet now plays a central role in mediating the transition of the UK plant science community from a focus on functional genomics to system/synthetic plant science and translational research. Indeed, GARNet played a key role in integrating research conducted on *Arabidopsis* (traditionally funded by the BBSRC and viewed as fundamental research with no immediate applicability) with research carried out on crops such as barley, maize, and wheat (traditionally funded by the Department for Environment, Food and Rural Affairs and viewed as applied biotechnology). The rapprochement of these two communities was needed and overdue: *Arabidopsis* research has advanced to yield precious insights for agriculture (e.g., how to increase plant yield) and emerging biofuels (e.g., how to increase cell metabolism to make plants produce more butanol). Additionally, crop science is realizing that *Arabidopsis* research provides excellent comparative tools for research across plant species. GARNet has taken the lead in coordinating meetings among investigators in both communities, resulting in the founding of the UK Plant Science Federation (Leonelli et al. 2012). GARNet has also strongly affected the provision of bioinformatic services to plant scientists and biologists interested in *Arabidopsis* data. In 2009, the National Science Foundation decided to dramatically cut funding to a key database, the Arabidopsis Information Resource (TAIR), due largely to a lack of long-term sustainability for such an infrastructure. GARNet organized two international workshops that gathered powerful PIs, information technology (IT) experts, and funders to discuss models for the long-term maintenance and development of databases in plant science, helped find an agreement for how TAIR was to survive and develop in the future, and provided guidance on how similar databases could be made more resilient and useful to researchers.

SELF-REGULATORY EFFORTS AS SOCIAL MOVEMENTS

Consortia and steering committees, exemplified in the previous cases, share a number of features. They are self-organized collectives, whose joint activities begin without a great deal of support from well-established institutions or even from the communities in which they operate. Individuals propose themselves as representative champions for their communities, with the duty to voice scientists' existing concerns and facilitate solutions to those problems. These collectives support a wider spectrum of values and ideals than the specific issues they emerged to tackle, such as fostering initiatives requiring broad changes in the governance of the social system within which they are working. Initially, these organized efforts were devised as provisional responses to a localized issue in data management and dissemination. They persist with minimal dedicated funding thanks to the voluntary support and contributions of members of the communities they represent. Despite a precarious status in the early stages of their operation, these self-organized collectives have garnered visibility and political power, building their credibility by strongly connecting to their communities and attempting to articulate scientists' concerns in a way that bridges communication gaps with relevant peers and other stakeholders. It is not a coincidence that the biological communities that managed to organize themselves in this way are among the largest and most successful today. As a consequence, the model organisms these groups have championed are currently recognized as the most important in experimental biology (Ankeny and Leonelli 2011), indeed exemplifying a specific mode of doing research that has come to define much of the field (Ankeny and Leonelli 2016). All this has happened within a relatively short period of time: both the GO Consortium and the GARNet steering committee have gone from outsider status to participating in the primary regulation of biological research within the space of ten years.

The scientists engaged in these efforts demonstrate an acute awareness of the deep ties between power and standardization and of the ways these ties affect day-to-day research practices. They have effectively created systems of governance via a complex web of activities (including sophisticated marketing strategies and enrollment techniques) within which the standards and norms they propose may help to address issues emerging from scientific work. How should we characterize these groups of scientists and their activities? What kind of collective agency is in operation, and how does it achieve both power and impact? One way to consider these questions is in light of discussions about the emergence and status of so-called new social and scientific/intellectual movements. Drawing from this literature is not a new idea, and I will refer to authors who have advanced similar views with respect to scientific agency. However, I believe this to be a powerful lens with which to analyze the development of contemporary biological knowledge, particularly the creation and implementation of standards and infrastructures to disseminate data. From this corpus of literature, I have extracted four characteristic features of social movements that can be readily observed in both case studies. I propose that we view these scientific consortia and steering committees as social movements because they exhibit these four characteristic features:

1. They emerge in response to changing research needs and landscapes.
2. They establish new practices.
3. They create a vision for how research should be conducted in the future.
4. They become political actors with the power to engender social, scientific, legal, and political shifts (e.g., data-sharing policies, rules for database access, publication strategies, or shifts to the credit system in science).

Movements as Reactions

Della Porta and Diani (1999, 6) define new social movements as

1. informal networks, based on
2. shared beliefs and solidarity, which mobilize about
3. *conflictual issues,* through
4. the frequent use of various forms of *protest.*

The emphasis within this definition is on the role of movements as *reactions* to the existing status quo. This is an important and suggestive intuition; the collective action characterizing consortia and steering committees is driven by the desire to resolve existing problems. For GO and GARNet, these problems emerge from scientific practice. To this end, a high level of epistemic and political agreement is required and must be targeted to specific issues. Consortia and steering committees are committed to using a rational, knowledge-based approach to reach such consensus; these are expert movements for an expert community and usage. This often means antagonizing the establishment, as in the case of many nonscientific social movements.

> A movement is a social/intellectual movement by our definition only if, at the time of its emergence, it significantly challenges received wisdom or dominant ways of approaching some problem or issue and thus encounters resistance. (Frickel and Gross 2005, 207)

For Frickel and Gross, the notion of "resistance," interpreted as opposition to a discriminating majority, is central. I agree that for cases of consortia and steering committees a degree of resistance and challenge to previous practices and normative demands that characterize a field or domain is involved. But although this provides a key motivation for collective action, there is another noteworthy goal central to the collective agency that initiates consortia. This is to draw attention to issues that have not been the focus of funding agencies or of the scientific community and yet have caused trouble for research (or are likely to do so in the future). In these cases a regulatory need is going unrecognized by regulatory bodies; thus, there is an opportunity to delegate decision-making power (and annexed responsibility) to a new form of agency or actor. If successful, some people or institutions are willing (or forced) to absorb the regulatory need, either because they are identified as likely candidates or because they are created for that purpose. Additionally, other scientists are happy to delegate responsibility to these new movements; they willingly give up their decisional power over the issues. A similar dynamic is currently seen in the rise of organizations such as the Research Data Alliance (2016), which started as a group of open science advocates and lobbyists in 2010 and within five years became a reference point for governments and funding agencies looking for guidance on how to collect and mobilize research data across all areas of society.

Movements as Collective Creation

Another significant feature of social movements is that they aim to create something new: "Temporary public spaces, movements of collective creation that provide societies with ideas, identities, and even ideals" (Eyerman and Jamison 1991, 4). GO is a good example of this kind of consortium, which is primarily geared toward the development of new tools and knowledge. GO managed to channel the creative energies of a number of prominent biologists and bioinformaticians into the development of a unique and highly popular database. At the same time, building the momentum and opportunity for such an endeavor is itself an imaginative and laborious act. Social movements have been defined as "luxury goods" because they need support in order to take off on the scale required for collective action to be effective. Thus, they are typically organized around "hot issues" most likely to attract the attention of funders and peers. (This is the case with both data infrastructure and synthetic and translational plant biology.) It is also critical to note the importance of the collective experience of unity through action as a means to form a social identity. The formation of a social nucleus with a distinct identity and sense of membership happens simultaneously with the focus on a common set of issues. Notably, the social unity or cohesion of the group is more important than agreement or consensus on the specifics of the issue itself; what matters is the individuals in the group's sense of agreement and belonging and their willingness to invest resources toward the same normative vision. Indeed, both GO and GARNet have contributed greatly to forming a well-defined research community bound together by similar worries and obligations.[4] Unavoidably, this has also involved conflicts over boundaries, the exclusion of individuals or groups for financial, geographical, or personal reasons (no matter how inclusive both the GO and the

GARNet groups strive to be), and the formation of other communities striving to counter or emulate their increasing visibility and resources.

Movements as Signs of Change

Melucci (1996, 1) proposes yet another definition of social movements:

> Movements are a sign; they are not merely an outcome of the crisis, the last throes of a passing society. They signal a deep transformation in the logic and processes that guide complex societies. Like the prophets, movements "speak before": they *announce what is taking shape* even before its direction and content has become clear.

Thus, according to Melucci, social movements have the key function of voicing a normative vision—in this sense they are "signs of change." This function is visible in both case studies in which the collectives in question have developed specific visions of what counts as good science (e.g., norms regulating standardization and data curation in databases; a commitment to enhancing research efficiency through collaboration and coordination in plant science). These visions play a key role in forming social identities (see section 2.2), but they also contribute to wider debates about the appropriateness of specific goals, norms, and methods in research at large and the changes that new technological and social developments foster. Another example can be seen in the ideas of Science 2.0 and Open Science, which the European Commission has used over the last decade to capture a perceived ongoing shift in the practice and results of science. This feature parallels the study of the formation of *communities of promise* with a common imagination, such as can be observed in the case of epistemic networks formed around stem cell research (Martin, Brown, and Kraft 2008), and, more generally, the study of the development and function of scientific *imaginaries* (Jasanoff and Kim 2015).

Notably, elaborating such a vision does not necessarily involve an explicit contrast between it and preexisting views. Making visions identifiable as new entities (i.e., as signs of change) is as important as building some continuity with the intellectual traditions characterizing the epistemic communities to which the vision is directed. A vision needs to be anchored somewhere in order to be understood. The language used to express the vision, the practices it involves, and the problems it is supposed to solve all need to be situated in specific contexts that coevolve with the vision itself. If spokespersons for a new vision cannot latch on to (and influence) one or more preexisting intellectual traditions, they will find it difficult, if not impossible, to enroll new participants in their movements (Frickel and Gross 2005, 221). This is on display in both GARNet and the GO Consortium because participants stress that these organizations championed existing understandings of good practice and reliable data sharing within model organism biology, particularly in the face of other ways of handling data preferred by other communities.

Movements as Rising Power

One final characteristic of social movements concerns the role of power dynamics in the emergence and operation of consortia and steering committees, including the importance of long-term influences on the environment.

> When backed by dense social networks and galvanised by culturally resonant, action-oriented symbols, contentious politics leads to *sustained* interaction with opponents. The result is a social movement. (Tarrow 1998, 2)

The actions of GO and GARNet (as well as other types of scientific organizations) result in the acquisition of political representation and agency on national and global agendas, even though their immediate target is primarily needs arising from day-to-day research practice. This large-scale political representation and agency often goes well beyond the resolution of the initial problems and can be referred to as these movements' *rising power*. Although often mentioned in sociological and anthropological studies of emerging fields, the ways in which such power is developed in and through scientific practice deserves much more research. For example, by what diverse paths does a movement quickly develop an internal hierarchy and administration in order to function, which in some cases transform into a semiofficial agency? The National Institute for Health and Care Excellence in the United Kingdom, which started as a grassroots movement of doctors trying to monitor the safety of guidelines provided by the National Health Service, is now a major evaluation agency with tremendous clout over government and patient organizations. A key element in this type of development is "access to key resources" (Frickel and Gross 2005, 214). These resources include (1) *organizational structures,* such as channels for information flow (e.g., conference venues and publications), frequently linked to epistemic cultures; (2) *intellectual power,* grown in parallel with the reputation and personal credibility of the movement's leaders and with the assessment of their vision and actions developed by peers over time; and (3) *long-term employment* within academia for at least some of the movement's leaders, which provides the stability and continuity necessary to the blossoming of collective agency on a large scale. Another important element is the ability to raise bottom-up support or *micromobilization* (Frickel and Gross 2005, 220). All of these display parallels to the situation outlined by Kaushik Sunder Rajan (2006, 52) in relation to what he calls "new corporate activism": corporations' political strategies for influencing the outcome of issues affecting their organizations.

ENTRENCHED CONFIGURATIONS AS SOURCES OF SOCIAL ROBUSTNESS

I have described four characteristics that social movements seem to have in common with scientists' attempts to regulate their own activities. Both ontology consortia and steering committees are instances of collective self-regulation stemming from perceived needs in a scientific field (e.g., conflict or lack of resources). They formulate creative solutions to such problems, which are developed and implemented by groups of individuals with the expertise to recognize the problems and to present them so that others within their field will recognize them as well. Additionally, these groups exhibit an entrepreneurial ability to devise ways in which risky, collective efforts can contribute to solving those problems. Focusing on these characteristics thus helps to explain how groups such as the GO Consortium and GARNet managed to evolve into well-recognized regulatory bodies.

The process of institutionalization at work in these groups relies heavily on widely entrenched forms of group socialization, which these organizations exploit in order to achieve two crucial and yet potentially contrasting goals. First, they maintain an enduring identity and some stability, which enables them to keep growing in scale, ambition, and visibility. Second, they retain the flexibility needed to fit the highly mutable and volatile nature of current research networks. The capacity to adapt to changes is crucial in the contemporary landscape of scientific funding, where intense competition for relatively small pots of money makes the majority of biological research dependent on collaborations around short-term projects. Collaborators, as well as the topics of interest, can and often do change radically from project to project. Scientists need to manage this environment in order to make interesting new links to people, fields, and topics, as well as maintain and develop existing interests and collaborations. Stability arises out of constant renewal; the necessity to enhance the robustness of social scaffolds in the face of environmental perturbations is one of the most fascinating aspects of these scientific initiatives.

A prime example showing how these groups depend on these forms of socialization is their reliance on charismatic individuals as group leaders who carry authority as well as recognition within the main communities of interest. In the case of GO, for instance, it is notable that the initial impetus toward the development of bio-ontologies was provided by key figures in model organism biology, such as Michael Ashburner, Suzanne Lewis, and Judith Blake, whose scientific authority among their peers was already established and well recognized. Building on existing credibility and reputation, these figures were able to attract the attention of their peers and the trust of funders, thereby creating a tidal wave of interest that culminated in the formation of a thriving community of developers and users of bio-ontologies. In the case of steering committees such as GARNet, we find similar dynamics; highly visible scientific figures in plant science and systems biology, such as Andrew Millar, loaned their credibility to the committee as it was being formed.

Given the responsibilities already weighing on the shoulders of these leading figures, individuals not well known for their scientific contributions, who nevertheless possessed the right set of competencies and skills, performed much of the actual legwork and coordination work. These individuals were willing to sacrifice time and resources toward making the enterprise successful at a time when resources allocated to the group were very scarce. In the case of both GO and GARNet, these turned out to be junior academics with broad-ranging scientific interests who were intrigued by the social organization of their communities. They had a strong drive toward promoting cooperative behaviors in science and often talked about the importance of "serving" the community of researchers by setting up useful data infrastructures. In some cases, family commitments made it difficult for these individuals to pursue a full-time career in research. Perhaps unsurprisingly, the majority of these individuals were women. Thus, to some extent, this embodied the well-established social configuration of womanhood as nurturing and service-oriented, providing colleagues and peers with trustworthy resources and highly skilled labor that did not fit the formal structures for scientific credit and measures of excellence.

Another form of socialization that features heavily in the history of these organizations (and also in the history of many social movements) is that of personal friendship. In both cases, the regulatory power of collective action was reinforced through informal networking, including late-night discussions, joint trips, and workshops involving a regular set of core attendees and the formation of strong personal bonds among some of them. This included a willingness to bring other friends and collaborators on board. These informal bonds became particularly important during times of trouble, when problems with the organization forced its members to regroup and rethink their strategies and general approach. GARNet faced such a moment at the end of its first ten years of funding, when it became apparent that its continuation would depend on its ability to (1) demonstrate the levels of support and appreciation for GARnet's work to the BBSRC and (2) formulate a vision for future work that embraced the whole of plant science, rather than only the *Arabidopsis* community, which tracked recent trends toward cross-species research (of the type that GARNet itself fostered, for instance, through helping to set up the UK Plant Science Federation). GARNet members appealed to prominent individuals in plant science with whom they had collaborated in the past and who were happy to testify to the usefulness of the organization and help articulate its vision for the next funding cycle.

These forms of socialization play the role of *core configurations,* which Linnda Caporael and colleagues (2014) have characterized as "subgroups of face-to-face interactions that are posited to recur in daily life, ontogeny, history, and plausibly, as part of human evolutionary history" (58). They can be identified and singled out on the basis of the specific functions they accomplish; indeed, their success in achieving a given purpose is what "explains their continued replication" (Caporael 1997, 282). Caporeal has focused on the size of groupings—the number of individuals involved—as a fundamental feature of core configurations, and my analysis of specific cases of collective agency in biology confirms her emphasis on groupings of a relatively small size, which enables strong personal relations and the ability to quickly reorganize in response to external challenges. Additionally, I have highlighted the distribution of the social roles required to spur a social movement to grow and become established, especially a scientific organization with these characteristics. Core configurations like personal friendship, by virtue of their proven track record in bringing and keeping individuals together, have become entrenched forms of socialization, which individuals fall back upon when attempting to achieve conceptual and institutional changes.[5] As such, these configurations provide stability and visibility to fledgling organizations, such as consortia and steering committees, while also enhancing their flexibility to changes in the environment. The result is robust social entities.

THE DISSEMINATION of scientific data relies on a great variety of material and social scaffolds, ranging from well-established institutions that determine data-sharing policies and related credit systems (funding agencies, policy bodies, academies, learned societies) to venues through which data can travel (annual conferences, data journals, repositories) and other types of organizations involved in the production and reuse of data (universities, networks). In this chapter I have considered two ways in which scientists have coordinated their actions and agendas to shape science governance and policy related to the means of data dissemination in biology. Both consortia and steering committees have played—and continue to play—crucial roles in supporting and structuring data curation practices, as well as making them visible and recognized by longstanding scientific institutions. In so doing, they have themselves acquired an institutional role and acted as key social scaffolds for the development and implementation of data-intensive biology. Looking at these organizations as social movements helps to identify some of the core strategies or configurations that helped to develop the ideas, values, and priorities of a few individual scientists on a large scale, thus shaping knowledge-making practices at the international level.

This analysis resonates with Wiebe Bijker's invitation to recognize that specific patterns of agency by groups of scientists play an important role in large technological systems (Bijker, Hughes, and Pinch 1987). It also shows why attention to social and institutional dynamics is critical to understanding scientific practices. Activities such as data sharing, data interpretation, publication patterns, the choice of topics for future research, and scientists' commitment to specific norms need to be analyzed with reference to their broad institutional and social contexts, especially in cases where scientists themselves play a key part in developing and shaping those contexts. In turn, social structures such as formal and informal committees and groups, often brought together by a common concern or goal, function as scaffolds for the development of new institutions (Gerson 2013; Wimsatt 2013; Caporeal et al. 2014). As illustrated by the speed with which both GARNet and GO have developed from a small group of scientists into large, influential organizations, an evaluation of the cultural role and impact of specific groups and associated norms and behaviors should take into account the highly dynamic context in which they operate. Different characteristics of social scaffolding help at different moments in the development of such institutions. For example, while imposing strong leadership may prove fatal when a feeling of community participation and engagement is required for social cohesion, it may well help when dynamics change, and social coordination is more effectively centered on the activities of a charismatic individual or subgroup. The same can be said for the extent to which norms of engagement are codified (e.g., participation in GARNet networks was voluntary but subject to specific rules of engagement—dictated by the broader funding structure through which it was supported—from the start), the choice to rely on given technologies versus attempting to develop new ones (GARNet drew its visibility from the former, while GO acquired social and political influence by virtue of the latter), and the choice to highlight existing "gaps" in governance versus trying to build new areas of influence (GO, notably, started with the former and ended up pursuing the latter).

In closing, it is critical to stress again that social scaffolds affect the production and transmission of knowledge through their tight interrelation with the development of material and infrastructural scaffolds. Indeed, the existence of organizations such as GARNet and GO has been strongly correlated with the development of computing facilities and data-extraction methods in molecular biology. The effective alignment of these material and social structures has made a significant difference to the methods and strategies for data production and interpretation currently in use within biology.[6] Philosophical research focused on the status of data in contemporary science, as well as the ways in which inferences are drawn and corroborated, needs to look beyond specific instances of data use and examine why certain configurations of norms, instruments, and methods become established and their implications for the development of knowledge-making practices.[7] The analysis herein points to an important direction for future work in the philosophy of science: the need to challenge the minimalist and asocial conceptualizations of scientific agency pervading much of contemporary philosophy. This type of work will help philosophers understand the material, social, conceptual, and institutional conditions for knowledge production as a necessarily interconnected and historically situated whole.

NOTES

Warm thanks to colleagues in GARNet and GO, Rachel Ankeny, Barry Barnes, Dario Castiglione, Elihu Gerson, James Griesemer, Alan Love, Bice

Maiguashca, Raffaele Marchetti, Hans Radder, Bill Wimsatt, and my colleagues at Egenis for helpful discussions on this topic. This research was funded by the European Research Council grant award 335925.

1. See the overviews of science and technology studies (STS) work on the social organization of scientific research provided in Bijker, Hughes, and Pinch (1987), Hackett et al. (2008), and Atkinson, Glasner, and Lock (2009).

2. For example, there are many STS analyses of standardization procedures, the role of standards as "coordination devices" for complex networks of actors (Latour 1987; Bowker and Star 1999), the relation between biomedical regulation and the production of "objective" knowledge (Cambrosio et al. 2009), and the way in which standards foster accountability and trust by facilitating the enactment of "rituals of verification" (Power 1997). The specific case of bioinformatic standards has also been subject to several studies (e.g., Hilgartner 1995, Bowker 2000, Hine 2006, Chow-White and Garcia-Sanchos 2012, Mackenzie 2012, and Lewis and Bartlett 2013, as well as my own work on the subject).

3. For analyses of the notions of scale at play in "big biology," see Davies, Frow, and Leonelli (2013).

4. For more detail on the ethos of model organism communities and the importance of repertoires in shaping research fields, see Ankeny and Leonelli (2011, 2016).

5. I am here thinking of simple entrenchment: "An evolving adaptive system with a recurring developmental trajectory, and differential entrenchment generating different degrees of evolutionary conservation" (Wimsatt 2013, 83).

6. For an expanded version of this argument, see Leonelli (2016).

7. To this aim, Rachel Ankeny and I have proposed to adopt the notion of *repertoires* as units of analysis for scientific organization and change over time (Ankeny and Leonelli 2016).

REFERENCES

Ankeny, Rachel A., and Sabina Leonelli. 2011. "What's So Special about Model Organisms?" *Studies in the History and the Philosophy of Science: Part A* 42 (2): 313–23.

Ankeny, Rachel A., and Sabina Leonelli. 2016. "Repertoires: A Post-Kuhnian Perspective on Scientific Change and Collaborative Research." *Studies in the History and the Philosophy of Science: Part A* 60:18–28.

Atkinson, Paul, Peter Glasner, and Margaret Lock, eds. 2009. *The Handbook for Genetics and Society*. Abingdon, UK: Routledge.

Beale, Michael, Paul Dupree, Kathryn Lilley, Jim Beynon, Martin Trick, Jonathan Clarke, Michael Bevan, Ian Bancroft, Jonathan Jones, Sean May, Karin van de Sande, and Ottoline Leyser. 2002. "GARNet, the Genomic Arabidopsis Resource Network." *Trends in Plant Science* 7 (4): 145–47.

Bijker, Wiebe E., Thomas P. Hughes, and Trevor Pinch, eds. 1987. *The Social Construction of Technological Systems: New Directions in the Sociology and History of Technology*. Cambridge, Mass.: MIT Press.

Bowker, Geoffrey C. 2000. "Biodiversity Datadiversity." *Social Studies of Science* 30 (5): 643–83.

Bowker, Geoffrey C., and Susan L. Star. 1999. *Sorting Things Out: Classification and Its Consequences*. Cambridge, Mass.: MIT Press.

Callebaut, W. 2012. "Scientific Perspectivism: A Philosopher of Science's Response to the Challenge of Big Data Biology." *Studies in the History and the Philosophy of the Biological and Biomedical Sciences* 43 (1): 69–80.

Cambrosio, Alberto, Peter Keating, Thomas Schlich, and George Weisz. 2009. "Biomedical Conventions and Regulatory Objectivity: A Few Introductory Remarks." *Social Studies of Science* 39 (5): 651–64.

Caporael, Linnda. 1997. "The Evolution of Truly Social Cognition: The Core Configurations Model." *Personality and Social Psychology Review* 1 (4): 276–98.

Caporael, Linnda, J. R. Griesemer, and W. C. Wimsatt, eds. 2014. *Developing Scaffolds in Evolution, Culture, and Cognition*. Cambridge, Mass.: MIT Press.

Castells, Manuel. 1996. *The Rise of the Network Society*. Oxford: Blackwell.

Chow-White, P., and Miguel Garcia-Sanchos. 2012. "Bi-directional Shaping and Spaces of Convergence: Interactions between Biology and Computing from the First DNA Sequencers to Global Genome Databases." *Science, Technology, and Human Values* 37 (1): 124–64.

Davies, Gail, Emma Frow, and Sabina Leonelli. 2013. "Bigger, Faster, Better? Rhetorics and Practices of Large-Scale Research in Contemporary Bioscience." *BioSocieties* 8 (4): 386–96.

Della Porta, Donatella, and Mario Diani. 1999. *Social Movements: An Introduction*. Oxford: Blackwell.

Eyerman, Ron, and Andrew Jamison. 1991. *Social Movements: A Cognitive Approach*. University Park, Pa.: Penn State University Press.

Floridi, Luciano. 2013. *The Philosophy of Information*. Oxford: Oxford University Press.

Frickel, Scott, and Neil Gross. 2005. "A General Theory of Scientific/Intellectual Movements." *American Sociological Review* 70 (2): 204–32.

Gerson, Elihu. 2013. "Integration of Specialties: An Institutional and Organizational View." *Studies in the History and the Philosophy of the Biological and Biomedical Sciences* 44:515–24.

Gibbons, Michael, Camille Limoges, Helga Nowotny, Simon Schwartzman, Peter Scott, and Martin Trow. 1996. *The New Production of Knowledge: The Dynamics of Science and Research in Contemporary Society.* London: Sage.

Hackett, Edward J., Olga Amsterdamska, Michael Lynch, and Judy Wajcman, eds. 2008. *The Handbook of Science and Technology Studies.* 3rd ed. Cambridge, Mass.: MIT Press.

Harvey, Mark, and Andrew McMeekin. 2007. *Public or Private Economies of Knowledge? Turbulence in the Biological Sciences.* Cheltenham, UK: Edward Edgar.

Hey, Tony, Stewart Tansley, and Kristine Tolle, eds. 2009. *The Fourth Paradigm: Data-Intensive Scientific Discovery.* Redmond, Wash.: Microsoft Research.

Hilgartner, Stephen. 1995. "Biomolecular Databases: New Communication Regimes for Biology?" *Science Communication* 17 (2): 240–63.

Hine, Christine. 2006. "Databases as Scientific Instruments and Their Role in the Ordering of Scientific Work." *Social Studies of Science* 36 (2): 269–98.

Jasanoff, Sheila, and Sang-Hyun Kim, eds. 2015. *Dreamscapes of Modernity: Sociotechnical Imaginaries and the Fabrication of Power.* Chicago: University of Chicago Press.

Jones, Katherine Maxon, Rachel A. Ankeny, and Robert Cook-Deegan. 2018. "The Bermuda Triangle: The Pragmatics, Policies, and Principles for Data Sharing in the History of the Human Genome Project." *Journal of the History of Biology,* online first.

Latour, Bruno. 1987. *Science in Action: How to Follow Scientists and Engineers through Society.* Cambridge, Mass.: Harvard University Press.

Leonelli, Sabina. 2009. "Centralising Labels to Distribute Data: The Regulatory Role of Genomicconsortia." In *The Handbook of Genetics and Society: Mapping the New Genomic Era,* edited by Paul Atkinson, Peter Glasner, and Margaret Lock, 469–85. Abingdon, UK: Routledge.

Leonelli, Sabina. 2010. "Documenting the Emergence of Bio-ontologies: Or, Why Researching Bioinformatics Requires HPSSB." *History and Philosophy of the Life Sciences* 32 (1): 105–26.

Leonelli, Sabina. 2012. "Classificatory Theory in Data-Intensive Science: The Case of Open Biomedical Ontologies." *International Studies in the Philosophy of Science* 26 (1): 47–65.

Leonelli, Sabina. 2016. *Data-centric Biology: A Philosophical Study.* Chicago: Chicago University Press.

Leonelli, Sabina, Berris Charnley, Alex R. Webb, and Ruth Bastow. 2012. "Under One Leaf: An Historical Perspective on the UK Plant Science Federation." *New Phytologist* 195 (1): 10–13.

Lewis, Jamie, and Andrew Bartlett. 2013. "Inscribing a Discipline: Tensions in the Field of Bioinformatics." *New Genetics and Society* 32 (3): 243–63.

Mackenzie, Adrian. 2012. "More Parts than Elements: How Databases Multiply." *Environment and Planning D: Society and Space* 30 (2): 335–50.

Martin, Paul, Nik Brown, and Alison Kraft. 2008. "From Bedside to Bench? Communities of Promise, Translational Research, and the Making of Blood Stem Cells." *Science as Culture* 17 (1): 29–41.

Melucci, Alberto. 1996. *Challenging Codes: Collective Action in the Information Age.* Cambridge: Cambridge University Press.

O'Malley, Maureen A., and Orkun S. Soyer. 2012. "The Roles of Integration in Molecular Systems Biology." *Studies in the History and the Philosophy of the Biological and Biomedical Sciences* 43 (1): 58–68.

Power, Michael. 1997. *The Audit Society: Rituals of Verification.* Oxford: Oxford University Press.

Ratti, Emanuele. 2015. "Big Data Biology: Between Eliminative Inferences and Exploratory Experiments." *Philosophy of Science* 82 (2): 198–218.

Research Data Alliance. 2016. "About Us." Accessed November 30, 2016. https://www.rd-alliance.org/about-rda.

Southan, Christopher, and Graham Cameron. 2009. "Beyond the Tsunami: Developing the Infrastructure to Deal with Life Sciences Data." In *The Fourth Paradigm: Data-Intensive Scientific Discovery,* edited by Tony Hey, Stewart Tansley, and Kristin M. Tolle, 117–23. Redmond, Wash.: Microsoft Research.

Sunder Rajan, Kaushik. 2006. *Biocapital: The Constitution of Postgenomic Life.* Durham, N.C.: Duke University Press.

Tarrow, Sidney G. 1998. *Power in Movement: Social Movements and Contentious Politics.* Cambridge: Cambridge University Press.

Wimsatt, William C. 2013. "Articulating Babel: An Approach to Cultural Evolution." *Studies in the History and the Philosophy of the Biological and Biomedical Sciences* 44:563–71.

3

CREATING COGNITIVE-CULTURAL SCAFFOLDING IN INTERDISCIPLINARY RESEARCH LABORATORIES

NANCY J. NERSESSIAN

THE BIOENGINEERING SCIENCES use a range of resources from various engineering fields and the biosciences to conduct groundbreaking basic biological research in the context of potential application. Pioneering university research laboratories in the bioengineering sciences are dynamic environments in which problems, methods, and technologies are continually undergoing development, and the primary researchers are students developing into full-fledged researchers. Research labs have long been sites for ethnographic research into social, cultural, and material practices of scientific research. Philosophers have only recently been attending to them as sites for developing fine-grained in situ analyses of the exploratory, incremental, nonlinear problem-solving practices of frontier science, their origins and evolution, and the epistemic principles guiding them.

For fourteen years I led a multidimensional, interdisciplinary research project funded by the U.S. National Science Foundation that, in addition to the usual kinds of questions that a philosopher and cognitive scientist might ask about reasoning, representation, problem-solving, and so forth, aimed to glean insights from our investigation of bioengineering sciences research labs to facilitate this kind of frontier interdisciplinary research. University research laboratories are highly significant contexts for making an impact on the research practices of a field because graduate students largely populate them, and these *researcher–learners* develop into the next generation of practitioners over the course of five to six years. In the context of this volume, I interpret "facilitating" as creating structures for research and learning. A premise of our research has been that to create such structures first requires examining the research practices and what already scaffolds them

64

in situ and then collaborating with the faculty researchers on developing scaffolding for learning and research appropriate to their objectives both within the lab and the larger educational ecosystem in which it is embedded. In the investigation discussed here, we had the opportunity to help build that educational ecosystem since there was no standard curriculum in biomedical engineering, and the faculty was engaged in creating a new department with a novel vision of what research in that field should be like.

From the outset we aimed at an *integrated* account of the lab research practices—one that would move beyond the perceived cognitive-cultural divide in science studies (Nersessian 2005). Specifically, because problem-solving processes comprise complex relations among researchers, technologies, and sociocultural practices, we adopted the framework of distributed cognition (d-cog) as a starting point of our analysis. The framework of distributed cognition developed from several strands of critique (see, especially, Hutchins 1995; Lave 1988) of both the context-free, body-independent "functionalist" construal of cognition by experimental psychology and artificial intelligence and the overly linguistic and thing-oriented construal of culture by cognitive anthropology. D-cog is part of a larger movement in the cognitive sciences, which I have called *environmental perspectives* (Nersessian 2005), that is grounded in empirical evidence from a range of disciplines and has persuasively argued that cognition and culture are mutually implicated. Concurring with the notions of cognition as embodied, situated, and enculturated and culture as a process, we framed problem-solving as situated within evolving distributed cognitive-cultural systems (Nersessian et al. 2003). In our case, the plural of the word "system" underscores that the lab and the multiple specific problem-solving processes within it can each be conceived as such a system, with the grain size dependent on the focus of an analysis. Our use of the hyphenated term "cognitive-cultural" is intentional.[1] It stresses that the systems comprising the humans and the artifacts investigated are simultaneously cultural and cognitive systems. The framework of distributed cognition is, itself, in need of development in several directions. On the one hand, as Georg Theiner (2013), among others, has underscored, the cultural dimensions of the framework have remained underdeveloped. On the other, the nature of cognitive contributions by human participants has also been underexamined at the expense of the cognitive affordances of technologies (Nersessian 2009). And, as will be discussed below, expanding the range of problem-solving tasks and environments studied from a distributed cognition framework contributes to further articulating the framework (Chandrasekharan and Nersessian 2015).

Recently, Edwin Hutchins has elaborated on the notion of distributed cognition in an effort to distinguish it from the "extended mind" thesis (Hutchins 2011, 2014). Hutchins stresses first that distributed cognition is not making an ontological claim but rather providing an analytical framework that attends explicitly to the cultural dimensions of cognition. "Distributed" signifies the spatiotemporal *process* nature of culture and cognition, or what he calls a "cultural-cognitive ecosystem" of people, artifacts, and embodied skills (Hutchins 2011, 440–41). Culture and cognition in this view are co-constructed and emergent from the dynamics of complex ecosystems; in our case, the research lab.[2] We contend, further, that the notion of culture-as-process implicates the history of the evolution of the lab in current practices (see, e.g., Kurz-Milcke, Nersessian, and Newstetter 2004; Nersessian, Kurz-Milcke, and Davies 2005).[3] Consonant with the notion of a dynamic ecosystem, we have analyzed both the evolving distributed cognitive-cultural system that is the laboratory and the specific problem-solving processes within it as comprising researcher–learners ("researchers"), technologies, practices, and research problems, all with evolving relational trajectories. As William Wimsatt (Wimsatt 2013b, chapter 1 of this book) points out, the fact that culture is dynamic means that elements will develop and change at different rates, and thus some elements provide *structuring constraints* for future development. As will be developed here, the evolving, historical nature of the system that is a research lab requires adding a new dimension of analysis—how the system builds itself, especially by means of the structuring constraints and affordances of its simulation technologies.

We have studied four labs, two in biomedical engineering (tissue engineering and neural engineering) and two in integrative systems biology (one, purely computational, that collaborates with external bioscientists and the other possessing a wet lab where researchers conduct their own experiments in service of modeling). Although the research in each of these pairs of labs is interdisciplinary, there are significant differences in the *kinds* of interdisciplinarity and thus in existing and needed structures for scaffolding the research of the labs. For this chapter I focus on one of the biomedical engineering labs, tissue engineering (Lab A), to dissect the processes by which problem-solving structure is created through the design of hybrid, bioengineered physical simulation technologies for investigating novel biological phenomena and through the lab's embedding in an educational context that has the goal of designing novel learning experiences to scaffold the development of hybrid biomedical engineers.

THE STUDY

The Interdiscipline of Biomedical Engineering

Biomedical engineering (BME) scientists are a breed of researcher whose aim is to make fundamental contributions to "basic science" and to create novel artifacts and technologies for medical applications. It is often the case that bioscientists have not conducted the basic biological research bioengineers need in order to make progress toward application. Indeed, they might not have even formulated the general problem, such as "what effects are the forces of blood flow having on the cardiovascular system." The basic science research in the labs we studied, such as on shear stresses in vascular biology or on learning in living neural networks, is approached largely from the perspective of engineering assumptions, principles, concepts, and values (Nersessian 2017). Most often, bioengineering scientists investigate real-world phenomena through designing, building, and running models—physical or computational. For engineers, "to engineer" means to conceive, design, and build artifacts in iterative processes. BME researchers extend this notion to making biological ("wet") artifacts through which to carry out research. In the BME labs we investigated, each laboratory engineers physical simulation models, locally called "devices." Devices are in vitro models that serve as sites of experimentation on selected aspects of in vivo phenomena of interest. They are *hybrid* artifacts where cells and cellular systems interface with nonliving materials in model-based physical simulations run under various experimental conditions (Nersessian 2008; Nersessian and Patton 2009). The devices participate in experimental research in various configurations of hybrid "model systems." As one researcher commented, they *"use that [notion] as the integrated nature, the biological aspect coming together with the engineering aspect, so it's a multifaceted model-system."*[4] In each BME lab, there is one or more *signature* device that plays a central role in evolving the research program. I call such devices "signature" because the lab is usually identified both internally and externally as "the lab that does X (device) studies."

As we will see, signature devices are *generatively entrenched* in that they provide structuring constraints for the evolution of a range of cognitive-cultural practices in the complex system that is "the lab." Jeffery Schank and

William Wimsatt (1986; Wimsatt 1986) introduced the notion of generative entrenchment to emphasize the role of specific entities in evolutionary processes in complex biological systems: "The generative entrenchment of an entity is a measure of how much of the generated structure or activity of a complex system depends upon the presence or activity of that entity. . . . The resulting picture suggests that generative entrenchment acts as a powerful and constructive developmental constraint on the course of evolutionary processes" (33). They also anticipated the potential to extend the notion to other kinds of evolving systems: "Since virtually any system exhibits varying degrees of generative entrenchment among its parts and activities, these studies and results have in addition broad potential application for the analysis of generative structures in other areas" (33). Wimsatt (2007, 2013b) has recently applied it to the role of certain elements in the cultural evolution of a complex system.

Devices, as physical simulation models, are experimental systems designed to function as analogical sources from which to draw inferences and make predictions regarding target in vivo phenomena. They are constructed so that experiments with them should enable the researcher *"to predict what is going to happen in a system [in vivo]. Like people use mathematical models . . . to predict what is going to happen in a mechanical system? Well, this [model system she was designing] is an experimental model that predicts—or at least you hope it predicts—what will happen in real life."* That is, research is conducted with these in vitro devices, and outcomes are transferred as candidate understandings and hypotheses to the corresponding in vivo phenomena. In effect, the researchers build parallel worlds in which devices mimic specific aspects of phenomena they cannot investigate directly due either to issues of control or ethics. In the philosophical literature on models, "simulation" is customarily reserved for computational modeling. However, respondents in our investigation variously use "simulate" and "mimic" in explaining how their physical models perform as they are "run" under experimental conditions. It is the dynamic nature of the models that makes them simulations. Thus, we use "physical simulation model" to refer to such bioengineered devices. Simulation by means of devices is an epistemic activity that comprises open exploration, testing and generating hypotheses, and inference. In our analysis, simulation devices are loci of integration of cognition and culture. They simultaneously constitute the "material culture" of the community, give rise to social practices, and perform as "cognitive artifacts" in their problem-solving practices. Understanding how these communities produce knowledge requires examining both aspects.

Early in our investigation, one researcher characterized the practice of building and experimenting with devices as *"putting a thought into the benchtop and seeing whether it works or not."* With respect to a researcher, as an instantiated thought, the device is a physically realized representation with correspondences to the researcher's mental model. As a tangible artifact, it evolves along with the researcher's understanding developed in the experimentation that the artifact makes possible. As a representation, it refers both to the in vivo phenomena and to the researcher's mental model. As such, the artifact is a site of simulation of not just some biological process but also the researcher's current understanding. The notion of *"the experimental model that predicts"* we encountered above is a distributed model-based reasoning system comprising researchers and simulation models. The "cognitive powers" created by constructing physical simulation models include enhanced ability for abstraction, for integrating knowledge and constraints from diverse domains, for conceptualization, and for changing representational format in ways that afford analogical, visual, and simulative reasoning (Nersessian 2008, 2009). Investigating scientific practice in the wild enables us to discern facets of how these powers are created in situ within the developing and evolving material and sociocultural environment.

BME labs can be cast broadly as what Karin Knor Cetina (1999) has characterized as *epistemic cultures*. These "are cultures that create and warrant knowledge" (1). Whereas the notions of discipline and specialty typically refer to the "differentiation of knowledge" and thus to the institutional organization of knowledge, epistemic culture shifts the focus of attention to "knowledge-in- action" (Cetina 1999, 3). As an approach to the study of science, the notion of an epistemic culture serves to focus on the differences of "knowledge-making machineries" in different subcultures. As Cetina details through case studies of experimental physics and molecular biology practices, these machineries comprise sociocultural structures as well as technologies of research. The analysis of epistemic cultures typically does not attend to the differences in epistemological assumptions underlying the practices of knowledge-making subcultures, which are important when considering interdisciplinary cultures since these assumptions often clash. As Evelyn Fox Keller has pointed out, there are differences in "the norms and mores of a particular group of scientists that underlie the particular meanings they give to words like theory, knowledge, explanation, and understanding, and even to the concept of practice itself" that are equally significant for individuating subcultures and understanding their practices (Keller 2002, 4). Accounting for the research in the labs we investigated requires both attending to the devices qua machineries of making knowledge, located in environments of "construction of the machineries of knowledge construction" (Cetina 1999, 3), and the devices qua artifacts that embed, and through which one can begin to discern, the epistemological assumptions, norms, and values of the culture of BME and its subdivisions (e.g., tissue and neural engineering).

In any analysis, researchers need to consider what constraints devices possess deriving from their design and construction (device qua device) and what limitations these impose on the simulation and subsequent inferences and interpretations (device qua model). Further, there is a tension between constraints on the design and functionality of a device (qua device) that derive from biology and from engineering. One respondent provided an example of this tension in the context of a problem in which the cells of a researcher kept dying in a simulation device, even though all the environmental conditions of media, temperature, and incubation seemed appropriate for sustaining them. The problem turned out to be the material from which the chamber holding the cells was built. As he recounted, "*His device was something he created and built based on the mechanical properties. But in the design process he did not take into account that maybe some of the materials used to build his device were toxic [to cells].*"

From a d-cog perspective, a device also has a dual nature: It serves as a site not only for the simulation of biological processes (machinery) but also for the researchers' understanding, epistemic norms, and epistemic values (model). The design of a device embeds norms and values primarily associated with the kind of quantitative analysis aimed at by engineers rather than biologists, such as approximation and simplification. Many of the researchers we interviewed characterized this difference as biologists focus on how "*everything interrelates to everything else,*" while engineers "*try to eliminate as many extraneous variables as possible so we can focus on the effect of one or perhaps two, such that our conclusions [qua model] can be drawn from the change of one variable.*" An overarching problem of the labs we studied is to determine the appropriate or best feasible abstraction of the in vivo phenomena to address their research questions. For instance, the endothelial cells lining an artery experience turbulent blood flows in vivo, but from an engineering perspective, it is desirable to begin with a first-order approximation (laminar flow).

Framing the lab activities as situated in complex cognitive-cultural systems provides a means of analyzing the problem-solving processes in a manner that integrates cognition and culture (Nersessian 2006). However, this analysis cannot be done simply by applying the current framework of d-cog (Hollan, Hutchins, and Kirsh 2000; Hutchins 1995; Kirsh and Maglio 1994). That framework was developed through studies of highly structured, dynamic problem-solving environments (plane cockpit, naval ship) in which participants carry out largely routinized tasks using existing technologies (such as the speed bug or the alidade), and the knowledge the pilot and crew bring to bear in those processes is relatively stable, even in novel situations. In contrast, the BME research lab is an innovation community where researchers often do not have established methods, technologies, and well-defined problems prior to beginning the research. Although loci of stability provide structuring constraints, equally important features of these labs include the evolution of technologies and the development of the researchers as learners in the processes of carrying out an overarching research agenda. The technological components of the research lab, for instance, evolve in unanticipated ways. We witnessed several cases in which the researcher had to design and construct new or redesign existing simulation devices in the course of problem-solving. At each slice in time, "the lab" comprises the current state of devices and research problems, students at various points of development into researchers, and a lab director at a stage of his or her research program. In effect, the lab *builds itself* as a cognitive-cultural system with specific affordances and limitations for problem-solving as it creates knowledge (Nersessian 2012). Examining how the lab builds itself provides significant insight into how researchers "create their cognitive powers by creating the environments in which they exercise those powers" (Hutchins 1995, xvi). We focus on designing and constructing physical simulation models because they are a major means through which engineering scientists build cognitive powers (Chandrasekharan and Nersessian 2011; Chandrasekharan and Nersessian 2015; Nersessian 2012).

University research labs are significant sites of learning, populated primarily with graduate students and, increasingly, undergraduates. It is a remarkable feature of BME labs that graduate students are simultaneously learners and pioneering researchers. Thus, the development of researchers as learners is a significant component of the evolution of the lab. Our labs reside in a BME community that places high value on what it calls "interdisciplinary integration" at the level of the individual researcher. For them this means moving beyond problematic collaborations that stem from the numerous differences between the practices and epistemic values of engineers and bioscientists, to the extent possible, and cultivating the individual researcher as a hybrid biomedical engineer from the outset, such as when the novice engineer learns to harvest and cultivate the cells she needs for research. The nature of the research requires lab members, who arrive predominantly with engineering backgrounds, to develop equal facility with wet-lab techniques, as well as engineering design and a selective deep knowledge of the biology of their research targets. These communities see themselves as cutting-edge, frontier researchers. The lab ethos is infused with an open-ended sense of possibility, as well as a tinge of anxiety about how little is known in their area and whether PhD research projects will work out. The researchers place a high value on innovation in methods, materials, and applications. Failure is omnipresent, as are lab-devised support structures for dealing with it. The social structure in each lab is largely *nonhierarchical*—a feature that in this case we attribute to the frontier and interdisciplinary nature of the research, where no one (including the director) considers herself or himself *the* expert, and requisite knowledge is distributed across the lab and wider community. Opportunities to innovate are provided to everyone, including a freshman who might have an interesting idea. In the period during which we conducted our investigation, we saw several instances in which "big gambles" led to high payoffs, sustaining this attitude, despite the fact that most of the researchers engaged in high-risk research were doing it for their dissertation projects. Projects can always be modified and scoped down to what is feasible in the time allotted. The sociocultural fabric each lab, and the community as a whole, builds has been highly successful in helping students to graduate and preparing them for excellent positions in academia and industry.

Method: Cognitive-Historical Ethnography

Our research group conducted an ethnographic study that sought to uncover the activities, artifacts, and sociocultural structures that constitute research as it is situated in the ongoing practices of the Lab A community.[5] We conducted two years of intensive data collection, followed by two years of targeted follow-up and, thereafter, limited tracking of students through to their graduation. We took field notes on our observations, audiotaped unstructured interviews (72), and video- and audiotaped research meetings (17). As a group (four ethnographers), we completed an estimated four hundred hours of field observations. Our "team ethnography" approach supplanted videotaping research activities in the lab. Although we were allowed to videotape research meetings, it did not prove feasible to videotape research as it was taking place. We used interpretive coding in analyzing interviews and field notes. Broadly consistent with the aims of "grounded theory" not to impose a specific theoretical perspective on the data (Glaser and Strauss 1967; Strauss and Corbin 1998), we approached coding to enable core categories and interpretations to emerge from the data and remain grounded in it while at the same time remaining guided by our initial research questions. We then developed case studies of specific researchers and deeper analyses of themes that had emerged from the data that were relevant to our research questions. We also examined findings with respect to pertinent philosophical, cognitive, and sociocultural theoretical frameworks.

Additionally, since these labs are evolving systems that reconfigure as the research program moves along and takes new directions in response to events occurring both in the lab and the broader community, there is a significant historical dimension to our analyses. As noted previously, the signature technologies of the cognitive-cultural systems are designed and redesigned in the context of research problems and projects, new methods are developed or adopted, and at any slice in time, the lab is populated by students at various points in their development into full-fledged researchers. To capture the historical dimension of these lab communities, we used interpretive methods of cognitive-historical analysis (Nersessian 1987, 1995, 2008). Coupling ethnography with cognitive-historical analysis affords examining how history is appropriated in the social, cultural, and material dimensions of practices, as these currently exist.

Specifically, cognitive-historical analysis enables following the trajectories of the human and technological components of a cognitive system on multiple levels, including the physical shaping and reshaping of artifacts in response to problems, their changing contributions to the models developed in the lab and the wider community, the nature of the concepts that are at play in the research activity at any particular time, and the development of learners as researchers.[6] Our cognitive-historical analysis uses the customary range of historical records to uncover how Lab A researchers have developed and used representational, methodological, and reasoning practices, as well as the histories of the research technologies. These practices can be examined over time spans of varying length, ranging from periods defined by the activity itself to decades or more.[7] In this context, the objective of cognitive-historical analysis is not to construct a historical narrative. Rather, it is to enrich our understanding of the cognitive-cultural system through examining how knowledge-producing practices originate, develop, and are used.

Making sense of the day-to-day practices and detailing the histories of researchers, artifacts, and practices are prima facie separate tasks. However, the research process within the labs we studied evolves at a fast pace, which necessitates integrating the two endeavors. With respect to the devices in particular, the ethnographic study of how they are understood and used by various lab members, coupled with ongoing interviews around research and learning, allows us to conjoin the cognitive-historical study of the developing lab members, the lab artifacts, and the lab itself, with an eye to the lab members' perception of these. Of particular note is our finding that device history, which chronicles the development of the current problem situation and what is known about the artifacts in question, is often appropriated hands-on. Since devices, inherited and new, need to be (re)designed for the current problem situation, avoiding past pitfalls requires, among other things, knowing why and how a certain problem situation has led to the realization of certain design options. The historicity of the artifacts becomes a resource for novel design options, though in practice it is not an easily accessible resource. However, it becomes more available as a researcher's membership in the community develops.

In the following sections, I will examine how the devices of Lab A provide not only scaffolding for current problem solving, but also create structural constraints and affordances for research potentialities not yet envisioned, and in so-doing build "the lab" itself. I will then discuss the wider ecosystem that has been "engineered" to provide scaffolding for researcher-learners to develop into the hybrid biomedical engineers envisioned to populate and build a novel twenty-first century version of the field of BME.

CREATING COGNITIVE-CULTURAL SCAFFOLDING: THE LABORATORY FOR TISSUE ENGINEERING

The laboratory for tissue engineering (Lab A) dates from 1987, when the director moved to a new university for the opportunity to begin research in the emerging area of tissue engineering. During our investigation the main members included the lab director, the lab manager, one postdoctoral researcher, seven PhD students, two master of science students, and four long-term undergraduates.[8] Several other undergraduates visited for a semester or a summer internship. Of the graduate students, two were male and seven were female, as was the postdoctoral researcher. All of these researchers came from engineering backgrounds, mainly mechanical or chemical engineering. The lab manager had a master's degree in biochemistry. The laboratory director's background was in aeronautical engineering, but he was by then a senior, highly renowned pioneer in the field of BME and the emerging subfield of tissue engineering.

Starting in the mid-1960s, the now director of Lab A had worked as an aeronautical engineer for the space program on how the effects of vibration along the axis of the Saturn launch vehicle *(pogo stick vibration)* affected the cardiovascular system of astronauts. He developed the hypothesis that the physical forces to which blood vessels are naturally exposed, such as pressure and forces associated with blood flow through the arteries, could adversely affect the blood vessels and thus be implicated in disease processes such as atherosclerosis. He embarked on a program of research into how and under what conditions the physical forces might create disease through *arterial shear* forces.

Building Simulation Devices and Model Systems

The future Lab A director decided early in this research to focus his efforts on the endothelial cells that line the arteries: "*It made sense to me that if there was this influence of flow on the underlying biology of the vessel wall, that somehow the cell type had to be involved, the endothelium.*" In the 1970s vascular biologists were focused on biochemical processes, and those he contacted were skeptical of the hypothesis. As a result, he ended up in the laboratory of a veterinary physiologist, where they surgically created animal models to induce pathologies in native arteries to investigate the nature and effects of arterial shear. Problems of control were significant in these modeling practices and led him to set out on a program of in vitro research that would require developing physical models through which to simulate biological phenomena with the desired experimental control. Thus, nearly all problem-solving activities in Lab A are model-based, in the sense that they require building (designing, constructing, redesigning) physical models, assembling them in various model-system configurations, and performing simulations under a variety of controlled experimental conditions.

The initial configuration of Lab A revolved around one physical model, the flow channel device, or *flow loop*, which is designed to enact selected in vivo blood flow conditions, normal and pathological. It consists of a flow channel (designed in a physiologically meaningful range) with accompanying flow-inducing components (such as a peristaltic pump, a pulse dampener, and liquid the viscosity of blood) designed to represent to a first-order approximation shear stresses that can occur during blood flow in an artery. It formed a model system with endothelial cell cultures on slides. When cells mounted on slides are "flowed" under different conditions, changes in cell morphology and proliferation can be related directly to the controlled shear stresses. This device started its life as a large, cumbersome artifact on a stand, for which contamination was a constant problem since it could not be assembled under the sterile workbench hood. Within a few years, it was re-engineered into a compact design that fits under the sterile hood, and experiments can be run in an incubator. This redesign process was chronicled for us in an interview with a recent graduate of the lab (see Kurz-Milcke et al. 2004).

After several years of research, the lab sought a better model system. Using cell cultures on slides provides only a limited understanding of arterial shear stress. Specifically, as the director noted, "*Putting cells in plastic and exposing them to flow is not a very good simulation of what is actually happening in the body. . . . If you look within the vessel wall you have smooth muscle cells and then inside the lining is [sic] the endothelial cells, but these cells types communicate with one another. So we had an idea: let's try to tissue-engineer a better model-system for using cell cultures.*" The idea was to create "*a more physiological model,*" where the effects of shear could be studied on more components of the blood vessel wall than with the endothelial cells in isolation. In principle this should help to better understand the functional properties related to arterial shear. To expand the possibilities for studying these properties, the director took "*the big gamble*"—to create a model of the blood vessel wall constructed from living tissue. If they were successful in building this model, it would open the possibility of turning it into a vascular graft for repairing diseased arteries. This model is variously referred to within the lab as the "construct," the "tissue-engineered blood vessel wall model," and, underscoring its application potential, the "tissue-engineered vascular graft." The flow loop and construct were the signature devices of the lab at the time of our investigation.

When we entered the lab, the construct was the focal simulation model. Because all the researchers would need to build their own constructs, cell culturing had supplanted learning the flow loop as the initial entry point into the lab culture. Learning to culture cells (bovine or porcine), as a senior researcher told us, is a "*baseline to everything.*" Whereas learning to manipulate the flow loop is a relatively easy task for an engineer, learning to culture cells and create constructs is not. The consequences of failure are high: when cell cultures die, experiments are ruined. As a result, much mentoring took place around learning to culture cells, starting with harvesting them from arteries donated from an animal lab at another institution. Although there are written protocols for the steps, we witnessed that these are learned in embodied apprenticeships over numerous sessions, first by hovering in close physical proximity as the mentor conducts the procedure under the sterile hood and then by the novice trying them herself. The discourse of the lab frequently centers on keeping the cells "happy," calling them "pets," bemoaning long weekends "babysitting" them, and sharing war stories about facing the recalcitrance of cells to respond in ways they desire. Novice researchers start to build up both tacit experimental know-how and resilience in the face of failure through this extended mentoring process.

Simultaneously, learning how to culture cells provides entre into the problem space and cognitive practices of the lab. Cell culturing is a prelude to building the construct models needed for most research projects. The in vivo blood vessel comprises several layers, and the in vitro construct device constitutes a family of models that can be built with different levels of approximation for simulating in vivo processes. The novel application goal that the construct gave rise to for Lab A—to tissue engineer a viable replacement blood vessel for human implantation—created the need to design and build other simulation models. To be either a functional model or an implant requires (among other things) that the cells embedded in the scaffolding material replicate the capabilities and behaviors of in vivo cells in order to achieve higher-level tissue functions, such as expressing the right proteins and genetic markers. Further, a vascular implant needs to be strong enough to withstand the in vivo blood forces. These problems, in turn, opened new lines of lab research and led to building new devices and model systems through which to manipulate and examine construct properties under various conditions. For example, the lab developed two devices to simulate mechanical forces (the pulsatile bioreactor and the equibiaxial strain device) and

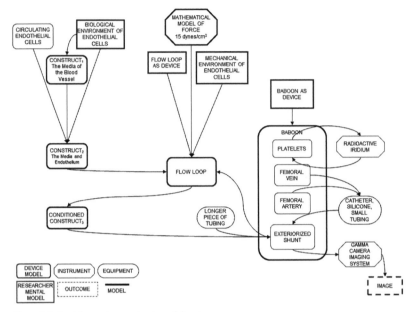

Figure 3.1. Partial vascular construct model system.

an "ex vivo" animal (baboon living in another lab) model system to investigate whether progenitor cells develop the ability to express anticoagulant proteins in response to shear (as do mature in vivo endothelial cells). A partial depiction of this animal model system (Figure 3.1) serves to illustrate how the distributed system *interlocks* a number of models, physical and mental.[9] It is "partial" because it can be extended out into a complex fabric of additional models and researchers that contribute to its functioning as a model-based reasoning system, through which inferences about the in vitro and the in vivo phenomena are made.

Building out the Cognitive-Cultural System

A glimpse of Lab A as a distributed system is provided in the representation created when we asked the director to draw a picture of the current lab research partway though our study (Figure 3.2). We gave no instructions for how to do this. His stated intention was to depict how his research "barriers" *(top section)*, researchers *(middle section)*, and technologies *(bottom section)* are interconnected. The diagram on paper is static, but the director's representation can be interpreted as providing a schematic of "the lab as an

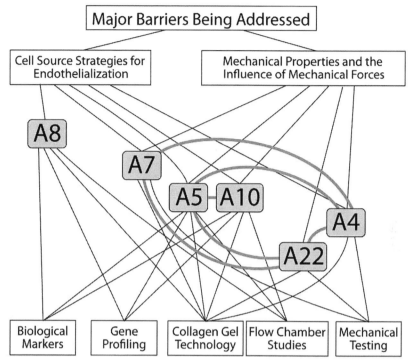

Figure 3.2. Lab A director's representation of the lab's research.

evolving distributed cognitive-cultural system"—a dynamic constellation of interrelated problems, researchers, simulation models, and other technologies. Although the director did not include himself on the diagram, he was, of course, an integral part of the system even though his visits to the physical space of the wet lab were rare. He spent a significant amount of time on the road promoting the research and obtaining funding, in addition to administering an interdisciplinary center. I next outline, briefly, how this system formed and evolved in the course of addressing the research problems through numerous iterations of designing, building, and experimenting with physical simulation models.

At the top of the diagram, the director categorized the "major barriers" with which the research dealt. From discussion with him about the diagram, it became clear that "barriers" are "addressed" by formulating and pursuing research problems that interconnect the basic biological research of the lab and its medical application aim. To address the barrier of "mechanical properties and the influence of mechanical forces" requires formulating research problems directed toward understanding the nature of arterial shear and its role in normal and disease processes. Solving these problems would further the application goal by bringing the research closer to creating an implant with the requisite mechanical properties to function within the body, such as shear strength. To address the barrier of "cell source strategies" requires research directed toward the problem of providing endothelial cells (among the most immune sensitive in the body) for a viable implant that would not be rejected by the recipient's body. This problem, in turn, opened lines of basic research for the members, such as the role of forces in stem cell differentiation (A8) and in the maturation of progenitor cells (A7).

The lab-built simulation models are designated by "collagen gel technology" (construct), "flow chamber studies" (flow loop), and "mechanical testing" (pulsatile bioreactor and equibiaxial strain device). The kinds of investigations along the bottom of the diagram implicate both the lab-built simulation models and the technologies through which simulation outcomes are examined. For instance, after a flow chamber study in which the construct would be subjected to controlled shear stresses ("conditioning"), the effects on the endothelial cells can be examined for various biological markers or gene profiling, which implicate a range of technologies, some external to the physical space of the lab, such as the confocal microscope to study morphology and migration or DNA microarray technology, used for studying gene expression. Also, "mechanical testing" implicates a lab-built instrument for testing the mechanical strength of a construct after conditioning.

The director intended that thick lines denote interconnections among the individual research projects with respect to the researcher designated to build the animal model system that would integrate these projects (A7). A postdoctoral researcher (A8) is represented as unconnected to the students because she started a new line of lab research into the possibility of stem cell differentiation by means of mechanical forces as a source of endothelial cells for the construct that only later (after success) became more central to the research. She did interact with other lab members about her and their research during conversations in the course of lab activities and at lab research meetings. Although research projects were carried out by individual lab members (sometimes assisted by an undergraduate or master of science student), we witnessed joint problem-solving episodes within the lab and during the lab meetings (held at varying intervals when the director was in town). Each individual research project and the problem-solving processes associated with it can be explicated as performed by a distributed cognitive-cultural system. But the lab's dual problems of understanding the effects of arterial shear and creating a viable implant built the lab itself into an evolving system that afforded and constrained the individual research. The diagram depicts the interconnected subsystems that contribute to building out the lab-as-distributed-cognitive-cultural-system.

When the researchers noted in Figure 3.2 entered the lab, the flow loop model was a well-established technology of research, but several formulated research problems that required some redesign of it. The construct model was a recent development, and all the researchers played significant roles in furthering its design in directions related to their specific projects. A5's research was directed toward correlating the development of arteriosclerosis with the genetic behavior of the endothelial cells and progenitor endothelial cells that circulate in the bloodstream by simulating various flow conditions. A10's research was investigating the effects of shear stress on aortic valve function, using valvular endothelial cells with a novel aortic construct that he designed. Mechanical integrity and strength were primary concerns for him, and, although he ended up not using it himself, he designed and built a new device for the lab: an equibiaxial strain device that simulates the strain (deformation from stress) experienced by vessels as blood flows through them (Nersessian, Kurz-Milcke, and Davies 2005). Components of it would be put to surprising use by a future researcher investigating stem cell differentiation (Harmon and Nersessian 2008). A4's research was to examine specific biological markers in relation to the controlled mechanical stimulation of constructs, as compared with their behavior in native tissue. A22's research focused on improving the mechanical strength of constructs. And, all of these projects are layers that undergird the baboon model system designed by A7 for the ex vivo experiment (Figure 3.1), intended to bring them closer to the medical application goal.

As indicated by the thick lines on the diagram, all of the system's components are connected to A7, who in an early interview noted that she had been designated as *"the person who would take the construct in vivo."* This meant that she would need to create a model system in which a construct would be connected to the vascular system of a living animal. To be successful, the project would need to *"obviously integrate the results of colleagues here in the lab."* At the start, she was quite unclear about just what she would study with the model. Once she decided on a specific animal, a lot of time was devoted to designing a means of connecting the fragile construct to the animal without it rupturing (and in a humane way). Her research project evolved into investigating whether shear stress conditioning of endothelial progenitor cells with the flow loop would make them function as mature endothelial cells in the production of thrombomodulin (a protein that prevents platelet formation) when attached to an animal circulatory system. For her investigation she designed a model system that could connect the construct to the bloodstream of a baboon by means of an exterior shunt between the femoral artery and the vein of the animal. The ex vivo simulation was designed to be run in real time through a gamma camera to provide functional imaging. Figure 3.1 provides a partial representation of this model system. Designing and running this model system with the requisite experimental controls was the most complex problem undertaken by the lab to date. As A7 noted, "*In the lab we can control . . . exactly what the flow is like. . . . But when we move to an animal model, it's more physiologic—the challenge then is that is a much more complex system.*" Despite the complexity, she was able to determine that preconditioning with flow loop shear at the normal human in vivo rate of arterial shear (15 dynes/cm^2) enhances the ability of progenitor cells to express anticoagulant proteins within the model system. This finding made a significant contribution to both the research community's understanding of the effects of arterial shear and the problem of endotheliel cell sources for implantation (see Nersessian 2009).

Discussion: Evolving the Cognitive-Cultural System

The few details of Lab A practices sketched in this chapter provide an illustration of how the devices that researchers build in the course of specific problem-solving efforts in a lab also participate in building and evolving a complex distributed cognitive-cultural system. These artifacts provide structuring constraints that create potentialities that researchers can exploit to evolve the system further. At the outset, this lab director did not envision his lab engaging in tissue engineering to make vascular construct models or conducting stem cell research and gene profiling. The initial understanding of arterial shear stemmed from his early mathematical modeling of vibratory forces on the human vascular system and experimental modeling with animals. The limitations of in vivo research with animal models led to taking the research in vitro, which afforded more control and opened the possibility of examining selected features of arterial shear in relation to endothelial cells. Building the flow loop model enabled them to focus largely on the structural properties and the proliferation behavior of cells under shear. But after several years of experimenting with endothelial cells, he realized the flow loop also offered the possibility of examining the relationships among different kinds of cells in the blood vessel wall if they could engineer a living three-dimensional model. Designing the construct family of models provided not only a range of more physiologically accurate models but also the potential to create a vascular graft for human implantation. Importantly, it afforded the possibility of investigating the functional properties of blood vessels in relation to shear (including building several new devices), which in turn led to the lab's ability to create a completely different kind of animal model system from those of the director's initial research. In sum, building physical simulation models has provided a platform for articulating a cognitive-cultural system comprising researchers, problems, models, technologies for experimentation, visualization, and analysis, and sociocultural practices that constitute Lab A as it evolves over time.

This brief glimpse of the evolving cognitive-cultural structures that enable research in Lab A calls into question the classic notion of a scaffold as something that falls away after it has served its purpose; cognitive-cultural scaffolding is *incorporated* into the evolving distributed cognitive-cultural system. The evolving technologies and researchers are interwoven into the fabric that *is* the system. This conception of scaffolding is exemplified by the construct model itself. The tissue-engineered vascular graft is built on a collagen scaffold that is incorporated into the fabric of the constructed vessel. The engineered vessel, if successful, would be incorporated into the in vivo vascular system. The notion of generative entrenchment helps to articulate this notion of scaffolding as incorporation in the evolution of the cognitive-cultural system of a research lab. Signature devices, in particular, and the cognitive-cultural practices associated with them provide constraints and affordances for evolving the research of the lab. They contain the potential for the development of future cycles of design and redesign, which often proceed in novel and unanticipated ways.

In an important sense, the core activity of the lab is building itself as a distributed cognitive-cultural system directed toward achieving the overarching goals of the research. The initial and persistent goal of Lab A has been to understand the role of physical forces on biological processes in the vascular system. The flow loop is particularly generatively entrenched in that it has served as an integral part of the research of the lab for all the years of its existence. It made possible taking the research in vitro because normal and pathological in vivo forces on cells could be replicated to a first-order approximation—and potentially could be redesigned for higher-order effects if necesssary. Its generative entrenchment can be seen on two levels. On a metalevel, through its design and use it has entrenched the practice of importing engineering concepts and methods of analysis pertaining to mechanical forces into the study of biological phenomena. On a physical level, as a device it has formed a component of most experimental model systems (see, e.g., Figure 3.1). Most importantly, its affordances and constraints led to the formation of new problems and novel technologies. The kinds of experimentation the researchers envisioned could be done with the flow loop led, for instance, to the novel construct family of models. The construct model needed to be designed to interlock with the flow loop in experimental situations, some of which, in turn, required modifications to its design. As we saw, the construct model provided the lab with a more physiologically realistic model. That, plus the potential it provided for a novel application, generatively entrenched the construct in the research program, thereby opening further lines of research, which led to the lab building several new simulation models and incorporating new technologies into its analyses.

Although I have focused on the hybrid simulation technologies, this analysis of Lab A as an evolving distributed cognitive-cultural system is, importantly, incomplete without at least briefly outlining, at the institutional level, the coevolution of educational cognitive-cultural scaffolding specifically developed and directed at creating a new kind of interdisciplinary researcher in BME—one designed toward evolving the research field beyond the problematic collaborations of researchers in different disciplines by creating hybrid BME researchers.

CREATING SCAFFOLDING FOR BME RESEARCHERS AS LEARNERS

University labs are populated largely by students, and the lab context and wider research communities are not sufficient to provide the cognitive-cultural scaffolding for students to develop fully as researchers. The Lab A director and other senior colleagues saw as their challenge the design and building of a new educational environment for developing their students into a new breed of researcher. This new breed would move beyond their own experiences of being educated as engineers who later moved into biomedical research by being educated as hybrid biomedical engineers from the outset. As a consequence they determined they would develop a pioneering educational program that would firmly establish BME as an "interdiscipline" that integrated all three components in its research and education.[10] The graduates of this program would be able to move into academia, medicine/public health, industry, or government and fluently collaborate with other hybrids or with disciplinary colleagues, thus mitigating much of the "interactional complexity" of interdisciplinarity (Wimsatt 1972). This was an explicit decision that had three main components: (1) two new buildings with architecture designed to promote interdisciplinarity among bioengineering, biosciences, and medicine, with one building dedicated entirely to the envisioned BME department; (2) a new joint department of BME across two universities with one university providing largely engineering expertise and the other medical expertise, with the biosciences drawn from each and with several new faculty lines for young hybrid researchers; and (3) a new educational program (starting at the graduate level but quickly adding an undergraduate degree) that would integrate the three components of the field throughout its curriculum and cultivate student identities as biomedical engineers. Together, these components would serve to articulate and institutionalize the kind of interdisciplinarity they broadly envisioned—hybridization (Gerson 2013; Wimsatt 2013b).[11]

When we became involved, the first two components were well underway and provided the institutional and material structures from which to develop an educational program. They had few ideas about how to construct an educational program, but in their estimation, there were no established curricula or textbooks that could be adapted to achieve their vision. Through a serendipitous circumstance, they became interested in understanding what cognitive science might have to offer as a resource. At that time the U.S. National Science Foundation (NSF) had a requirement that any grant that included an educational program also had to include a cognitive science dimension. The leaders of the BME initiative were applying for an engineering research center that would include graduate training. I was director of the program in cognitive science, so they contacted me and asked if I could explain why the NSF would have such a requirement (i.e., what did cognitive science have to offer education). This created a partnership between us and my colleague Wendy Newstetter, whom they would hire into the new department and who became the co-principal investigator on our NSF-funded research. Our NSF funding, in turn, led to our creating a research group for conducting investigations of the cognitive and learning practices in the labs. Creating what they called "a cognitively informed educational program" was a novel conception consonant with novel objectives.

If successful, it would put them on the map as leaders in education as well as research.[12]

Much cognitive science research has established that making students active participants in their learning is more effective than simply lecturing to them, and in the sciences, this holds particularly true when they are engaged in attempting to solve authentic problems.[13] In the K–12 area, there was by then a long history of learning initiatives based on *problem-based learning* (PBL) activities. We were thus predisposed to find a way to make PBL central to the developing curriculum. Our choice was further reinforced by the fact that the method is widely used in medical education as a means of preparing students for the clinic and thus familiar to the medical faculty. With medical PBL, small groups are presented with problems—rich and complex real-world medical cases—that enable them to engage in the authentic practices of the field, with "scaffolding" created by the teachers (who act as "facilitators" to student problem-solving) to support their developing expertise in diagnostic practices. In the course of problem-solving, they develop a deep understanding of the human body, diagnostic capabilities, and an identity as medical problem solvers. PBL, as used in medical schools, scaffolds the kind of hypothetical deductive and inductive reasoning needed for diagnosing ailments. We argued that the model-based reasoning (Nersessian et al. 2003; Nersessian 2008, 2009; Nersessian and Patton 2009) used in engineering problem-solving requires a different kind of scaffolding that needs to be developed with the faculty. To distinguish the practices, we called the new PBL-informed method for engineering education *problem-driven learning* (PDL). Over time, through several iterations, this method has been woven into the BME curriculum. At the graduate level, there are two core PDL classes, and at the undergraduate level, there are three core PDL courses, two classes, and one instructional lab, all created collaboratively with the faculty (Newstetter 2006; Newstetter et al. 2010). Notably, as the undergraduate level developed, it began to provide an additional pool of lab researchers. Much of the rest of the curriculum at both levels has evolved to contain significant PDL elements developed by individual faculty members who were inspired by their experiences as facilitators of the introductory PDL course (all faculty facilitate). Thus, PDL, as a method, has become generatively entrenched, providing structuring constraints for course design in this and other BME programs our curriculum has influenced.

The introductory course is taken by all incoming students, who work in groups of eight on the problem outside of class and with one faculty or postdoc facilitator during the class periods.[14] The problems tackled are carefully designed by the faculty to present complex, ill-structured health-care problems drawn from the real world, which encourage students to integrate and anchor their developing bioscience and engineering knowledge. For example, in a problem about cancer screening, student teams must formulate and address questions concerning the biology of cancer, current screening technologies (e.g., CT scans or MRI), and future screening strategies (e.g., at the nanoscale) and develop statistical models, among other topics. There is now a substantial repository of problems that faculty can draw from and modify to keep up to date.

It is important to underscore that the curriculum development was not a linear process. Hutchins has characterized learning as "adaptive reorganization in a complex system" (Hutchins 1995, 289). We (my research group) and the BME faculty were also learners, and much "adaptive reorganization" took place in the early years of this curriculum development. One important dimension was that prior to our reseach, scant research had been conducted on the cognitive practices of biomedical engineering (or any field of engineering). Further, although university research laboratories are the main training grounds for future researchers, they have rarely served as sites for studying situated learning. We proposed a program of "translational research" that focused on translating insights about the nature of cognitive practices and effective strategies for supporting learning and problem-solving in research labs into the instructional setting. A major goal was to infuse the curriculum development with insights drawn from studying the situated research practices of BME, such as those of Lab A briefly outlined in this chapter.

Our philosophical and cognitive science research objective was to illuminate the ways in which the social, cultural, material, and cognitive aspects of practice and learning are intertwined in the research setting. We analyzed the ecological features of the research labs—the cognitive, investigational, and interactive practices—that invite and support complex learning and used them to guide design principles for instructional settings. Our findings led us to characterize the research labs as *agentive learning environments,* where researcher–learners are made agents of their own learning, unlike traditional passive instruction via lecture and the canned, recipe-driven instructional lab. These findings reinforced our initial choice of PDL as a pedagogical method. Now, learning scientists and experienced faculty work with incoming faculty,[15] which, together with the repository of PDL problems, constitute a "faculty incubator" that provides cognitive-cultural saffolding for their rapid participation in what for them is usually a novel pedagogical method and learning-centered BME ecosystem. Finally, through the outreach efforts of our learning researchers, the BME faculty, and the PhD students of the program who have gone on to university appointments, significant elements of our PDL approach have become generatively entrenched in other BME programs in the United States and internationally.

THIS CHAPTER has provided a glimpse into the complex processes through which cognitive-cultural scaffolding for research is created in the process of the lab articulating itself as an evolving distributed cognitive-cultural system. Although I have looked only at the tissue-engineering lab here, the highlighted features of these processes transferred robustly across the other BME labs we studied. Designing, redesigning, building, and experimenting with hybrid physical models that simulate selected aspects of in vivo phenomena is a central practice of these communities. Simulation models are the signature technologies of a lab and provide the structuring constraints that afford ways of furthering the research program without rigidly specifying in advance what moves can be made. One model can provide possibilities for creating other models that can interlock with it in extended experimental model systems. A major reconfiguration took place shortly before we entered Lab A because of the researchers' felt need to have a better model of the blood vessel wall than could be provided by their practice of studying the reactions of *"cells in plastic to fluid flow."* As we saw, the existing flow loop device facilitated and constrained the lab's development of the construct model, which in turn opened a range of new possibilities for research, including the creation of a novel animal model system and a line of stem cell research.

What this brief analysis establishes is that cognitive-cultural scaffolding becomes generatively entrenched in the evolving research program as "the lab" builds itself in the course of building the research technologies and the researchers. A highly significant part of the educational scaffolding directed toward creating hybrid researchers is the "repeated assembly" (Caporael 1996, 2014) of a PDL method through the years of the curriculum and of lab researchers who have at least the introductory PDL course in common. Further, the development of educational scaffolding for facilitating BME research provides a demonstration of "the manner in which epistemic integration interacts with organizations and institutions" (Gerson 2013, 515). Existing institutions adopted the notion that innovative research required a richer epistemic integration of BME, which in turn required the creation of new institutions and modes of organization, including a novel educational program, generatively entrenched in new kinds of buildings designed specifically to foster interdisciplinarity and a new kind of cross-university department aimed at creating hybrid researchers, themselves poised to work at the forefront of BME and to extend the frontiers for the next generation.

NOTES

I am grateful to the U.S. National Science Foundation for the funding required to carry out a project of this scope (REC0106733 and DRL0411825). I appreciate also the support of the Center for Philosophy of Science, University of Pittsburgh, where the final version of the chapter was written. I thank my fellow fellows, postdocs, and John Norton for their helpful comments on a draft discussed in our reading group. The chapter has also benefited from comments by participants during the "Beyond the Meme" workshop from which this volume derives. I thank the editors for inviting me to participate and for their comments on the penultimate draft, which significantly improved this article.

The research on which this analysis is based was a joint undertaking with my research group. I thank the members of the Cognition and Learning in Interdisciplinary Cultures (clic.gatech.edu) research group for their extensive and creative contributions to data collection, analysis, and interpretation; notably, for Lab A: Wendy Newstetter (co-PI), Elke Kurz-Milcke, Lisa Osbeck, Ellie Harmon, and Jim Davies. We are grateful to the lab director and researchers for welcoming us into their work space, granting us numerous interviews, and being so generous with time.

1. I use *cognitive-cultural* as an abbreviation for "cognitive, social, cultural, and material" dimensions as understood in science studies fields and for what Hutchins termed "socio-technical systems."

2. Gerson (2013) rightly cautions against the unreflective appropriation of biological metaphors for analyzing culture. However, Wimsatt (2013a) argues for the appropriateness of the ecosystem analogy for characterizing cultural evolution "because of the multiple evolving and interdependent lineages acting on different time and size scale" (564), which, as we will see, aptly characterizes the complexity of the evolution of the lab technologies, people, and practices as distributed cognitive-cultural systems.

3. See also Gerson (2013), Shore (1997), and Wimsatt (2013a).

4. All italicized quotes are taken from interviews with the researchers.

5. Carrying out this project required building our own interdisciplinary research group (with expertise in philosophy of science, history of science, anthropology, linguistics, psychology, artificial intelligence, human-centered computing, public policy, design, and qualitative methods) and developing innovative methods of data collection and analysis.

6. For a comparison of cognitive-historical analysis to other methodologies—laboratory experiments, observational studies, computational modeling—employed in research on scientific discovery, see Klahr and Simon (1999).

7. It is through fine-grained analysis of the processes of the coevolution of cognitive and cultural practices in science that we come to understand how, contra Kuhn, scientific change is continuous but not simply cumulative development (Nersessian 1984). Importantly, such analysis enables us to reveal the structuring constraints of prior conceptual structures and theories that contribute to the creation of genuinely novel concepts and theories (Nersessian 2008); see also chapter 4 of this volume.

8. Of the PhD students, three graduated while we were in the lab, and four graduated after we concluded formal data collection. We attended the dissertation defenses of the latter students and obtained their dissertations for our archive.

9. The notion of "interlocking models" we have developed is complex and cannot be developed fully within the confines of this chapter. Briefly, it is a multidimensional notion that serves to articulate relations among the components of the laboratory cast as a distributed cognitive-cultural system. Exemplars of interlocking models are physical models that interlock biology and engineering components, researcher mental models that interlock with artifact models in simulation processes, and configurations of models that interlock in experimental situations (Nersessian et al. 2003; Nersessian 2009; Nersessian and Patton 2009).

10. As a field, BME had been in existence since at least the early 1960s, but there were a few established departments, most notably at Johns Hopkins University (est. 1962). Since this is not a historical account of the field, I will present the situation as the founders of this new department expressed it to us.

11. Although I cannot articulate it in the confines of this chapter, there are significant insights to be gained regarding the explicit formation of interdisciplinary research fields—characteristic of much late twentieth- and early twenty-first-century science and engineering—by examining them

through the lens of research on "social movements" (see chapter 2). In the case at hand, the call of these researchers for a new breed of biomedical engineer suited for the twenty-first century came well in advance of an articulated means to carry out the objective, but it was a collective normative vision, the broad outlines of which were announced to the administrations of the schools involved, the wider intellectual community, funding agencies, and prospective donors. Importantly, it made a bid to reshape the knowledge-producing practices of a field, which the leaders felt needed to move beyond collaboration to hybridization in order to meet specified goals for twenty-first century BME.

12. Interestingly, they did not get the NSF engineering research center on that round but decided to proceed with what they dubbed "a cognitively informed" educational program anyway. The gamble paid off in that in approximately five years they went from nonexistent to the number two BME department in the *U.S. News and World Report* rankings, recently taking the number one spot over such rivals as the long-established Johns Hopkins department, the Massachusetts Institute of Technology, and Stanford. Twelve years after the program started, they won the state regents' award for the best educational program in the state. The program has been awarded the 2019 Bernard M. Gordon Prize for Innovation in Engineering and Technology Education by the National Academy of Engineering. These prestigious awards also provide validation for the "translational approach" pioneered in our research.

13. In cognitive science the notion that problem-solving is central to scientific thinking stems from the work of one of its founders, Herbert Simon, who traces his intellectual roots to the Würtzburg school of psychology, as does Karl Popper, for whom problem solving is the generator of scientific progress (Berkson and Wettersten 1984).

14. Because the BME-dedicated building was under construction as we began to plan the implementation of the PDL approach, five specially designed classrooms were constructed with seating and wall-to-ceiling whiteboards surrounding the room to facilitate interaction among the participants. Since we were doing research on the courses, two rooms were equipped with a separate observational window and recording compartment. The plan for students to work in groups of eight with a facilitator was recognized as costly from the outset, but the educational experiment is seen as so successful by the administration that it has continued to support the model despite significant growth in the student population. In recent years more than 160

undergraduate students are enrolled per semester, with facilitators needed for more than twenty teams, plus graduate courses.

15. The department has hired its own cognitive and learning scientists to provide support for ongoing curriculum development.

REFERENCES

Berkson, W., and J. Wettersten. 1984. *Learning from Error: Karl Popper's Psychology of Learning.* LaSalle, Ill.: Open Court.

Caporael, L. R. 1996. "Repeated Assembly." *Psycoloquy* 7 (4).

Caporael, L. R. 2014. "Evolution, Groups, and Scaffolded Minds." In *Developing Scaffolds in Evolution, Culture, and Cognition,* edited by Linnda R. Caporael, James R. Griesemer, and William C. Wimsatt, 57–76. Cambridge, Mass.: MIT Press.

Cetina, K. K. 1999. *Epistemic Cultures: How the Sciences Make Knowledge.* Cambridge, Mass.: Harvard University Press.

Chandrasekharan, S., and N. J. Nersessian. 2011. "Building Cognition: The Construction of External Representations for Discovery." *Proceedings of the Cognitive Science Society* 33:264–73.

Chandrasekharan, S., and N. J. Nersessian. 2015. "Building Cognition: The Construction of External Representations for Discovery." *Cognitive Science* 39:1727–63.

Gerson, E. M. 2013. "Integration of Specialties: An Institutional and Organizational View." *Studies in History and Philosophy of Science Part C: Studies in History and Philosophy of Biological and Biomedical Sciences* 44 (4): 515–24.

Glaser, B., and A. Strauss. 1967. *The Discovery of Grounded Theory: Strategies for Qualitative Research.* Piscataway, N.J.: Aldine Transaction.

Harmon, E., and N. J. Nersessian. 2008. "Cognitive Partnerships on the Bench Top: Designing to Support Scientific Researchers." *Proceedings of the 7th ACM Conference on Designing Interactive Systems.* New York: ACM.

Hollan, J., E. Hutchins, and D. Kirsh. 2000. "Distributed Cognition: Toward a New Foundation for Human-Computer Interaction Research." *ACM Transactions on Computer-Human Interaction* 7 (2): 174–96.

Hutchins, E. 1995. *Cognition in the Wild.* Cambridge, Mass.: MIT Press.

Hutchins, E. 2011. "Enculturating the Supersized Mind." *Philosophical Studies* 152 (3): 437–46.

Hutchins, Edwin. 2014. "The Cultural Ecosystem of Human Cognition." *Philosophical Psychology* 27 (1): 34–49.

Keller, E. F. 2002. *Making Sense of Life: Explaining Biological Development with Models, Metaphors, and Machines.* Cambridge, Mass.: Harvard University Press.

Kirsh, D., and P. Maglio. 1994. "On Distinguishing Epistemic from Pragmatic Action." *Cognitive Science* 18:513–59.

Klahr, D., and H. A. Simon. 1999. "Studies of Scientific Discovery: Complimentary Approaches and Divergent Findings." *Psychological Bulletin* 125:524–43.

Kurz-Milcke, E., N. J. Nersessian, and W. Newstetter. 2004. "What Has History to Do with Cognition? Interactive Methods for Studying Research Laboratories." *Journal of Cognition and Culture* 4:663–700.

Lave, J. 1988. *Cognition in Practice: Mind, Mathematics, and Culture in Everyday Life.* New York: Cambridge University Press.

Nersessian, N. J. 1984. *Faraday to Einstein: Constructing Meaning in Scientific Theories.* Dordrecht, The Netherlands: Martinus Nijhoff/Kluwer Academic.

Nersessian, N. J. 1987. "A Cognitive-Historical Approach to Meaning in Scientific Theories." In *The Process of Science,* 161–79. Dordrecht, The Netherlands: Kluwer Academic.

Nersessian, N. J. 1995. "Opening the Black Box: Cognitive Science and the History of Science." *Osiris: Constructing Knowledge in the History of Science* 10:194–211.

Nersessian, N. J. 2005. "Interpreting Scientific and Engineering Practices: Integrating the Cognitive, Social, and Cultural Dimensions." In *Scientific and Technological Thinking,* edited by M. Gorman, R. D. Tweney, D. Gooding, and A. Kincannon, 17–56. Hillsdale, N.J.: Lawrence Erlbaum.

Nersessian, N. J. 2006. "The Cognitive-Cultural Systems of the Research Laboratory." *Organization Studies* 27:125–45.

Nersessian, N. J. 2008. *Creating Scientific Concepts.* Cambridge, Mass.: MIT Press.

Nersessian, N. J. 2009. "How Do Engineering Scientists Think? Model-Based Simulation in Biomedical Engineering Research Laboratories." *Topics in Cognitive Science* 1:730–57.

Nersessian, N. J. 2012. "Engineering Concepts: The Interplay between Concept Formation and Modeling Practices in Bioengineering Sciences." *Mind, Culture, and Activity* 19 (3): 222–39.

Nersessian, N. J. 2017. "Hybrid Devices: Embodiments of Culture in Biomedical Engineering." In *Cultures without Culturalism,* edited by K. Chemla and E. F. Keller, 117–44. Durham, N.C.: Duke University Press.

Nersessian, N. J., E. Kurz-Milcke, and J. Davies. 2005. "Ubiquitous Computing in Science and Engineering Research Laboratories: A Case Study from

Biomedical Engineering." In *Knowledge in the New Technologies,* edited by G. Kouzelis, M. Pournari, M. Stöppler, and V. Tselfes, 167–95. Berlin: Peter Lang.

Nersessian, N. J., E. Kurz-Milcke, W. Newstetter, and J. Davies. 2003. "Research Laboratories as Evolving Distributed Cognitive Systems." In *Proceedings of the Cognitive Science Society 25,* edited by D. Alterman and D. Kirsch, 857–62. Hillsdale, N.J.: Lawrence Erlbaum.

Nersessian, N. J., and C. Patton. 2009. "Model-Based Reasoning in Interdisciplinary Engineering: Two Case Studies from Biomedical Engineering Research Laboratories." In *Philosophy of Technology and Engineering Sciences,* edited by A. Meijers, 678–718. Amsterdam: Elsevier Science.

Newstetter, W. C. 2006. "Fostering Integrative Problem Solving in Biomedical Engineering: The PBL Approach." *Annals of Biomedical Engineering* 34 (2): 217–25.

Newstetter, W. C., E. Behravesh, N. J. Nersessian, and B. B. Fasse. 2010. "Design Principles for Problem-Driven Learning Laboratories in Biomedical Engineering Education." *Annals of Biomedical Engineering* 38 (10): 3257–67.

Schank, J. C., and W. C. Wimsatt. 1986. "Generative Entrenchment and Evolution." *PSA: Proceedings of the Biennial Meeting of the Philosophy of Science Association* 1986:33–60.

Shore, B. 1997. *Culture in Mind: Cognition, Culture, and the Problem of Meaning.* New York: Oxford University Press.

Strauss, A., and I. Corbin. 1998. *Basics of Qualitative Research Techniques and Procedures for Developing Grounded Theory.* 2nd ed. London: Sage.

Theiner, G. 2013. "Onwards and Upwards with the Extended Mind: From Individual to Collective Epistemic Action." In Caporael, Griesemer, and Wimsatt 2013, 191–208.

Wimsatt, W. C. 1972. "Complexity and Organization." *Biennial Meeting of the Philosophy of Science Association* 1972: 67–86.

Wimsatt, W. C. 1986. "Developmental Constraints, Generative Entrenchment, and the Innate-Acquired Distinction." In *Integrating Scientific Disciplines,* edited by William Bechtel, 185–208. Dordrecht, The Netherlands: Kluwer Academic.

Wimsatt, W. C. 2007. *Re-engineering Philosophy for Limited Beings: Piecewise Approximations to Reality.* Cambridge, Mass.: Harvard University Press.

Wimsatt, W. C. 2013a. "Articulating Babel: An Approach to Cultural Evolution." *Studies in History and Philosophy of Science Part C: Studies in History and Philosophy of Biological and Biomedical Sciences* 44 (4): 563–71.

Wimsatt, W. C. 2013b. "Entrenchment and Scaffolding: An Architecture for a Theory of Cultural Change." In Caporael, Griesemer, and Wimsatt 2013, 77–105.

4

ARCHES AND SCAFFOLDS
Bridging Continuity and Discontinuity in Theory Change

MICHEL JANSSEN

I. SYNOPSIS

In principle, new theoretical structures in physics, unlike arches and other architectural structures, could be erected without the use of any scaffolds. After all, that is essentially how the four-dimensional formalism of special relativity, the curved space-times of general relativity, and the Hilbert-space formalism of quantum mechanics are introduced in modern textbooks. Historically, however, such structures, like arches, were originally erected on top of elaborate scaffolds provided by the structures they eventually either partially or completely replaced. The metaphor of arches and scaffolds highlights the remarkable degree of continuity in instances of theory change that, at first sight, look strikingly discontinuous. After putting to rest some historiographical worries about the metaphor and presupposing as little knowledge of the relevant physics and mathematics as possible, I describe how some key steps in the development of relativity and quantum theory in the early decades of the twentieth century can be captured quite naturally in terms of arches and scaffolds. Given how easy it is to find examples of this kind, I argue that it may be worthwhile to further analyze this pattern of theory change with the help of some of Stephen Jay Gould's ideas about evolutionary biology, especially his notion of constraints. In honor of Gould, I have tried to write this paper as a Gouldian pastiche.[1]

II. METAPHORS FOR THEORY CHANGE

In the section "Plans for Research" of a 1953 application for a Guggenheim Fellowship, Thomas S. Kuhn outlined two book projects that would eventually result in *The Copernican Revolution* (Kuhn [1957] 1999) and *The Structure of Scientific Revolutions* (Kuhn [1962] 2012). He already had the title of *Structure* but not the terms *paradigm* and *paradigm shift*. Comparing science to architectural structures, he wrote:

> Science, then, does not progress by adding stones to an initially incomplete structure, but by tearing down one habitable structure and rebuilding to a new plan with the old materials and, perhaps, new ones besides. (Hufbauer 2012, 459)[2]

The "adding stones" metaphor with which Kuhn contrasts the "tearing down" metaphor can be found, for instance, in the preface of Rudolf Carnap's *Aufbau,* one of the central texts of logical positivism, the philosophical program that Kuhn was reacting against. In philosophy, Carnap wrote, one ought to proceed as in the natural sciences, where "one stone gets added to another, and thus is gradually constructed a stable edifice, which can be further extended by each following generation" (Carnap 1928; quoted in Sigmund 2017, 137).[3]

Neither of these building metaphors for how old theories get to be replaced by new ones does justice to all or even most instances of theory change. When building a new theory, one tends to neither simply *add to* nor simply *tear down* an old theory. The old cumulative picture may be wrong but so is the alternative picture of a new theory or paradigm built on the burning embers of the old one, a picture conjured up and reinforced by the way in which Kuhn exploited the political connotations of his revolution metaphor in *Structure.*[4] It is good to remind ourselves right at the outset of this paper that "the price of metaphor is . . . eternal vigilance" (Lewontin 1963, 230).[5]

It has widely been accepted that neither the transition from geocentric to heliocentric astronomy nor the transition from nineteenth-century ether theory to special relativity fits the mold of a Kuhnian paradigm shift in the sense of *tearing down* one structure and replacing it by another (Swerdlow and Neugebauer 1984; Janssen 2002). In *The Copernican Revolution,* Kuhn ([1957] 1999, 182) himself used the completely different metaphor of a "bend in an otherwise straight road" to characterize the "shift in . . . direction in . . . astronomical thought" marked by Copernicus's *De Revolutionibus.* Both this "bend in the road" metaphor and another metaphor Kuhn was fond of using, that of a *gestalt switch* (Kuhn [1962] 2012, 85), are incompatible with the metaphor of tearing down some old structure and erecting a new one.

Does the metaphor of Kuhn's Guggenheim application at least capture the major theoretical upheaval of the mid-1920s known as the quantum revolution? In his book on the Bohr model of atomic structure, historian of science Helge Kragh suggests it does. He writes that matrix mechanics, the earliest incarnation of the new quantum theory, "grew out of what little was left" of the old quantum theory of Niels Bohr and Arnold Sommerfeld—"its ruins" (Kragh 2012, 368). The preface of a popular undergraduate physics textbook gives a similar impression:

> Quantum mechanics is not, in my view, something that flows smoothly and naturally from earlier theories. On the contrary, it represents an abrupt and revolutionary departure from classical ideas. (Griffiths 2005, viii)

The Dutch physicist Hendrik B. G. Casimir, who studied with some of the quantum revolutionaries of the mid-1920s, likewise emphasized the disruptive nature of the quantum revolution. "Between 1924 and 1928," he wrote in his autobiography, based in part on six lectures at the University of Minnesota in 1980, the development of a new quantum mechanics "swept physics like an enormous wave, tearing down provisional structures, stripping classical edifices of illegitimate extensions, and clearing a most fertile soil" (Casimir 1983, 51). Casimir's mixed metaphor, however, still leaves room for continuity in the quantum cataclysm. His tidal wave did not level the classical building in its entirety but only washed away parts of it.

One will search Kuhn's writings in vain for an account of the quantum revolution in which the new paradigm was erected on the ruins of the old one. And this is not just because, confounding some of his commentators,[6] Kuhn avoided the terminology of *Structure* in his historical writings.[7] In "Reflections on my Critics," his contribution to the proceedings of the 1965 conference in London that pitted him against Karl Popper, Imre Lakatos, Paul Feyerabend, and others, Kuhn (1970, 256–59) sketched how he saw the transition from the old quantum theory to matrix mechanics. He put great emphasis on what he saw, with considerable justification, as the crisis of the old quantum (calling it a "case book example" of this key concept of *Structure*[8]) but characterized the way out of this crisis as "a series of connected steps too complex to be outlined here" and criticized Lakatos for introducing, in his account of the same episode, "the crisis-resolving innovation . . . like a magician pulling a rabbit from a hat" (Kuhn 1970, 256–57). So, for Kuhn, a paradigm shift following a crisis did not necessarily have to be a wholesale and abrupt break with the past.

In fact, many elements of continuity in the transition from classical to quantum physics are on display in Max Jammer's (1966) classic, *The Conceptual Development of Quantum Mechanics,* the closest thing we have to a canonical account of this transition. As Jammer says in the preface, one of his main objectives was to show

> how in the process of constructing the conceptual edifice of quantum mechanics each stage depended on those preceding it without necessarily following from them as a logical consequence. (Jammer 1966, vii)

Olivier Darrigol more explicitly focused on continuities rather than discontinuities in his book *From c-Numbers to q-Numbers.* In the introduction, he expressed his conviction that "to obtain new theories" modern physicists "extend, combine, or transpose available pieces of theory" (Darrigol 1992, xxii). Jürgen Renn (2006) has introduced the notion of "Copernicus processes" to make a similar point.[9] One of the mottos Darrigol chose for his book has quantum architect Paul Dirac, in unpublished lecture notes of 1927, directly contradicting the assessment of the modern textbook writer quoted above:

> The new quantum theory requires very few changes from the classical theory . . . so that many of the features of the classical theory to which it owes its attractiveness can be taken over unchanged into the quantum theory. (Darrigol 1992, xiii)

The clause I left out—"these changes being of a fundamental nature"—gives Dirac's statement a paradoxical flavor. The key to the resolution of this paradox will be given in section IV (see note 47). Following Darrigol's lead, more recent work on the early history of quantum physics has highlighted a variety of continuities.[10]

To sum up: as long as the concept of a paradigm shift includes the "tearing down" element emphasized in Kuhn's Guggenheim application, neither the Copernican revolution, nor the relativity revolution, nor the quantum revolution fits the bill. Some of the revolutionaries in these cases can fairly be labeled iconoclasts, but none of them simply smashed the icons of the old guard.

I therefore propose—with some trepidation—a different building metaphor for theory change, one that involves both *adding to* and *tearing down* old structures and one that captures both continuities and discontinuities. I will present five examples from the early history of relativity and quantum theory, my area of expertise as a historian of science, in which a new theoretical structure can be seen, or so I will argue, as an arch built on top of a scaffold provided by an older theoretical structure discarded (at least in part) once the arch was finished.[11]

In four of these examples, the scaffold was discarded in its entirety once physicists recognized that the arch could support itself. In my last example, the second one from the history of quantum theory, only part of the scaffold was dismantled while other parts stayed in place long after the arch was finished. In this case, both arch and scaffold became part of the edifice of quantum theory as physicists use and teach it today. In this example, the relation between arch and scaffold is considerably more complicated than in the first four. However, to the extent that the arch-and-scaffold metaphor can profitably be used to characterize other instances of theory change at all, I expect the messy complicated cases to be more typical than the clean, simple ones.

As my two examples from the history of quantum theory will illustrate, the theoretical structure that plays the role of the arch in one instance of theory change can play the role of the scaffold in the next. This observation helps explain why scaffolds are sometimes only partially discarded after they have served their purpose in the building of an arch. What is merely a scaffold for some theorists may have been an arch for some of their predecessors and continue to be seen and treated as such by part of the relevant community.

The arch-and-scaffold metaphor can be broken down into specific elements with the help of Figure 4.1. This figure shows the construction of one of a total of nine arches, each 120 feet wide, of a bridge over the Thames in London in the 1810s. Originally called the Strand Bridge, it was renamed in 1817 to commemorate the Battle of Waterloo. It was demolished and replaced by a new one in the 1940s.

Figure 4.1 shows various components of the arch-and-scaffold metaphor that I will be using in my examples, mindful of the old adage that nothing kills a metaphor faster than the attempt to formalize it. The foundation on which both scaffold and arch are built is called the *tas-de-charge*. Then there

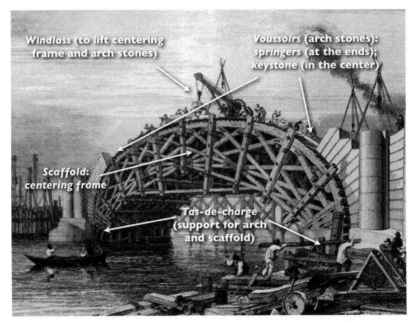

Figure 4.1. Elements of the arch-and-scaffold metaphor illustrated by the construction of the Strand Bridge over the Thames in London (renamed the Waterloo Bridge in 1817). Cropped version of *Print of the Strand Bridge (taken in the Year 1815)*, drawn by Edward Blore, engraved by George Cooke, and published by William Bernard Cooke (London, 1817). British Museum, museum number 1880,1113.1403.

is the scaffold or *centering frame*. The stones placed on this scaffold to make up the arch are called the *voussoirs*. Those at the ends are called the *springers;* the one in the center is called the *keystone*. The distinction between springers and keystones can meaningfully be made for the theoretical structures playing the role of an arch in my metaphor as well. For instance, the initial conception of a new theoretical structure can be seen as springers; the finishing touches as the keystone. Or, combining the two examples I will discuss from the history of special relativity, one can think of the new space-time structure as springers and of the new formalism for the physics of systems in that new space-time as the keystone. In the first of my two examples from the history of quantum theory (the fourth case study in section IV), we will even encounter an element (Niels Bohr's correspondence principle as it was used by several physicists in 1924–1925) corresponding to the final element labeled in Figure 4.1, the *windlass*, the instrument used to lift both the scaffold and the arch stones.

A nice feature of the example of the scaffolding of a stone-arch bridge is that both scaffold and arch can serve as a bridge, though the latter will make for a sturdier one than the former. In the case of a bridge, the relation between arch and scaffold is thus similar in this respect to that between a scaffolded and a scaffolding theory. In fact, the scaffolding shown in Figure 4.1 has basically the same structure as the wooden footbridge in Cambridge known as the Mathematical Bridge shown in Figure 4.2. The same relation is illustrated by two incarnations of the arched Walton Bridge across the Thames: a wooden bridge was completed in 1750 (and painted by Canaletto in 1754); a stone bridge opened in 1788 (and was painted by J. M. W. Turner in 1805).[12]

The basic idea behind the arch-and-scaffold metaphor is hardly new. For instance, in his contribution to the proceedings of the 1965 conference in London mentioned above, Lakatos noted that

> some of the most important research programs in the history of science were grafted on to older programs with which they were blatantly inconsistent. For instance, Copernican astronomy was "grafted" on to Aristotelian physics, Bohr's programme on to Maxwell's . . . As the young grafted program

Figure 4.2. The "Mathematical Bridge" in Cambridge. Picture by Joseph D. Martin.

strengthens, the peaceful co-existence comes to an end, the symbiosis becomes competitive and the champions of the new programme try to replace the old programme altogether. (Lakatos 1970, 142)

Whereas Lakatos used the term *grafting,* Werner Israel, in a paper on the prehistory of black holes, actually uses the term *scaffolding* ("The old and discarded is often scaffolding for the new") and explicitly offers the mechanism for theory change his metaphor is supposed to capture as an alternative to "the Kuhnian cycle of paradigm and revolution" (Israel 1987, 200). He gives two examples, both from nineteenth-century physics: "Faraday's concept of field grew out of the aether, Carnot's thermodynamics from the notion of a caloric fluid" (Israel 1987, 200). Unlike Lakatos and Israel, I will work out my examples in considerable detail.

There is some family resemblance between the arch-and-scaffold metaphor and two other well-known metaphors for the construction of knowledge, Sextus Empiricus's and Wittgenstein's ladders and Neurath's ship. In the penultimate proposition of the *Tractatus,* Wittgenstein (1921, 89, Proposition 6.54) noted that his reader "must, so to speak, throw away the ladder after he has climbed up it." This ladder metaphor has a rich history going back at least to Plato (Agassi 1975, 456–57; Gakis 2010). To mention just one precedent, Francis Bacon (1620, 14), in his *Novum Organum,* talked about "The Ladder of the Intellect." He compared the way in which authors suppress some of the evidence for their claims to "what men do in building, namely after completion of the building, remove the scaffolding and ladders from sight" (Bacon 1620, 97, Aphorism 125).

Neurath used his ship metaphor in several publications. In a booklet written in response to the first volume of Oswald Spengler's *Decline of the West,* he wrote:

> We are like sailors who on the open sea must reconstruct their ship but are never able to start afresh from the bottom. Where a beam is taken away a new one must at once be put there, *and for this the rest of the ship is used as support.* (Neurath 1921, 199 [emphasis added]; quoted and discussed in Sigmund 2017, 88)

Substituting *scaffold* for *support* in the italicized clause, we see that Neurath's metaphor is closer to the arch-and-scaffold metaphor than the proverbial ladders of Wittgenstein and others.[13] This was recognized by Wimsatt and Griesemer (2007, 300), who refer to Neurath's metaphor approvingly in a paper on scaffolding. The italicized clause, however, does not return in a later version of the metaphor. As Neurath put it in *Erkenntniss,* the house journal of the Vienna Circle in the early 1930s (Sigmund 2017, 223): "We are like sailors, who have to rebuild their ship on the open sea, without ever being able to dismantle it in dry-dock and reconstruct it from the best components" (Neurath 1932–1933, 92).[14]

The metaphor of a scaffold has also been used by several mathematicians. Carl Friedrich Gauss, we read in reminiscences recorded by a friend shortly after his death, "never gave a piece of work to the public until he had given it the perfection of form he desired for it. A good building should not show its scaffolding when completed, he used to say" (Sartorius von Waltershausen 1856, 67). Some of Gauss's contemporaries therefore compared him to a fox erasing its tracks in the sand with its tail.[15] Historian of astronomy Curtis Wilson quotes a much earlier example of this use of the scaffolding metaphor. Wilson (2001, 168–69) is relating how Newton's early devotion to algebra gave way to a strong preference for geometry. He recalls how Newton himself once claimed that he had constructed the proofs of most propositions in his *Principia* analytically but presented them in geometrical terms and thus, to use the Gaussian metaphor, removed the analytical scaffolding. This would have made for a nice illustration of the metaphor had not Tom Whiteside, editor of Newton's mathematical papers, shown Newton's claim to be false. In a footnote to this passage, however, Wilson quotes Thomas Hobbes actually using the metaphor in his attack on John Wallis's algebra: "Symbols are poor, unhandsome, though necessary, scaffolds of demonstration; and ought no more to appear in public, than the most deformed necessary business which you do in your chambers" (Hobbes 1656, 248; quoted in Wilson 2001, 185n36). This use of the scaffolding metaphor by Hobbes and Gauss is close to mine. The more common usage among mathematicians, however, appears to be quite different.

As David Hilbert liked to point out, the mathematician "erects living quarters *before* turning to the foundations" (Rowe 1997, 548). This is a paraphrase of the following passage in (unpublished) notes taken by Max Born, another quantum architect, of lectures Hilbert delivered in Göttingen in 1905:

> The buildings of science are not erected the way a residential property is, where the retaining walls are put in place before one moves on to the construction

and expansion of living quarters. Science prefers to get inhabitable spaces ready as quickly as possible to conduct its business. Only afterwards, when it turns out that the loosely and unevenly laid foundations cannot carry the weight of some additions to the living quarters, does science get around to support and secure those foundations. This is not a deficiency but rather the correct and healthy development. (Peckhaus 1990, 51; quoted in a different translation in Corry 1999, 163–64)

This attitude is clearly in evidence in Hilbert's early work on general relativity. Whereas Einstein only threw in his fate with the elegant mathematics of Riemannian geometry once he had convinced himself that the blueprint it provided for a spacious new theory of gravity included a solid grounding in physics, Hilbert was happy to move into these new quarters without inspecting the foundations first (see note 36).

The same attitude can be found in the methodological reflections of other mathematicians and mathematical physicists. My first example comes from a 1900 textbook by Paul Volkmann, professor of physics in Königsberg and an associate of Hilbert:

The conceptual system of physics should not be conceived as one which is produced bottom-up like a building. Rather it is a thorough system of cross-references, which is built like a vault or the arch of a bridge, and which demands that the most diverse references must be made in advance from the outset, and reciprocally, that as later constructions are performed the most diverse retrospections to earlier dispositions and determinations must hold. Physics, briefly said, is a conceptual system which is consolidated retroactively. (Volkmann 1900, 3–4; quoted in Corry 2004, 61)

My second example comes from an address to the London Mathematical Society, delivered in 1924 by its outgoing president, William H. Young, and published two years later. I am not aware of any direct connection between Young and Hilbert, but Young's wife, Grace Chisholm Young, a mathematician in her own right, earned her doctorate in Göttingen working with Hilbert's most famous colleague, Felix Klein. Echoing the observation attributed to Gauss in the passage from a friend's reminiscences quoted above, Young began his presidential address by reminding his audience of the common view that "all scaffolding is futile, because no scaffolding is to appear on the finished edifice" (Young 1926, 421). Young took exception to this view and argued that it had been harmful to "The Progress of Mathematical Analysis in the Twentieth Century," the topic of his presidential address. "It is essential," he insisted, "to go to a higher discipline [in this case: set theory] in order to master a lower one [in this case: analysis]" (Young 1926, 427). More advanced mathematics—to paraphrase Young's point—may be needed to prove results in more basic mathematics, which lesser mortals can then use without having to worry about the higher mathematics ever again. So it is the higher rather than the lower mathematics that plays the role of the scaffolding for Young. Hilbert expressed the same idea in a 1917 lecture. Hilbert, to paraphrase again, "described the axiomatic method as a process analogous to constructing ever deeper foundations to support a building still under construction" (Rowe 1997, 548). In my final case study in section IV, I will present Jordan's transformation theory as the scaffold on which von Neumann built the arch of his Hilbert space formalism. This case study, however, can alternatively be cast in terms of von Neumann deepening the foundation of Jordan's theory, in which case von Neumann's theory would provide the scaffold (in Young's sense) built to secure Jordan's mathematically unsound arch.

In his 1917 book, *The Electron,* American physicist Robert A. Millikan used a metaphor that combines my arch-and-scaffold metaphor and Hilbert's building-and-foundation metaphor to describe the relation between Einstein's formula for the photoelectric effect, which Millikan had experimentally verified the year before, and the controversial light-quantum hypothesis from which Einstein had derived this formula in 1905. Millikan wrote:

> Despite . . . the apparently complete success of the Einstein equation, the physical theory of which it was designed to be the symbolic expression is found so untenable that Einstein himself, I believe, no longer holds to it, and we are in the position of having built a very perfect structure and then knocked out entirely the underpinning without causing the building to fall. It stands complete and apparently well tested, but without any visible means of support. These supports must obviously exist, and the most fascinating problem of modern physics is to find them. (Millikan 1917, 230; quoted and discussed by Stuewer 2014, 156)

Millikan, in other words, saw Einstein's formula for the photoelectric effect as an arch in danger of collapsing and thus in need of a scaffold to support it. As we will see in section IV, we can likewise think of von Neumann bringing in Hilbert space as a scaffold to prevent Jordan's arch from collapsing (see note 37 for another, more clear-cut, example).

On the face of it, Heinrich Hertz's attitude toward Maxwell's theory, expressed in his famous line that "Maxwell's theory is Maxwell's system of equations" (Hertz 1893, 21), may look similar to Millikan's attitude toward the theory from which Einstein derived the formula for the photoelectric effect. Hertz, on this reading, took Maxwell's "perfect structure and then knocked out entirely the underpinning without causing the building to fall." Hertz himself, however, saw his contribution to Maxwell's theory quite differently. In a lecture at the 1889 *Naturforscherversammlung* (the annual meeting of the German Society of Natural Scientists and Physicians) held that year in Heidelberg, Hertz used the metaphor of a bridge, similar to the one used by Volkmann in the passage quoted above, to explain the importance of his experimental demonstration of electromagnetic waves for the further development of Maxwell's theory. In his commemorative speech at the German Physical Society in Berlin, shortly after Hertz's early death, Max Planck referred to this lecture and further elaborated on Hertz's metaphor. In his 1889 lecture, Planck recalled, Hertz

> compared Maxwell's theory to a bridge that, with a bold arch, spans across the wide ravine between the regions of optical and electromagnetic phenomena, characterized by molecular and cosmic wavelengths, respectively. Because of these fast electrical vibrations, he explained at the time, new land had been gained in the middle of this ravine and a firmly planted pillar now stood on it providing additional support for the bridge. Since that time, various kinds of expert craftsmanship have made this pillar taller and broader, ensuring that today the bridge stands stronger and prouder than ever. It no longer just serves, as it did in the past, the occasional forays of the odd bold adventurer. No, it can already carry the heavy trucks of research in the exact sciences, constantly shipping the land's treasures from one region to another, thereby enriching both. (Planck 1894, 283)

Planck's elaboration of this metaphor reminds us that bridges are not built to be admired but used. The same is true for the metaphorical arches and bridges to be examined in this paper.

III. HISTORIOGRAPHICAL SCRUPLES

Whiggishness: Three Lines of Defense

Before I present some concrete examples of theory change that, I argue, fit my metaphor of arches and scaffolds, I need to address the obvious worry that any historical narrative deploying this metaphor is intrinsically Whiggish.[16] After all, we can only see in hindsight that one scientist's arch was the next scientist's scaffold. I have three lines of defense against this charge of Whiggishness, and I am prepared to make my stand on the third.

My first line of defense is that scientists sometimes *do* realize that they are building a scaffold and not an arch. Throughout the reign of the old quantum theory, for instance, Bohr was keenly aware of its provisional character. In a letter to Sommerfeld of April 30, 1922, he described his work on the theory as a "sincere effort to obtain an inner connection such that one can hope to create a valid fundament for further construction" (Sommerfeld 2004, doc. 55; translation from Eckert 2013, 126). In the early 1920s, Göttingen emerged as another center for work on the old quantum theory, alongside Bohr's Copenhagen and Sommerfeld's Munich. Born, the leader of this third center, clearly saw the limitations of the theory too. In the preface of his book, *Atomic Mechanics,* he cautioned:

> The work is deliberately conceived as an attempt . . . to ascertain the limits within which the present principles of atomic and quantum theory are valid and . . . to explore the ways by which we may hope to proceed . . . To make this program clear in the title, I have called the present book "Vol. I;" the second volume is to contain a closer approximation to the "final" atomic mechanics . . . The second volume may, in consequence, remain for many years unwritten. In the meantime let its virtual existence serve to make clear the aim and spirit of this book. (Born 1925, v)[17]

The sequel would be written much sooner than Born anticipated. It was published only five years later (Born and Jordan 1930). First, however, in 1927, an English translation of Born's 1925 book appeared. Given the rapid developments since 1925, the decision to republish his treatise in translation without any substantive changes or additions could be called into question. Born tries to preempt criticism on this score in the preface. The first argument he gives in his defense is that "it seems to me that the time is not yet arrived when the new mechanics can be built up on its own foundations, without any connection with classical theory" (Born 1927, xi). Without denying the self-serving purpose of this preface, we can say that Born clearly recognized the role of the old theory in building up the new one.

Einstein used a building metaphor, suggestive at least of an arch and a scaffold, to describe the sense in which he considered general relativity, his theory of gravity, to be preliminary. In the fall of 1915, Einstein and Hilbert found themselves in a race to complete the theory (Janssen and Renn 2015). They eventually arrived at the same field equation. This equation determines what gravitational field a given matter distribution will produce. The two men agreed that the left-hand side of this equation describes the curvature of space-time, reflecting the central idea of the theory that gravity is part of the fabric of space-time. They did not agree, however, on the interpretation of the right-hand side, describing the matter responsible for this space-time curvature. Hilbert endorsed the view of Gustav Mie, a late representative of the so-called electromagnetic worldview (see the second case study in section IV), that all matter ultimately consists of electromagnetic fields satisfying some nonlinear generalization of Maxwell's equations. Einstein briefly flirted with this idea but rejected it. For him the right-hand side of his field equation was just a placeholder for whatever would be supplied later by a more satisfactory theory of matter. Two decades after he had introduced general relativity, Einstein was still waiting for such a theory. General relativity, he wrote in 1936,

> is similar to a building, one wing of which is made of fine marble (left part of the equation), but the other wing of which is built of low grade wood (right side of equation). (Einstein 1936, 312; quoted and discussed in van Dongen 2010, 62)

The wing made out of wood can be thought of as a scaffold awaiting the arrival of the marble for the completion of this part of the building.

Robert Hooke is an early example of a natural philosopher with the humility to recognize that he was building a scaffold rather than an arch. Hooke even used the arch-and-scaffold metaphor, although he cast it in somewhat different terms.[18] In the preface of his *Micrographia*, he wrote: "If I have contributed the meanest foundations whereon others may raise nobler Superstructures, I am abundantly satisfied" (Hooke 1665, xii–xiii). In a similar vein, Einstein (1917b, 91) wrote in his popular book on relativity that "the most beautiful fate of a physical theory is to point the way to the establishment of a more inclusive theory, in which it lives on as a limiting case" (translation from Holton 1981, 101).[19]

One would expect it to be easier to find examples of scientists recognizing the preliminary character of theories put forward by their contemporaries and calling the arches erected by them mere scaffolds. Despite some crowdsourcing, however, I have only been able to find two such examples so far. Only in one of those is the term *scaffold* used explicitly.

The first example comes from Arthur Stanley Eddington's Gifford Lectures in Edinburgh in 1927. Eddington prefaced his discussion of wave mechanics with the following warning:

> Schrödinger's theory is now enjoying the full tide of popularity . . . Rather against my better judgment I will try to give a rough impression of the theory. It would probably be wiser to nail up over the door of the new quantum theory a notice, "Structural alterations in progress—No admittance except on business," and particularly to warn the doorkeeper to keep out prying philosophers. I will, however, content myself with the protest that, whilst Schrödinger's theory is guiding us to sound and rapid progress in many of the mathematical problems confronting us and is indispensable in its practical utility, I do not see the least likelihood that his ideas will survive long in their present form. (Eddington 1928, 210–11)

The second example comes from an article in *The Electrician* in 1893, in which John Henry Poynting criticized the mechanical model of the ether that Oliver Lodge (1889) had promoted in his bestseller *Modern Views of Electricity*. This model is known as the cogwheel ether (see Figure 4.3).[20] Pierre Duhem famously and disapprovingly compared Lodge's book to a factory:

> Here is a book intended to expound the modern theories of electricity and to expound a new theory. In it there are nothing but strings which move around pulleys, which roll around drums, which go through pearl beads, which carry weights; and tubes which pump water while others swell and contract; toothed wheels which are geared to one another and engage hooks. We thought we were entering the tranquil and neatly ordered abode of reason, but we find ourselves in a factory. (Duhem 1914, 70–71; quoted in Hunt 1991, 87)

Figure 4.3. Lodge's "cogwheel ether" (Lodge 1889, 207).

Poynting had more sympathy for Lodge's project. "We are all looking forward," he wrote, "to the time when by mechanical explanation of electromagnetism, light shall once more become mechanical" (Poynting 1893, 635). Yet, he cautioned, such explanations

> are solely of value as a scaffolding enabling us to build up a permanent structure of facts, i.e., of phenomena affecting our senses. And inasmuch as we may at any time have to replace the old scaffolding by new, more suitable for new parts of the building, it is a mistake to make the scaffolding too solid, and to regard it as permanent and of equal value with the building itself. (Poynting 1893, 635)

The problem with Lodge's work, according to Poynting (1893), was that he made the "scaffolding . . . as important as the building" (635).

As striking as these passages from the writings of Bohr, Born, Einstein, Eddington, Hooke, and Poynting are, I do not expect to find too many scientists characterizing the theories of their contemporaries, let alone their own, as mere scaffolds for future theories. These examples thus only go so far in deflecting the charge of Whiggishness against the arch-and-scaffold metaphor.

My second line of defense is that structures often *inadvertently* serve as scaffolds for other structures. So the use of the arch-and-scaffold metaphor does not *automatically* imply the kind of teleology responsible for the odium of Whig history. A good example—and the one that originally inspired my exploration of the metaphor—is the proposal by Alexander Graham Cairns-Smith (1985) in *Seven Clues to the Origin of Life* that the complex nucleotides of RNA and DNA were first assembled on a scaffold of minute clay crystals. Cairns-Smith (1985) asks: "How can a complex collaboration between components evolve in small steps?" (58). The answer, he suggests, is "that every so often an older way of doing things will be displaced by a newer way that depends on a new set of subsystems. It is then that seemingly paradoxical collaborations may come about" (Cairns-Smith 1985, 59). He uses the arch-and-scaffold metaphor, complete with two simple diagrams (see Figure 4.4), to illustrate how this can happen:

> Consider this very simplified model—an arch of stones. This might seem to be a paradoxical structure if you had been told that it arose from a succession of small modifications, that it had been built one stone at a time. How can you build any kind of arch *gradually?* The answer is with a supporting scaffolding. In this case you might have used a scaffolding of stones. First you would build a wall, one stone at a time. Then you would remove stones to leave the "paradoxical" structure. Is there any other way than with scaffolding of *some* sort? Is there any other way to explain the kind of complex leaning together of subsystems that one finds in organisms, when each of several things depends on each other, than that there had been earlier pieces, now missing? (Cairns-Smith 1985, 59–60)

John Norton (2014, 685–87) used essentially the same analogy (including a drawing of an arch) that Cairns-Smith used for the origin of life to explain

Figure 4.4. A wall of stones as a scaffold for an arch of stones (Cairns-Smith 1985, 59–60).

how the long sequence of inductive inferences that eventually got us to to-day's scientific knowledge could ever have gotten off the ground.

Cairn-Smith's example—and, for that matter, Norton's—clearly shows that the arch-and-scaffold metaphor can be used without implying that the earlier structure was intentionally built as a scaffold. Yet it still does not quite put the worry about Whiggishness to rest. Unlike Norton, I am starting in *media res* and not at the mythical dawn of humanity. In that case, a historical narrative in which a structure is called a *scaffold* before the construction of the arch built on top of it has even begun at least strongly *suggests* an element of teleology. Such a narrative is bound to lend an air of linearity and inevita-bility to the transition from the *scaffolding* to the *scaffolded* theoretical struc-ture. Fortunately, spelling out the danger of the metaphor in this way points to an obvious way to avoid it: tell the story backward! In other words—as the order of the terms already happens to suggest—have narratives using the arch-and-scaffold metaphor introduce the arches before the scaffolds.

This third line of defense against the charge of Whiggishness is both more effective than the other two and more natural than it may initially sound. A narrative that moves from arches to scaffolds neither implies nor suggests teleology. Instead it answers a question one naturally asks when standing in awe in front of an arch or some other architectural marvel: How did they build that? A satisfactory answer to such questions will often involve the identification of an earlier structure that served as the scaffold for the archi-tectural structure and was then partially or completely taken down. Similar questions can be asked about theoretical structures in physics, such as the four-dimensional formalism of special relativity, the curved space-times of general relativity, and the Hilbert-space formalism of quantum mechanics. How did physicists come up with these formalisms to describe and explore nature? In at least some instances, as the examples discussed below will il-lustrate, such questions can be answered by the identification of some ear-lier formalism that scaffolded the new formalism and was discarded either in whole or in part once it was recognized that the new formalism no longer needed extraneous support.

Following Hilbert's lead (see section II), mathematicians might want to put the metaphor on its head and have the new formalism play the role of the scaffold built retroactively to support earlier formalism playing the role of the arch, built hastily and in danger of collapsing. Two of the examples I will be discussing—both, unsurprisingly, involving mathematicians (Minkowski and von Neumann)—can be used to illustrate this "inverted" use of the metaphor. When the metaphor is used in this way, we do not run afoul of the charge of Whiggishness because the scaffold is built after the arch.

When I use the metaphor without this kind of inversion (i.e., when the scaffold is built before the arch), I hope to steer clear of any pernicious Whiggishness by making the kind of question I am asking explicit in the way indicated above. The arch-and-scaffold narratives offered in response to such questions neither imply nor suggest that the progression from the earlier to the later theoretical structure was linear or inevitable. This line of defense, however, does place an important constraint on the construction of such narratives. To bring out and reinforce the analogy with the "How did they build that?" question we ask when we happen upon a puzzling architectural structure, such narratives should all start with at least a preliminary characterization of the theoretical structure that plays the role of the arch in the story. Ideally, one then proceeds to tell the story backward. This kind of time reversal has been attempted in many movies and TV shows—with limited success.[21] Given this checkered track record, it may be wiser to stick to the tried-and-true strategy of presenting a new historical account as an alternative to some canonical received view. I will at least try to combine this safe standard approach with the challenging one of telling the story backward, which is clearly the more elegant way of constructing a historical narrative based on the arch-and-scaffold metaphor.

Some historians may still not be satisfied. Even if they grant that I am not giving Whiggish answers, they may still object that I am asking Whiggish questions. Asking how something came about, after all, obviously presupposes that it was there to begin with. This residual charge of Whiggishness does not bother me. I make no excuse for asking questions informed by present-day knowledge and for making decisions about what source material to look at based on what I expect to shed light on the development of currently accepted theories. Which is not to say that there are no dangers and pitfalls to this approach (see, e.g., note 43). Still, this kind of Whiggishness strikes me as relatively benign. In fact, it is implied by the commonly accepted names for the subfields I take myself to be working in—*history of relativity theory* and *history of quantum mechanics*. My examples of arches and scaffolds are all based on my earlier research in these two areas. The papers in which I published this research implicitly answer questions similar to those I explicitly raise when I recast parts of their narratives in terms of arches and scaffolds. Hence, far from compounding my Whiggish sins, I am actually atoning for them by adopting the arch-and-scaffold metaphor!

No Actors' Categories: Two Lines of Defense

A related objection to the arch-and-scaffold metaphor is that it is not an actors' category. I can offer two rejoinders to defuse that charge. The first is that, even though there may not have been many, several historical actors did use the language of arches (or, at least, buildings) and their scaffolds. We have already seen passages from (or attributed to) several scientists (Poynting, Millikan), philosophers (Bacon, Hobbes, Wittgenstein), and mathematicians (Gauss, Hilbert), in which they use the term *scaffold* or terms like it *(ladder, support, foundations).*

We saw Poynting, in his 1893 critique of Lodge's theory, warn his readers not to mistake the scaffold for the building in science. Six years later, in his 1899 presidential address to the Section of Mathematical and Physical Science of the British Association for the Advancement of Science, he issued a more general version of this warning. "To give the hypotheses equal validity with facts," he said on that occasion, "is to confuse the temporary scaffolding with the building itself" (Poynting 1899, 620). The paragraph concluding with this sentence suggests that "the building itself" refers to nature itself. This paragraph also yields two more occurrences of the term *scaffolding* and one of *ladder* (cf. the passage from Bacon's *Novum Organum* quoted in section II):

> While the building of Nature is growing spontaneously from within, the model of it, which we seek to construct in our descriptive science, can only be constructed by means of scaffolding from without, a scaffolding of hypotheses. While in the real building all is continuous, in our model there are detached parts which must be connected with the rest by temporary ladders and passages, or which must be supported till we can see how to fill in the understructure. (Poynting 1899, 620; quoted in Freund [1904] 1968, 227)

On the next page, Poynting alerted his readers to the danger of mistaking a preliminary theory for a definitive one: "It is necessary to bear in mind what part is scaffolding, and what is the building itself, already firm and complete" (Poynting 1899, 621; quoted in Freund [1904] 1968, 227). Here the contrast between scaffolding and building does not refer to the contrast between our scientific models of nature and nature itself but to that between preliminary partial theories (the "detached parts") and more comprehensive and secure theories (the "understructure").

A few years before Poynting, the American paleontologist Edward Drinker Cope had issued a similar warning. In a letter to the editor of *Science,* Cope used an analogy to buildings and scaffolds to defend the value of hypotheses as long as judgments about their truth or falsity are suspended:

> Builders generally know the difference between the scaffolding and the building. And a builder will value the indication of faults in his scaffolding rather than general disquisitions on the uselessness of scaffolds in general. (Cope 1895, 522)

Cope and Poynting were by no means the first to use the analogy to buildings and scaffolds to make these points. In 1820, Humphry Davy had already voiced some of the same concerns they raised in the 1890s, using remarkably similar language. In his address to the Royal Society upon taking up its presidency, Davy had cautioned his audience to "attach no importance to hypotheses" and to treat "them rather as part of the scaffolding of the building of science, than as belonging either to its foundations, materials, or ornaments" (Davy 1820, 14).

On the European continent, Johann Wolfgang von Goethe made essentially the same point in a passage that only seems to have been published posthumously:

> Hypotheses are scaffoldings which one puts up before building and which one tears down once the building is complete. They are indispensable for the worker: only one should not take the scaffolding for the building. (Beutler 1949, 9:653; translation from Agassi 1975, 457; see also Agassi 2013, 118)

The sentiment expressed here by Davy, Goethe, Cope, and Poynting can also be found in *Preliminary Discourse on the Study of Natural Philosophy* by the British astronomer and philosopher of science avant la lettre John Herschel: "To lay any great stress on hypotheses . . . except in as much as they serve as a scaffold for the erection of general laws, is to 'quite mistake the scaffold for the pile'" (Herschel [1830] 1966, 204; see Agassi 1975, 457).

In an essay in *Quarterly Reviews* in 1840, Herschel provided a more elaborate and more eloquent statement of Poynting's observation, quoted above, that "while the building of Nature is growing spontaneously [our] model of it . . . can only be constructed by means of scaffolding." This is how Herschel put it:

> In erecting the pinnacles of this temple [of science], the intellect of man seems quite as incapable of proceeding without a scaffolding or circumstructure foreign to their design, and destined only for temporary duration, as in the rearing of his material edifices. A philosophical theory does not shoot up like the tall and spiry pine in graceful and unencumbered natural growth, but, like a column built by men, ascends amid extraneous apparatus and shapeless masses of materials. (Herschel 1857, 67)

As the quotations from Planck in section II and Poynting in this section suggest, the history of electromagnetism may make for good hunting grounds for scaffolding metaphors. Here is one from the preserves of James Clerk Maxwell's *Treatise on Electricity and Magnetism*:

> We can scarcely believe that Ampère really discovered the law of action by means of the experiments which he describes. We are led to suspect, what, indeed, he tells us himself, that he discovered the law by some process which he has not shewn us, and that when he had afterwards built up a perfect demonstration he removed all traces of the scaffolding by which he has raised it. (Maxwell 1873, 162–63)

Maxwell's observation about Ampère is reminiscent of Abel's (or Jacobi's) comparison of Gauss to a fox erasing its tracks (see note 15).

In his *History of the Theories of Aether and Electricity,* Sir Edmund Whittaker used a scaffolding metaphor to describe some of Maxwell's own work. After discussing the paper in which Maxwell (1861–1862) first published what in hindsight we recognize as Maxwell's equations of electrodynamics (see Figure 4.6 in section V for the mechanical model Maxwell used to derive those equations), he wrote:

> Maxwell's views were presented in a more developed form in a memoir . . . read to the Royal Society in 1864 [Maxwell 1865]; in this the architecture of the system was displayed, stripped of the scaffolding by aid of which it had been first erected. (Whittaker [1951–1954] 1987, 1:255)

At least one later commentator on Maxwell (1865) used the same metaphor: "The scaffolding could now be kicked away from the edifice" (Kargon 1969, 434).[22] To give another example of a historian of science using the metaphor, Owen Gingerich (1989, 69), in an article about Kepler, wrote that "both

Ptolemy in the *Almagest* and Copernicus in *De Revolutionibus* had carefully concealed the scaffolding by which they had erected their mathematical models."

For my final example of scientists themselves using the metaphor, I return to Einstein. In 1953, the year before he died, Einstein compared the extraction of concepts from experience to the construction of houses and bridges with the help of a scaffold. In a letter to Maurice Solovine, his old friend and fellow member of the mock *Olympia Academy* of his halcyon days as a patent clerk in Berne, he wrote:

> Concepts can never be derived logically from experience and be above criticism. But for didactic and also heuristic purposes such a procedure is inevitable. Moral: Unless one sins against logic, one generally gets nowhere; or, one cannot build a house or construct a bridge without using a scaffold which is really not one of its basic parts. (Einstein 1987, 147; cf. Agassi 1975, 456–59)

The point for which Einstein used this metaphor is rather different from the one pursued in this paper. It is closer, actually, to the point Bacon, Wittgenstein, and others wanted to make with their ladder metaphor (Agassi 1975, 456–58; cf. section II).

Despite these intriguing passages obtained largely through crowdsourcing (see the acknowledgments), I must admit that I have only found a handful of examples so far of historical actors using scaffolding metaphors for the development of science. Moreover, I am not aware of any of the actors in my five examples using this kind of language. Let me use another analogy to explain why I ultimately see this not as a weakness but as a strength of my project.[23]

Historians are in the business of selecting parts of invariably incomplete source material and carefully arranging it in historical narratives to make their audience look at it from their point of view. In this respect, historical narratives are not unlike museum exhibits of dinosaurs (cf. Figure 4.5).

When mounting a dinosaur for exhibit, curators use several devices (often in combination) to create the illusion of a complete animal. First, they restore missing or damaged pieces of fossil bone with plaster. Next, they combine complementary specimens of the same species to form a composite skeleton. Finally, they make sculptures or casts of any missing elements to create a whole animal. To hold the fossil bones in lifelike positions, they always use some kind of metal armature, or scaffold. Normally, the armature is designed to be as unobtrusive as possible to give the museum visitor the

Figure 4.5. Apatosaurus under construction at the Chicago Field Museum in 1908. Courtesy of the Field Museum, Chicago. ID No. CSGEO23972. This picture is also reproduced in Brinkman (2010, 244).

impression that the specimen is self-supporting. Eugene S. Richardson, a curator at the Field Museum in Chicago, wrote a laudatory poem about an exhibit of Gorgosaurus, which ends with the following lines: "Here he stands without a scaffold! Gorgosaur is self-supporting!" (Brinkman 2013, 223–25).

There is some controversy about the practice of disguising these curatorial devices to enhance the illusion. At some museums, restored pieces of fossil bone, or plaster casts of fossils, have been carefully painted to match the original fossil material and to hide its artificial origins. Other museums, by contrast, have made their restorations in a different color so that the visitor can easily distinguish them from original fossils. Museums will sometimes exhibit only original fossil material, restoring the complete animal in a background mural or a drawing.[24]

The latter approach, where the curator's role in reconstructing the animal is explicitly acknowledged, corresponds to the approach to writing history of science that I am adopting in this project. (The approach in the papers that I draw on for this project was closer to the former.) Using a metaphor that is clearly my own and not an actors' category to present my source material to my readers, I am like the curator whose use of elements that will not be mistaken for parts of the original specimen forcefully reminds visitors of their own roles in the reconstruction of the specimen.

IV. FIVE EXAMPLES FROM THE DEVELOPMENT OF RELATIVITY AND QUANTUM THEORY

I have identified five instances of major theory change in the history of relativity and quantum theory that fit the arch-and-scaffold metaphor. In these examples, we will encounter two kinds of relations between arch and scaffold, which will be examined more generally in sections V–VI. In this section I present my five case studies, explain how they fit the metaphor, and indicate what we gain, in terms of our historical understanding of these episodes, by recasting their narratives in terms of it.[25] To do so, I need to cover these examples in some detail. At sufficiently low resolution just about any episode in the history of science can be made to fit just about any metaphor. It is thus imperative to show that the arch-and-scaffold metaphor captures such episodes at a much more fine-grained level.

I have tried to write this section without presupposing any knowledge of the relevant physics. Even so, readers who are familiar with (the history of) relativity and quantum theory will undoubtedly find this section much easier to read than readers who are not. Those without a background in physics are encouraged to read as much as they can stomach of the first, the third, and the fourth case study and skim or skip only the mathematically more demanding (parts of) the second and the fifth. A detailed understanding of section IV is not required to appreciate the more general points about the use of the arch-and-scaffold metaphor in the history of science in sections V and VI.

How Minkowski Space-Time Was Scaffolded by Lorentz's Theorem of Corresponding States[26]

The natural starting point for a history of special relativity in terms of arches and scaffolds is the lecture "Space and Time" by the Göttingen mathematician Hermann Minkowski (1909). Minkowski gave this lecture at the 1908 *Naturforscherversammlung* held that year in Cologne. It was published posthumously the following year. Minkowski began his lecture by proclaiming that henceforth space by itself, and time by itself, are doomed to fade away into mere shadows, and only a kind of union of the two will preserve an independent reality. (Minkowski 1909, 75)

He proceeded to develop the now familiar geometry of what has come to be known as Minkowski space-time, the space-time structure of the special theory of relativity. Although it only got its name later, this is the theory introduced in the most famous paper of Einstein's annus mirabilis (Einstein 1905). Minkowski showed that the transformations that relate the space-time coordinates of one reference frame in Minkowski space-time to the space-time coordinates of another reference frame in uniform motion with respect to the first are completely analogous to the transformations that relate the Cartesian coordinates with respect to one set of orthogonal axes in ordinary Euclidean space to the Cartesian coordinates with respect to another set of orthogonal axes rotated with respect to the first. In fact, such rotations in three-dimensional Euclidean space can be subsumed under Lorentz transformations in four-dimensional Minkowski space-time.

How did Minkowski build this magnificent arch? As he made clear in his lecture, he used a scaffold provided by recent work in electrodynamics. However, he also imagined a scenario in which the arch would have been built without a scaffold or, better perhaps, a different scaffold, provided by Newtonian mechanics rather than Maxwellian electrodynamics.[27] The equations of Newtonian mechanics, he noted, "exhibit a two-fold invariance" (Einstein 1905, 75). They do not change when we rotate the axes of our spatial coordinate system or when we set that spatial coordinate system in motion with a constant velocity. The latter invariance expresses the principle of relativity in Newtonian mechanics (Einstein extended this principle from mechanics to all of physics, especially electrodynamics). In Newtonian theory, Minkowski noted, these two operations

lead their lives entirely apart. Their utterly heterogeneous character may have discouraged any attempt to compound them. But it is precisely when they are compounded that the complete group, as a whole, gives us to think. (Minkowski 1909, 76)

"The thought might have struck some mathematician," Minkowski mused, that maybe Newton's theory, invariant under rotations and under transformations from one inertial frame to another (now called Galilean transformations to distinguish them from Lorentz transformations), ought to be replaced by a theory based on invariance under Lorentz transformations. In this way mathematicians might have anticipated special relativity. "Such a premonition," Minkowski continued, "would have been an extraordinary triumph of pure mathematics." Alas, this possibility had not occurred to any mathematician, including Minkowski himself, before physicists had recognized the importance of Lorentz transformations in electrodynamics. He consoled himself with the thought that "mathematics, though it now can display only staircase-wit,[28] has the satisfaction of being wise after the event" (Minkowski 1909, 79).

In the opening sentence of his lecture, Minkowski had already identified the actual source of his insight into the importance of Lorentz transformations: "The views of space and time which I wish to lay before you have sprung from the soil of experimental physics, and therein lies their strength." "They are radical," he continued, before warning his audience that the old concepts of space and time were "doomed to fade away into mere shadows." Minkowski thus emphasized the discontinuity in the transition from the old to the new views of space and time. At the same time, a certain continuity is suggested by his acknowledgment that these new views sprang "from the soil of experimental physics."

At the end of the lecture, he returned to this point, locating the germ of his new views in the theoretical tools for cultivating the "soil of experimental physics" developed by Einstein and the Dutch physicist Hendrik Antoon Lorentz. In the course of his lecture, Minkowski had introduced what he called the "world postulate," which basically says that we live in a four-dimensional world that can be described in infinitely many equivalent space-time coordinate systems all related to each other via Lorentz transformations. In the conclusion, he wrote:

> The validity without exception of the world-postulate, I like to think, is the true nucleus of an electromagnetic image of the world, [a nucleus[29]] which, discovered by Lorentz and further revealed by Einstein, now lies open in the full light of day. (Minkowski 1909, 91)

Lorentz had been the first to show that Maxwell's equations for electric and magnetic fields are invariant under Lorentz transformations (Janssen 2017). Initially, he could only do this to a good approximation and for a restricted class of charge distributions acting as sources of the fields (Lorentz 1895).

Eventually, he could do it exactly and for arbitrary charge distributions (Lorentz 1904, 1916). Lorentz referred to this result as his "theorem of corresponding states." A pair of corresponding states consists of two physical systems in the ether, the nineteenth-century medium serving as the carrier of light waves and electric and magnetic fields, one at rest, the other in uniform motion. The quantities pertaining to the system in motion are related to those pertaining to the system at rest by a Lorentz transformation, the name given to these transformations by the French mathematician Henri Poincaré.

Poincaré and Einstein recognized that quantities for the system in motion in Lorentz's theorem of corresponding states are the space-time coordinates and the components of the electric and magnetic fields measured by an observer moving with the system. Before the advent of special relativity, Lorentz himself saw these quantities as nothing but auxiliary variables in terms of which the system in motion could be described in the same way as the corresponding system at rest in terms of the real quantities. Lorentz could show that the same measurement performed on two systems forming a pair of corresponding states would give the same result in a broad class of optical experiments. On the further assumption that, when set in motion with respect to the ether, the system at rest turns into the corresponding moving system, Lorentz could use his theorem to explain the negative results of many experiments aimed at detecting the earth's motion through the ether, including the famous 1887 Michelson-Morley experiment.

This additional assumption, however, is not as innocuous as it looks. It boils down to the assumption that the laws governing the material objects with which light waves interact in optical experiments (mirrors, lenses, screens, etc.) are all invariant under Lorentz transformations, just as Maxwell's equations governing the electric and magnetic fields making up the light waves themselves. As long as Lorentz invariance was restricted to the laws governing the fields, one could think of it as a special property of Maxwell's equations. This, of course, was precisely how Lorentz had discovered Lorentz invariance in the first place. The negative results of many ether-drift experiments suggested that it was a much more general property, common to all laws of physics. One way to avoid this conclusion was to assume that all laws of physics could eventually be reduced to the laws of electrodynamics. This view was promoted by German physicists such as Wilhelm Wien and Max Abraham in the early years of the twentieth century (Janssen and

Mecklenburg 2007). For several years this so-called electromagnetic view of nature was seen as the cutting edge in theoretical physics research. This is what Minkowski is referring to as the "electromagnetic image of the world" in the passage from his 1908 Cologne lecture quoted above. Minkowski had been an early supporter of the electromagnetic worldview but by 1908, as this same passage shows, he had distanced himself from it, supporting Einstein's special theory of relativity instead (Janssen 2009, 39).

Although the title of his paper, "On the Electrodynamics of Moving Bodies," suggests otherwise, Einstein, like Minkowski, recognized that Lorentz invariance has nothing to do with the particulars of electromagnetism but reflects a new space-time structure. In a letter of February 19, 1955, to Carl Seelig, one of his early biographers, he succinctly described the main novelty of special relativity as his "realization that the Lorentz transformation transcends its connection with Maxwell's equations and has to do with the nature of space and time in general" (Janssen 2009, 41). Minkowski, as we saw, reached the same conclusion and rephrased it in geometrical terms, identifying his "world-postulate" as "the true nucleus" of the electromagnetic world picture.

The relation between arch and scaffold is pretty straightforward in this case. Lorentz invariance is the key structural element shared by arch and scaffold. In the scaffold, Lorentz invariance is tied to electromagnetism. We get from scaffold to arch in this case by recognizing, as Einstein and Minkowski did, that Lorentz invariance has nothing to do with electromagnetism per se but is a property of all laws governing systems in Minkowski space-time, the new space-time structure of special relativity. The Lorentz invariance of the laws reflects the symmetry of this space-time structure. It took a few years for physicists to distinguish between the Lorentz and the Einstein-Minkowski interpretation of Lorentz invariance (the two interpretations are empirically equivalent), but eventually, the latter prevailed. For physicists using Minkowski space-time today, the only thing left to remind them that it was scaffolded by Lorentz's theorem of corresponding states is that the transformations between different perspectives on the arch are named after the man responsible for the scaffold.

Note how easy it was to tell this story backward in time, starting with Minkowski's geometrical interpretation of Lorentz invariance and ending with Lorentz's original interpretation of it in the context of electrodynamics. Also note that Minkowski's contribution might alternatively be characterized in terms of Hilbert's "build first and worry about the foundations later" metaphor (see section II). In that version of the story, Minkowski would be the mathematician who provided more secure foundations for the electromagnetic worldview of the physicist Abraham.

How Laue's Relativistic Continuum Mechanics Was Scaffolded by Abraham's Electromagnetic Mechanics[30]

Work in the history of special relativity typically focuses on how the theory changed our concepts of space and time. Special relativity, however, involved much more than the introduction of a new space-time structure and some minor adjustments to the laws of Newtonian mechanics to make them Lorentz invariant (such as the insertion of factors of $\sqrt{1-v^2/c^2}$ in various equations, where v is the relative velocity of two inertial frames, and c is the velocity of light). For one thing, the theory required the general relation between energy and mass (or inertia) expressed in its most famous equation, $E = mc^2$. It also required a "mechanics"—in the sense of a general framework for doing physics (cf. the term *quantum mechanics*)—of fields rather than particles.

In the years just prior to the arrival of special relativity, Abraham and others had developed such a mechanics for the special case of electromagnetic fields (Abraham 1903). It provided the foundation of the electromagnetic worldview, the attempt to reduce all of physics to electrodynamics. Insofar as the electromagnetic worldview is covered at all in histories of special relativity, it is typically presented, implicitly or explicitly, as a research program that was briefly considered cutting edge at the beginning of the twentieth century but was then vanquished by special relativity.[31]

However, as Einstein, for one, clearly recognized, it is more accurate to say that it was *co-opted* by special relativity. Within a few years of the introduction of special relativity, the electromagnetic mechanics of Abraham had morphed into the relativistic continuum mechanics presented in the first textbook on relativity (Laue 1911). Max Laue basically obtained his relativistic mechanics for fields by taking Abraham's electromagnetic mechanics, rewriting it in terms of the four-dimensional formalism developed by Minkowski and Sommerfeld, and stripping it of its electromagnetic particulars.

The arch-and-scaffold metaphor captures this development in a natural way. It underscores the importance of the formulation of relativistic continuum mechanics in the development of special relativity by making it—to use the terms introduced in Figure 4.1—the *keystone* of the arch for which Minkowski had provided the *springers*.

The backward-looking perspective is especially important in this case. If the developments are covered forward in time, it is hard to bring out those features of the electromagnetic program that proved most relevant for the formulation of relativistic continuum mechanics without the account becoming blatantly Whiggish. Put differently, and as illustrated by the historiographically impeccable coverage of this episode by Richard Staley (2008, chapters 6–7), Whiggishness can be avoided only at the price of obscuring what in hindsight were the most salient elements of the electromagnetic program and how they were incorporated into special relativity.

The clearest way to bring out these elements is to rewrite some of the equations of the electromagnetic program in terms of the formalism of Minkowski, Sommerfeld, and Laue. The proponents of the electromagnetic program wrote their equations in terms of vectors, with the usual three (spatial) components and tensors one can think of as three-by-three matrices, and they handled spatial and temporal derivatives separately. The relativists grouped quantities into vectors with four space-time components and tensors one can think of as four-by-four matrices, and they put spatial and temporal derivatives on equal footing.

It will be instructive to look at this in a little more detail. Abraham replaced the energy and momentum of particles of Newtonian mechanics, with definite positions in space, by the energy *density* and momentum *density* of electromagnetic fields, spread out all over space. The x-, y-, and z-components of the electromagnetic momentum density he introduced are proportional to the electromagnetic energy flow density in the x-, y-, and z-direction. The electromagnetic energy flow density in turn is given by the Poynting vector, the cross product of the electric and the magnetic field and the main claim to fame of the physicist we already encountered in section III.

Central to Abraham's efforts to reduce mechanics to electrodynamics was his attempt to eliminate the Newtonian concept of mass by replacing the inertial force on a massive particle by the force exerted on a massless charge distribution by the electromagnetic field generated by that charge distribution itself. The interaction of massless charges with these self-fields thus mimics inertial mass (those with limited tolerance for equations can skip ahead to the quotation following Equation (5)).

Abraham showed that the components $f^i_{\text{self}_{\text{EM}}}$ (with $i = 1, 2, 3$ labeling x-, y-, and z-components) of the density of the electromagnetic force exerted by a charge distribution's self-field can be written as minus the time derivative of the components of its electromagnetic momentum density and the divergence (a sum of spatial derivatives) of its stress-energy density, given by the so-called Maxwell stress tensor. This tensor can be thought of as a three-by-three matrix. Its nine components represent the flow of the x-, y-, and z-components of the electromagnetic momentum density in the x-, y-, and z-direction. Like the electromagnetic momentum density, the components of the Maxwell stress tensor are functions of the components of the electric and magnetic fields.

In modern notation, f^i_{EM} can be written as

$$f^i_{\text{self}_{\text{EM}}} = -\frac{\partial}{\partial t}\left(p^i_{\text{self}_{\text{EM}}}\right) - \sum_{j=1}^{3}\frac{\partial}{\partial x^j}\left(T^{ij}_{\text{self}_{\text{EM}}}\right), \tag{1}$$

where $p^i_{\text{self}_{\text{EM}}}$ is the i-component of the electromagnetic momentum of the charge distribution's self-field, and $T^{ij}_{\text{self}_{\text{EM}}}$ is the ij-component of its Maxwell stress tensor.[32]

The *energy density,* the *three* components of the *momentum density,* the *three* components of the *energy flow density,* and the *nine* components of the *momentum flow density* make for a total of sixteen components. In special relativity, they are combined (with an extra factor of c here and there) to give the *sixteen* components of the *energy-momentum tensor $T^{\mu\nu}$* (with $\mu, \nu = 0, 1, 2, 3$ corresponding to one time and three spatial components).[33] This tensor can be thought of as a four-by-four matrix, with the first index labeling the rows and the second index labeling the columns. Its components are:

$$T^{\mu\nu} = \begin{pmatrix} \textbf{00 component} & \textbf{0j components} \\ \text{energy density} & \text{energy flow density} \div c \\ & \text{in } x, y, \text{ and } z \text{ direction} \\[1em] \textbf{i0 components} & \textbf{ij components} \\ \text{momentum density} \times c & \text{momentum flow density} \\ (x, y, \text{ and } z \text{ components}) & (x, y, \text{ and } z \text{ components}) \\ & \text{in } x, y, \text{ and } z \text{ direction} \end{pmatrix}. \tag{2}$$

The energy-momentum tensor is symmetric in its indices—that is, $T^{\mu\nu} = T^{\nu\mu}$ In other words, the matrix representing $T^{\mu\nu}$ stays the same if we switch rows and columns. Focusing on the first row and the first column, we have $T^{i0} = T^{0i}$, from which it follows that

$$\text{momentum density} = \text{energy flow density} \div c^2.$$

As was first realized by Planck (1906), this is an elegant way of expressing $E = mc^2$ in the four-dimensional formalism.

Equation (1) for the components $f^i_{\text{self}_{\text{EM}}}$ of the electromagnetic force density, the rate of change of the momentum density of the self-field, can be combined with a similar equation for $f^0_{\text{self}_{\text{EM}}}$, the rate of change of the energy density of the self-field. This combination can be written compactly in terms of the components of the energy-momentum tensor, $T^{\mu\nu}_{\text{self}_{\text{EM}}}$, for the charge distribution's self-field:

$$f^\mu_{\text{self}_{\text{EM}}} = -\frac{1}{c}\frac{\partial}{\partial t}\left(T^{\mu 0}_{\text{self}_{\text{EM}}}\right) - \sum_{j=1}^{3}\frac{\partial}{\partial x^j}\left(T^{\mu j}_{\text{self}_{\text{EM}}}\right) = -\sum_{v=0}^{3}\frac{\partial}{\partial x^v}\left(T^{\mu\nu}_{\text{self}_{\text{EM}}}\right), \qquad (3)$$

where the four components of the electromagnetic four-force density $f^\mu_{\text{self}_{\text{EM}}}$ are $f^0_{\text{self}_{\text{EM}}}$ and $f^i_{\text{self}_{\text{EM}}}$ ($i = 1, 2, 3$) and where $x^\mu \equiv (ct, x, y, z)$. The final expression in Equation (3) is called the four-divergence of the energy-momentum tensor $T^{\mu\nu}_{\text{self}_{\text{EM}}}$.

In relativistic continuum mechanics, this equation is generalized from the electromagnetic field to arbitrary spatially extended systems. The subscript "EM" in Equation (3) can thus be dropped. Using the so-called Einstein summation convention, which says that any index occurring twice in the same expression (once "upstairs" and once "downstairs") is summed over, we can also drop the summation sign. Finally, we introduce the abbreviation $\partial_\mu \equiv \partial/\partial x^\mu$ and arrive at:

$$f^\mu_{\text{self}} = -\partial_v T^{\mu\nu}_{\text{self}}. \qquad (4)$$

In general there will be an external four-force density with components f^μ_{ext} acting on the system as well as the four-force density with components f^μ_{self} that the system exerts on itself. The basic equation of motion for the system says that the sum of these force densities vanishes everywhere—that is, $f^\mu_{\text{ext}} + f^\mu_{\text{self}} = 0$. Using Equation (4) for f^μ_{self}, we can write this equation as

$$f^\mu_{\text{ext}} = \partial_v T^{\mu\nu}_{\text{self}}. \qquad (5)$$

This is the fundamental law of relativistic continuum mechanics.

As Einstein wrote in an unpublished review article on special relativity in 1912:

> The general validity of the conservation laws [of energy and momentum] and the law of the inertia of energy [$E = mc^2$] suggests that [the energy-momentum tensor (2) and the force equation (5)] are to be ascribed a general significance, even though they were obtained in a very special case [i.e., electrodynamics]. We owe this generalization, which is the most important new advance in the theory of relativity, to the investigations of Minkowski, Abraham, Planck, and Laue. (Einstein 1987–2018, vol. 4:92; cf. Janssen and Mecklenburg 2007, 110)

Note that Abraham, the undisputed leader of the electromagnetic program and a staunch opponent of special relativity, is mentioned here in the same breath as Minkowski, Planck, and Laue. Abraham richly deserved to be mentioned alongside this trio of enthusiastic supporters of special relativity. His electromagnetic mechanics provided the scaffold on which Laue built the arch of relativistic continuum mechanics.

The relation between scaffold and arch in this case is the same as in my first example. The scaffold exhibits the structure of the arch for the special case of electromagnetism. Once again, the arch is thus obtained by stripping the scaffold of its electromagnetic particulars. The comparison of the equations of the electromagnetic view of nature and relativistic continuum mechanics also highlights a different aspect of the relation between arch and scaffold in both these examples. The step from scaffold to arch in both cases also involved grouping various quantities defined in three-dimensional space (scalars, vectors, and tensors) into quantities defined in four-dimensional space-time.

Another passage from Einstein's 1912 review article provides the natural starting point for telling the story about the transition from Abraham's electromagnetic mechanics to Laue's relativistic continuum mechanics backward in time. Einstein gave the following concise characterization of how one applies the latter:

> To every kind of material process we want to study, we have to assign a symmetric tensor ($T_{\mu\nu}$), whose components have the physical meaning given in [Equation (2)]. [Equation (5)] must always be satisfied. The problem to be

solved always consists in finding out how ($T_{\mu\nu}$) is to be formed from the variables characterizing the processes under consideration. If several processes take place in the same region that can be isolated in the energy-momentum balance, we have to assign to each individual process its own stress-energy tensor ($T_{\mu\nu}^{(1)}$) etc., and set ($T_{\mu\nu}$) equal to the sum of these individual tensors. (Einstein 1987–2018, vol. 4:92)

The question to which the arch-and-scaffold narrative then provides the answer is: How did physicists go from Newtonian particle mechanics to relativistic continuum mechanics as a general framework for doing physics. In other words, how did they go from writing down the forces acting on some collection of bodies and using Newton's second law, $\mathbf{F} = m\mathbf{a}$, to solve for the motion of these bodies to writing down the energy-momentum tensors for various processes occurring in some region of space-time and setting the four-divergence of their sum, $\partial_\nu(T_{(1)}^{\mu\nu} + T_{(2)}^{\mu\nu} + \cdots)$, equal to the external four-force density, f_{ext}^μ?

One final question that needs to be answered about this episode is why the new relativistic mechanics, despite being identified by Einstein as "the most important new advance in the theory of relativity," did not catch on in the physics community of the early 1910s. The short answer: because of quantum theory. Only two years after Laue published his relativity textbook, in which relativistic continuum mechanics takes center stage, Niels Bohr (1913) proposed his model of the hydrogen atom based on nonrelativistic Newtonian particle mechanics. This model was further developed in the old quantum theory of Bohr and Sommerfeld (Kragh 2012; Eckert 2014). The old quantum theory gave Newtonian mechanics, especially Newtonian celestial mechanics, a new lease on life. Sommerfeld used relativistic particle mechanics to explain the fine structure of spectral lines, but the old quantum theory had no use for relativistic continuum mechanics. The latter only played a role in the development of general relativity, the topic of my next example.

How the Field Equations of General Relativity Were Scaffolded by the Field Equations of the Earlier Entwurf Theory[34]

In 1907, only two years after he published the special theory of relativity, Einstein, still a patent clerk in Berne, started thinking about a generalization of the theory that would incorporate gravity. By 1911, when he was appointed full professor in Prague, he had arrived at the basic idea of his general theory of relativity. Contrary to what its name suggests, this theory does not extend the principle of relativity from uniform to nonuniform motion (Janssen 2014), but it does weave gravity into the fabric of space-time. In general relativity, gravity is represented by space-time curvature. Free-falling bodies (i.e., particles subject to no other forces than gravity) will follow the straightest possible paths in these curved space-times. Such paths are called *geodesics* and satisfy the geodesic equation.[35]

This much was already becoming clear to Einstein when he returned from Prague to Zurich in 1912 and began working in earnest on his new theory of gravity with the help of his new colleague at the Federal Institute of Technology (ETH), Marcel Grossmann. Grossmann introduced Einstein to the elegant mathematics of Bernard Riemann, Elwin Bruno Christoffel, Gregorio Ricci-Curbastro, and others needed for the formulation of the kind of theory Einstein was after. The two of them had been classmates at what had then still been called the Federal Polytechnic back in the late 1890s. Grossmann had become professor of mathematics at their alma mater in 1907. Their collaboration, recorded in Einstein's famous Zurich notebook (Renn 2007, vols. 1–2), resulted in a joint paper published in the spring of 1913 (Einstein and Grossmann 1913). Its title modestly announced an "outline" *(Entwurf)* of a new theory of gravity and a generalized theory of relativity. Historians refer to it as the *Entwurf theory.*

The *Entwurf* theory already put in place most of the formalism of the general theory of relativity, which Einstein completed two and a half years later, in November 1915, in four short communications to the Prussian Academy in Berlin (Einstein 1915a, 1915b, 1915c, 1915d). In March 1914, he had left Zurich to take up a prestigious appointment in the German capital. In his papers of November 1915, Einstein basically changed only one important element of the *Entwurf* theory: its field equations.

The field equations, as the reader may recall from section III, determine how matter curves space-time. (It is customary to use the plural *equations* even though they can be written as one equation because this one equation has several components.) It was clear that matter had to be represented by its energy-momentum tensor (cf. the preceding case study). This tensor appears on the right-hand side of the field equations. The difference between the field equations of 1913 and 1915 was the left-hand side. In his first November paper, Einstein claimed that he and Grossmann had already considered the new candidate for the left-hand side three years earlier. The Zurich notebook confirms this. The notebook shows how mathematical considerations, supplied by Grossmann, had led Einstein to this elegant candidate and how he had abandoned it because it looked as if the resulting field equations did not reduce to the equations of Newtonian theory in the appropriate limit and were incompatible with energy-momentum conservation. Einstein thereupon switched from a mathematical to a physical strategy. Exploiting the analogy with the Maxwell-Lorentz theory for the electromagnetic field (rewritten in the four-dimensional formalism of Minkowski, Sommerfeld, and Laue), he constructed field equations for the gravitational field guaranteed to satisfy energy-momentum conservation and to have the correct Newtonian limit. These are the field equations published in the *Entwurf* paper.

In the introduction of his first paper of November 1915, Einstein made it sound as if he had suddenly turned his back on the physical strategy that had led him to the *Entwurf* field equations and gone back to the mathematical strategy that had originally led him to the field equations with which he now proposed to replace them. Subsequent sections of the paper make it clear that this was, at best, an exaggeration. Einstein relied heavily on lessons learned pursuing the physical strategy over the preceding two and a half years to show that these resurrected field equations passed muster on the counts of energy-momentum conservation and the Newtonian limit on which they had failed earlier. To put it in terms of the arch-and-scaffold metaphor: Einstein may already have envisioned the arch in 1913, but the confidence to put weight on it only came in November 1915.[36]

Closer examination of both the first November paper and Einstein's correspondence at the time makes it doubtful that there was an eleventh-hour return to the mathematical strategy and strongly suggests that, instead, it was his dogged pursuit of the physical strategy that led Einstein back to the field equations to which the mathematical strategy had already led him in the Zurich notebook. As Jürgen Renn and I have argued in detail, Einstein used the *Entwurf* field equations as a scaffold to construct the field equations with which he replaced them in his first paper of November 1915 (Janssen and Renn 2015, 2019).

By late 1914, Einstein had perfected the analogy between the Maxwell-Lorentz theory for the electromagnetic field and the *Entwurf* theory for the gravitational field. He convinced himself that the formalism he had developed relying on this analogy uniquely determined the field equations and that these were the *Entwurf* field equations. Satisfied that his arch was now complete, he published a lengthy review article on his theory. The title no longer talks about an "outline" of a "generalized" theory of relativity but promises nothing less than the "formal foundation of the general theory of relativity" (Einstein 1914).

In October 1915, Einstein discovered, to his dismay, that his uniqueness argument for the *Entwurf* field equations was fallacious. Rather than abandoning his general formalism—or tearing down his house, to use the metaphor of Kuhn's Guggenheim application quoted in section II—Einstein merely replaced the definition of the gravitational field in the *Entwurf* theory by what seemed to be the only other physically plausible candidate, the so-called Christoffel symbols. Inserting this new definition into his general formalism, he ended up with the same equations that he had rejected in the Zurich notebook, thereby reestablishing the connection between his theory and the elegant mathematics that, as Einstein (1915a, 778) noted in his first November 1915 paper, he and Grossmann had abandoned "with a heavy heart" in 1913. The general formalism developed for the *Entwurf* theory provided a number of relations, the counterparts of similar relations for the electromagnetic field, that the gravitational field had to satisfy to be acceptable from a physics point of view. These relations continued to hold when the old definition of the gravitational field was replaced by the new one. This, then, is how Einstein got from scaffold to arch. By changing the definition of the gravitational field, he swapped out one building block for another, confident that the structure he had erected with the old building blocks would remain stable upon this substitution.[37]

Einstein himself identified this as the crucial step in the transition from the *Entwurf* field equations to the field equations of his first November paper. In the paper, he called the new definition of the gravitational field "the key to the solution." In the letter to Sommerfeld from which I already quoted in note 36, he called the old definition "a fateful prejudice" (Janssen and Renn 2007, 859, 875–79).

Worried that Hilbert might beat him to the punch, Einstein rushed his new field equations into print. Over the next three weeks, he continued to tweak them. Throughout this period Einstein was laboring under the misconception that the extent to which his theory generalized the principle of relativity from uniform to nonuniform motion is directly related to the degree of *covariance* of its equations—that is, to the size of the class of coordinate transformations under which the equations retain their form. The covariance of the field equations of his first November paper was much broader than that of the *Entwurf* field equations, but they were still not *generally* covariant—that is, they do not retain their form under arbitrary coordinate transformations. Einstein quickly realized, however, that with relatively minor modifications these new field equations could be turned into generally covariant field equations that just happened to be written in a special form in which their general covariance is not immediately apparent.

In his second November paper, he proposed one such modification, only to replace it with another, more satisfactory, one in the fourth (Einstein 1915b, 1915d). The latter modification was to add a term with the so-called *trace* of the energy-momentum tensor (the sum of the terms on the diagonal in Equation (2)) on the right-hand side of the field equations. This trace term is necessary to ensure that the quantity representing the energy-momentum density of the gravitational field enters the field equations in the exact same way as the energy-momentum tensor for matter. Since this was a requirement that had been one of Einstein's guiding principles, he was now confident that no further corrections would be needed. The amended equations of this fourth and final communication to the Berlin Academy of November 1915 are the Einstein field equations used to this day.[38] The trace term formed the keystone of what is widely admired as the most marvelous arch Einstein left us.

Einstein still had to write his field equations in a form in which they only retain their form under a restricted class of coordinate transformations, as this was the only way in which he could connect them to the general formalism for the *Entwurf* theory and show that they were compatible with energy-momentum conservation. As a result, the arch unveiled in Einstein's papers of November 1915 still showed clear traces of the scaffold used to build it. The same is true for the section on the field equations in the review article that Einstein (1916a) published in May the following year to replace the premature review article of late 1914. It was only in a short paper published in November that year that Einstein (1916b) finally removed all traces of the *Entwurf* scaffold.

A natural starting point for telling this story backward in time is to begin with Einstein's later recollections of how he found the field equations of general relativity. This is the approach Jürgen Renn and I took in a talk we have given in various places based on our article in *Physics Today* (Janssen and Renn 2015).[39] The older Einstein routinely claimed that he had found the Einstein field equations following the mathematical strategy. Some commentators, notably John Norton (2000), have taken him at his word. Renn and I concur with the conclusion of Jeroen van Dongen's (2010) study of Einstein's unified field theory program that these statements should be seen, first and foremost, as propaganda for this program. In his ultimately fruitless pursuit of a classical field theory unifying general relativity and electromagnetism, Einstein relied on a purely mathematical strategy. It served his purposes to suggest that this approach could boast of at least one spectacular success, the discovery of the Einstein field equations.

An arch-and-scaffold narrative working backward from these later pronouncements by Einstein reveals that, while supported by several passages in his writings and correspondence from the gestation period of general relativity, they do not square with the full range of textual evidence available. The arch-and-scaffold metaphor can be used to put the physical strategy, suppressed in Einstein's later recollections, in sharp relief, which makes it easier to compare competing accounts of how Einstein found his field equations in the fall of 1915, the account of Janssen and Renn (2007), in which Einstein stuck to the physical strategy, and the classic account of Norton (1984), in which he switched to the mathematical strategy.[40]

The clarification of the difference between the two strategies Einstein used in his search for the field equations of general relativity may shed light on at least two other issues in modern history and philosophy of physics. The first concerns the interpretation of general relativity (Lehmkuhl 2014). Should one think of it in analogy with electrodynamics as the theory of a particular field, as the physical strategy suggests, or should one think of it as a theory about geometry, as the mathematical strategy suggests?[41] The second issue is about methodology in physics. On the face of it, Einstein's own later account of how he found the field equations of general relativity in 1913–1915 is strong testimony in support of a purely mathematical approach to theory construction. This approach remains popular to this day in certain quarters of the physics community. An account of this episode that emphasizes the importance of the physical strategy, conveyed concisely with the help of the arch-and-scaffold metaphor, can likewise serve as powerful counter-testimony.

How the Matrix Mechanics of Heisenberg, Born, and Jordan Was Scaffolded by the Kramers Dispersion Formula[42]

The paper known as the *Umdeutung* (Reinterpretation) paper, with which Werner Heisenberg (1925) laid the foundation for matrix mechanics, draws on an earlier paper he wrote as the junior coauthor of Bohr's right-hand man in Copenhagen at the time, the Dutch physicist Hans Kramers. This paper by Kramers and Heisenberg (1925) provides a detailed derivation and further exploration of a new quantum formula for optical dispersion that Kramers (1924a, 1924b) had proposed in two short notes in *Nature* the year before. Max Dresden (1987, 275), Kramers's biographer, goes so far as calling this paper "the direct, immediate, and exclusive precursor to the Heisenberg paper on matrix mechanics." Recent work by Alex Blum, Martin Jähnert, Christoph Lehner, and Jürgen Renn (2017) at the Max Planck Institute for the History of Science in Berlin suggests that the *Umdeutung* paper owes as much, if not more, to work on intensities in multiplet spectra.[43] The work on dispersion theory, however, remains an important strand in the genealogy of the *Umdeutung* paper. This strand is nicely captured by the arch-and-scaffold metaphor. Blum et al. (2017, 4n3) find this to be true for the multiplet strand as well.

Optical dispersion is the phenomenon, familiar from rainbows and prisms, that the refraction of light in an optical medium depends on its color. Although it already occupied the minds of Descartes and Newton, it was not until two centuries later that a halfway satisfactory theory of the phenomenon was formulated.[44] Particularly challenging was a puzzling feature discovered by early pioneers of photography in the 1840s. Normally, the index of refraction increases with the frequency of the refracted light. Blue light is refracted more strongly than red light. However, in narrow frequency bands around the absorption frequencies of an optical medium, the index of refraction decreases with increasing frequency in some materials. This is called *anomalous dispersion*. In the 1870s, Wolfgang Sellmeier and others introduced a new generation of dispersion theories that could account for this phenomenon. The characteristic feature of this new class of theories is that optical media contain small oscillators with resonance frequencies at the absorption frequencies of the material. These theories correctly predict that dispersion becomes anomalous in the vicinity of these resonance frequencies. In the 1890s, Hermann von Helmholtz, Lorentz, and Paul Drude reformulated these originally purely mechanical theories in terms of electromagnetic waves interacting with electrically charged oscillators, soon to be identified as electrons. Such harmonically bound electrons were sometimes called *dispersion electrons*.

These classical dispersion theories were incompatible with Bohr's (1913) atomic model. In this model, electrons orbit the nucleus the way planets orbit the sun, except that in the atom only a discrete set of orbits are allowed, labeled by integer-valued quantum numbers. A straightforward adaptation of the classical dispersion formula to Bohr's new atomic model would have been to replace the *oscillation* frequencies of the harmonically bound dispersion electrons in the former by the *orbital* frequencies of the planetary electrons in the latter. This was not an option. The problem is that, in general, these orbital frequencies differ sharply from the atom's absorption frequencies. Light is absorbed or emitted in a Bohr atom when an electron jumps from one orbit to another. The frequency of the absorbed or emitted radiation is determined not by the orbital frequencies of either of these orbits but by the energy difference between them. Only in the limit in which the quantum numbers labeling the orbits get very large do the radiation frequencies coincide with the orbital frequencies. Simply replacing oscillation frequencies by orbital frequencies would thus lead to a theory predicting anomalous dispersion at the wrong frequencies.

While this posed a serious problem for Bohr's atomic model and the old quantum theory that grew out of it, the classical dispersion theory also faced a serious problem, which could actually be solved with some of the resources provided by the old quantum theory. Experimentalists had found puzzling values for an important set of free parameters of the classical theory, the so-called oscillator strengths. In the classical theory, the oscillator strength for a particular resonance frequency is the number of dispersion electrons per atom with that resonance frequency. Intuitively, one would expect these numbers to be in the single digits—a few dispersion electrons with the same resonance frequency per atom—but the values giving the best fit with the data tended to be much lower. It was not uncommon to find values as low as one dispersion electron with a particular resonance frequency per hundreds or even tens of thousands of atoms.

The German experimentalist Rudolf Ladenburg (1921) reinterpreted these parameters in a way that such low values were only to be expected. The oscillator strength does not, Ladenburg suggested, represent the number of electrons with a particular resonance frequency but the number of electron *jumps* between two orbits associated with the absorption of radiation at that frequency. Ladenburg set the number of jumps equal to the product of the occupation number of the initial orbit (the fraction of the total number of electrons in that orbit) and the probability that an electron in that initial orbit would jump to the final orbit. For these probabilities he used the probability coefficients for transitions between different quantum states introduced by Einstein (1917a). Replacing numbers of electrons by products of occupation numbers and transition probabilities, Ladenburg turned the classical dispersion formula into a new quantum formula.

The formula still had two limitations. First, it was restricted to situations in which electrons would jump to and from their ground state, the orbit of lowest energy. Second, Ladenburg could still not explain why dispersion is anomalous at the absorption frequencies. He just retained this feature of the classical formula, as it was clearly borne out by the experimental data. In a follow-up paper, Ladenburg and Fritz Reiche, a theoretical physicist, introduced the notion of *substitute oscillators (Ersatzoszillatoren)* operating between two orbits and with resonance frequencies equal to the absorption frequencies associated with transitions between them (Ladenburg and Reiche 1923). If one thought of these substitute oscillators as the conduits of dispersion, one at least had some way of understanding why dispersion is anomalous at these transition frequencies.

This is where matters stood when Kramers entered the fray. Most likely at the instigation of Bohr (whose doctoral adviser at the University of Copenhagen, Christian Christiansen, had done important work on optical dispersion), Kramers tried to derive a dispersion formula in the old quantum theory modeled on the one given by Ladenburg. The central tool he used for this derivation was Bohr's correspondence principle. In the hands of Kramers (1924a, 1924b), Born (1924), and the American theoretical physicist John H. Van Vleck (1924a, 1924b), this principle turned into a powerful instrument for generating quantum formulae designed to merge with their classical counterparts in the limit of high quantum numbers.[45]

Using canonical perturbation theory in special momentum and position variables known as *action-angle variables,* a technique originating in celestial mechanics, Kramers first derived a formula for dispersion in classical mechanics. He then made three substitutions to turn this classical formula into a quantum formula. First, he expanded the orbital motion into a Fourier series and replaced the squares of the amplitudes of the various Fourier components by the Einstein coefficients for transition probabilities. Second, he replaced orbital frequencies by radiation frequencies corresponding to transitions between orbits. Third, he replaced derivatives with respect to action variables by difference quotients. The basic quantization conditions of the old quantum theory, which select the allowed electron orbits in an atom, set such action variables equal to an integral multiple of Planck's constant h. In the limit of high quantum numbers N, where the allowed orbits are getting closer and closer together, one can thus approximate a derivative of a quantity with respect to an action variable by subtracting that quantity's value at the Nth orbit from its value at the $(N + 1)^{th}$ orbit and dividing the result by h.

With these three substitutions, the classical dispersion formula Kramers had derived turned into a quantum formula. Because of the third substitution, this formula is the difference of two terms. Both have the same structure as Ladenburg's formula. As long as electrons only jump to and from their ground state, Kramers's formula reduces to Ladenburg's. Kramers's formula, however, applies to all possible transitions between orbits. Its construction guarantees that it merges with the well-tested classical formula in the limit of high quantum numbers. It still required a leap of faith that the formula would continue to hold all the way down to the smallest quantum numbers, but its agreement with Ladenburg's formula for the ground state was reassuring on that score. In hindsight, Kramers's faith in his formula was well placed. It carries over completely intact to modern quantum mechanics and has been fully confirmed experimentally.

The Kramers dispersion formula was incorporated into a short-lived but influential quantum theory of radiation proposed by Bohr, Kramers, and Slater (1924) and known as the BKS theory. John C. Slater was an American postdoc visiting Copenhagen at the time. The substitute oscillators introduced by Ladenburg and Reiche (1923) return in the BKS theory under the name *virtual oscillators*. The BKS theory thus introduces a dual representation of atoms. To the set of orbits of Bohr's original theory, the BKS theory adds—to use a term introduced by another early quantum theorist, Alfred Landé (1926, 456)—an orchestra of virtual oscillators associated with every possible transition between those orbits. All information about observable quantities—that is, frequencies and intensities of spectral lines, is contained in the latter. The Kramers dispersion formula nicely illustrates this: it only contains quantities referring to transitions between orbits and makes no reference whatsoever anymore to individual orbits.

After the examples given in the first three case studies in this section, I trust that the reader will have no trouble seeing in the sequence of dispersion theories outlined above (Sellmeier, Helmholtz-Lorentz-Drude, Ladenburg-Reiche, Kramers) how the later theory was scaffolded by the earlier one. But how did the Kramers dispersion formula (partly) scaffold Heisenberg's *Umdeutung* paper? As sketched above, Kramers derived his quantum dispersion formula in two steps. First, he derived a formula in classical mechanics. Then he used the correspondence principle to translate the result into a quantum formula. The fundamental idea of *Umdeutung* is to use the correspondence principle to translate the input rather than the output of such classical derivations and do the entire derivation in terms of the new quantum language. This strategy is not limited to the derivation of a formula for dispersion. Heisenberg realized that it could serve as a new framework for all of physics. A little more concretely, the basic idea is to take positions and momenta, the fundamental variables of classical mechanics, in terms of which Kramers had derived his classical dispersion formula, translate them according to "the scheme of the dispersion theory," as Heisenberg himself put it in an interview for the Archive for the History of Quantum Physics (AHQP) in the early 1960s (cf. note 8),[46] into quantum variables and calculate with those new variables *on the assumption that they satisfy the same laws as their classical counterparts*. Hence, the term *Umdeutung*: rather than repealing the laws of classical mechanics, Heisenberg sought to reinterpret them.[47]

In Heisenberg's *Umdeutung* or reinterpretation scheme, quantities associated with a single orbit get replaced by quantities associated with a transition between two orbits. Electron orbits are eliminated altogether. Heisenberg formulated his theory entirely in terms of transition quantities without answering the obvious question "transitions between *what?*" These transition quantities have two indices, referring to an initial and a final state, but Heisenberg had nothing whatsoever to say about the nature of those states. Multiplication of his two-index objects, Heisenberg found, is noncommutative: $A \times B \neq B \times A$. In their elaboration of Heisenberg's *Umdeutung* paper, Born and his former student Pascual Jordan identified these two-index objects as matrices, their rows and columns labeled by Heisenberg's two indices (Born and Jordan 1925). This showed that Heisenberg's strange noncommutative multiplication rule is nothing but the standard multiplication rule for matrices.[48]

The relation between arch and scaffold in this example is a combination of those encountered in the relativity examples. First, the way in which Heisenberg, with help from Born and Jordan, generalized Kramers's theory for a specific phenomenon (dispersion) to a new framework for all of physics (matrix mechanics) is reminiscent of the way in which Laue generalized Abraham's electromagnetic mechanics to a new framework for all of physics (relativistic continuum mechanics). Second, the way in which Heisenberg replaced classical quantities by two-index objects soon to be recognized as matrices while keeping the structure of classical mechanics intact is reminiscent of the way in which Einstein replaced the definition of the gravitational field in the *Entwurf* theory by a new definition while keeping the formalism developed for the *Entwurf* field equations intact.

The example also illustrates an element of the arch-and-scaffold metaphor identified in Figure 4.1 that we did not encounter in the relativity examples. Kramers built his quantum formula and Heisenberg built his quantum theory on the foundation—the *tas-de-charge* in terms of Figure 4.1—of classical mechanics. The instrument they used to erect their quantum constructions, the *windlass* in terms of Figure 4.1, was the correspondence principle, in the specific guise of the three substitutions listed above.

As part of his *Umdeutung* project, Heisenberg also had to bring the quantization conditions of the old quantum theory, formulated in terms of individual orbits, into his new framework. Translating these conditions according to the "scheme of the dispersion theory" (as he put it in his AHQP interview), Heisenberg arrived at a corollary of the Kramers dispersion formula found independently by Werner Kuhn (1925) and Willy Thomas (1925; see also Reiche and Thomas 1925). This Thomas-Kuhn sum rule served as the quantization condition in the *Umdeutung* paper. It thus has nothing to do with Thomas S. Kuhn, who is the only one I know who refers to it as the Kuhn-Thomas sum rule (Duncan and Janssen 2007, 594). Born and Jordan (1925) showed that this sum rule can be rewritten as

$$\hat{q}_j \, \hat{p}_j - \hat{p}_j \, \hat{q}_j = i\hbar \qquad (6)$$

($j = 1, 2, 3$), where $\hat{q}_j$ and $\hat{p}_j$ are the components of position and momentum (with the "hats" to indicate that these quantities are not numbers but matrices), i is the imaginary unit, and $\hbar$ is Planck's constant divided by 2π (Duncan and Janssen 2007, 659–60). Equation (6) gives the diagonal elements ($j = k$) of

$$\hat{q}_j \, \hat{p}_k - \hat{p}_k \, \hat{q}_j = i\hbar \, \delta_{jk} \qquad (7)$$

($j, k = 1, 2, 3$; $\delta_{jk} = 1$ for $j = k$, and $\delta_{jk} = 0$ for $j \neq k$), the familiar commutation relations for position and momentum that serve as the basic quantization conditions in matrix mechanics.

As in the case of the November 1915 papers in which Einstein first presented the Einstein field equations, the *Umdeutung* paper, the harbinger of matrix mechanics, still showed clear traces of the scaffold on which it was built. Heisenberg's two-index objects satisfying a peculiar noncommutative multiplication rule are still somewhere in between the transition amplitudes and transition frequencies of the Kramers dispersion formula and the matrices introduced by Born and Jordan. The Thomas-Kuhn sum rule, the quantization condition of the *Umdeutung* paper, comes straight out of dispersion theory.

Most importantly, perhaps, the notion of a virtual oscillator that Bohr, Kramers, and Slater (1924) had taken over from Ladenburg and Reiche (1923) served as a placeholder until a more satisfactory way had been found to represent the states that the systems studied in matrix mechanics were transitioning between. New and better representations would soon be provided, be it Schrödinger's wave functions or von Neumann's vectors in Hilbert space (see the next case study). Virtual oscillators could now be identified either as Fourier components of a wave function (Duncan and Janssen 2007, 617) or as matrix elements of position (Casimir 1973, 492). However, in their follow-up to the *Umdeutung* paper, written before these contributions by Schrödinger and von Neumann, Born and Jordan (1925, 884) still talked about "substitute oscillators," Ladenburg and Reiche's original term for virtual oscillators. Although today it is used in connection with the BKS theory, Landé (1926) actually introduced his "orchestra of virtual oscillators" to describe matrix mechanics. At least one popular book continued to use closely related imagery—a "band" *(Kapelle)* of "assistant musicians" *(Hilfsmusiker)*—to explain matrix mechanics to a lay audience long after the concept of a quantum state had been incorporated into it (Zimmer 1934, 161–62; quoted in Duncan and Janssen 2007, 616).

In Duncan and Janssen (2007), we already indicated how to tell this story backward in time. Our starting point was exactly the kind of wonder one experiences upon first seeing an improbable architectural structure. One of those left wondering how Heisenberg built his arch is particle physicist and Nobel laureate Steven Weinberg. Talking about the *Umdeutung* paper in *Dreams of a Final Theory,* he wrote:

> If the reader is mystified at what Heisenberg was doing, he or she is not alone. I have tried several times to read the paper that Heisenberg wrote on returning from Helgoland [where he had gone to seek relief from an attack of hay fever], and, although I think I understand quantum mechanics, I have never understood Heisenberg's motivations for the mathematical steps in his paper. (Weinberg [1992] 1994, 67; cf. Duncan and Janssen 2007, 559)

This same quote is used to motivate at least two other studies of the *Umdeutung* paper (Aitchison, MacManus, and Snyder 2004; Blum et al. 2017). This underscores the point I made in section III that the arch-and-scaffold metaphor, far from compounding the historiographical sin of Whiggishness, can be seen as an attempt to legitimize a common and benign form of it, even if one has to remain vigilant (see note 43).

How von Neumann's Hilbert-Space Formalism Was Scaffolded by the Dirac-Jordan Statistical Transformation Theory[49]

Quantum theory developed rapidly in the years 1925–1927. By the middle of 1926, four different versions were in circulation: the Göttingen matrix mechanics of Heisenberg, Born, and Jordan; the wave mechanics of Austria's Erwin Schrödinger; the q-number theory of Cambridge's Paul Dirac; and, though more problematic and less influential than the other three, the operator calculus of Born and the American mathematician Norbert Wiener. Schrödinger (1926) had shown that wave mechanics and matrix mechanics always give the same empirical predictions. Born (1926a, 1926b) had shown that Schrödinger's wave functions call for a probabilistic interpretation. A general formalism tying the four different versions together, however, had yet to be found. Then, in late 1926, independently of one another, Jordan (1927a) and Dirac (1927) submitted papers proposing essentially the same overarching formalism along with its probabilistic interpretation. It became known as the *Dirac-Jordan statistical transformation theory*, or *transformation theory* for short. I focus here on Jordan's formulation, though I will borrow some of Dirac's vastly superior notation. For a comparison of Jordan's approach to Dirac's—widely disseminated through his influential textbook on quantum mechanics (Dirac 1930)—see Duncan and Janssen (2013, 185–90).

Statistical transformation theory can be seen as an arch built on a scaffold constructed out of the four related yet different theories it unified. The arch that Heisenberg (1925) had built on the scaffold of the Kramers dispersion formula (see the preceding case study) thus became part of the scaffold on which Jordan (1927a, 1927b) and Dirac (1927) erected their arch. Within a few months, the Hungarian polymath John von Neumann (1927a, 1927b, 1927c) would use Jordan and Dirac's arch as a scaffold to build an arch of his own, his Hilbert-space formalism for quantum mechanics, although one could also say, in the spirit of Hilbert and Young (see section II), that von Neumann produced a scaffold to prop up Jordan's arch. Like Dirac's paper, von Neumann's papers were later expanded into a book (von Neumann 1932).

I will not even attempt to characterize the relation between arch and scaffold in the transition from the four early versions of quantum theory to transformation theory, other than to say that it is considerably more complicated than in the examples analyzed so far. It will be difficult enough to decide which formalism played the role of the arch and which that of the scaffold in the transition from Jordan's version of transformation theory to von Neumann's Hilbert-space formalism. Either way, it is a challenge to precisely characterize the relation between these two formalisms.

As mentioned in the introduction, another important difference between this case study and the other four in this section is that in the transition from matrix and wave mechanics to transformation theory as well as in the subsequent transition from transformation theory to Hilbert space, the scaffold was not dismantled once the arch had been built. Elements of all four of these formalisms continue to be used to this day. While many philosophers of quantum mechanics use vectors in Hilbert space, quantum chemists for the most part get by with Schrödinger wave functions. This is true both in research and in teaching. In introductory physics courses, quantum mechanics is typically presented in the guise of wave mechanics, while for some problems techniques from matrix mechanics are used (e.g., raising and lowering operators to find the energy spectrum of a simple harmonic oscillator). More advanced courses typically present a blend of von Neumann's Hilbert-space formalism and Dirac's version of transformation theory. As we will see, this blend depends, for its mathematically cogent formulation, on advances made long after the period under consideration here, the late 1920s, to which I now return.

JORDAN'S NEW FOUNDATION FOR QUANTUM THEORY. The new foundation (*Neue Begründung*) of quantum theory that Jordan (1927a) announced in the title of his paper is based on two fundamental ideas. First, quantum mechanics is ultimately a theory about conditional probabilities $\Pr(\tilde{A} = a \mid \tilde{B} = b)$ that some physical (i.e., observable or measurable) quantity $\tilde{A}$ has the value a given that another physical quantity $\tilde{B}$ has the value b (the tildes indicate that these physical quantities are quantum variables; q-numbers in Dirac's terminology). Second, such conditional probabilities are given by the absolute square of corresponding complex probability amplitudes, $\varphi(a, b)$. I use the notation of Duncan and Janssen (2013), which follows Dirac rather than Jordan, whose notation is a veritable nightmare.[50]

Examples of probability amplitudes are the familiar energy eigenfunctions $\psi_n(x)$ of Schrödinger's wave mechanics, where n refers to the eigenvalue E_n and where, for convenience, we restrict ourselves to one-dimensional problems. The absolute square of this function, $|\psi_n(x)|^2 = \psi_n(x)^* \psi_n(x)$ (where the star denotes complex conjugation), multiplied by the infinitesimal distance dx, gives the probability that the position $\tilde{x}$ of the system is somewhere in the narrow interval $(x, x+dx)$ given that its energy $\tilde{E}$ is equal to E_n:

$$\Pr(\tilde{x} \in (x,\ x+dx) \mid \tilde{E} = E_n) = |\psi_n(x)|^2\ dx. \tag{8}$$

Though eventually named after Born (1926a, 1926b), the probabilistic interpretation of $\psi_n(x)$ in this particular form is due to Wolfgang Pauli, a quantum theorist of the same generation as Heisenberg and Jordan, who was in close contact with all three founders of matrix mechanics and made several key contributions himself (Duncan and Janssen 2013, 182–83). Jordan (and Dirac) generalized Equation (8) for position and energy to arbitrary quantities $\tilde{A}$ and $\tilde{B}$:

$$\Pr(\tilde{A} \in (a, a+da) \mid \tilde{B} = b) = |\varphi(a,b)|^2\ da. \tag{9}$$

In Jordan's formalism, the energy eigenfunction $\psi_n(x)$ thus becomes the probability amplitude $\varphi(x, E_n)$.

In Jordan's first paper on his new formalism, only quantities with continuous spectra are considered. When, in a second paper, Jordan (1927b) tried to generalize his formalism to quantities with wholly or partly discrete spectra (such as, typically, the energy), he ran into serious difficulties, which mercilessly exposed the limitations of his approach.

Jordan's approach, reflecting his mathematical training in Göttingen, was axiomatic (Lacki 2000). He started from a set of postulates for his probability amplitudes and then looked for a realization of these postulates. As Hilbert, von Neumann, and Lothar Nordheim, one of Hilbert's assistants at the time, put it in a joint paper on Jordan's new formalism:

> One imposes certain physical requirements on these probabilities, which are suggested by earlier experience and developments, and the satisfaction of which calls for certain relations between the probabilities. Secondly, one searches for a simple analytical apparatus in which quantities occur that satisfy these relations exactly. (Hilbert, von Neumann, and Nordheim 1928, 2–3; cf. Lacki 2000, 296)

The number of Jordan's postulates in various expositions of his formalism fluctuates between two and six (Duncan and Janssen 2013, 199). I will use a version here based on three postulates.

JORDAN'S POSTULATES. The first postulate gives the probability amplitude for the basic variables, a generalized coordinate $\tilde{q}$ and its conjugate momentum $\tilde{p}$ (again, we will restrict ourselves to one-dimensional problems):

$$\varphi(p, q) = e^{-ipq/\hbar}. \tag{10}$$

This postulate takes the place of the commutation relations in Equation (7) for position and momentum as the basic quantization condition in Jordan's formalism. Since $|\varphi(p, q)|^2 = 1$, he concluded (ignoring the issue of how to normalize his probabilities) that "for a given value of $[\tilde{q}]$ all possible values of $[\tilde{p}]$ are *equiprobable*" (Jordan 1927a, 814). Jordan's formalism thus contains the kernel of the uncertainty principle, which Heisenberg (1927), drawing on Jordan's work, would publish later that year.

The second postulate says that the amplitude $\varphi(b, a)$ is the complex conjugate of the amplitude $\varphi(a, b)$:

$$\varphi(b, a) = \varphi(a, b)^*. \tag{11}$$

For example, $\varphi(q, p) = \varphi(p, q)^* = e^{ipq/\hbar}$, from which it follows that for a given value of $\tilde{p}$ all values of $\tilde{q}$ are equiprobable.

The basic amplitude in Equation (10) trivially satisfies the following pair of differential equations:

$$\left(p + \frac{\hbar}{i} \frac{\partial}{\partial q} \right) \varphi(p, q) = 0, \quad \left(q + \frac{\hbar}{i} \frac{\partial}{\partial p} \right) \varphi(p, q) = 0. \tag{12}$$

Jordan thought that Equations (10)–(12) sufficed to find the probability amplitudes for any pair of quantities $\tilde{A}$ and $\tilde{B}$ related to $\tilde{q}$ and $\tilde{p}$ by a so-called canonical transformation.

Canonical transformations belong to the bag of tricks the old quantum theory had borrowed from celestial mechanics. Closely related techniques were central to the derivation of the Kramers dispersion formula (see the preceding case study). Born, Heisenberg, and Jordan (1926) had imported canonical transformations into matrix mechanics in their famous *Dreimännerarbeit*. Before he worked out his new foundation for quantum mechanics, Jordan (1926a, 1926b) had published two papers on how to implement canonical transformations in matrix mechanics (Lacki 2004; Duncan and Janssen 2009). Asked about the use of canonical transformations in the *Dreimännerarbeit* in an interview for the AHQP (cf. note 8) in the early 1960s, Jordan said:

> Canonical transformations in the sense of Hamilton-Jacobi [theory in celestial mechanics] were . . . our daily bread in the preceding years, so to tie in the new results with those as closely as possible—that was something very natural for us to try. (Duncan and Janssen 2009, 355)

Canonical transformations, however, proved ill-suited to the task Jordan assigned to them in his new formalism. They are both too restrictive and too permissive for his purposes. They are too restrictive because quantities related by a canonical transformation always have the same spectrum (Duncan and Janssen 2013, 216). A canonical transformation can thus never take us from a quantity with a continuous spectrum to a quantity with a (partly) discrete spectrum. As Jordan (1927b) eventually had to concede, this means that there is no canonical transformation that takes us from the basic amplitude $\varphi\,(p, q)$ satisfying the pair of differential equations (12) to the new amplitude $\varphi\,(x, E_n) = \psi_n\,(x)$ satisfying a transformed version of this pair of differential equations, one of which would have to be equivalent to the time-independent Schrödinger equation.

Canonical transformations are also too permissive. Jordan's realization of his postulates turned on identifying probability amplitudes with quantities characterizing associated canonical transformations. Unfortunately, as we will see, there are many canonical transformations giving probability amplitudes that do not satisfy Equation (11), Jordan's second postulate. Jordan thus had to artificially restrict the class of allowed canonical transformations.[51] In hindsight, we can see that Jordan was stretching the classical formalism beyond its breaking point in trying to make it work for his new quantum formalism (Duncan and Janssen 2013, 188–91, 253–54).

Jordan's third postulate, to which we now turn, also has its share of problems, though these are not fatal to his project. This postulate is about how to combine probability amplitudes for different pairs of quantities. It states that in quantum mechanics the usual rules of probability theory, the addition rule for the disjunction and the multiplication rule for the conjunction of two outcomes, apply to the probability amplitudes rather than to the probabilities themselves. Following Born and Pauli, Jordan (1927a, 812) called this the "interference of probabilities."

The famous double-slit experiment illustrates that this is a sensible name. Let φ_1 be the amplitude for the conditional probability that an electron strikes a screen at position x given that it went through the first slit. Let φ_2 be the amplitude that the electron strikes at x given that it went through the second slit. According to Jordan's addition rule for probability amplitudes, the probability of the electron striking at x if it could have gone through either slit is then given by

$$|\varphi_1 + \varphi_2|^2 = (\varphi_1 + \varphi_2)(\varphi_1 + \varphi_2)^* = |\varphi_1|^2 + |\varphi_2|^2 + \varphi_1\varphi_2^* + \varphi_2\varphi_1^*. \tag{13}$$

The first two terms in the final expression give the probability that the electron strikes the screen at x if it went through one of the slits. The last two terms give the interference effects if the electron could have gone through both.

In the paper introducing the uncertainty principle, Heisenberg (1927) took Jordan to task for his third postulate, arguing that the laws of probability are what they are independently of the laws of physics. Even quantum mechanics cannot change them. While most modern commentators would agree with this criticism, it does not affect Jordan's formalism. Jordan only used his dubious new quantum probability laws to derive two conditions, which in the further elaboration of the formalism took over the role of those new probability laws as the third postulate. These two conditions are eminently reasonable whether or not one accepts Jordan's derivation of them. They both continue to hold in modern quantum mechanics.

The first of these two conditions says that the probability amplitudes $\varphi\,(a, b)$, $\varphi\,(b, c)$, and $\varphi\,(a, c)$ involving the physical quantities $\tilde{A}$, $\tilde{B}$, and $\tilde{C}$ should satisfy the relation

$$\varphi(a, c) = \int db\ \varphi(a, b)\ \varphi(b, c). \tag{14}$$

In the example of the double-slit experiment, $\tilde{A}$ is the position where the electrons hit the screen (with a continuum of values a), $\tilde{C}$ is the position of the source of the electrons (with some fixed value c), and $\tilde{B}$ is the position of the slits (with two possible values b_1 and b_2). The integral in Equation (14) then reduces to a sum of two terms,

$$\varphi\,(a, c) = \varphi\,(a, b_1)\ \varphi\,(b_1, c) + \varphi\,(a, b_2)\ \varphi\,(b_2, c).$$

These two terms are more explicit expressions for the amplitudes φ_1 and φ_2 in Equation (13) (Duncan and Janssen 2013, 186–87n38).

The second of the two conditions effectively serving as Jordan's third postulate says that if $\tilde{A} = \tilde{C}$ in Equation (14), it should be the case that

$$\varphi(a, a') = \int db\ \varphi(a, b)\ \varphi(b, a') = \delta(a - a'), \tag{15}$$

where $\delta(x)$ is defined as "vanishing everywhere except at $x = 0$ where it is infinite." I put this definition in scare quotes to flag its gross mathematical sloppiness. Dirac (1927) introduced this notorious delta function in his version of transformation theory. Jordan used it implicitly. Equation (15) expresses the obvious requirement that the probability of finding the value a for a quantity $\tilde{A}$ given the value a' for that same quantity should be zero unless those two values are the same.

If $\tilde{A}$ has a fully discrete spectrum, its possible values can be labeled with a discrete index, and the requirement (15) can be formulated in mathematically unobjectionable fashion as:

$$\varphi(a_i, a_j) = \int db \, \varphi(a_i, b) \varphi(b, a_j) = \delta_{ij}.$$

If $\tilde{A}$ has a fully continuous spectrum, the Kronecker delta δ_{ij} (see Equation (7) for its definition) has to be replaced by the Dirac delta function.

A REALIZATION OF JORDAN'S POSTULATES USING CANONICAL TRANS-FORMATIONS. Jordan's three postulates boil down to the requirement that his probability amplitudes satisfy the relations (10), (11), (14), and (15). All that is left to do at this point is to find a mathematical representation of these probability amplitudes such that these four relations are guaranteed to hold (see the description of Jordan's approach by Hilbert, von Neumann, and Nordheim above). Jordan does not tell us how he arrived at this representation. He just states his choice and shows that with that choice his postulates are satisfied. Jordan's choice, however, is a natural one.

Consider the familiar result that an energy eigenfunction $\psi_n(p)$ in momentum space is the Fourier transform of that energy eigenfunction $\psi_n(q)$ in position space:

$$\psi_n(p) = \int dq \, e^{-ipq/\hbar} \, \psi_n(q). \tag{16}$$

Using notation introduced by Dirac (1927), we can write this transformation of ψ_n from q-space to p-space as

$$\psi_n(p) = \int dq \, (p/q) \, \psi_n(q). \tag{17}$$

If p and q were discrete indices, the integral would turn into a sum, and the equation would express that a vector with components $\psi_n(p)$ is equal to the product of a matrix with components (p/q), where p labels rows and q labels columns, and a vector with components $\psi_n(q)$. Equation (17) can be seen as the generalization of this relation to a situation in which p and q are continuous variables. Neither Jordan nor Dirac was overly concerned with the mathematical niceties of this generalization.

Comparison between Equations (16)–(17) and Equation (10) suggests that the basic probability amplitude for momentum and position be identified with the "matrix" (more accurately: the integral kernel) for the transformation from position space to momentum space:

$$\varphi(p, q) = (p/q) = e^{-ipq/\hbar}. \tag{18}$$

This in turn suggests that the probability amplitude for an arbitrary pair of physical quantities $\tilde{A}$ and $\tilde{B}$ be identified with the "matrix" for the transformation from b-space to a-space,

$$\varphi(a, b) = (a/b). \tag{19}$$

This, of course, is why the Dirac-Jordan formalism is called statistical *transformation* theory.

Equation (18) shows that the first postulate (i.e., Equation (10)) is satisfied. As long as the transformation "matrix" (a/b) is *unitary*—which means that its inverse $(a/b)^{-1} = (b/a)$ is given by its complex conjugate $(a/b)^*$—the second postulate (i.e., Equation (11)) is also satisfied. Alas, not all canonical transformations are unitary, which is why Jordan somewhat artificially had to restrict the class of allowed transformations (see note 51).

Substituting $\psi_n(p) = \varphi(p, E_n) = (p/E_n)$, $\psi_n(q) = \varphi(q, E_n) = (q/E_n)$ and $e^{-ipq/\hbar} = (p/q)$ into Equation (16), we arrive at

$$(p/E_n) = \int dq\, (p/q)(q/E_n). \tag{20}$$

This shows that Equation (14), one of the two conditions effectively playing the role of Jordan's third postulate, is satisfied in the special case that the quantities $\tilde{A}$, $\tilde{B}$, and $\tilde{C}$ are $\tilde{p}$, $\tilde{q}$, and $\tilde{E}$, respectively. To show that this is true for any triplet of quantities, consider some eigenfunction ψ of the energy or some other quantity. Its transformation from c-space to a-space is given by

$$\psi(a) = \int dc\, (a/c)\, \psi(c). \tag{21}$$

Its transformation from c-space to a-space via b-space is given by

$$
\begin{aligned}
\psi(a) &= \int db \; (a/b) \; \psi(b) \\
&= \int db \, (a/b) \left(\int dc \, (b/c) \; \psi(c) \right) \\
&= \int dc \left(\int db \; (a/b) \, (b/c) \right) \psi(c).
\end{aligned}
\tag{22}
$$

Comparison of these two transformations shows that (a/c) in Equation (21) is equal to the expression in large parentheses in the last line of Equation (22). This is just as it should be according to Equation (14) (Duncan and Janssen 2013, 185).

To verify that Equation (15), the other half of Jordan's third postulate, is also satisfied, compare the final expression for $\psi(a)$ in Equation (22) for $c = a'$ to

$$
\psi(a) = \int da' \; \delta(a - a') \; \psi(a'),
\tag{23}
$$

which holds on the basis of the defining equation for the delta function (i.e., for any function $f(x)$, $\int dx' \; \delta(x - x') f(x') = f(x)$). This comparison shows that

$$
\int db \; (a/b)(b/a') = \delta(a - a'),
\tag{24}
$$

in accordance with Equation (15).

A REALIZATION OF JORDAN'S POSTULATES USING HILBERT SPACE. In the first installment of a trilogy of papers that would provide the backbone of his famous 1932 book, von Neumann (1927a) introduced the Hilbert-space formalism of quantum mechanics. With the help of a modern version of this formalism, a new realization of Jordan's postulates can be given. In this new realization, integral kernels of canonical transformations, which Jordan used to represent his probability amplitudes, are replaced by "inner products" of "vectors" in Hilbert space. I use scare quotes to indicate that the justification for treating the relevant quantities as vectors and inner products of vectors turns on results in mathematics only found much later, in particular the theory of distributions and the theory of rigged Hilbert space, both developed in the 1950s. These developments are beyond the level of this paper—and beyond my command of mathematics. They nicely illustrate the point William Young (1926) made in his presidential address to the London Mathematical Society (see section II). Sometimes more sophisticated mathematics can be used to shore up more basic mathematics. Once that has been done, one can use the latter without worrying about the former. A modern student of quantum mechanics will hardly ever go wrong envisioning elements in Hilbert space as vectors in a finite-dimensional vector space. This section is written in that spirit. From now on, I will talk about vectors and inner products without using scare quotes, even though I will remind the reader at several junctures of the mathematical difficulties lurking just below the surface. With that preamble, let me introduce the Hilbert-space formalism and sketch how it can be used to construct a realization of Jordan's postulates.

In the Hilbert-space formalism, physical quantities, $\tilde{A}$, are represented by certain linear operators mapping vectors onto other vectors in a complex, infinite-dimensional vector space known as Hilbert space: $\hat{A}: |f\rangle \rightarrow |g\rangle$ (a "hat" denotes an operator; $|\ \rangle$ is the standard modern notation, due to Dirac, for a vector in Hilbert space). That $\hat{A}$ is linear means that

$$\hat{A}\left(\lambda\,|f_1\rangle + \mu|f_2\rangle\right) = \lambda\,\hat{A}|f_1\rangle + \mu\hat{A}|f_2\rangle \qquad (25)$$

for any vectors $|f_1\rangle$ and $|f_2\rangle$ and any complex numbers λ and μ. If a vector $|a\rangle$ satisfies

$$\hat{A}|a\rangle = a|a\rangle, \qquad (26)$$

it is called an *eigenvector* of $\hat{A}$, and the (in general, complex) number a is called an *eigenvalue* of $\hat{A}$. Physical quantities are represented by so-called self-adjoint (or Hermitian) operators. Their eigenvalues are always real numbers. The (infinite) set of all eigenvectors of any self-adjoint operator forms an orthogonal basis for Hilbert space.

The standard notation, again due to Dirac, for the inner product of two arbitrary vectors, $|f\rangle$ and $|g\rangle$, in Hilbert space is $\langle f|g\rangle$. Since this will in general be a complex number, the order matters:

$$\langle g|f\rangle = \langle f|g\rangle^*. \qquad (27)$$

It thus makes a difference whether $\hat{A}|\ \rangle$ enters an inner product $\langle\ |\ \rangle$ on the right, as a vector $|\ \rangle$, or on the left, as a dual vector $\langle\ |$. The dual vector of $\hat{A}|\ \rangle$ is $\langle\ |\hat{A}^\dagger$, where $\hat{A}^\dagger$ is called the adjoint of $\hat{A}$. For self-adjoint operators, $\hat{A}^\dagger = \hat{A}$.

The energy $\tilde{E}$ is represented by a self-adjoint operator $\hat{E}$ with normalized eigenvectors $|E_n\rangle$ and eigenvalues E_n. The position $\tilde{x}$ is likewise represented by a self-adjoint operator $\hat{x}$ with normalized eigenvectors $|x\rangle$ and eigenvalues x. The normalization is mathematically more problematic in the case of continuous spectra than in the case of discrete spectra. For systems with a fully discrete energy spectrum, for instance, we can simply use the Kronecker delta: $\langle E_{n_i} | E_{n_j} \rangle = \delta_{ij}$. For quantities such as position with fully continuous spectra, we need the Dirac delta function: $\langle x | x' \rangle = \delta(x - x')$. The inner products $\langle x | E_n \rangle$ of these normalized eigenvectors give the familiar energy eigenfunctions $\psi_n(x)$ of wave mechanics. As we saw above (cf. Equations (8)–(9)), these are also the probability amplitudes $\varphi(x, E_n)$.

This is true in general. Jordan's three postulates are satisfied if the probability amplitude $\varphi(a, b)$ for any pair of physical quantities $\tilde{A}$ and $\tilde{B}$ is set equal to the inner product $\langle a | b \rangle$ of the normalized eigenvectors $|a\rangle$ and $|b\rangle$ of the corresponding self-adjoint operators $\hat{A}$ and $\hat{B}$.

It will be instructive to explicitly verify this for Jordan's second and third postulates. The second postulate (i.e., Equation (11)) follows directly from the definition of the inner product in Hilbert space: $\varphi(b, a) = \varphi(a, b)^*$ because $\langle b | a \rangle = \langle a | b \rangle^*$ (see Equation (27)). There is no need for the kind of restrictions on $\langle a | b \rangle$ that Jordan had to impose on (a/b).

The third postulate (i.e., Equations (14)–(15)) holds by virtue of a key result of von Neumann's Hilbert-space formalism, his famous spectral theorem for self-adjoint operators. We need not worry about the proof of this theorem, but we do need to understand at least roughly what it says.

Consider a discrete orthonormal basis $\{|e_i\rangle\}$ (with $\langle e_i | e_j \rangle = \delta_{ij}$) in a finite dimensional complex vector space. Any vector $|f\rangle$ in that space can be written in terms of its components with respect to this basis:

$$|f\rangle = \sum_{i=1}^{N} |e_i\rangle \langle e_i | f\rangle. \tag{28}$$

The complex number $\langle e_i | f \rangle$ gives the component of $|f\rangle$ in the direction of $|e_i\rangle$. Equation (28) can also be parsed in a different way. We can identify the expression $|e_i\rangle \langle e_i|$ as a projection operator,

$$\hat{P}_{e_i} = |e_i\rangle \langle e_i|, \tag{29}$$

that maps any vector $|f\rangle$ onto the part of $|f\rangle$ in the direction of $|e_i\rangle$ ($\hat{P}_{e_i}$ is a self-adjoint operator). Equation (28) then expresses that the sum of these projection operators is the identity operator

$$\hat{1} = \sum_{i=1}^{N} |e_i\rangle \langle e_i| = \sum_{i=1}^{N} \hat{P}_{e_i}, \tag{30}$$

which maps any vector $|f\rangle$ back onto itself. Equation (30) is called the resolution of unity corresponding to the orthonormal basis $\{|e_i\rangle\}$.

Von Neumann's spectral theorem sanctions the generalization of Equations (28)–(30) from finite-dimensional complex vector spaces to infinite-dimensional complex Hilbert space with both discrete and continuous orthonormal bases. The analogue of Equation (28) in Hilbert space with an orthonormal basis consisting of normalized eigenvectors $|a\rangle$ of the self-adjoint operator $\hat{A}$ is

$$|f\rangle = \int da\, |a\rangle\langle a|f\rangle, \tag{31}$$

where the integral is to be taken over all eigenvalues of $\hat{A}$. Using the decomposition of $|f\rangle$ in Equation (31), the definition of the eigenvectors $|a\rangle$ in Equation (26), and the linearity of the operator $\hat{A}$, we can write the action of $\hat{A}$ on $|f\rangle$ as

$$\hat{A}|f\rangle = \int da\left\{\hat{A}|a\rangle\right\}\langle a|f\rangle = \int da\, a|a\rangle\langle a|f\rangle. \tag{32}$$

It follows that $\hat{A}$ can be written as

$$\hat{A} = \int da\, a|a\rangle\langle a| = \int da\, a\hat{P}_a, \tag{33}$$

where, in analogy to $\hat{P}_{e_i}$ in Equation (29), the projection operator $\hat{P}_a$ is given by

$$\hat{P}_a = |a\rangle\langle a|. \tag{34}$$

This operator maps any vector $|f\rangle$ in Hilbert space onto the part of $|f\rangle$ in the direction of $|a\rangle$. In analogy to Equation (30), the integral of $\hat{P}_a$ over all eigenvalues of $\hat{A}$ is the identity operator,

$$\hat{1} = \int da\, |a\rangle\langle a| = \int da\, \hat{P}_a. \tag{35}$$

Equation (33) gives the spectral decomposition of the self-adjoint operator $\hat{A}$. Equation (35) gives the corresponding resolution of unity.

Once the hard work of proving the spectral theorem is done, it is easy to show that Equations (14)–(15) (and thereby Jordan's third postulate) are satisfied if probability amplitudes $\varphi(a, b)$ are identified with inner products $\langle a|b\rangle$. Equation (14) requires that

$$\langle a|c\rangle = \int db\, \langle a|b\rangle\langle b|c\rangle, \tag{36}$$

where $|a\rangle$, $|b\rangle$, and $|c\rangle$ are the normalized eigenvectors of the self-adjoint operators $\hat{A}$, $\hat{B}$, and $\hat{C}$, representing the quantities $\tilde{A}$, $\tilde{B}$, and $\tilde{C}$. This relation holds by virtue of the resolution of unity corresponding to the spectral decomposition of $\hat{B}$, which allows us to rewrite the right-hand side as $\langle a|\hat{1}|c\rangle = \langle a|c\rangle$ Equation (15) requires that

$$\langle a|a'\rangle = \int db\, \langle a|b\rangle\langle b|a'\rangle = \delta(a-a').\tag{37}$$

This relation holds by virtue of the spectral decomposition of $\hat{B}$ and the normalization $\langle a\,|\,a'\rangle = \delta(a-a')$ of the eigenvectors of $\hat{A}$.

HOW VON NEUMANN DID NOT BUILD HIS ARCH ON JORDAN'S SCAFFOLD AND WHY NOT.

The preceding two subsections suggest that we have found another picture-perfect example of my arch-and-scaffold metaphor in the history of early twentieth-century physics. The relation between arch and scaffold in this case is reminiscent of the general-relativity example discussed earlier. In both cases, swapping out one building block for another while leaving the structure built with them intact resulted in a new building exhibiting the splendor of a magnificent mathematical formalism that had been waiting in the wings. In the case of general relativity, the building blocks were two different definitions of the gravitational field, and the mathematical formalism was the differential geometry of Riemann and others. In this case, the building blocks are two different realizations of Jordan's probability amplitudes— $\varphi(a, b) = (a/b)$ and $\varphi(a, b) = \langle a\,|\,b\rangle$ —and the mathematical formalism is Hilbert's spectral theory of operators as generalized by von Neumann.

Historically, however, this is *not* how von Neumann got from the Jordan-Dirac transformation theory to his own Hilbert-space formalism. Even in the historical literature, von Neumann's formalism is not always clearly distinguished from Dirac's. In the classic book on the conceptual development of quantum mechanics mentioned in section II, for instance, Jammer (1966, 307–22) gave the section dealing with von Neumann (1927a, 1927b, 1932) the misleading title "The Statistical Transformation Theory in Hilbert Space" (Duncan and Janssen 2013, 193n51). This is what von Neumann had to say about Dirac in the preface of the book that grew out of his 1927 papers:

> Dirac's method does not meet the demands of mathematical rigor in any way—not even when it is reduced in the natural and cheap way to the level that is common in theoretical physics . . . the correct formulation is not just a matter of making Dirac's method mathematically precise and explicit but right from the start calls for a different approach related to Hilbert's spectral theory of operators. (von Neumann 1932, 2)

Rather than using Hilbert space to provide a new realization of probability amplitudes, von Neumann wanted to avoid probability amplitudes altogether. One of his major objections against the Dirac-Jordan formalism was its reliance on the Dirac delta function. This is not a well-defined function, and von Neumann (1927a, 2) dismissed it as simply "absurd." He also objected to the basic probability amplitude $\varphi(p, q) = e^{-ipq/\hbar}$ for position and momentum introduced by both Dirac and Jordan (with the latter even elevating it to the status of a postulate; see Equation (10)). Although this is at least a well-defined function, the integral of its absolute square diverges. That means that it is not an element of the space of square-integrable functions, which is one instantiation of Hilbert space.

As mentioned above, the mathematics needed to solve these problems (the theory of distributions and the theory of rigged Hilbert space) was not developed until the 1950s. Using these new tools, we can replace transformation "matrices" $(a|b)$ by "inner products" $\langle a|b \rangle$ in a mathematically rigorous way. So, contrary to what von Neumann believed in 1927 and 1932, it is possible to make "Dirac's method mathematically precise and explicit." I already alluded to the continued use of the resulting formalism, blending elements of Dirac and von Neumann, in more advanced courses on quantum mechanics, although textbook writers and instructors typically (and understandably!) only gesture at the mathematics needed for its rigorous formulation.

Given the familiarity of this formalism, modern readers may be tempted to read it back into Dirac's original paper of 1927 on transformation theory—that is, to read his "brackets" $(a|b)$ as inner products $\langle a|b \rangle$ and then break those up into "bra"-s $\langle a|$ and "ket"-s $|b \rangle$, the now familiar names, due to Dirac, for vectors (kets) and their duals (bras) in Hilbert space. Although Dirac (1930) made use of the Hilbert-space formalism in his book, it was only in 1939 that he himself first split "brackets" into "bras" and "kets" (Borrelli 2010).

How von Neumann did introduce his formalism in response to Jordan's. There is no doubt that von Neumann introduced his Hilbert-space formalism in response especially to Jordan's version of the Dirac-Jordan statistical transformation theory. What is not clear, as I mentioned at the beginning of this section, is whether von Neumann's formalism is best understood as an arch built on top of the scaffold provided by Jordan's formalism or as a scaffold built to support Jordan's mathematically unsound arch. I will return to this ambiguity at the end of this section, after I have gone over the steps actually taken by von Neumann in 1927.

I already mentioned the paper on Jordan's formalism by Hilbert, von Neumann, and Nordheim (1928), submitted in April 1927 but published only the following year. The authors emphasized the mathematical problems with Jordan's formalism and referred to a forthcoming paper by von Neumann that would address them. Rather than confronting these problems head-on, however, von Neumann (1927a) avoided them by deviating from Jordan's approach almost from the start. He only took over the two basic ideas on which Jordan had built his formalism: first, that quantum mechanics is a theory about conditional probabilities; second, that these probabilities satisfy some peculiar rules.

As Hilbert and his coauthors had written approvingly about Jordan's third postulate, "axiom IV" in their exposition:

> This requirement is obviously analogous to the addition and multiplication theorems of ordinary probability calculus, except that in this case they hold for the amplitudes rather than for the probabilities themselves. (Hilbert, von Neumann, and Nordheim 1928, 5)

In his own paper, von Neumann reiterated that, in Jordan's formalism,

> the multiplication law of probabilities does not hold in general (what does hold is a weaker law corresponding to Jordan's "combining of probability amplitudes"). (von Neumann 1927a, 46)

Instead of introducing Jordan's probability amplitudes, however, von Neumann constructed a formula for conditional probabilities out of projection operators in Hilbert space—*Einzeloperatoren,* as he called them, or *E.op.*s for short. Using the notation for projection operators introduced in Equation (34), we can write von Neumann's formula as (Duncan and Janssen 2013, 242–44)

$$\Pr(\tilde{A} \in (a, a+da) \,|\, \tilde{B}=b) = \mathrm{Tr}(\hat{P}_a \, \hat{P}_b)\, da. \tag{38}$$

where the *trace* $\mathrm{Tr}(\hat{A})$ of any operator $\hat{A}$ is defined with the help of an arbitrary discrete orthonormal basis $\{|e_i\rangle\}$ of Hilbert space:

$$\mathrm{Tr}(\hat{A}) \equiv \sum_{i=1}^{\infty} \langle e_i | \hat{A} | e_i \rangle. \tag{39}$$

It is easily shown that the result does not depend on which orthonormal basis we use to evaluate the trace.

Using this definition and using Equation (34) for the projection operators, we verify that Equation (38) reduces to Equation (9), Jordan's formula for the same conditional probability:

$$\text{Tr}(\hat{P}_a\,\hat{P}_b)\,da = \sum_{i=1}^{\infty}\langle e_i\,|\,a\rangle\langle a\,|\,b\rangle\langle b\,|\,e_i\rangle\,da$$

$$= \sum_{i=1}^{\infty}\langle b\,|\,e_i\rangle\langle e_i\,|\,a\rangle\langle a\,|\,b\rangle\,da = |\langle a\,|\,b\rangle|^2\,da. \tag{40}$$

In the last step we used that $\sum_i |e_i\rangle\langle e_i|$ is the identity operator and that $\langle b\,|\,a\rangle\langle a\,|\,b\rangle = \langle a\,|\,b\rangle^*\langle a\,|\,b\rangle = |\langle a\,|\,b\rangle|^2$. It is important to note, however, that the projection operators were the fundamental quantities for von Neumann. Expressing them in terms of "bras" and "kets" reintroduces some of the mathematical objections that he got around by using projection operators instead of probability amplitudes.

Using the resolution of unity the same way as in Equation (40), one readily verifies that $\text{Tr}(\hat{A}\hat{B}) = \text{Tr}(\hat{B}\hat{A})$ for arbitrary operators $\hat{A}$ and $\hat{B}$. Von Neumann's formalism thus reproduces the relation $\text{Pr}(\tilde{B}=b\,|\,\tilde{A}=a) = \text{Pr}(\tilde{A}=a\,|\,\tilde{B}=b)$ that follows directly from Jordan's second postulate (see Equation (11)).

It is also easy to verify that the relation

$$\text{Tr}(\hat{P}_a\,\hat{P}_c) = \int db\,\text{Tr}(\hat{P}_a\,\hat{P}_b\,\hat{P}_c) \tag{41}$$

in von Neumann's formalism is the equivalent of Equation (14), which expresses the "interference of probabilities" in Jordan's formalism. If the projection operators are written in terms of bras and kets, the left-hand side of Equation (41) reduces to $\langle c\,|\,a\rangle\langle a\,|\,c\rangle$ (cf. Equation (40)). The right-hand side can similarly be written as

$$\int db\,\sum_i \langle e_i\,|\,a\rangle\langle a\,|\,b\rangle\langle b\,|\,c\rangle\langle c\,|\,e_i\rangle = \int db\,\langle c\,|\,a\rangle\langle a\,|\,b\rangle\langle b\,|\,c\rangle. \tag{42}$$

It follows that

$$\langle a\,|\,c\rangle = \int db\,\langle a\,|\,b\rangle\langle b\,|\,c\rangle, \tag{43}$$

which is Equation (14) for Jordan's probability amplitudes if these amplitudes are identified with inner products in Hilbert space (see Equation (36)). It is probably no coincidence that Equation (41) is nowhere to be found in von Neumann (1927a). Von Neumann was interested in the outcome of an actual measurement of one quantity given the outcome of a prior measurement of another quantity. In the type of situation involving three quantities considered by Jordan in Equation (14), it is critical, modern quantum mechanics tells us, that the quantity $\hat{B}$ is not actually measured.

Getting from the formalism of Jordan (1927a) to the formalism of von Neumann (1927a) is clearly not as straightforward as replacing one building block by another. It is true that projection operators replaced probability amplitudes as the basic elements but, unlike the substitution of inner products for transformation "matrices," this replacement was accompanied by invasive structural changes in the edifice built out of these elements.

We can distinguish three layers in Jordan's formalism: basic ideas, postulates expressing those ideas, and a realization of those postulates. Von Neumann only took over the first of these layers. His paper with Hilbert and Nordheim, however, shows that he had carefully examined Jordan's entire building. While this inspection had revealed it to be a rickety mathematical structure from top to bottom, it at least had given him a good idea as to what a general formalism for quantum mechanics would have to deliver to be viable as a new framework for doing physics.

Von Neumann recognized that a generalization of Hilbert's spectral theory of operators was much more appropriate for the purposes of Jordan and Dirac than the theory of canonical transformations that they themselves had pressed into service. The Hilbert-space formalism thus freed quantum mechanics from some of the vestiges of classical mechanics that can clearly be recognized in transformation theory (in the original form in which Jordan and Dirac presented it) and its progenitors, matrix mechanics, and q-number theory.

We already saw that Jordan's strong reliance on canonical transformations created a number of serious problems. Many of these are specific to Jordan's axiomatic formulation of transformation theory and do not affect Dirac's formulation. In other respects, however, both Jordan and Dirac were handicapped by their commitment to canonical transformations. A feature of canonical transformations that I did not emphasize so far is that it is always a transformation from a *pair* of variables, some generalized coordinate q and its conjugate momentum p, to another such pair. As long as probabilities are defined in terms of canonical transformations, all physical quantities thus need to be sorted in terms of such conjugate variables. Part of von Neumann's new way of defining these same probabilities in terms of projection operators was the recognition that physical quantities can be represented by self-adjoint operators acting in Hilbert space. With that recognition, the need to group quantities in pairs of conjugate variables evaporated: one no longer had to mind one's p's and q's (Duncan and Janssen 2013)

Von Neumann's Hilbert-space formalism also brought the definitive clarification of the relation between matrix mechanics and wave mechanics. The two theories correspond to different instantiations of Hilbert space. Wave mechanics works in the space of square-integrable complex functions; matrix mechanics in the space of square-summable complex sequences. Von Neumann (1927a) referred to theorems by Parseval, Fischer, and Riess—theorems mathematicians had known about for at least two decades—proving the isomorphism of these two infinite-dimensional complex vector spaces (Duncan and Janssen 2013, 238–39).

Von Neumann (1927a) submitted the paper in which he introduced his Hilbert-space formalism for quantum mechanics in May 1927. Six months later, he submitted another paper in which he distanced himself even further from Jordan's approach than he had in May. In Duncan and Janssen (2013, 187n39), we conjectured that this was in response to Heisenberg's uncertainty paper published in late March. In that paper, Heisenberg (1927) criticized Jordan's idea that quantum mechanics called for a modification of the basic rules of probability theory. In April (in his paper with Hilbert and Nordheim) and in May, von Neumann had endorsed Jordan's position (see the quotations above). In November, however, he unequivocally rejected it. One of the shortcomings of his earlier paper, he wrote, was that

> the relation to the ordinary probability calculus was not sufficiently clarified: the validity of its basic rules (addition and multiplication law of the probability calculus) was not sufficiently stressed. (von Neumann 1927b, 246)

The title of von Neumann's new paper accordingly promised a new "probability-theoretical construction" *(Wahrscheinlichkeitstheoretischer Aufbau)* of quantum mechanics. Von Neumann was familiar with the work on probability theory that Richard von Mises (1928) would publish in book form the following year. To define the probability that a particular property of a system has a particular value, von Neumann, following von Mises, imagined an ensemble of a large number of copies of the system and asked about the relative frequency with which a copy randomly drawn from this ensemble would have that value for that property. He introduced the as yet unknown function $\mathcal{E}(\dots)$ for the expectation value of a property in such ensembles. Assuming that properties are represented by self-adjoint operators acting in

Hilbert space and imposing some seemingly innocuous conditions on the function $\mathcal{E}(\ldots)$, von Neumann was able to derive a unique expression for it (Duncan and Janssen 2013, 247–50; Bub 2010; Dieks 2017).

In modern terms, von Neumann's formula for the expectation value of a property of a system, a property represented by some self-adjoint operator $\hat{A}$, in an ensemble of a great many copies of this system, an ensemble characterized by a density operator $\hat{\rho}$, can be written as

$$\mathcal{E}(\hat{A}) = \operatorname{Tr}(\hat{\rho}\,\hat{A}). \tag{44}$$

For a uniform ensemble consisting of identical copies of the system, the density operator, von Neumann showed, is just the projection operator $\hat{P}_{\psi}$ onto the unit vector $|\psi\rangle$ representing the state of all members of the ensemble. Inserting

$$\hat{\rho}_{\text{uniform}} = \hat{P}_{\psi} = |\psi\rangle\langle\psi| \tag{45}$$

for $\hat{\rho}$ in Equation (44), we recover the more familiar expression for the expectation value of the property represented by $\hat{A}$ in a system in the state $|\psi\rangle$,

$$\langle A \rangle_{\psi} = \operatorname{Tr}(|\psi\rangle\langle\psi|\hat{A}) = \sum_i \langle e_i|\psi\rangle\langle\psi|\hat{A}|e_i\rangle = \langle\psi|\hat{A}|\psi\rangle. \tag{46}$$

For a nonuniform ensemble, the density operator $\hat{\rho}$ is a weighted sum of projection operators $\hat{P}_{\psi_k}$ onto unit vectors in the set $\{|\psi_k\rangle\}$ representing the various states of the members of the ensemble:

$$\hat{\rho}_{\text{nonuniform}} = \sum_k \alpha_k \hat{P}_{\psi_k} = \sum_k \alpha_k |\psi_k\rangle\langle\psi_k|, \tag{47}$$

where the α_k's are real numbers such that $\sum_k \alpha_k = 1$. Inserting Equation (47) for $\hat{\rho}$ in Equation (44), we find that the expectation value in this nonuniform ensemble is given by

$$\langle A \rangle_{\{\psi_k\}} = \sum_k \alpha_k \operatorname{Tr}(|\psi_k\rangle\langle\psi_k|\hat{A}) = \sum_k \alpha_k \langle\psi_k|\hat{A}|\psi_k\rangle. \tag{48}$$

Von Neumann's dissatisfaction with Jordan's uncritical introduction of probabilities thus led him to the important distinction between uniform and nonuniform ensembles or, in modern terms, between *pure states* and *mixed states*. Thermal states are represented by mixed states in quantum mechanics. Before the end of the year, von Neumann (1927c), using his density operators to describe various ensembles, published yet another paper, the final installment of his 1927 trilogy, that helped lay the foundations for quantum-statistical mechanics.

JORDAN AND VON NEUMANN: ARCH OR SCAFFOLD? If we look at the sequence of general formalisms for quantum mechanics in *Neue Begründung* (Jordan 1927a, 1927b), *Mathematische Begründung* (von Neumann 1927a), and *Wahrscheinlichkeitstheoretischer Aufbau* (von Neumann 1927b), we can clearly see how important elements of the earlier formalisms were retained in the later ones while others were dropped (such as, for instance, the need to group quantities in pairs of conjugate variables). However, if we try to characterize this progression in terms of arches and scaffolds, it is not clear which version of the metaphor we should use. Was the earlier formalism used as a scaffold to facilitate the construction of the arch of the later formalism, or was the later formalism used as a scaffold to prevent the earlier arch from collapsing? We can tell the story using either version of the metaphor. The best way to tell it may be by mixing the two. In any event, this case calls for a loosening of the metaphor. No matter which formalism played the role of the scaffold and which one that of the arch, the fact remains that the scaffold was never taken down. Instead we are left with a composite of arch and scaffold.

V. THE ARCH-AND-SCAFFOLD METAPHOR AND EVOLUTIONARY BIOLOGY

In the introduction to his magnum opus, *The Structure of Evolutionary Theory*, Stephen Jay Gould (2002, 1–6) used an architectural metaphor to describe the development of evolutionary theory that fits nicely with the arch-and-scaffold metaphor, even though they work on different scales. Gould compared a sequence of closely related theories to a cathedral; I compare pairs of adjacent terms in such a sequence to arches and scaffolds.[52]

Gould took his metaphor from the Scottish geologist, botanist, and paleontologist Hugh Falconer. In an 1863 paper on Darwin's theory of descent with modification from a common ancestor through natural selection, Falconer suggested that the further development of the theory would end up resembling the building of the *Duomo* in Milan. This cathedral was built over several centuries and combines conflicting Gothic and baroque styles. Gould contrasted Falconer's view that Darwin had laid the foundations for a building bound to be built according to plans very different from Darwin's original ones with Darwin's own view, expressed in his response to Falconer that the latter would continue to govern the construction of the entire building (or, in Darwin's own terms, that the whole "framework will stand," not just the foundations).

In Gould's view, the actual development of evolutionary theory has been much closer to what Falconer than to what Darwin expected (Grantham 2004, 30). In other words, the way Gould saw it, contemporary evolutionary theory resembles the *Duomo* in just the way Falconer envisioned. Whittaker made a similar observation about the development of Maxwell's theory of electromagnetism. After discussing the elaboration of Maxwell's theory by J. J. Thomson, George Francis FitzGerald, Oliver Heaviside, Poynting, and others, he noted: "Maxwell's theory was now being developed in ways which could scarcely have been anticipated by its author. But although every year added something to the superstructure, the foundations remained much as Maxwell had laid them" (Whittaker [1951–1954] 1987, 1:318). My first stab at an analysis of the development of quantum theory in terms of arches and scaffolds in section IV suggests that similar observations can be made about quantum theory.

Talking about an early stage in the development of quantum theory, physicist and philosopher Henry Margenau used the same building metaphor as Falconer: "Bohr's atom sat like a baroque tower upon the Gothic base of classical electrodynamics" (Margenau 1950, 311; quoted in Lakatos 1970, 142). Unlike Falconer, however, he considered this "a malformation in the theory's architecture" (Margenau 1950, 311; quoted in Lakatos 1970, 142). In a lecture at Keio University in 1989, condensed-matter icon Philip W. Anderson also compared science to a cathedral but did so to emphasize science's beauty. After a brief sketch of various important contributions to physics that build on the 1957 paper by John Bardeen, Leon Cooper, and John Robert Schrieffer introducing the BCS theory of superconductivity named after them, Anderson (1994, 239) asked: "Where does the beauty reside?" The best answer he could come up with is that it resides in the network of citations connecting the relevant papers. He then added:

> Science has the almost unique property of collectively building a beautiful edifice: perhaps the best analogue is a medieval cathedral like Ely or Chartres . . . where many dedicated artists working with reference to each other's work jointly created a complex of beauty. (Anderson 1994, 239)

The same metaphor has been used to describe technological developments. Discussing the question "who deserves the most credit for inventing the internet" in his bestseller *The Innovators,* Walter Isaacson (2014, 260)[53] quotes pioneer Paul Baran "using a beautiful image that applies to all innovation":

> The process of technological development is like building a cathedral. Over the course of several hundred years new people come along and each lays down a block on top of the old foundations. (Hafner and Lyon, 1996, 79)

Another comparison of science to a cathedral can be found in the preface of a book on thermodynamics by Gilbert Lewis and Merle Randall (1923). Their image of the cathedral of science under construction is reminiscent of the factory that Duhem saw in Lodge's *Modern Views of Electricity* (see section III). Unlike Duhem, however, Lewis and Randall saw this as a good thing. The awe inspired by science's cathedrals should not get in the way of its day-to-day business.

> There are ancient cathedrals which, apart from their consecrated purpose, inspire solemnity and awe . . . The labor of generations of architects and artisans has been forgotten, the scaffolding erected for their toil has long since been removed, their mistakes have been erased, or have become hidden by the dust of centuries. Seeing only the perfection of the completed whole, we are impressed as by some superhuman agency. But sometimes we enter such an edifice that is still under construction; then the sound of hammers, the reek of tobacco, the trivial jests bandied from workman to workman, enable us to realize that these great structures are but the result of giving to ordinary human effort a direction and a purpose.
>
> Science has its cathedrals, built by the efforts of a few architects and of many workers. In these loftier monuments of scientific thought, a tradition has arisen whereby the friendly usages of colloquial speech give way to a certain severity and formality. While this may sometimes promote precise thinking, it more often results in the intimidation of the neophyte. Therefore, we have attempted, while conducting the reader through the classic edifice of thermodynamics into the workshops where construction is now in progress, to temper the customary severity of the science insofar as is compatible with clarity of thought. (Lewis and Randall 1923, vii)

In the preface of his textbook on special relativity, J. L. Synge similarly wrote: "My ambition has been to make [Minkowski] space-time a real workshop for physicists, and not a museum visited occasionally with a feeling of awe" (Synge [1955] 1972, vii).

After these examples of physicists comparing the development of their field to the building of cathedrals, I return to Gould's discussion of Falconer's use of the metaphor. After contrasting the different ways in which Darwin and Falconer expected the cathedral of evolutionary theory to be built, he noted parenthetically that

> no one has suggested the third alternative, often the fate of cathedrals— destruction, either total or partial, followed by a new building of contrary or oppositional form, erected over a different foundation. (Gould 2002, 6)

As I pointed out in section II, the original Waterloo Bridge did suffer the fate of Gould's third alternative, which corresponds to the metaphor Kuhn used in his Guggenheim application of "tearing down one habitable structure and rebuilding to a new plan." Neither the development of evolutionary theory nor the development of quantum and relativity theory fits this metaphor.

Since I brought up evolutionary theory, the question naturally arises how these architectural metaphors for theory change (arches and scaffolds, building a cathedral) relate to accounts of theory change modeled on biological evolution. Toward the end of *Structure,* Kuhn ([1962] 2012) used an analogy "that relates the evolution of organisms to the evolution of scientific ideas," albeit with the caveat that the analogy "can easily be pushed too far" (171). The evolutionary biology Kuhn had in mind was almost certainly the population genetics of the Modern Synthesis, which reigned supreme in the early 1960s (Bowler 2003, chapter 9).

The analogy between population genetics and cultural evolution is best known through the last chapter of Richard Dawkins's *The Selfish Gene,* in which selection of *memes,* units of culture, takes the place of selection of genes (Dawkins 1976, chapter 11). Dawkins does not apply his model for cultural evolution to science but gives no indication that it could not be applied there as well.

One key difference between the evolution of theories and the evolution of species, however, is that modifications of theories, unlike variations in species that form the input for natural selection, are anything but generated at random. Kuhn has little to say about where new theories come from,[54] and some of what he does say might give comfort to those tempted to push the analogy beyond its breaking point. Consider, for instance, the following passage in *Structure:* "The new paradigm, or a sufficient hint to permit later articulation, emerges all at once, sometimes in the middle of the night, in the mind of a man deeply immersed in crisis" (Kuhn [1962] 2012, 90). Combining statements such as these with Kuhn's emphasis on the proliferation of different articulations of a paradigm in a period of crisis— both in general (Kuhn [1962] 2012, chapter 7) and more specifically (Kuhn 1970, 257; "more and wilder versions of the old quantum theory than before")—one may come away with the impression that modifications of theories, not unlike variations in species, are typically generated in great profusion and in no particular direction and that the way in which modifications of theories compete for acceptance by a given scientific community is not dissimilar to the way variations in species compete for a given ecological niche.

Incidentally, in his critique of Lodge mentioned in section III, Poynting used the biological metaphor suggested by Kuhn toward the end of *Structure* to characterize the tradition of constructing mechanical models for the ether. Lodge, he wrote,

> uses the main idea of Maxwell's model [see Figure 4.6] but replaces Maxwell's duality of magnetic wheels and electric "idle" wheels by a duality of electric wheels. It is, perhaps, an open question whether this is really a simplification, but the attempt was well worth making, for it is only by variation and natural selection that the mechanical model will be suited to its environment in the electric world. (Poynting 1893, 635)

The random proliferation of ether models in great profusion suggested by this analogy hardly does justice to the development of ether theory by late-Victorian Maxwellians. The arch-and-scaffold metaphor fits this development much better. Lodge's ether was scaffolded by Maxwell's, as a superficial comparison of Figures 4.3 and 4.6 already suggests.

Gould was among those leading the charge against the hard-line version of the Modern Synthesis.[55] In his attack on this hardening orthodoxy, he used an architectural metaphor that has become more popular than the

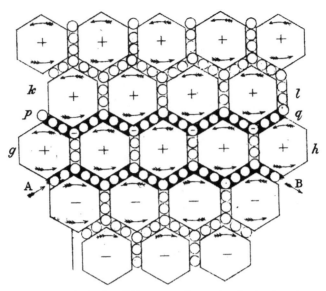

Figure 4.6. The "honeycomb ether" scaffolding Maxwell's equations for electric and magnetic fields (Maxwell 1861–1862, pt. II, plates following 488).

one by Falconer he unearthed. Gould's own metaphor involves the *Basilica di San Marco* in Venice rather than the *Duomo* in Milan (see Figure 4.7 below). In "The Spandrels of San Marco and the Panglossian Paradigm," Gould teamed up with Richard Lewontin—he who warned that the price of metaphor is eternal vigilance (see section II)—to offer, as they announced in the subtitle of their paper, "a critique of the adaptationist programme." *Panglossian* refers to Dr. Pangloss, Voltaire's caricature of Leibniz in *Candide*, who sees adaptation everywhere. In the abstract, the authors wrote:

> An adaptationist programme has dominated evolutionary thought in England and the United States during the past forty years. It is based on faith in the power of natural selection as an optimizing agent. It proceeds by breaking an organism into unitary "traits" and proposing an adaptive story for each considered separately ... We criticize this approach and attempt to reassert a competing notion (long popular in continental Europe) that organisms must be analyzed as integrated wholes, with *baupläne* so constrained by phyletic heritage, pathways of development, and general architecture that *the*

Figure 4.7. Spandrels in the *Basilica di San Marco*. Courtesy of the Procurator of San Marco.

constraints themselves become more interesting and more important in delim-
iting pathways of change than the selective force that may mediate change
when it occurs. (Gould and Lewontin 1979, 581 [my emphasis]; see also 593–94
as well as Gould 1980, 39–40)

Gould and Lewontin thus championed an approach to biological evolution
that de-emphasizes the agent of evolutionary change (natural selection) and
puts the emphasis on the role of constraints instead. In his bestseller *Your
Inner Fish*, Neil Shubin (2008) shows what an account of evolutionary change
along these lines looks like. Taking the same backward-looking perspective
that I argued we should strive for in arch-and-scaffold narratives (see section III), Shubin traces the evolution of various parts of the human body back
along our branch of the evolutionary tree. Although he clearly acknowledges
that natural selection is the mechanism that "mediate[s] change when it
occurs," I do not recall coming across the term *natural selection* even once
when reading his book and a search in an electronic version did not return
a single instance of the term. Shubin's emphasis is on Gould and Lewontin's

constraints. He even uses the term *scaffolding* at one point: "The scaffolding of our entire body originated in a surprisingly ancient place: single-celled animals" (Shubin 2008, 123).

A full-blown version of the approach advocated by Gould and Lewontin, which goes by the acronym *evo-devo* (for evolution and development), has become popular in biology (Sansom and Brandon 2007). Evo-devo fits much better with the arch-and-scaffold metaphor for the evolution and development of scientific theories than the population genetics of the Modern Synthesis. An arch-and-scaffold narrative for an episode of theory change highlights how structures in a later theory can be traced to structures in an earlier one. It brackets the question of how the new theory displaced the old one and instead focuses on how the new theory grew out of the old one. The parallels to Gould and Lewontin's "Spandrels of San Marco" or Shubin's *Your Inner Fish* should be obvious. They are interested in tracing structures in later species to structures in earlier species and less interested in spelling out the details of the selection process through which the former displaced the latter.

The concept of constraints can fruitfully be used both in evo devo–type accounts of biological evolution and in arch-and-scaffold-type accounts of theory change, even though the forces behind the constraints are different. With biological organisms, as with architectural structures, the constraints ultimately come from limitations of the malleability of (the arrangement of) the components out of which the organism or the architectural structure are made. These components and arrangements can only be tweaked so much before the creature ceases to be viable or the building collapses. Over time, natural selection can change one creature into a radically different creature, but it cannot get there in one fell swoop. In a mature science, it turns out, a theory that has proved its mettle by accounting for a wide array of empirical data can likewise only be tweaked so much before it ceases to be *empirically viable*. In this case, it is conceivable, of course, that an exceptionally imaginative scientist dreams up a radically different theory that accounts for an even more impressive array of data than the prevailing one. At first sight, it may even look as if that is essentially how relativity theory and quantum theory burst upon the scene. On closer examination (see section IV), the founding fathers of these theories arrived at them by tweaking existing theories under the tight constraints imposed by empirical viability. But whereas nature tweaks species at random, scientists tweak theories by design—or so one would hope! Because of this key difference, constraints play an even larger role in the evolution of theories than in the evolution of species. It is only natural then for modern historians of science looking for help from evolutionary biology in their studies of theory change to turn to Gould and evo-devo and to forget about Dawkins and population genetics. Of course, fifty years from now this new alliance might mainly show us that we were prisoners of our time, just as Kuhn was of his.

Be that as it may, it seems to me that the concept of constraints, as developed by Gould and others, has the potential to help us overcome the limitations of the arch-and-scaffold metaphor that we ran into at various junctures in section IV. In particular, it may help us articulate relations between the scaffolded and the scaffolding theory that do not naturally fit the basic architectural metaphor of arches and scaffolds. To illustrate this potential, I close this section with a first exploration of the possible applications of Gould's ideas about constraints to the evolution and development of scientific theories.

In a paper titled "The Evolutionary Biology of Constraint," written around the same time as "The Spandrels of San Marco," Gould distinguishes two kinds of constraints that the adaptationist tends to neglect:

> One is that the possible routes of selection are channeled by inherited morphology, building material, and the amount and nature of variation itself. Though selection moves organisms down the channels, the channels themselves . . . impose primary constraints on the direction of change. The second is that selection on one part of a structure may impose a set of correlated and nonadaptive modifications of other parts . . . Many features, even fundamental ones, may be nonadaptive (though not, to be sure, strongly unadaptive) either as developmental correlates of primary adaptations or as "unanticipated" structural consequences of primary adaptations themselves. (Gould 1980, 44)

The structural constraints most relevant to my arch-and-scaffold metaphor for theory change are of the first kind. The transition from the old quantum theory to matrix mechanics provides a nice example of this kind (cf. section IV). Consider perturbation theory in matrix mechanics developed in the famous *Dreimännerarbeit* (Born, Heisenberg, and Jordan 1926). Perfectly adapted to the task at hand, one might think that it was especially developed for matrix mechanics. It was not. The perturbation techniques of celestial mechanics had been transferred to atomic mechanics in the old quantum theory of Bohr and Sommerfeld. Heisenberg's (1925) *Umdeutung,* or "reinterpretation," of classical mechanics to what would become matrix mechanics essentially also dictated how these perturbation techniques had to be "reinterpreted." Recognition of this state of affairs helps us pinpoint what accounts for the continuity in the transition from the old quantum theory to modern quantum mechanics (cf. section II). Suman Seth (2010, 266) quotes Sommerfeld emphasizing that continuity in 1929: "The new development does not signify a revolution, but a joyful advancement of what was already in existence, with many fundamental clarifications and sharpenings." Seth focuses on a continuity of scientific *practice*. What made this continuity of practice possible, however, was a continuity of mathematical structure and technique (Midwinter and Janssen 2013, 146–47, 198).

The spandrels of San Marco from the title of Gould and Lewontin's article are constraints of the second kind—more specifically, "'unanticipated' structural consequences of primary adaptations." As they explain in the introduction of their paper,

> [s]pandrels—the tapering triangular spaces formed by the intersection of two rounded arches at right angles [see Figure 4.7]—are necessary architectural by-products of mounting a dome on rounded arches. Each spandrel contains a design admirably fitted into its tapering space. An evangelist sits in the upper part flanked by the heavenly cities . . . The design is so elaborate, harmonious, and purposeful that we are tempted to view it as the starting point of any analysis, as the cause in some sense of the surrounding architecture. But this would invert the proper path of analysis. The system begins with an architectural constraint: the necessary four spandrels and their tapering triangular form. (Gould and Lewontin 1979, 581–82, see also Gould 2002, 1249–53)

As a simple example of "spandrels" in biological evolution, Gould and Lewontin point to the tiny front legs of *Tyrannosaurus rex*. One could try to give an adaptationist account of this odd feature—maybe they developed "to help the animal rise from a lying position" (Gould and Lewontin 1979, 587)—but, given the homologies between *T. rex* and its ancestors (i.e., every bone in the skeleton of one corresponds to a bone in the skeleton of any other), it is more likely that it was "a developmental correlate of allometric fields for relative increase in head and hindlimb size" (Gould and Lewontin 1979, 587). Simply put: with a limited supply of bone material, for some parts to get bigger, other parts had to get smaller. The tiny front legs would then be an automatic by-product of evolution driven by constraints. As with the spandrels of San Marco, this seemingly useless feature was subsequently given some purpose.[56]

This simple example can be used to give a (rough) definition of the concept of a "spandrel" independently of its architectural origin. A spandrel is a feature that initially looks specifically designed for a particular purpose but that on closer examination is an inessential but inevitable by-product of a highly constrained development that was only subsequently given some purpose. Defined in this way, "spandrels" can also be recognized in instances of scientific theory change. I can think of at least one example in the history of quantum theory.

On first encountering Bohr's theory of the atom and the old quantum theory of Bohr and Sommerfeld that grew out of it, one might think that the notion of electron orbits is perfectly adapted to the job at hand—that is, the explanation of atomic spectra. Electron orbits represent the different energy states of electrons in atoms, and jumps between those energy states are associated with the spectral lines that were the main object of study in the old quantum theory. Electron orbits are so central to the old quantum theory that they came to dominate the theory's iconography (Schirrmacher 2009). Using Gould and Lewontin's metaphor, however, one can say that electron orbits were nothing but "spandrels" arising as by-products of the use of mathematical techniques borrowed from celestial mechanics in atomic physics. These techniques were very effective in determining the energy levels of an electron in an atom. It was therefore only natural that elements associated with these techniques got transferred along with them. Planets orbiting the sun in the solar system thus became electrons orbiting the nucleus in an atom. This element, it seemed, could be put to good use. Energy states of electrons were represented in the old quantum theory by definite electron orbits. This representation, however, turned out to be highly problematic and was abandoned in the transition from the old to the new quantum theory (see the fourth case study in section IV; Duncan and Janssen 2007, 2014, 2015).[57]

VI. THE ARCH-AND-SCAFFOLD METAPHOR AND SCIENTIFIC INNOVATION

In this chapter, I showed that a metaphor of arches and scaffolds can be used to capture both continuities and discontinuities in various episodes in the early history of special relativity, general relativity, and quantum theory. The metaphor thus helps stake out a middle ground between the traditional cumulative picture of theory change and the discontinuous picture of paradigm shifts made popular by Kuhn's *Structure* (though its author, as we saw in section II, vacillated between different and not necessarily compatible metaphors for theory change). In four of my five examples, I indicated how the narrative could be constructed backward in time, which, I argued, is the most effective defense against the obvious charge of Whiggishness against the metaphor (section III).

I identified two specific ways in which a scientist can get from the theory playing the role of the scaffold in the metaphor to the theory playing the role of the arch and gave concrete examples of each in the five case studies in section IV.

1. **Generalization.** A scientist recognizes that a structure exhibited by the scaffold for a special case has broader significance.

   1a. Einstein and Minkowski realized that the Lorentz invariance of Lorentz's theory of electromagnetism transcends its connection with electromagnetism and reflects a symmetry of a new space-time structure (first case study).

   1b. Laue developed relativistic continuum mechanics by stripping Abraham's electromagnetic mechanics of its electromagnetic particulars (second case study).

   1c. Heisenberg recognized that the way in which Kramers had used Bohr's correspondence principle to construct a new formula for optical dispersion could be generalized to construct a new framework for all of physics (fourth case study).

   1d. Von Neumann unified wave mechanics and matrix mechanics by showing that their mathematical formalisms are different instantiations of a more general formalism that he called Hilbert space (fifth case study).

2. **Substitution.** A scientist replaces the basic building blocks of the scaffold with new ones while leaving the structure built out of them

intact. An example of this in the evolution of technology would be the replacement of vacuum tubes by transistors in a logic board.

2a. Einstein arrived at equations within hailing distance of the Einstein field equations of general relativity by changing the definition of the gravitational field in the formalism he had developed around the older *Entwurf* field equations (third case study).[58]

2b. The central idea of Heisenberg's *Umdeutung* paper was to replace classical quantities by two-index quantum objects soon to be recognized as matrices without changing the relations between those quantities given by the laws of classical mechanics (fourth case study).

2c. One obtains the Hilbert-space incarnation of the Dirac-Jordan transformation theory by replacing transformation matrices by inner products of vectors in Hilbert space (fifth case study).

In this last example, however, I also noted that this is *not* how von Neumann introduced Hilbert space. Like John Stachel's (2007) Newstein fable (see note 35), this arch-and-scaffold story provided a counterfactual history that could be used as a foil for the actual history. The actual history in this case can also be captured in terms of the arch-and-scaffold metaphor. It is not clear, however, whether the best way to do so is in terms of a scaffold built *before* the arch, discarded when the arch could support itself, or in terms of a scaffold built *after* the arch, left in place to prevent the arch from collapsing. The former use of the metaphor is mine; the latter is a tweaked version of Hilbert's metaphor of building a house before laying its foundations (see section II). In my version of the metaphor, Jordan's transformation theory is the scaffold, and von Neumann's Hilbert space-formalism is the arch. In the Hilbert-inspired version, it is just the other way around. In this final case study, I ended up mixing these two metaphors in my attempt to characterize the relation between these two general frameworks for quantum mechanics.

This should serve as a reminder that the arch-and-scaffold metaphor is an expository device—a gimmick, some might say—not an analytical tool. As an expository device, it does useful work, as is perhaps best illustrated by the general-relativity example (Janssen and Renn 2015). Both Einstein himself and later commentators have suggested that he found the Einstein field equations in November 1915 by switching from physics to mathematics at the eleventh hour. The arch-and-scaffold metaphor helped counter this dramatic but highly misleading account by putting the alternative account, with Einstein doggedly pursuing his physics, in sharper relief.

At a more basic level, the arch-and-scaffold metaphor served to bring out common patterns in different instances of theory change that would have been much harder to spot without it. In the introduction, I broke down the metaphor into specific elements using a picture of the construction of the Waterloo Bridge (Figure 4.1). These elements worked well to draw special attention to certain features of the examples I presented in section IV. The metaphor of Minkowski providing the *springers* and Laue providing the *keystone* of the arch of special relativity helped underscore the importance of relativistic continuum mechanics. The metaphor of Kramers and Heisenberg using the correspondence principle as their *windlass* nicely brought out the way in which several physicists used this principle in the period right around *Umdeutung*. But breaking down the metaphor in this way should not be mistaken for turning it into a philosophical tool for further analysis, either of the general pattern or of the individual examples. It remains an expository device similar to the curatorial devices used in museum exhibits of dinosaurs (see section III).

For analytical tools we need to look elsewhere. As I suggested in section V, they may be found in evolutionary biology, not in the population genetics of the Modern Synthesis but in the more recent tradition known as evo-devo. The concept of constraints looks especially promising, but additional concepts will undoubtedly be needed. In the development of special relativity, for instance, we saw that the transition from scaffold to arch involved grouping various quantities defined in three-dimensional space into new quantities defined in four-dimensional space-time. Can analogous processes be identified in evolutionary biology? If so, can the concepts developed to deal with those processes be customized to deal with their possible analogues in the evolution of theories? Such concepts could then be used to bring features glimpsed through the lens of the arch-and-scaffold metaphor into sharper focus. In this way, my project could support broader efforts, already underway, to develop a new framework for cultural evolution, including the evolution of science, that draws on advances made in evolutionary biology over the past few decades (Caporael, Griesemer, and Wimsatt 2014; Laubichler and Renn 2015; Renn 2019). In the spirit of Hooke (see section III), I would be satisfied if the arch-and-scaffold metaphor were to help scaffold this new framework and were then thrown away like Wittgenstein's ladder.[59]

NOTES

1. *Acknowledgments.* I am grateful for helpful feedback from audiences in Washington, DC, Minneapolis, Pittsburgh, Jerusalem, Berlin, and Paris, where I gave talks on my arch-and-scaffold project between 2011 and 2016; participants in a summer school in Tübingen, August 2014; participants in the workshop "Beyond the Meme" at the University of Minnesota, October 2014; members of a reading group at the Center for Philosophy of Science at the University of Pittsburgh, fall 2015; and students in and visitors to the graduate seminar "The Evo-Devo of Theories" I cotaught with Mark Borrello at the University of Minnesota, fall 2016. I am indebted to Tony Duncan, Anne Kox, John Norton, and Jürgen Renn, with whom I have worked over the years on the episodes in the history of relativity and quantum theory from which I drew my examples of theory change. I thank Rich Bellon, Victor Bo-antza, Paul Brinkman, Tony Duncan, Sam Fletcher, Clayton Gearhart, Marco Giovanelli, Cameron Lazaroff-Puck, Joe Martin, David Rowe, Jim Smoak, and Andy Zangwill for supplying me with examples of scientists, mathematicians, and historians using the term *scaffold* or related terms to describe theory development. I thank John Eade for establishing the provenance of the drawing I used for Figure 4.1 and Pietro Omodeo for helping me obtain a picture of the spandrels of San Marco (see Figure 4.7). Bill Wimsatt first suggested to me that the pattern of theory change that I am trying to capture with my metaphor fits with the general model of cultural evolution that he and his collaborators have been trying to work out, drawing on ideas from evo-devo in biology. Jim Griesemer, one of these collaborators, made me see that evo-devo, rather than population genetics, makes for a fruitful comparison between evolutionary biology and theory development in science. Jim thus took me "beyond the meme." I have benefited from further discussion of this comparison with Mark Borrello, Max Dresow, Manfred Laubichler, Jürgen Renn, and Gunter Wagner. I gratefully acknowledge support for work on this project from the *Max-Planck-Institut für Wissenschaftsgeschichte* and the *Alexander von Humboldt Stiftung*. Special thanks to Jim Smoak, veteran of the Vietnam War, for his heroic efforts in checking the page proofs for this paper.

2. This statement comes from a section of the application called "Plans for Research," which is included in the appendix of an article on Kuhn's education and early career by Karl Hufbauer (2012).

3. John Earman (1993) shows that Carnap and Kuhn have much more in common than these two quotations suggest. In line with what I will argue

about the development of modern physics, Earman (1993, 9) sees evolution rather than revolution in going from Carnap to Kuhn.

4. In February 1947, responding to the hype in various newspaper reports about a new theory by Erwin Schrödinger, Albert Einstein released a press statement saying that "the reader gets the impression that every five minutes there is a revolution in science, somewhat like the *coups d'état* in some of the smaller unstable republics" (Klein 1975, 113). See also the chapter on Einstein in Cohen (1985, chapter 28, 435–45).

5. The origin of this warning remains unclear (like the original about liberty rather than metaphor, which is often but wrongly attributed to Thomas Jefferson). Lewontin put it in quotation marks but did not give a source. In a book review decades later, Lewontin (2001) wrote: "As Arturo Rosenblueth and Norbert Wiener once noted, 'The price of metaphor is eternal vigilance.'" This may be why the warning is often attributed to Rosenblueth and Wiener (1945), which is cited in Lewontin (1963), though not for this warning, which is nowhere to be found in it. I am grateful to Kris Fowler for her help in trying to track down the source of this warning.

6. See Klein, Shimony, and Pinch (1979, 437).

7. See Kuhn (1984, 363), reprinted as a new afterword in the second edition of Kuhn (1978).

8. See also an unpublished essay on the "crisis of the old quantum theory" (Kuhn 1966) and the videotape of a 1980 lecture at Harvard based on this essay. In the proceedings of the 1965 London conference, Kuhn (1970, 258) wrote: "History of science, to my knowledge, offers no equally clear, detailed, and cogent example of the creative functions of normal science and crisis." In the Q&A following his 1980 lecture at Harvard, he reiterated that the crisis of the old quantum theory is "a textbook example . . . as described in *Structure*," adding: "I don't think there are many if any that are that good" (transcribed from the videotape of the lecture). In his interviews in the early 1960s with surviving members of the first generation of quantum physicists for the *Archive for History of Quantum Physics* (AHQP) (Kuhn et al. 1967), Kuhn routinely asked his subjects (leading) questions about their awareness of this crisis at the time (Seth 2010, 265).

9. See also Renn and Rynasiewicz (2014, 38) and a more programmatic earlier paper, Renn (1993, 312–13).

10. See, e.g., Duncan and Janssen (2007, 2013, 2015); Joas and Lehner (2009); Seth (2010); Midwinter and Janssen (2013); James and Joas (2015); Jähnert (2016); Jordi Taltavull (2017); Blum et al. (2017).

11. A new history of quantum mechanics of which I am a coauthor (Duncan and Janssen, in preparation) will also make use of this metaphor as is reflected in the subtitles of its two volumes. Following Kuhn's example, however, we largely refrain from explicitly using the metaphor in the text (cf. note 7).

12. I am grateful to John Eade, who maintains a website about the Thames, for drawing my attention to these bridges.

13. Neurath may have drawn inspiration from another ship metaphor: Does the ship of Theseus remain the same when all its parts are replaced? For discussion of this conundrum in the philosophy of identity, see, e.g., Pesic (2002, 15–23). I am grateful to Alexander Greff for this suggestion.

14. See Rabossi (2003, section II, 176–78) for a discussion of how W. V. O. Quine used Neurath's ship metaphor in several places (e.g., Quine 1960, 3–4) and combined it with his own metaphor of a "web of belief." In the paragraph that ends with the ship metaphor in his book against Spengler, Neurath (1921, 198–99) actually uses language suggestive of Quine's "web of belief" ("We always have to do with a whole network of concepts"), and the ship metaphor is introduced as a metaphor for the kind of holism found in Duhem, whom Neurath explicitly mentions at this point.

15. This oft-repeated but never properly sourced comparison is attributed to Niels Henrik Abel in some versions of the story and to Carl Gustav Jacob Jacobi in others.

16. See Michael Gordin's (2014) review of Chang (2012) for some interesting musings on Whiggishness, anti-Whiggishness, and anti-anti-Whiggishness.

17. See Midwinter and Janssen (2013, 162–63) for further discussion of this passage.

18. Ofer Gal (2002) used Hooke's terms as the title for a book on Hooke and Newton.

19. Einstein (1917b, 91) gives two examples: electrostatics and Maxwellian electrodynamics and special and general relativity.

20. For discussion of this model and Lodge's book, see Hunt (1991, 87–95). Figure 4.3 is reprinted as Figure 4.7 on p. 92 of Hunt's book.

21. See, e.g., the episode "The Betrayal" of the sitcom *Seinfeld,* which first aired November 20, 1997.

22. Cameron Lazaroff-Puck (2015) has shown that the characterization of the relation between Maxwell's 1864–1865 and 1861–1862 papers by Whittaker and Kargon is misleading, but even on Lazaroff-Puck's alternative

account, the relation between the two can still be captured in terms of arches and scaffolds.

23. I am grateful to Paul Brinkman, a leading historian of vertebrate paleontology, for helping me develop this analogy. One could develop a similar one about ancient sculptures.

24. For a discussion of composite dinosaur displays and the metal armatures used to support them, see Brinkman (2010, especially 237–46).

25. See my home page for links to the slides of my lectures at a summer school in Tübingen in 2014 on all five examples and to the papers on which these lectures and section IV are based.

26. Based on Janssen (1995, 2002, 2009, 2017).

27. Jon Dorling (1976) showed how, in principle, Euclid could have arrived at Minkowski space-time by dropping one of the axioms of his geometry (Janssen 2009, 49). Dorling's analysis beautifully brings out the relation between Euclidean geometry and the pseudo-Euclidean geometry of Minkowski space-time. At the same time, it serves as a reductio of the notion that special relativity could have arisen prior to the development of electrodynamics in the nineteenth century.

28. The German original has *Treppenwitz*, which is based on the French idiom *l'esprit de l'escalier*, meaning "thinking of the perfect retort too late." Here is an example in honor of singer-songwriter Glenn Frey (1948–2016), cofounder of the Eagles. On a flight from San Diego to Minneapolis in February 2013, the pilot told the passengers over the intercom that we could see Winslow, Arizona, from the plane. Upon arrival, the pilot joined the flight attendants saying their goodbyes as we got off the plane. I was already in the terminal when I realized what I should have said to him: "Take it easy."

29. In the German original, it is unambiguous that *which* [*der*] refers to *nucleus* [*der Kern*] rather than to *image of the world* [*das Weltbild*].

30. Based on Janssen and Mecklenburg (2007) and Janssen (2009).

31. See, e.g., Miller ([1981] 1988, sections 1.8–1.14, 7.4, and 12.4) and Kragh (1999, chapter 8, "A revolution that failed").

32. When Equation (1) is integrated over all of space, the second term on the right-hand side vanishes (as long as $T^{ij}_{\text{self}_{\text{EM}}}$ drops off fast enough if we go to infinity), and what is left can be written in vector form as

$$\mathbf{F}_{\text{self}_{\text{EM}}} = -\frac{d\mathbf{P}_{\text{self}_{\text{EM}}}}{dt}.$$

The total force on the charge distribution is the sum of this force and the force $\mathbf{F}_{\text{ext}_{EM}}$ coming from the external field. Since the Newtonian mass m_N of the charge distribution is assumed to be zero, it follows from Newton's second law, $\mathbf{F}_{\text{tot}} = m_N\mathbf{a}$, that the total force, $\mathbf{F}_{\text{tot}} = \mathbf{F}_{\text{self}_{EM}} + \mathbf{F}_{\text{ext}_{EM}}$, vanishes. Using the expression for $\mathbf{F}_{\text{self}_{EM}}$ above, we then find that

$$\mathbf{F}_{\text{ext}_{EM}} = \frac{d\mathbf{P}_{\text{self}_{EM}}}{dt},$$

which has the same form as the Newtonian law, $\mathbf{F} = d\mathbf{p}/dt = m_N\mathbf{a}$ (where we used that momentum is the product of mass and velocity, $\mathbf{p} = m_N\mathbf{v}$, and that acceleration is the time derivative of velocity, $\mathbf{a} = d\mathbf{v}/dt$). This, then, is how Newton's second law is recovered in Abraham's electromagnetic mechanics (Janssen and Mecklenburg 2007, 108–10).

33. The energy-momentum tensor is sometimes called the stress-energy tensor or the stress-energy-momentum tensor. As Joe Martin once observed (private communication), it is the Crosby, Stills, Nash, and Young of the tensors.

34. Based on Janssen and Renn (2007). See also Janssen (2005) and Renn (2006). In an article in *Physics Today* to mark the centenary of general relativity, the two of us explicitly used the arch-and-scaffold metaphor to tell the story of how Einstein found the field equations of general relativity (Janssen and Renn 2015). An expanded version of this article will serve as the introduction of a sourcebook we are preparing on the subject (Janssen and Renn 2019).

35. Although I will not attempt to do so here, the early phase of the development of general relativity can also be captured quite naturally in terms of arches and scaffolds. Einstein essentially generalized the metric field of a flat Minkowski space-time to the metric field of curved space-times, identifying *paths of extremal length* as the trajectories of free-falling bodies. Such extremal paths are called *metric geodesics* to distinguish them from *straightest paths,* which are called *affine geodesics.* In the pseudo-Riemannian geometry of general relativity (*pseudo* in the same sense that the geometry of Minkowski space-time is pseudo-Euclidean), metric and affine geodesics coincide. The concept of an *affine connection* used to characterize affine geodesics was only introduced a few years after Einstein completed general relativity by the mathematicians Gerhard Hessenberg, Tullio Levi-Civita, and Hermann Weyl. John Stachel (2007) has written a counterfactual history of general relativity in which a fictitious nineteenth-century mathematician,

Weylmann (a composite of Weyl and Grossmann), rewrote Newton's gravitational theory in terms of an affine connection in Newtonian space-time (a reformulation actually provided in 1923 by the French mathematician Élie Cartan), which a fictitious early twentieth-century physicist, Newstein (a composite of Newton and Einstein), then reworked in a relativistic space-time. Stachel's counterfactual history, which helps put various aspects of the actual history in sharp relief, can be recast in terms of an arch (general relativity) built on a scaffold (Newton-Cartan theory). A clear exposition of the mathematics needed for such a recasting can be found in Fletcher (2017).

36. Hilbert had no such compunctions. He was, metaphorically speaking (see section II), ready to move into new dwellings without checking the foundations first. In the fall of 1915, as I mentioned in section III, Einstein and Hilbert found themselves in a race for the field equations (Janssen and Renn 2015). In a letter to Sommerfeld of November 28, 1915, Einstein gave a detailed account of his route to these equations. That he did so in a letter to Sommerfeld, who knew both Einstein and Hilbert well, was probably at least in part to secure his priority. "It is easy," Einstein told Sommerfeld, clearly referring to Hilbert, "to write down these generally-covariant field equations but difficult to see that they are a generalization of the Poisson equation [of Newtonian theory] and not easy to see that they satisfy the conservation laws" (Einstein 1987–2018, vol. 8, doc. 153).

37. A similar mechanism can be discerned in Planck's attempts to find the formula for the spectral distribution of blackbody radiation. Blackbody radiation is an ideal kind of heat radiation. The formula for its spectral distribution should tell us how much energy is emitted at each frequency given the temperature of the emitting body. In the late 1890s, Planck developed a framework that allowed him to derive this formula from an expression for the entropy of a *resonator* (which can be thought of as a charge on a spring with a particular resonance frequency) in interaction with the radiation. Initially, Planck (1900a) convinced himself that the second law of thermodynamics uniquely determines the expression for this entropy, which when inserted into his general formalism gives the formula for the spectral distribution of blackbody radiation proposed in 1896 by Wien. When the empirical adequacy of the Wien law was called into question shortly thereafter, Planck (1900b) discovered that this uniqueness claim was in error. The second law of thermodynamics is compatible with a range of expressions for

the entropy of his resonators. A few months later, Planck (1900c) used this wiggle room to cook up a new expression for resonator entropy, which when inserted into the same general formalism gives a new formula for the spectral distribution of blackbody radiation. This Planck law, as it came to be called, was in excellent agreement with the experimental data. Planck now found himself in the same predicament as Millikan a decade and a half later (see section II). His new formula for blackbody radiation, like Einstein's formula for the photoelectric effect, stood "complete and apparently well tested, but without any visible means of support" (Millikan 1917, 230). Planck immediately set out to find such support. Supplying a derivation of the expression for the entropy of his resonators that led to his new formula for blackbody radiation, Planck (1900d, 1901) took the first steps toward quantizing the energy of these resonators (Kuhn 1978).

38. In 1917, Einstein added another term with the infamous cosmological constant (Janssen 2014).

39. My home page has links to the video and the slides of the version presented at the symposium "General Relativity at 100" at the Institute for Advanced Studies in Princeton in 2015.

40. See the introduction of Janssen and Renn (2019) for such a comparison.

41. Lehmkuhl (2014, 317) cites a passage from a 1926 letter to Hans Reichenbach, in which Einstein uses language that is suggestive of the arch-and-scaffold metaphor: "It is wrong to think that 'geometrization' is something essential. It is only a kind of crutch [*Eselsbrücke*] for the discovery of numerical laws" (Einstein, 1987–2018, vol. 15, doc. 249).

42. Based on Duncan and Janssen (2007). For a concise version of this story, see Midwinter and Janssen (2013, 156–62).

43. It is beyond the scope of this paper to evaluate the argument of Blum et al. (2017). I will just note that *time* is on their side. The Kramers-Heisenberg paper on dispersion theory was written around Christmas 1924. An important letter from Heisenberg to Ralph Kronig, documenting key steps toward the *Umdeutung* paper, was not written until early June 1925. It is implausible, on the face of it, that nothing of consequence would have happened between January and June. Heisenberg may thus already have been rewriting history when he gave dispersion pride of place in his *Umdeutung* paper (just as Einstein, as we saw in the preceding case study, was already rewriting history when he suggested in November 1915 that an

eleventh-hour switch from physics to mathematics had led him to the field equations of general relativity). This illustrates the residual dangers of the benign Whiggishness still lurking in my use of the arch-and-scaffold metaphor (see section III).

44. For the history of dispersion theory in the period of interest here, roughly from 1870 to 1925, see Jordi Taltavull (2017).

45. Duncan and Janssen (2007) and Midwinter and Janssen (2013) focus on these particular correspondence-principle arguments. See Rynasiewicz (2015) and Jähnert (2016) for broader accounts of the correspondence principle and its evolution.

46. Quoted and discussed in Duncan and Janssen (2007, section 3.5, 593–97; see also section 4.3, 613–17).

47. This is the key to the resolution of the paradoxical statement by Dirac quoted in section II.

48. An instructive application of the *Umdeutung* strategy outlined in these last two paragraphs is Jordan's derivation of a formula for the mean-square fluctuation of the energy in blackbody radiation in Born, Heisenberg, and Jordan (1926), the sequel to Heisenberg (1925) and Born and Jordan (1925), known as the *Dreimännerarbeit*. For a detailed reconstruction of Jordan's derivation, see Duncan and Janssen (2008).

49. Based on Duncan and Janssen (2009, 2013).

50. I also suppress Jordan's notion of a "supplementary amplitude" [*Ergänzungsamplitude*] (Duncan and Janssen 2013, 189 and section 2.4, 217–21).

51. The "supplementary amplitude" (see note 50) was an (unsuccessful) attempt to avoid such restrictions.

52. Norton (2014, 685–87) uses the arch-and-scaffold and cathedral metaphors on an even larger scale to capture the construction of the totality of our empirical knowledge (cf. section III).

53. Isaacson's overall account of the digital revolution fits nicely with the accounts of the relativity and quantum "revolutions" sketched in this paper. This is obscured by the unfortunate choice of the book's subtitle, the negation of which would actually have provided a more accurate characterization of its contents: "how [it was not just] a group of hackers, geniuses, and geeks [who] created the digital revolution."

54. The same can be said about the bold conjectures of Popperian falsificationism or the hypotheses tested according to the rules of hypothetico-deductivism or Bayesianism.

55. Gould's dissatisfaction with the dominant selectionist paradigm was fueled by outrage over its shoddy applications to human populations (Gould 1981; cf. Kevles 1985, 284).

56. Many more and much better examples can be found in Shubin (2008).

57. Quantum mechanics, as it is taught and practiced today, may provide another example of a spandrel. In the old quantum theory, the spandrel was the misleading visualization of energy levels as electron orbits that was imported into the theory along with the mathematical techniques Schwarzschild, Sommerfeld, and others borrowed from celestial mechanics. One could argue (see, e.g., Bub 2019) that wave mechanics, which remains a popular form of quantum mechanics, likewise provides a misleading visualization of quantum states as wave functions, which was imported into the theory along with the mathematical techniques Schrödinger borrowed from wave optics and analytical mechanics to develop his optical-mechanical analogy (Joas and Lehner 2009).

58. I briefly described a similar example in Planck's work on blackbody radiation (see note 37).

59. The arch-and-scaffold metaphor will also have done its job if it helps put to rest the question of whether science develops continuously or discontinuously. That question, in the end, only distracts from a more fundamental one: Where is scientific novelty coming from? This is where population genetics seems to have seriously tripped up Kuhn. Evo-devo should provide a much better guide in the search for answers to this question.

REFERENCES

Aaserud, Finn, and Helge Kragh, eds. 2015. *One Hundred Years of the Bohr Atom.* Copenhagen: Det Kongelige Danske Videnskabernes Selskab.

Abraham, Max. 1903. "Prinzipien der Dynamik des Elektron." *Annalen der Physik* 10:105–79.

Agassi, Joseph. 1975. *Science in Flux.* Dordrecht: Reidel.

Agassi, Joseph. 2013. *The Very Idea of Modern Science: Francis Bacon and Robert Boyle.* Dordrecht: Springer.

Aitchison, Ian J. R., David A. MacManus, and Thomas M. Snyder. 2004. "Understanding Heisenberg's 'Magical' Paper of July 1925: A New Look at the Calculational Details." *American Journal of Physics* 72:1370–79.

Anderson, Philip W. 1994. "Some Ideas on the Aesthetics of Science." In the 1st ed. of *A Career in Theoretical Physics,* 570–83. Singapore: World Scientific. (698–711 in 2nd ed., 2004). Reprinted in *Nishina Memorial Lectures. Creators of Modern Physics*, 235–43. Berlin: Springer, 2008. Page references are to the Springer edition.

Bacon, Francis. 1620. *Novum Organum*. Part 2 of *Francisci de Verulamio, Summi Angliae Cancellarii, Instauratio magna*. London: [Bonham Norton and] Ioannem Billium typographum regium. Page references are to *The New Organon*, edited by Lisa Jardine and Michael Silverthorne. Cambridge: Cambridge University Press, 2000.

Badino, Massimiliano, and Jaume Navarro, eds. 2013. *Research and Pedagogy: A History of Early Quantum Physics through Its Textbooks*. Berlin: Edition Open Access.

Beutler, Ernst, ed. 1949. *Johann Wolfgang von Goethe: Gedenkausgabe der Werke, Briefe, und Gespräche*. 24 vols. Zurich: Artemis.

Blum, Alexander, Martin Jähnert, Christoph Lehner, and Jürgen Renn. 2017. "Translation as Heuristics: Heisenberg's Turn to Matrix Mechanics." *Studies in History and Philosophy of Modern Physics* 60:3–22.

Bohr, Niels. 1913. "On the Constitution of Atoms and Molecules (Part I)." *Philosophical Magazine* 26:1–25.

Bohr, Niels, Hendrik Anthony Kramers, and John Clarke Slater. 1924. "The Quantum Theory of Radiation." *Philosophical Magazine* 47:785–802. Reprinted in van der Waerden 1968, 159–76.

Born, Max. 1924. "Über Quantenmechanik." *Zeitschrift für Physik* 26:379–95. English translation in van der Waerden 1968, 181–98.

Born, Max. 1925. *Vorlesungen über Atommechanik*. Berlin: Springer.

Born, Max. 1926a. "Zur Quantenmechanik der Stoßvorgänge. Vorläufige Mitteilung." *Zeitschrift für Physik* 37:863–67.

Born, Max. 1926b. "Zur Quantenmechanik der Stoßvorgänge." *Zeitschrift für Physik* 38:803–27.

Born, Max. 1927. *The Mechanics of the Atom*. London: Bell. Translation of Born 1925.

Born, Max, Werner Heisenberg, and Pascual Jordan. 1926. "Zur Quantenmechanik II." *Zeitschrift für Physik* 35:557–615. English translation in van der Waerden 1968, 321–85.

Born, Max, and Pascual Jordan. 1925. "Zur Quantenmechanik." *Zeitschrift für Physik* 34:858–88. English translation (except for chapter 4) in van der Waerden 1968, 277–306.

Born, Max, and Pascual Jordan. 1930. *Elementare Quantenmechanik*. Berlin: Springer.

Borrelli, Arianna. 2010. "Dirac's Bra-ket Notation and the Notion of a Quantum State." In *Styles of Thinking in Science and Technology*, edited by Hermann

Hunger, Felicitas Seebacher, and Gerhard Holzer, 361–71. Vienna: Verlag der Österreichischen Akademie der Wissenschaften.

Bowler, Peter J. 2003. *Evolution: The History of an Idea.* 3rd ed. Berkeley: University of California Press.

Brinkman, Paul D. 2010. *The Second Jurassic Dinosaur Rush: Museums and Paleontology in America at the Turn of the Twentieth Century.* Chicago: University of Chicago Press.

Brinkman, Paul D. 2013. "Red Deer River Shakedown: A History of the Captain Marshall Field Paleontological Expedition to Canada, 1922." *Earth Sciences History* 32:204–34.

Bub, Jeffrey. 2010. "Von Neumann's 'No Hidden Variables' Proof: A Re-appraisal." *Foundations of Physics* 40:1333–40.

Bub, Jeffrey. 2019. "Interpreting the Quantum World: Old Questions and New Answers." *Historical Studies in the Natural Sciences.* Forthcoming.

Cairns-Smith, Alexander Graham. 1985. *Seven Clues to the Origin of Life: A Scientific Detective Story.* Cambridge: Cambridge University Press.

Caporael, Linnda R., James R. Griesemer, and William C. Wimsatt. 2014. *Developing Scaffolds in Evolution, Culture, and Cognition.* Cambridge, Mass.: MIT Press.

Carnap, Rudolf. 1928. *Der logische Aufbau der Welt.* Frankfurt: Meiner. Page reference is to the English translation, *The Logical Structure of the World,* Berkeley: University of California Press, 1967.

Casimir, Hendrik B. G. 1973. "Kramers, Hendrik Anthony." In Vol. 7 of *Dictionary of Scientific Biography,* edited by Charles Coulston Gillispie, 491–94. New York: Charles Scribner's Sons.

Casimir, Hendrik B. G. 1983. *Haphazard Reality: Half a Century of Science.* New York: Harper and Row.

Chang, Hasok. 2012. *Is Water H_2O? Evidence, Realism, and Pluralism.* Dordrecht: Springer.

Cohen, I. Bernard. 1985. *Revolution in Science.* Cambridge, Mass.: Harvard University Press.

Cope, Edward Drinker. 1895. "Professor Brooks on Consciousness and Volition." *Science,* new series, vol. 2, no. 42 (October 18): 521–22.

Corry, Leo. 1999. "Hilbert and Physics (1900–1915)." In *The Symbolic Universe: Geometry and Physics 1890–1930,* edited by Jeremy J. Gray, 145–88. Oxford: Oxford University Press.

Corry, Leo. 2004. *David Hilbert and the Axiomatization of Physics (1898–1918).* Dordrecht: Kluwer.

Darrigol, Olivier. 1992. *From c-Numbers to q-Numbers: The Classical Analogy in the History of Quantum Theory.* Berkeley: University of California Press.

Davy, Humphry. 1820. "Address of the President on Taking the Chair of the Royal Society, for the First Time; December 7th, 1820.—On the Present State of That Body, and on the Prospects and Progress of Science." Reprinted in *The Collected Works of Humphry Davy,* edited by John Davy, vol. 7, *Discourses Delivered before the Royal Society; and Agricultural Lectures,* part I, 5-15. London: Smith, Elder, and Cornhill, 1840. Page reference is to the Smith, Elder edition.

Dawkins, Richard. 1976. *The Selfish Gene.* Oxford: Oxford University Press.

Dieks, Dennis. 2017. "Von Neumann's Impossibility Proof: Mathematics in the Service of Rhetorics." *Studies in History and Philosophy of Modern Physics* 60:136-48.

Dirac, Paul Adrien Maurice. 1927. "The Physical Interpretation of the Quantum Dynamics." *Proceedings of the Royal Society of London: Series A* 113:621-41.

Dirac, Paul Adrien Maurice. 1930. *Principles of Quantum Mechanics.* Oxford: Clarendon.

Dorling, Jon. 1976. *Special Relativity out of Euclidean Geometry.* Unpublished manuscript.

Dresden, Max. 1987. *H. A. Kramers: Between Tradition and Revolution.* New York: Springer.

Duhem, Pierre. 1914. *La Théorie Physique: Son Objet, Sa Structure.* Paris: Marcel Rivière and Cie. Page reference is to the English translation, *The Aim and Structure of Physical Theory,* Princeton, N.J.: Princeton University Press, 1954.

Duncan, Anthony, and Michel Janssen. 2007. "On the Verge of *Umdeutung* in Minnesota: Van Vleck and the Correspondence Principle." 2 pts. *Archive for History of Exact Sciences* 61:553-624, 625-71.

Duncan, Anthony, and Michel Janssen. 2008. "Pascual Jordan's Resolution of the Conundrum of the Wave-Particle Duality of Light." *Studies in History and Philosophy of Modern Physics* 39:634-66.

Duncan, Anthony, and Michel Janssen. 2009. "From Canonical Transformations to Transformation Theory, 1926-1927: The Road to Jordan's *Neue Begründung.*" *Studies in History and Philosophy of Modern Physics* 40:352-62.

Duncan, Anthony, and Michel Janssen. 2013. "(Never) Mind Your p's and q's: Von Neumann versus Jordan on the Foundations of Quantum Theory." *European Physical Journal H* 38:175-259.

Duncan, Anthony, and Michel Janssen. 2014. "The Trouble with Orbits: The Stark Effect in the Old and the New Quantum Theory." *Studies in History and Philosophy of Modern Physics* 48:68–83.

Duncan, Anthony, and Michel Janssen. 2015. "The Stark Effect in the Bohr-Sommerfeld Theory and in Schrödinger's Wave Mechanics." In Aaserud and Kragh 2015, 217–71.

Duncan, Anthony, and Michel Janssen. (In preparation). *Constructing Quantum Mechanics*. Vol. 1. *The Scaffold, 1900–1923* (expected: 2019); Vol. 2. *The Arch, 1923–1927* (expected: 2022).

Earman, John. 1993. "Carnap, Kuhn, and the Philosophy of Scientific Methodology." In *World Changes: Thomas Kuhn and the Nature of Science,* edited by Paul Horwich, 9–36. Cambridge, Mass.: MIT Press.

Eckert, Michael. 2013. "Sommerfeld's *Atombau und Spektrallinien.*" In Badino and Navarro 2013, 117–35.

Eckert, Michael. 2014. "How Sommerfeld Extended Bohr's Model of the Atom." *European Physical Journal H* 39:141–56.

Eddington, Arthur S. 1928. *The Nature of the Physical World.* New York: MacMillan.

Einstein, Albert. 1905. "Zur Elektrodynamik bewegter Körper." *Annalen der Physik* 17:891–921. English translation in Einstein et al. 1952, 37–65; Stachel 2005, 123–59.

Einstein, Albert. 1914. "Die formale Grundlage der allgemeinen Relativitätstheorie." *Sitzungsberichte der Königlich Preußische Akademie der Wissenschaften.* Berlin, 1030–85. Reprinted in Einstein 1987–2018, vol. 6, doc. 9.

Einstein, Albert. 1915a. "Zur allgemeinen Relativitätstheorie." *Sitzungsberichte der Königlich Preußische Akademie der Wissenschaften.* Berlin, 778–86. Reprinted in Einstein 1987–2018, vol. 6, doc. 21.

Einstein, Albert. 1915b. "Zur allgemeinen Relativitätstheorie. (Nachtrag)." *Sitzungsberichte der Königlich Preußische Akademie der Wissenschaften.* Berlin, 799–801. Reprinted in Einstein 1987–2018, vol. 6, doc. 22.

Einstein, Albert. 1915c. "Erklärung der Perihelbewegung des Merkur aus der allgemeinen Relativitätstheorie." *Sitzungsberichte der Königlich Preußische Akademie der Wissenschaften.* Berlin, 831–39. Reprinted in Einstein 1987–2018, vol. 6, doc. 23.

Einstein, Albert. 1915d. "Die Feldgleichungen der Gravitation." *Sitzungsberichte der Königlich Preußische Akademie der Wissenschaften.* Berlin, 844–47. Reprinted in Einstein 1987–2018, vol. 6, doc. 25.

Einstein, Albert. 1916a. "Die Grundlage der allgemeinen Relativitätstheorie." *Annalen der Physik* 49:769–822. Reprinted in Einstein 1987–2018, vol. 6, doc. 30. English translation in Einstein et al. 1952, 111–64.

Einstein, Albert. 1916b. "Hamiltonsches Prinzip und allgemeine Relativitätstheorie." *Sitzungsberichte der Königlich Preußische Akademie der Wissenschaften.* Berlin, 1111–16. Reprinted in Einstein 1987–2018, vol. 6, doc. 41. English translation in Einstein et al. 1952, 167–73.

Einstein, Albert. 1917a. "Zur Quantentheorie der Strahlung." *Physikalische Zeitschrift* 18:121–128. Reprint from *Physikalische Gesellschaft Zürich. Mitteilungen* 18:47–62, reprinted in Einstein 1987–2018, vol. 6, doc. 38. English translation in van der Waerden 1968, 63–77.

Einstein, Albert. 1917b. *Über die spezielle und allgemeine Relativitätstheorie.* Braunschweig: Vieweg. Reprinted in Einstein 1987–2018, vol. 6, doc. 42. Page reference is to *Relativity: The Special and The General Theory. 100th Anniversary Edition,* edited by Hanoch Gutfreund and Jürgen Renn. Princeton, N.J.: Princeton University Press, 2015.

Einstein, Albert. 1936. "Physik und Realität." *Journal of the Franklin Institute* 221:313–47. Page reference is to the translation in *Ideas and Opinions.* New York: Bonanza Books, n.d.

Einstein, Albert. 1987. *Letters to Solovine.* New York: Philosophical Library.

Einstein, Albert. 1987–2018. *The Collected Papers of Albert Einstein.* 15 vols. Edited by John Stachel, Martin J. Klein, Robert Schulmann, Diana Kormos Buchwald et al. Princeton, N.J.: Princeton University Press.

Einstein, Albert et al. 1952. *The Principle of Relativity.* New York: Dover.

Einstein, Albert, and Marcel Grossmann. 1913. *Entwurf einer verallgemeinerten Relativitätstheorie und einer Theorie der Gravitation.* Leipzig, Germany: Teubner. Reprinted in Einstein 1987–2018, vol. 4, doc. 13.

Fletcher, Samuel C. 2017. *On the Reduction of General Relativity to Newtonian Gravitation.* Unpublished manuscript.

Freund, Ida. (1904) 1968. *The Study of Chemical Composition: An Account of Its Method and Historical Development.* Cambridge: Cambridge University Press. Reprint, New York: Dover. Page reference is to the 1968 edition.

Gakis, Dimitris. 2010. "Throwing Away the Ladder before Climbing It." In *Bild und Bildlichkeit in Philosophie, Wissenschaft und Kunst/Image and Imaging in Philosophy, Science, and the Arts,* edited by Richard Heinrich, Elisabeth Nemeth, and Wolfram Pichler, 98–100. Kirchberg am Wechsel: Austrian Ludwig Wittgenstein Society.

Gal, Ofer. 2002. *Meanest Foundations and Nobler Superstructures: Hooke, Newton, and the "Compounding of the Celestiall Motions of the Planetts."* Dordrecht: Kluwer.

Gingerich, Owen. 1989. "Johannes Kepler." In *Planetary Astronomy from the Renaissance to the Rise of Astrophysics*, part A: *Tycho Brahe to Newton*, edited by René Taton and Curtis Wilson, 54–78. Cambridge: Cambridge University Press.

Gordin, Michael D. 2014. "The Tory Interpretation of History." Review of Chang 2012. *Historical Studies in the Natural Sciences* 44:413–23.

Gould, Stephen Jay. 1980. "The Evolutionary Biology of Constraint." *Daedalus* 109 (2): 39–52.

Gould, Stephen Jay. 1981. *The Mismeasure of Man*. New York: Norton.

Gould, Stephen Jay. 2002. *The Structure of Evolutionary Theory*. Cambridge, Mass.: Harvard University Press.

Gould, Stephen Jay, and Richard C. Lewontin. 1979. "The Spandrels of San Marco and the Panglossian Paradigm: A Critique of the Adaptationist Programme." *Proceedings of the Royal Society of London B* 205:581–98.

Grantham, Todd A. 2004. "Constraints and Spandrels in Gould's *Structure of Evolutionary Theory*." *Biology and Philosophy* 19:29–43.

Griffiths, David J. 2005. *Introduction to Quantum Mechanics*. 2nd ed. Upper Saddle River, N.J.: Pearson Prentice Hall.

Hafner, Katie, and Matthew Lyon. 1996. *Where Wizards Stay Up Late: The Origins of the Internet*. New York: Simon & Schuster. Page reference is to paperback edition (2006).

Heisenberg, Werner. 1925. "Über die quantentheoretische Umdeutung kinematischer und mechanischer Beziehungen." *Zeitschrift für Physik* 33:879–93. English translation in van der Waerden 1968, 261–76.

Heisenberg, Werner. 1927. "Über den anschaulichen Inhalt der quantentheoretischen Kinematik und Mechanik." *Zeitschrift für Physik* 43:172–98.

Herschel, John Frederick William. (1830) 1966. *A Preliminary Discourse on the Study of Natural Philosophy*. London: Longman, Rees, Orme, Brown, and Green. Reprinted in facsimile as *The Sources of Science*, no. 17. New York: Johnson Reprint.

Herschel, John Frederick William. 1857. *Essays from the Edinburgh and Quarterly Reviews with Addresses and Other Pieces*. London: Longman, Rees, Orme, Brown, and Green.

Hertz, Heinrich. 1893. *Electric Waves: Being Researches on the Propagation of Electric Action with Finite Velocity through Space*. London: Macmillan.

Hilbert, David, John von Neumann, and Lothar Nordheim. 1928. "Über die Grundlagen der Quantenmechanik." *Mathematische Annalen* 98:1–30.

Hobbes, Thomas. 1656. *Six Lessons to the Professors of Mathematics*. London: Andrew Crook. Page reference is to vol. 7 of *The English Works of Thomas Hobbes of Malmesbury*, edited by Sir William Molesworth. 11 vols. Darmstadt, Germany, 1966. Reprint of the 1839 edition.

Holton, Gerald. 1981. "Einstein's Search for the *Weltbild*." *Proceedings of the American Philosophical Society* 125:1–15. Page reference is to reprint in *The Advancement of Science and Its Burdens*. 1st paperback ed. Cambridge, Mass.: Harvard University Press, 1998.

Hooke, Robert. 1665. *Micrographia: Or, Some Physiological Descriptions of Minute Bodies Made by Magnifying Glasses. With Observations and Inquiries Thereupon*. London: Royal Society.

Hufbauer, Karl. 2012. "From Student of Physics to Historian of Science: T. S. Kuhn's Education and Early Career, 1940–1958." *Physics in Perspective* 14: 421–70.

Hunt, Bruce J. 1991. *The Maxwellians*. Ithaca, N.Y.: Cornell University Press.

Isaacson, Walter. 2014. *The Innovators. How a Group of Hackers, Geniuses, and Geeks Created the Digital Revolution*. New York: Simon & Schuster. Page reference is to paperback edition (2015).

Israel, Werner. 1987. "Dark Stars: The Evolution of an Idea." In *Three Hundred Years of Gravitation,* edited by Stephen W. Hawking and Werner Israel, 199–276. Cambridge: Cambridge University Press.

Jähnert, Martin. 2016. *Practicing the Correspondence Principle in the Old Quantum Theory: A Transformation through Implementation*. PhD diss., Technische Universität, Berlin.

James, Jeremiah, and Christian Joas. 2015. "Subsequent and Subsidiary? Rethinking the Role of Applications in Establishing Quantum Mechanics." *Historical Studies in the Natural Sciences* 45:641–702.

Jammer, Max. 1966. *The Conceptual Development of Quantum Mechanics*. New York: McGraw-Hill.

Janssen, Michel. 1995. *Lorentz's Ether Theory and Einstein's Special Theory of Relativity in the Light of the Experiments of Trouton and Noble*. PhD diss., University of Pittsburgh, Pa.

Janssen, Michel. 2002. "Reconsidering a Scientific Revolution: The Case of Lorentz versus Einstein." *Physics in Perspective* 4:421–46.

Janssen, Michel. 2005. "Of Pots and Holes: Einstein's Bumpy Road to General Relativity." *Annalen der Physik* 14 (Suppl.): 58–85.

Janssen, Michel. 2009. "Drawing the Line between Kinematics and Dynamics in Special Relativity." *Studies in History and Philosophy of Modern Physics* 40:26–52.

Janssen, Michel. 2014. "'No Success Like Failure . . .': Einstein's Quest for General Relativity, 1907–1920." In Janssen and Lehner 2014, 167–227.

Janssen, Michel. 2017. "How Did Lorentz Find His Theorem of Corresponding States?" *Studies in History and Philosophy of Modern Physics.* http://dx.doi.org/10.1016/j.shpsb.2017.06.006.

Janssen, Michel, and Christoph Lehner, eds. 2014. *The Cambridge Companion to Einstein.* Cambridge: Cambridge University Press.

Janssen, Michel, and Matthew Mecklenburg. 2007. "From Classical to Relativistic Mechanics: Electromagnetic Models of the Electron." In *Interactions: Mathematics, Physics, and Philosophy, 1860–1930,* edited by Vincent F. Hendricks, Klaus Frovin Jørgensen, Jesper Lützen, and Stig Andur Pedersen, 65–134. Berlin: Springer.

Janssen, Michel, and Jürgen Renn. 2007. "Untying the Knot: How Einstein Found His Way Back to Field Equations Discarded in the Zurich Notebook." In Renn 2007, vol. 2:839–925.

Janssen, Michel, and Jürgen Renn. 2015. "Arch and Scaffold: How Einstein Found His Field Equations." *Physics Today,* November, 30–36.

Janssen, Michel, and Jürgen Renn. Forthcoming 2019. *How Einstein Found His Field Equations: A Source Book.* Heidelberg: Springer.

Joas, Christian, and Christoph Lehner. 2009. "The Classical Roots of Wave Mechanics: Schrödinger's Transformations of the Optical-Mechanical Analogy." *Studies in History and Philosophy of Modern Physics* 40:338–51.

Jordan, Pascual. 1926a. "Über kanonische Transformationen in der Quantenmechanik." *Zeitschrift für Physik* 37:383–86.

Jordan, Pascual. 1926b. "Über kanonische Transformationen in der Quantenmechanik II." *Zeitschrift für Physik* 38:513–17.

Jordan, Pascual. 1927a. "Über eine neue Begründung der Quantenmechanik." *Zeitschrift für Physik* 40:809–38.

Jordan, Pascual. 1927b. "Über eine neue Begründung der Quantenmechanik II." *Zeitschrift für Physik* 44:1–25.

Jordi Taltavull, Marta. 2017. *Transformation of Optical Knowledge from 1870 to 1925: Optical Dispersion between Classical and Quantum Physics.* PhD diss., Humboldt Universität, Berlin.

Kargon, Robert. 1969. "Model and Analogy in Victorian Science: Maxwell's Critique of the French Physicists." *Journal of the History of Ideas* 30:423–36.

Kevles, Daniel J. 1985. *In the Name of Eugenics: Genetics and the Uses of Human Heredity*. Berkeley: University of California Press.

Klein, Martin J. 1975. "Einstein on Scientific Revolutions." *Vistas in Astronomy* 17:113–33.

Klein, Martin J., Abner Shimony, and Trevor Pinch. 1979. "Paradigm Lost? A Review Symposium." *Isis* 70:429–40.

Kragh, Helge. 1999. *Quantum Generations: A History of Physics in the Twentieth Century*. Princeton, N.J.: Princeton University Press.

Kragh, Helge. 2012. *Niels Bohr and the Quantum Atom: The Bohr Model of Atomic Structure, 1913–1925*. Oxford: Oxford University Press.

Kramers, Hendrik Anthony. 1924a. "The Law of Dispersion and Bohr's Theory of Spectra." *Nature* 113:673–76. Reprinted in van der Waerden 1968, 177–80.

Kramers, Hendrik Anthony. 1924b. "The Quantum Theory of Dispersion." *Nature* 114:310–11. Reprinted in van der Waerden 1968, 199–201.

Kramers, Hendrik Anthony, and Werner Heisenberg. 1925. "Über die Streuung von Strahlung durch Atome." *Zeitschrift für Physik* 31:681–707. English translation in van der Waerden 1968, 223–52.

Kuhn, Thomas S. (1957) 1999. *The Copernican Revolution: Planetary Astronomy in the Development of Western Thought*. Cambridge, Mass.: Harvard University Press. Page reference is to the 1999 20th printing.

Kuhn, Thomas S. (1962) 2012. *The Structure of Scientific Revolutions*. Chicago: University of Chicago Press. Page references are to the 2012 4th edition.

Kuhn, Thomas S. 1966. *The Crisis of the Old Quantum Theory, 1922–1925*. Unpublished manuscript. Cambridge, Mass.: Massachusetts Institute of Technology Archives.

Kuhn, Thomas S. 1970. "Reflections on My Critics." In Lakatos and Musgrave 1970, 231–78.

Kuhn, Thomas S. 1978. *Black-Body Theory and the Quantum Discontinuity, 1894–1912*. Oxford: Oxford University Press. 2nd ed., with a reprint of Kuhn 1984 as an afterword, Chicago: University of Chicago Press, 1987.

Kuhn, Thomas S. 1984. "Revisiting Planck." *Historical Studies in the Physical Sciences* 14:231–52. Page references are to Kuhn 1978 2nd ed. of 1987.

Kuhn, Thomas S., John L. Heilbron, Paul Forman, and Lini Allen. 1967. *Sources for the History of Quantum Physics: An Inventory and Report*. Philadelphia: American Philosophical Society.

Kuhn, Werner. 1925. "Über die Gesamtstärke der von einem Zustande ausgehenden Absorptionslinien." *Zeitschrift für Physik* 33:408–12. English translation in van der Waerden 1968, 253–57.

Lacki, Jan. 2000. "The Early Axiomatizations of Quantum Mechanics: Jordan, von Neumann, and the Continuation of Hilbert's Program." *Archive for History of Exact Sciences* 54:279–318.

Lacki, Jan. 2004. "The Puzzle of Canonical Transformations in Early Quantum Mechanics." *Studies in History and Philosophy of Modern Physics* 35:317–44.

Ladenburg, Rudolf. 1921. "Die quantentheoretische Deutung der Zahl der Dispersionselektronen." *Zeitschrift für Physik* 4:451–68. English translation in van der Waerden 1968, 139–57.

Ladenburg, Rudolf, and Fritz Reiche. 1923. "Absorption, Zerstreuung, und Dispersion in der Bohrschen Atomtheorie." *Die Naturwissenschaften* 11:584–98.

Lakatos, Imre. 1970. "Falsification and the Methodology of Scientific Research Programmes." In Lakatos and Musgrave 1970, 91–196.

Lakatos, Imre, and Alan Musgrave, eds. 1970. *Criticism and the Growth of Knowledge.* Cambridge: Cambridge University Press.

Landé, Alfred. 1926. "Neue Wege der Quantentheorie." *Die Naturwissenschaften* 14:455–58.

Laubichler, Manfred D., and Jürgen Renn. 2015. "Extended Evolution: A Conceptual Framework for Integrating Regulatory Networks and Niche Construction." *Journal of Experimental Zoology. Part B: Molecular and Developmental Evolution* 324B:565–77.

Laue, Max. 1911. *Das Relativitätsprinzip.* Braunschweig: Vieweg.

Lazaroff-Puck, Cameron. 2015. "Gearing Up for Lagrangian Dynamics: The Flywheel Analogy in Maxwell's 1865 Paper on Electrodynamics." *Archive for History of Exact Sciences* 69:455–90.

Lehmkuhl, Dennis. 2014. "Why Einstein Did Not Believe that General Relativity Geometrizes Gravity." *Studies in History and Philosophy of Modern Physics* 46:316–26.

Lewis, Gilbert N., and Merle Randall. 1923. *Thermodynamics and the Free Energy of Chemical Substances.* New York: McGraw-Hill.

Lewontin, Richard C. 1963. "Models, Mathematics, and Metaphors." *Synthese* 15:222–44.

Lewontin, Richard C. 2001. "In the Beginning Was the Word." *Science* 291 (February 16): 1263–64.

Lodge, Sir Oliver. 1889. *Modern Views of Electricity.* New York: Macmillan.

Lorentz, Hendrik Antoon. 1895. *Versuch einer Theorie der electrischen und optischen Erscheinungen in bewegten Körpern.* Leiden: Brill.

Lorentz, Hendrik Antoon. 1904. "Electromagnetic Phenomena in a System Moving with Any Velocity Smaller than That of Light." *Koninklijke Akademie van Wetenschappen te Amsterdam, Section of Sciences, Proceedings* 6 (1903–1904): 809–31. Reprinted in Einstein et al. 1952, 11–34.

Lorentz, Hendrik Antoon. 1916. *The Theory of Electrons and Its Applications to the Phenomena of Light and Radiant Heat.* 2nd ed. Leipzig: Teubner.

Margenau, Henry. 1950. *The Nature of Physical Reality.* New York: McGraw-Hill.

Maxwell, James Clerk. 1861–1862. "On Physical Lines of Force." 4 pts. *Philosophical Magazine.* Pt. I: 21, no. 139 (March 1861): 161–75; Pt. II: 21, no. 140 (April 1861): 281–91; (continued:) 21, no. 141 (May 1861): 338–45; Pt. III: 23, no. 151 (January 1862): 12–24; Pt. IV: 23, no. 152 (February 1862): 85–95. Reprinted in Maxwell 1890, vol. 1:451–513.

Maxwell, James Clerk. 1865. "A Dynamical Theory of the Electromagnetic Field." *Philosophical Transactions of the Royal Society of London.* 155 (January 1865): 459–512. Reprinted in Maxwell 1890, vol. 1:526–97.

Maxwell, James Clerk. 1873. *A Treatise on Electricity and Magnetism.* 1st ed. 2 vols. Cambridge: Cambridge University Press.

Maxwell, James Clerk. 1890. *The Scientific Papers of James Clerk Maxwell.* 2 vols. Edited by W. D. Niven. Cambridge: Cambridge University Press.

Midwinter, Charles, and Michel Janssen. 2013. "Kuhn Losses Regained: Van Vleck from Spectra to Susceptibilities." In Badino and Navarro 2013, 137–205.

Miller, Arthur I. (1981) 1998. *Albert Einstein's Special Theory of Relativity: Emergence (1905) and Early Interpretation (1905–1911).* Reading, Mass.: Addison-Wesley. Reprint, New York: Springer.

Millikan, Robert A. 1917. *The Electron: Its Isolation and Measurement and the Determination of Some of Its Properties.* Chicago: University of Chicago Press.

Minkowski, Hermann. 1909. "Raum und Zeit." *Physikalische Zeitschrift* 20:104–11. Page references are to the English translation in Einstein et al. 1952, 75–91.

Neurath, Otto. 1921. *Anti-Spengler.* Munich: Callwey. Reprinted in Neurath 1981, vol. 1:139–96. Page reference is to the English translation in Neurath 1973, 158–213.

Neurath, Otto. 1932–1933. "Protokollsätze." *Erkenntnis* 3:204–14. Reprinted in Neurath 1981, vol. 2:577–85. Page reference is to the English translation in Neurath 1983, 91–99.

Neurath, Otto. 1973. *Empiricism and Sociology.* Edited by Robert S. Cohen and Marie Neurath. Dordrecht: Reidel.

Neurath, Otto. 1981. *Gesammelte philosophische und methodologische Schriften.* Edited by Rudolf Haller and Heiner Rutte. Vienna: Hölder-Pichler-Tempsky.

Neurath, Otto. 1983. *Philosophical Papers, 1913–1946.* Edited by Robert S. Cohen and Marie Neurath. Dordrecht: Reidel.

Norton, John D. 1984. "How Einstein Found His Field Equations, 1912–1915." *Historical Studies in the Physical Sciences* 14:253–316. Reprinted in *Einstein and the History of General Relativity,* edited by Don Howard and John Stachel, 253–316. Boston: Birkhäuser.

Norton, John D. 2000. "'Nature Is the Realisation of the Simplest Conceivable Mathematical Ideas': Einstein and the Canon of Mathematical Simplicity." *Studies in History and Philosophy of Modern Physics* 31:135–70.

Norton, John D. 2014. "A Material Dissolution of the Problem of Induction." *Synthese* 191: 671–90.

Peckhaus, Volker. 1990. *Hilbertprogramm und Kritische Philosophie. Das Göttinger Modell interdisziplinärer Zusammenarbeit zwischen Mathematik und Philosophie.* Göttingen: Vandenhoeck and Ruprecht.

Pesic, Peter. 2002. *Seeing Double: Shared Identities in Physics, Philosophy, and Literature.* Cambridge, Mass.: The MIT Press.

Planck, Max. 1894. "Heinrich Rudolf Hertz. Rede zu seinem Gedächtnis am 16.2.1894." *Deutsche Physikalische Gesellschaft. Verhandlungen* 13:9–29. Reprinted in Planck 1958, vol. 3:268–88.

Planck, Max. 1900a. "Über irreversible Strahlungsvorgänge." *Annalen der Physik* 1:69–122. Reprinted in Planck 1958, vol. 1:614–67.

Planck, Max. 1900b. "Entropie und Temperatur strahlender Wärme." *Annalen der Physik* 1:719–37. Reprinted in Planck 1958, vol. 1:668–86.

Planck, Max. 1900c. "Über eine Verbesserung der Wien'schen Spectralgleichung." *Deutsche Physikalische Gesellschaft. Verhandlungen* 2:202–4. Reprinted in Planck 1958, vol. 1:687–89.

Planck, Max. 1900d. "Zur Theorie des Gesetzes der Energieverteilung im Normalspektrum." *Deutsche Physikalische Gesellschaft. Verhandlungen* 2:237–45. Reprinted in Planck 1958, vol. 1:698–706.

Planck, Max. 1901. "Über das Gesetz der Energieverteilung im Normalspektrum." *Annalen der Physik* 4:553–63. Reprinted in Planck 1958, vol. 1:717–27.

Planck, Max. 1906. "Das Prinzip der Relativität und die Grundgleichungen der Mechanik." *Deutsche Physikalische Gesellschaft. Verhandlungen* 8:136–41. Reprinted in Planck 1958, vol. 2:115–20.

Planck, Max. 1958. *Physikalische Abhandlungen und Vorträge.* 3 vols. Braunschweig: Vieweg.

Poynting, John Henry. 1893. "An Examination of Prof. Lodge's Electromagnetic Hypothesis." *Electrician* 31:575–77, 606–8, 635–36.

Poynting, John Henry. 1899. "Address by the President of the Section, September 14, 1899." *British Association for the Advancement of Science. Transactions of the Sciences. Section A. Mathematical and Physical Science,* 615–24.

Quine, Willard Van Orman. 1960. *Word and Object.* Cambridge, Mass.: MIT Press.

Rabossi, Eduardo. 2003. "Some Notes on Neurath's Ship and Quine's Sailors." *Principia* 7:171–84.

Reiche, Fritz, and Willy Thomas. 1925. "Über die Zahl der Dispersionselektronen, die einem stationären Zustand zugeordnet sind." *Zeitschrift für Physik* 34: 510–25.

Renn, Jürgen. 1993. "Einstein as a Disciple of Galileo: A Comparative Study of Concept Development in Physics." In *Einstein in Context,* edited by Mara Beller, Robert S. Cohen, and Jürgen Renn, 311–41. Cambridge: Cambridge University Press.

Renn, Jürgen. 2006. *Auf den Schultern von Riesen und Zwergen: Einsteins unvollendete Revolution.* New York: Wiley.

Renn, Jürgen, ed. 2007. *The Genesis of General Relativity.* 4 vols. New York: Springer.

Renn, Jürgen. 2019. *The Evolution of Knowledge: Rethinking Science for the Anthropocene.* Princeton: Princeton University Press.

Renn, Jürgen, and Robert Rynasiewicz. 2014. "Einstein's Copernican Revolution." In Janssen and Lehner 2014, 38–71.

Rosenblueth, Arturo, and Norbert Wiener. 1945. "The Role of Models in Science." *Philosophy of Science* 12:316–21.

Rowe, David E. 1997. "Perspective on Hilbert." Review of books by Herbert Mehrtens, Volker Peckhaus, and Michael-Markus Toepell. *Perspectives on Science* 5:533–70.

Rynasiewicz, Robert. 2015. "The (?) Correspondence Principle." In Aaserud and Kragh 2015, 175–99.

Sansom, Roger, and Robert N. Brandon, eds. 2007. *Integrating Evolution and Development: From Theory to Practice.* Cambridge, Mass.: MIT Press.

Sartorius von Waltershausen, Wolfgang. 1856. *Gauß zum Gedächtniss: Biographie Carl Friedrich Gauß.* Leipzig, Germany: S. Hirzel. Page reference is to the 1966 English translation by Helen Worthington Gauss, privately printed in Colorado Springs.

Schirrmacher, Arne. 2009. "Bohrsche Bahnen in Europa: Bilder und Modelle zur Vermittlung des modernen Atoms." In *Atombilder: Ikonographien des Atoms in Wissenschaft und Öffentlichkeit des 20. Jahrhunderts,* edited by Charlotte Bigg and Jochen Hennig, 73–82. Göttingen: Wallstein.

Schrödinger, Erwin. 1926. "Über das Verhältnis der Heisenberg-Born-Jordanschen Quantenmechanik zu der meinen." *Annalen der Physik* 79:734–56.

Seth, Suman. 2010. *Crafting the Quantum: Arnold Sommerfeld and the Practice of Theory, 1890–1926.* Cambridge, Mass.: MIT Press.

Shubin, Neil. (2008) 2009. *Your Inner Fish: A Journey into the 3.5-Billion-Year History of the Human Body.* New York: Pantheon Books. Reprint, New York: Vintage Books. Page reference is to the 2009 edition with a new afterword.

Sigmund, Karl. 2017. *Exact Thinking in Demented Times: The Vienna Circle and the Epic Quest for the Foundations of Science.* New York: Basic Books.

Sommerfeld, Arnold. 2004. *Wissenschaftlicher Briefwechsel. Band 2: 1919–1951.* Edited by Michael Eckert and Karl Märker. Munich: Deutsches Museum and GNT-Verlag.

Stachel, John, ed. 2005. *Einstein's Miraculous Year: Five Papers That Changed the Face of Physics.* Centenary ed. Princeton, N.J.: Princeton University Press.

Stachel, John. 2007. "The Story of Newstein or Is Gravity Just Another Pretty Force?" In Renn 2007, vol. 4:1041–78.

Staley, Richard. 2008. *Einstein's Generation: The Origins of the Relativity Revolution.* Chicago: University of Chicago Press.

Stuewer, Roger H. 2014. "The Experimental Challenge of Light Quanta." In Janssen and Lehner 2014, 143–66.

Swerdlow, Noel M., and Otto Neugebauer. 1984. *Mathematical Astronomy in Copernicus's De Revolutionibus.* 2 vols. New York: Springer.

Synge, John Lighton. (1955) 1972. *Relativity: The Special Theory.* Amsterdam: North-Holland. Page reference is to the 1972 3rd printing.

Thomas, Willy. 1925. "Über die Zahl der Dispersionselektronen, die einem stationären Zustande zugeordnet sind (Vorläufige Mitteilung)." *Die Naturwissenschaften* 13:627.

van der Waerden, Bartel Leendert, ed. 1968. *Sources of Quantum Mechanics*. New York: Dover.

Van Dongen, Jeroen. 2010. *Einstein's Unification*. Cambridge: Cambridge University Press.

Van Vleck, John Hasbrouck. 1924a. "The Absorption of Radiation by Multiply Periodic Orbits, and Its Relation to the Correspondence Principle and the Rayeigh-Jeans Law. Part I. Some Extensions of the Correspondence Principle." *Physical Review* 24:330–46. Reprinted in van der Waerden 1968, 203–22.

Van Vleck, John Hasbrouck. 1924b. "The Absorption of Radiation by Multiply Periodic Orbits, and Its Relation to the Correspondence Principle and the Rayeigh-Jeans Law. Part II. Calculation of Absorption by Multiply Periodic Orbits." *Physical Review* 24:347–65.

Volkmann, Paul. 1900. *Einführung in das Studium der theoretischen Physik, insbesondere das der analytischen Mechanik mit einer Einleitung in die Theorie der Physikalischen Erkentniss*. Leipzig: Teubner.

Von Mises, Richard. 1928. *Wahrscheinlichkeit, Statistik, und Wahrheit*. Vienna: Springer.

von Neumann, John. 1927a. "Mathematische Begründung der Quantenmechanik." *Königliche Gesellschaft der Wissenschaften zu Göttingen. Mathematisch-physikalische Klasse. Nachrichten*, 1–57.

von Neumann, John. 1927b. "Wahrscheinlichkeitstheoretischer Aufbau der Quantenmechanik." *Königliche Gesellschaft der Wissenschaften zu Göttingen. Mathematisch-physikalische Klasse. Nachrichten*, 245–72.

von Neumann, John. 1927c. "Thermodynamik quantenmechanischer Gesamtheiten." *Königliche Gesellschaft der Wissenschaften zu Göttingen. Mathematisch-physikalische Klasse. Nachrichten*, 273–91.

von Neumann, John. 1932. *Mathematische Grundlagen der Quantenmechanik*. Berlin: Springer.

Weinberg, Steven. (1992) 1994. *Dreams of a Final Theory*. New York: Pantheon. Reprint, New York: Vintage Books. Page reference is to the 1994 edition.

Whittaker, Sir Edmund. (1951–1954) 1987. *A History of the Theories of Aether and Electricity*. 2 vols. London: Nelson. Reprint, New York: Tomash/American Institute of Physics.

Wilson, Curtis. 2001. "Newton on the Moon's Variation and Apsidal Motion: The Need for a Newer 'New Analysis.'" In *Isaac Newton's Natural Philosophy*, edited by Jed Z. Buchwald and I. Bernard Cohen, 139–88. Cambridge, Mass.: MIT Press.

Wimsatt, William C., and James R. Griesemer. 2007. "Reproducing Entrenchments to Scaffold Culture: The Central Role of Development in Cultural Evolution." In Samson and Brandon 2007, 227–323.

Wittgenstein, Ludwig. 1921. "Logisch-Philosophische Abhandlung." *Annalen der Naturphilosophie* 14:185–262. Page reference is to the English translation, *Tractatus Logico-Philosophicus*, New York: Routledge, 2001.

Young, William Henry. 1926. "The Progress of Mathematical Analysis in the Twentieth Century." *Proceedings of the London Mathematical Society* (Series 2) 24:421–34.

Zimmer, Ernst. 1934. *Umsturz im Weltbild der Physik*. Munich: Knorr and Hirth.

5

PROMISCUOUS INVENTIONS
Modeling Cultural Evolution with Multiple Inheritance

JACOB G. FOSTER AND JAMES A. EVANS

"Trillian, this is my semi-cousin Ford, who shares three of the same mothers as me . . ."

—ZAPHOD BEEBLEBROX, THE HITCHHIKER'S GUIDE TO THE GALAXY

In THIS CHAPTER, we argue that ideas and inventions—like Zaphod—can have many mothers. This is not always the default assumption. Powerful tools from macroevolution have been used to reconstruct cultural phylogenies ("trees") in a variety of spheres (O'Brien et al. 2013), including language (Gray, Drummond, and Greenhill 2009), crafts (Tehrani and Collard 2002), and lithic technology (O'Brien, Darwent, and Lyman 2001). In order for these tools to retrieve accurate genealogies, however, the underlying patterns of cultural evolution must fit the assumptions of the biological methods—above all, the predominance of vertical information transfer (from "mother" to "child") and tree-like branching. Such phylogenetic methods treat the horizontal transfer of information from other lineages as contaminating noise. For this reason, the application of phylogenetic methods to prokaryotic taxa has been challenged. Far from being "noise," horizontal transfer is common among these organisms (Doolittle and Bapteste 2007).

Horizontal transfer is common in human culture, too, thanks to our rich communicative capacity and increasingly frequent population movement. In fact, scholars of modern technology often assume recombinant processes and hence substantial horizontal transmission (Arthur 2009; Wimsatt 2013b). This common assumption suggests that phylogenetic methods are not always applicable to cultural data, just as they have limited application to prokaryotes. In a further contrast with biology, contemporary cultural evolution often leaves a detailed and relatively complete "fossil record" of past forms. Standard phylogenetic methods are designed to work without fossil data and typically only use them, if at all, for calibration (Felsenstein 2004; Gray and Atkinson 2003). More sophisticated methods that fully incorporate available fossil data (Fisher 2008; Huelsenbeck and Rannala 1997) exist but are rarely deployed.

In this chapter, we sketch an exploratory framework that learns from this rich historical data to infer the possible histories and patterns of cultural inheritance, beginning with which and how many "parents" contribute to each offspring. Simple vertical transfer is now treated as a special case of combinatorial evolution (Arthur 2009; Wimsatt 2013a), in which one or more parents from distinct lineages provide the raw materials involved in spawning a new "child," which could be an invention, an organization, or a literal developing human being. We focus on the directed acyclic graphs (DAGs) that best trace the inheritance of known features and explain their observed distribution across cultural types. Temporal and geographical constraints on the space of plausible histories allow us to enforce the time directedness and spatial localization of inheritance. We describe a formalism that assumes independence in the choice of parents, but this can be relaxed to allow nonindependence and structured parent choice. Groups of parents can "mate" with each other (or be "chosen" by offspring) according to a range of criteria, including unobserved but inferred fitness (i.e., appeal), past fecundity (i.e., number of offspring), or population structure (e.g., different disciplines or craft traditions that limit interbreeding between lineages). We describe both parsimony-based and probabilistic, generative approaches. Probabilistic methods are especially valuable when dealing with historical phenomena because they allow us to encode our assumptions about the underlying process and then reason rigorously from the data to a universe of plausible historical trajectories. In other words, these methods demand and leverage "new conceptual frameworks" to tackle the "massive new data sets" that are often available to trace the trajectories of cultural evolution (see chapter 1).

We show that our approach can apply to a wide range of cultural phenomena, including the evolution of technology, organizations, genres, and art forms, as well as the changing cultural constitution of individual human beings—that is, it can be used to model the "sequential dependencies in the acquisition of cultural traits during development" (see the introduction). We conclude by describing the relationship between the modes of cultural inheritance revealed by this approach (branching or reticulate) and the mixture of transmission-isolating mechanisms (TRIMs) (Durham 1991) and transmission-accelerating mechanisms (TRAMs) that together shape inheritance and pattern cultural evolution.

This chapter responds to the key question raised by Love and Wimsatt in their introduction to this volume: "How to characterize cultural heredity with multiple parents." In answering that question, we embrace core insights about the distinctive internal and external structures (Wade 2016) that influence cultural evolution: "Sequential dependencies in the acquisition of cultural traits" and "the roles of external structure" like institutions, organizations, and infrastructures in setting up the population structure that shapes cultural evolution (see the introduction by Love and Wimsatt; chapter 1 by Wimsatt).

Our approach incorporates several of the elements of an "adequate theory of cultural evolution" outlined by Wimsatt (chapter 1). It explicitly analyzes the complex lineages of "ideational, behavioral, and material items, which are capable of being modularly decomposed or chunked and black boxed hierarchically." It can be used to model the complex cultural growth of "developing biological individuals" as well as organizations. Finally, in our analysis of TRIMs and TRAMs, we show how institutions and infrastructures can work together to produce and maintain "cultural breeding populations" and structure the processes of inheritance and invention by which culture evolves. In other words, it represents a sustained conceptual and mathematical effort to think "beyond the meme."

THE QUESTION CONCERNING PHYLOGENIES

Are cultural phylogenies possible? In other words, are there cultural units whose evolution traces tree-like topologies? As recently as 1997, this was an open question (Boyd et al. 1997). Less than twenty years later, it has been answered decisively in the affirmative. Formal methods for phylogenetic inference, including parsimony, maximum likelihood, and Bayesian inference, have been applied to a range of cultural forms, from languages (Gray, Drummond, and Greenhill 2009) to projectile points (O'Brien, Darwent, and Lyman 2001) and textiles (Tehrani and Collard 2002); see O'Brien et al. (2013) for an extensive list. Explicit tests using a standard goodness-of-fit metric (the retention index) suggest that trees fit cultural data just as well as they do biological data (Collard, Shennan, and Tehrani 2006). What is the alternative to the branching, tree-like pattern of cultural inheritance? Reticulation: a topology in which lineages not only split but blend, join, and recombine. Given that cultural phylogenies are indeed possible, why did we (and should we) consider reticulation?

Reticulation is plausible for a simple reason: the capacity to generate particular cultural traits moves with relative ease from one living individual to another. In other words, such traits or *transmissable elements* (Wimsatt 2013a) are capable of horizontal transmission between cultural lineages. A small note on terminology: when discussing the transmission of genetic information, biologists typically refer to *transfer*. In vertical transfer, genetic information flows from parent to offspring via reproduction. This preserves the integrity of lineages and ultimately builds up tree-like, branching topologies. In horizontal transfer, genetic information flows nonreproductively from one individual to another, potentially between distinct lineages. This breaks down the integrity of lineages and, if common, produces highly reticulate, recombinant topologies.[1] In both cases, the underlying genetic information is assumed to transfer unaltered, although under special circumstances it may undergo simultaneous mutation or recombination.

When discussing the flow of cultural traits (Mesoudi 2011), evolutionary anthropologists often refer to *transmission* rather than transfer, making an analogy to epidemiology. In vertical transmission, cultural traits flow from parent to child. In horizontal transmission, cultural traits flow between a pair of individuals, who may be unrelated.[2] Vertical transmission helps to preserve the integrity of cultural lineages. Horizontal transmission can produce distinct and well-separated lineages, as long as the reach of horizontal transmission is limited—for example, via TRIMs (Durham 1991) that maintain separate cultural breeding populations (see chapter 1). In contrast to the genetic case, however, the "flow" of cultural traits can be much more complex than the "flow" of genetic information. While genetic information can be copied with minimal error, transmitting the capacity to manifest particular cultural traits is nontrivial. Far from mere copying (as in simple memetic pictures), it often involves detailed reconstruction and reverse engineering (Claidière, Scott-Phillips, and Sperber 2014) and can depend on the prior acquisition of other cultural traits that scaffold sequential acquisition (see the introduction by Love and Wimsatt, as well as chapter 1).

Despite the need for reconstruction and reverse engineering, the horizontal transmission of cultural traits from person to person can be relatively easy, and hence the transmission of cultural traits from one cultural "lineage" to another becomes possible. As Stephen J. Gould (2010) quipped in an oft-cited quote, "Five minutes with a wheel, a snowshoe, a bobbin, or a bow and arrow may allow an artisan of one culture to capture a major achievement of another." While Gould underestimates the difficulty of inferring a generative procedure from an artifact, five *years* as apprentice to an artisan from another culture probably suffices for the horizontal, cross-cultural transmission of many major technical achievements (see chapter 8). Given the right scaffolding[3] and enough time, a novice will acquire the skills, knowledge, and practices that make her "infectible" by a new technology (Wimsatt 2013a).[4] The same is true for other cultural traits, like complex beliefs. For this reason, proponents of cultural phylogenetics are careful to emphasize that tree-like structures may not always be appropriate (Cochrane and Lipo 2010). As we noted above, tree-like structures are not always suitable for biological evolution either, as in the case of bacteria, with their rampant horizontal transfer of genetic information via plasmids, transformation, or transduction (Gogarten and Townsend 2005). Whether phylogenetic trees can accurately represent a particular evolutionary history is therefore an empirical question, not a theoretical one. Trees provide a reasonable representation of particular histories of cultural transmission when TRIMs (Durham 1991) limit or prevent cross-lineage transmission for the trait in question and thus maintain branching as the dominant mode of cultural macroevolution (at a certain level of analysis).

As Mesoudi (2011) notes, however, the TRIMs that apply to projectile points and textiles (e.g., language, limited intergroup contact, and ethnocentrism) are unlikely to apply directly to the evolution of scientific ideas and technological inventions—although related social mechanisms might, along with the need for scaffolded skill acquisition (Goodwin 2017).[5] Instead, the picture presented by the literature on science and technology is positively promiscuous, with recombination an essential and often primary process (Fleming and Sorenson 2001, 2004; Uzzi et al. 2013; Arthur 2009; Foster, Rzhetsky, and Evans 2015). This implies that reticulation should be common; that ideas and inventions—like Zaphod in our epigraph—can have many mothers.

Before proceeding, an important caveat. We note that vertical and horizontal transmission at the level of people is logically independent from vertical and horizontal transmission at the level of specific cultural traits or products. For example, the degree to which a biological individual's repertoire of cultural traits or products (e.g., ideas, beliefs, practices, technologies) emerges within a social lineage or across them is independent of whether specific, novel instances of cultural traits or products are produced through conservative, vertical tinkering or liberal, horizontal recombination. We focus here not on the vertical or horizontal transmission of cultural traits from person to person (though see section 7) but on the vertical or horizontal transmission of elements from one cultural trait to another. Still, the two may be empirically related. If the capacity for horizontal transmission from person to person were limited, then possibilities for the combinatorial generation of new culture would be highly constrained. As a result, promiscuous horizontal transmission between people is a necessary but not sufficient condition for promiscuous *combinatorial invention* in which cultural artifacts have multiple parents.

TACKLING MULTIPLE INHERITANCE

This picture of promiscuous combinatorial invention suggests that standard methods of phylogenetic inference will often produce distorted pictures of the pattern of cultural evolution in contemporary science and technology.[6] It is worth noting that some extensions of phylogenetic methods permit horizontal transmission between lineages (Nicholls and Gray 2006). Inference can be robust to moderate levels of horizontal transmission (Greenhill, Currie, and Gray 2009), and network-based methods can detect signals of reticulation directly (Lipo 2006; Gray, Bryant, and Greenhill 2010; Huson Rupp, and Scornavacca 2010). Ignoring reticulation, however, runs the risk of distorting a true history by forcing multiple inheritance into a branching tree. Network-based methods, on the other hand, are essentially exploratory. Because they lack an underlying generative model (but see Wen, Yu, and Nakhleh 2016), they can neither draw on existing knowledge about the inventive process nor represent uncertainty or improve with additional data (Ghahramani 2015). Nevertheless, cultural evolution is sufficiently complex (see chapter 1) that all models are distortions (Wimsatt 2002), and there is no one "true" representation for every trajectory. As Doolittle and Bapteste (2007) note, "Different evolutionary models and representations of relationships will be appropriate, and true, for different taxa or at different scales or for different purposes." We thus embrace pattern and process pluralism in cultural just as in biological evolution.

Pattern pluralism aside, traditional phylogenetic methods do not take advantage of a distinctive feature of contemporary technological evolution: the incredibly rich "fossil" record of past forms, which often possesses detailed information about timing, sequence, and spatial location (Evans and Foster 2011). In biology, such information is often sparse and always hard to come by. It usually involves a lot of digging and scraping. For this reason, phylogenetic methods are generally designed to make inferences in the *absence* of extensive evidence about past forms. Available information can be used to "root" trees with an out-group or to calibrate particular branching points (Felsenstein 2004). The latter application is relatively common in the reconstruction of linguistic phylogenies, where fossil traces, for example, written materials that have persisted to the present, are exceedingly rare (see, for example, Gray and Atkinson 2003). Biologists have also developed parsimony (Fisher 2008) and likelihood-based (Huelsenbeck and Rannala 1997) methods that penalize trees if they infer ancestral states with no trace in the fossil record.

We set these methods aside for two reasons. First, they are fundamentally tree-based and hence suffer the same problem of forcing multiple inheritance onto branching topologies.[7] Second, they depend on the inference of past forms. Yet past forms are densely documented in scientific and technological data sets (see chapter 6), thanks to ongoing incentives to publish (Merton 1973) and patent (Owen-Smith and Powell 2001). This rich history is ripe for analysis, thanks to the increasing electronic availability of data (Evans and Foster 2011). Scholars of technological evolution should use it, and phylogenetic methods are simply not designed for situations in which history is as richly and densely documented as it is in science and technology. Taking this record into account will dramatically improve our characterization of the process that generated it, as well as our prediction of what will come next.

Finally, ancestral cultural forms can influence the present in a way that ancestral biological forms cannot. When a species goes extinct, its distinct genetic information is lost forever, along with its phenotype, behavior, and entailed ecological interactions, the loss of which can tip other species toward extinction. When an idea or technology "goes extinct"—in the sense that it no longer occupies any minds or has a physical presence in the contemporary population—it can nevertheless contribute to a new technology or idea. If an artifactual or textual trace of the extinct idea remains, contemporary inventors can draw components, features, or inspiration from it (Tëmkin and Eldredge 2007). Because any act of cultural transmission always involves some inference and reconstruction (Claidière, Scott-Phillips, and Sperber 2014), even limited traces of an earlier idea or technology can contribute ingredients to a novel cultural unit decades or even centuries later.

These considerations suggest that if we take the possibility of multiple inheritance seriously, we should develop models that allow both reticulate patterns of multiple inheritance and tree-like patterns of descent with modification. In other words, we should go beyond the inference of phylogenetic trees. In the rest of this chapter, we introduce exploratory and model-based methods that can detect and describe multiple inheritance in densely sampled cases of cultural evolution. These methods themselves have multiple parents; in addition to the phylogenetic tradition, they draw on ideas from latent variable modeling (Blei 2014), Bayesian nonparametric models (Gershman and Blei 2012; Ghahramani 2013), probabilistic machine learning (Ghahramani 2015), and network analysis (Newman 2003). In the following sections, we provide a high-level description of these methods and the underlying ideas and intuitions. The appendix gives specific mathematical descriptions of several methods. We also discuss the philosophy behind inference using probabilistic models. We close with a reflection on the role of entrenchment and scaffolding in the tempo and mode of technological evolution.

Before proceeding, we describe in words and a little notation our general picture of the evolution of ideas and technologies. Imagine that we observe an invention at time t_j. Given our observation, we know that at some earlier time $(t_j - \epsilon)$ a "creative unit" (which could be an individual inventor/ scientist or a team) must have assembled (consciously or not) a set of influences P_j. For example, the Bessemer process of steel production involved removing impurities from pig iron with oxidation by blowing air through the molten metal (Birch 1967). "Parental influences" here include pig iron, the oxidation process, and, ultimately, the use of dolomite or limestone linings for the Bessemer converter. Taken together, these influences P_j provide the raw material from which the invention was assembled; hence the set P_j contains the parents of the new invention j. If cultural evolution in this particular domain is dominated by vertical transmission and descent with modification (i.e., tinkering), then P_j may only have one member, and the invention j only has one parent (for example, the inventor slightly adjusts the technique typically used to process a particular stone). If cultural evolution is dominated by horizontal transmission and is combinatorial, then P_j may have several members, and the invention will have multiple parents. Note that, in principle, any invention that precedes invention j in time is a possible parent. The set of all possible parents is denoted P_{t_j}. It is time ordered (each earlier invention is time stamped). Depending on the typical length of the inventive process and the typical difficulty of mastering a new invention, it may take some time before a given invention p can become a parent. This implies a lower bound on the difference between the time of invention t_j and the time of observation of an allowable parent t_p, such that some time Δt_{jp} must have passed before p is a possible parent of j.

Any idea or technology may be coarsely characterized by its elementary building blocks; as Wimsatt notes in his discussion of transmissible or replicable elements (chapter 1), TREs can be "modularly decomposed." The outcome of this modular decomposition might be the components that make up an invention or the concepts that make up the idea of a scientific paper. We refer to these parts as *features,* and the set of all features as $\mathcal{F}$.[8] This characterization is a necessary precondition for analysis, but it is more than a useful fiction: for any particular community of practice, the relevant coarse-graining—the "principles of vision and division" (Bourdieu 1990; Foster, Rzhetsky, and Evans 2015)—will be relatively consistent.[9]

For any invention j, each of its parents $p \in P_j$ is characterized by its own set of features $F_p \in \mathcal{F}$. To create a new invention j, the inventor selects its features from the set of features possessed by its parents. This set is simply the union of all the parental feature sets: $\bigcup_{p \in P_j} F_p$. Occasionally, an invention introduces an entirely novel feature rather than just drawing on the features of the past. At other times, an invention may "bundle" together several preexisting features (from one or several parents) into an effective, integrated unit. This new unit becomes a "feature" available to future inventions—its constituent subfeatures are henceforth sampled together. This process is called *black boxing* (Latour 1987). To embrace these generative possibilities, our model must allow inventors to black box or introduce a novel feature with some probability (which will typically be small in cases where invention is largely combinatorial).

Note that this inventive process can generate a range of inheritance pathways and hence topologies. It can describe unilineal inheritance, in which a single (cultural) parent is selected and (perhaps) slightly modified. It can also describe multilineal inheritance, in which multiple (cultural) parents are selected and their features recombined. Starting from a picture of invention that is entirely agnostic about tinkering versus recombination is essential if we are to let the rich traces of inventive activity reveal the underlying modes of inheritance and pattern(s) of technological evolution. Such an agnostic analysis can also provide data-driven hints as to the modes and mechanisms of cultural evolution.

Before defining a model-based approach for studying multiple inheritance, we describe some exploratory methods for tracing multiple inheritance in densely sampled, time-ordered data. These methods are much simpler than the model-based approach but require further assumptions about the inventive process.

GENERALIZING PARSIMONY

Parsimony-based methods provide powerful exploratory tools for constructing possible phylogenies (O'Brien et al. 2013). Here we describe methods than can reconstruct possible reticulated histories (directed graphs) in the case of multiple inheritance. As a modeling strategy, parsimony emphasizes simplicity; it seeks the minimal explanation for the observed facts. In phylogenetic reconstruction, the present distribution of features (genetic or morphological) provides the observed facts; the phylogenetic tree provides a potential explanation for those facts. In seeking the simplest explanation, parsimony methods minimize the number of genetic changes implied by the proposed tree. A tree that can explain the present distribution of features with five changes is preferable to a tree that requires six. The intuition underlying this simplicity criterion is quite plausible: not only are fewer changes "simpler" in an absolute sense, but every change (mutation) is a low-probability event. Hence, we should generally seek explanations for present facts (i.e., trees) that minimize the number of such events (note that this basic idea is also used in some approaches to phylogenetic networks; see Huson and Scornavacca [2011]).

What is the analog of this parsimony principle in our generalized model of inheritance? Recall that in our model, new inventions sample over the features of past inventions. All else being equal, it probably takes more time and effort to sample from three past inventions than from two. Hence, the most parsimonious or simplest explanation might minimize the number of past inventions needed to account for the features of the present invention. On the other hand, consider the following scenario. A new invention has six features. Four of those six features can be found in a single predecessor; the remaining two can be found separately in several possible predecessors. This history would lead to three "parents." Alternately, three of the features can be found in a single predecessor and the remaining three in another predecessor.

This history would lead to two parents. Which history provides the simplest explanation? Naive parsimony (i.e., minimizing the number of parents) ignores the fact that a single parent can account for the majority of features in the new invention; in such ambiguous cases, it's quite likely that there are many ways to account for the residual features, whereas there might be only one history that splits all features across two parents. We balance these various nuances of the parsimony principle through a *greedy* inventive process. In essence, we assume that at any stage of invention inventors draw as many features from a parent as possible.[10] In particular, inventors tend to draw a large number of features from one parent and a small number from several others, rather than drawing a moderate number from two or three. This greedy assumption also makes the problem more computationally tractable, as we do not have to search over different combinatorial histories.[11] At every step, we pick the simplest explanation—the parent that accounts for the most features.[12]

We now describe how to implement this parsimony principle in practice. Before implementing the following algorithm, we first reduce the feature sets of all inventions by removing any novel features—features that appear for the first time in that particular invention. These features cannot be accounted for by the past. In cases where the same novel feature appears in multiple inventions in the same time slice, we treat those features as novel for all inventions. Now, we execute the following algorithm for each invention *j*. Note that this algorithm can be executed independently for every invention.

PARSIMONY ALGORITHM
1. Establish the set of possible parents. This may be the set of all earlier inventions, or it may have some restrictions (e.g., all inventions more than six months older than the focal invention).
2. For each possible parent, count up the number of features in the focal invention that could have been inherited from that parent.
3. Add to the set of *j*'s parents whichever prior invention explains the most features (and remove that invention from the set of possible parents). If there are multiple equally explanatory inventions, we can implement additional principles of simplicity as desired. For example, we can prefer the most recent ancestor or the ancestor with the smallest spatial distance, social distance (e.g., as computed in a social network), or cognitive distance (e.g., as computed in a network or space of skills,

ideas, etc.). If *all* parsimony principles have been exhausted and multiple possible parents remain, choose at random.

4. Eliminate from the feature set of the focal invention all features that have been explained by the parent set.

5. If features remain to be explained, go back to step 2. Otherwise, stop.

We repeat this procedure for all j to reconstruct a parsimonious history for our observed technologies. This history will be a directed acyclic graph (DAG); we establish the convention that arcs run *to* an invention j from its parents p, k, m, and so on to represent the flow of ideas from past to present.[13] This directed graph can be weighted, with arc weights counting the number of features inherited from each parent. Because of the random steps in the construction of the DAG, we should construct multiple complete histories for a given data set and look at properties of the ensemble, which is only necessary if the random number generator is called. Because the reconstruction process is independent for each distinct invention, it can be trivially parallelized. See appendix for mathematical details.

Parsimonious Insights

What can we learn from the DAGs reconstructed via parsimony? First, remember that we are trying to make inferences about an unobserved history of invention and inheritance from richly sampled but incomplete evidence. In general, we will not have traces of the inheritance process (i.e., that the inventors of technology j drew on technologies p, k, and m); even when we do, those traces are incomplete and potentially biased. What we do have is a record of what technologies with what features exist at what times. Our inferences are also shaped by assumptions about the inventive process that has generated this record—namely, that it is a greedy local search, in which inventors sequentially sample the space of possible parents and prefer to extract as much as possible from each parent along the way, until their new invention is complete.[14]

We can mine several insights from the weighted DAG that represents a parsimonious reconstruction of cultural inheritance under this model of technological evolution and the inventive process. First, we can ask what fraction of "explicable" features in each invention is inherited from each parent. For a particular invention, this tells us whether most of its features come from a single parent or whether its features can be better explained by even sampling from several parents. Consider the largest such fraction for each invention; call this the *primary inheritance*. The frequency distribution of primary inheritance reveals how many inventions are largely explicable with a single parent and how many require multiple parents to explain their features (see Bedau, chapter 6, for an empirical analysis of multiple parentage in U.S. patents). From this distribution, we can obtain a good guess at the dominant mode of cultural inheritance in a given domain and make inferences about the dominant mode of cultural evolution. If the distribution is peaked at large values (close to one), then the mode of inheritance is primarily unilineal, and evolution proceeds via descent with modification. We can turn the DAG into a tree by retaining the highest-weight incoming arc for each node. This tree likely represents a good first approximation of the inheritance pattern. At the very least, it suggests that vertical transfer and descent with modification together provide a parsimonious explanation for the facts. If the distribution is spread out across possible fractions—or even peaked at lower values—then we have evidence that the inheritance pattern is reticulate, involving multiple parents, and that the mode of evolution may be combinatorial. Given that our greedy reconstruction process is biased toward trees, a broad distribution of primary inheritance provides substantial evidence for reticulate cultural evolution and multiple parentage.

Second, consider the number of features present in a given invention p. These are the features that could be passed on to any descendants. In a given reconstruction, we can compute the fraction of such features that are actually passed on to each descendant; call this the *primary contribution*. The mean and mode of this quantity, computed over all descendants, can tell us whether the features of p are typically inherited as a block or whether they are separable and used as a selective smorgasbord. While block inheritance of features happens in biology (e.g., genes are bundled together into chromosomes), cultural and especially technological evolution is distinctive in its capacity to create such building blocks from more primitive pieces; this is a key part of the internal or *endogenetic* structure of cultural inheritance (see the introduction by Love and Wimsatt).[15] We briefly discuss how such black-boxing events can be detected in parsimonious DAGs.

Black Boxing

We can use the two measures described above to identify potential moments of modularization or black boxing (Latour 1987). In practice, black boxing may involve miniaturization (allowing a bundle of features to "fit" in new places); compression (simplifying components, removing redundancies, and integrating parts to maximize efficiency); autocatalysis (relations of mutual dependence across parts that sustain and reproduce coparticipation; see chapter 11 of this book); and the streamlining and/or standardizing of input and output (making it easier for the set of features to recombine; see chapter 2 for the importance of standardization to combinatorial processes in genomics and proteomics). The key signature of black boxing for a particular invention p is the relative size of its (average) primary contribution, compared to all other inventions. If recombinant evolution is typical, the average size of the average primary contribution will be relatively low. When an invention has an above-average primary contribution across descendants, this strongly suggests that its components are black boxed and drawn upon as whole units rather than as a set of parts. Now consider the primary inheritance of a specific invention k. If the primary inheritance is low, then invention k has sampled from several sources. When the primary inheritance is low and the primary contribution is high, this suggests that k has drawn on several parents and bundled the parts together into a unit with emergent value.[16] There is a synergistic, nonadditive, epistatic interaction among the parts, which leads others to select the whole black box. We can validate this intuition using related traces. For example, black-boxing events will likely correspond to cases in which the citations to a black-boxing patent supersede and largely replace citations to the patents (and separable components) on which it draws (Funk and Owen-Smith 2016).

PROBABILISTIC MODELS, POSSIBLE HISTORIES

Before discussing model-based approaches to the study of cultural genealogies, we pause to discuss the role of models, uncertainty, and evidence. Despite the rich electronic record of inventive activity in science, technology, and other cultural domains, much remains unknown. Specific influence pathways may be discoverable, but only after considerable effort—for example, using traditional historical methods. Hence, the principled integration of model and evidence is important and the rigorous representation of uncertainty essential. Probabilistic models provide a comprehensive framework for such integration (Ghahramani 2015).

Insights from the previous section were limited in two ways. First, our model of the discovery process was implicit and narrow: greedy search. While this model provided a useful parsimony principle allowing us to construct well-defined, parsimonious "explanations" of observed histories (i.e., DAGs), it may distort inference insofar as it misspecifies the inventive process. By biasing reconstruction toward tree-like structures, parsimony provides a conservative test for multiple inheritance. Data that support a parsimonious explanation with multiple inheritance are quite likely to have been generated by some kind of recombinant process, but the details are likely to be wrong, and we learn nothing about the inventive process from the data. This leads to the second limitation. We have no idea how much certainty to have in our reconstructed cultural genealogy.

Probabilistic model-based inference has neither of these difficulties. First, we can construct a much more flexible model than greedy search. This flexible model allows us to specify what we know about cultural inheritance (e.g., from qualitative or historical investigations of innovation, of which there are several examples in this book)—and what we do not know. This lack of knowledge is represented by parameters in the model: we may have a general sense of the underlying generative process, but different model parameters realize different generative scenarios. Any hunches we have about the generative process can be further specified through priors on the parameters. For example, our model might have a parameter controlling the average number of parents that contribute to a new invention (it will). If we have a strong reason to suspect that the average number of parents is two, then we can put a prior on that parameter concentrated around two. If not, we may choose a totally uninformative prior to represent our uncertainty about its value. But most of the action in probabilistic modeling does not take place in the priors; it takes place in inference. Inference is simply a process of learning from the data. The rules of probability (specifically, Bayes' rule) allow us to use data to update our uncertainty. Doing so avoids the second limitation of parsimony methods: we can precisely quantify our certainty in the reconstructed genealogy.

Flexibility is important when reconstructing cultural genealogies, because of our agnostic position on patterns and processes of cultural evolution. It is very likely that cultural evolution follows different patterns in different domains. We know that stone tools and some features of language (to pick two examples) follow branching patterns. We strongly suspect that some areas of high technology follow combinatorial, reticulate patterns (Fleming and Sorenson 2001, 2004; Arthur 2009). Model-based inference allows us to discover different patterns and processes of cultural evolution in different domains. We do not claim that our models perfectly describe the world (even after inference). We do claim, however, that they give a relatively precise sense of the generative processes and historical trajectories that could explain available evidence. Crucially, our models can focus attention on the most plausible or informative influence pathways that merit detailed, costly historical or ethnographic investigation (Wimsatt 2013b); in other words, they can provide structure to the larger problem agenda of understanding cultural evolution in specific domains and guide the attention of relevant disciplinary partners to maximize the value of their contributions (see the introduction by Love and Wimsatt).

LEARNING ABOUT MULTIPLE INHERITANCE

With model-based inference, we allow the data to reduce our uncertainty about the nature of the inheritance process (Ghahramani 2015). As input data, we again have a set of types ordered in time. These types could be patents, publications, products, or other complex cultural entities (e.g., organizations or people—anything decomposable into documented building blocks; see chapter 1). Each type j is characterized by a unique set of features.[17] A type may correspond to multiple entities, insofar as these entities are "indistinguishable" from the perspective of these features. The more refined the set of elementary features (i.e., the larger the number of distinct features), the more types there will be. Consider, for example, the description of patents using a few classification codes, as opposed to more detailed descriptions extracted and normalized from full text. In the former case, many patents might correspond to the same type; in the later case, a single type might correspond to just a few patents, or even a unique one.

Types are ordered in time; we can retain time as a component of the model to account for time intervals, as in parsimony. We can also use temporal information to model the probability that the recent and the ancient past are considered as sources of potential parents. If we have information about the spatial, social, and cognitive "place" of invention, we can learn from the data whether there are similar "local" biases (Adams 2002). For example, we might have detailed information about the time and place of invention. Since recent inventions are generally easier to retrieve than much older inventions and local knowledge is easier to access than distant knowledge (e.g., due to the institutional or organizational structuring of cultural breeding populations), we might modulate the probability of choosing a particular parent by a decaying function of temporal separation and geographic distance. But since we do *not* know how much more likely inventors are to retrieve recent or local knowledge over ancient or distant knowledge, we characterize that function with unknown parameters. We can learn from the data a reasonable range of possible parameter values.

For simplicity and concreteness, we describe the model as a generative process. From the generative model, we can construct a joint probability distribution over the observed types **F**, the DAG of parentage assignments **P**, and the parameters Θ that control the number of parents and the sampling of features from parents. Given the joint distribution, we can construct the posterior probability over DAGs and parameters conditional on the observations using Bayes' rule. The posterior probability is our ultimate target: given our modeling assumptions, our priors, and (most importantly) our available evidence, we can sample from the posterior probability distribution to discover which DAGs are more (and less) likely and which parameter values are more probable, given the evidence.

This is conceptually identical to the standard Bayesian approach to phylogenetic tree reconstruction (Felsenstein 2004; Bergstrom and Dugatkin 2012). In that case, we want to construct (or at least sample from) the posterior distribution over trees, conditioned on available data D. In principle, we may have a prior over trees; in practice, a flat prior is usually used so that every tree is equally probable, a priori. Using Bayes' rule, we can construct the posterior:

$$\text{Pr(Tree}\,|\,D) = \frac{\text{Pr}(D\,|\,\text{Tree})\text{Pr(Tree})}{\text{Pr}(D)} \tag{1}$$

where parameters of the model of character or sequence evolution have been suppressed. The likelihood $\text{Pr}(D\,|\,\text{Tree})$ is well defined and can be easily computed; it is the probability of the observed data given a particular tree, a particular model of evolution, and particular parameter values characterizing that model.

Our generative model for multiple inheritance can be quickly summarized by listing its steps. The model begins at the earliest observation and iterates over the following:

1. Choose the number of parents.
2. Choose the identity of the parents.
3. Choose features from the set of parents.

This is, of course, the same basic picture that guided our parsimony method above. In a fully Bayesian approach, we would begin the generative process by drawing the parameters from prior distributions (Gershman and Blei 2012). Given these parameters, we would then iterate the steps above. Note that even here some assumptions are baked into the model; for example, the number of parents is not influenced by the identity of the parents nor can the features selected from one particular parent influence the selection of subsequent parents.[18] We now (briefly) describe each step; we provide mathematical details in the appendix.

Number of Parents

For a given observation, the generative process begins by choosing the number of parents. The number of parents is drawn from a distribution controlled by one or more parameters θ_p. The simplest such distribution would be a Poisson, in which case the parameter would control the average number of parents. This picture is similar to the so-called Indian buffet process, in which customers sample dishes from the buffet until they have chosen a number of dishes drawn from a Poisson distribution (Griffiths and Ghahramani 2011). Using a Poisson distribution, however, assumes that there is a typical number of parents and that the distribution of parents is tightly peaked around that number. That might not be the case—another instance in which model specification will shape inference. Ideally, one would explore models with alternative distributions (and mixtures of distributions) and check them using techniques for model criticism, as through predictive sample reuse or posterior predictive checks (Blei 2014). In full generality, one might permit the parameter(s) controlling the number of parents to change over the course of evolutionary history, allowing one mode of cultural evolution (e.g., branching) to dominate earlier portions of the DAG and another mode (e.g., reticulation) to dominate later parts. See Silvestro et al. (2014) for an inspiring approach to capturing such shifts.

Choosing Parents

Once the number of parents has been selected—equivalently, once we have selected the in-degree of the node j in the directed acyclic graph representation—we must choose *specific* parents. There are many ways to formalize this choice process. For simplicity, we assume that each parent is chosen independently. If each parent, in turn, has an equal probability of being chosen (a highly unrealistic assumption), then each invention will have an asymptotically Poisson number of offspring (i.e., out-degree). A slightly more complicated model, imitating the Indian buffet process, assumes that inventors start with the most recent potential parents and then work backward in time. Each potential parent is considered; it is selected as an ancestor with a probability proportional to its popularity (i.e., its current number of offspring or, equivalently, out-degree). This process repeats until the full complement of parents is chosen. In this model, preferential attachment (which asymptotically produces a power-law distribution of out-degrees) competes with recency bias. Although older nodes may have given birth several times, they are less likely to be selected as they get older; more nodes must be "skipped" to get to them. In general, the probability of choosing a particular parent can depend on many different factors. Parents might have an intrinsic "fitness." This fitness could be drawn from some distribution when the parent is initially created. More realistically, the fitness could be determined by the constellation of features present in the invention (thus allowing for inventions with similar features to have similar fitness). Preferential attachment (rich-get-richer dynamics) could play a role, reflecting prestige bias, conformist bias, or both (Boyd and Richerson 1988; Mesoudi 2011). Parent choice may be shaped by explicit markers of social identity, such as disciplinary, professional, or institutional affiliation (see chapters 1 and 12), as well as by temporal, spatial, social, and conceptual distance. Finally, we could explicitly model choice-set formation so that inventors make a cognitively plausible choice across a small number of possibilities, rather than implicitly considering the entire universe of possible parents (Swait and Ben-Akiva 1987; Bruch, Feinberg, and Lee 2016). These more complex models of parent choice would allow researchers to test important assumptions. For example, we could discover that inventors are more likely to select a set of parents from the "same" cultural breeding population (e.g., scientific discipline or technology area).

Number of Features

We assume that the number of features sampled from the parents is independent of the number or identity of the parents. This is, again, a simplifying assumption; it could be that inventions with more parents tend to sample more features or that inventions with high-fitness or popular parents sample more. As with the number of parents, the simplest choice for this distribution is Poisson, though this could be generalized to admit more complex distributions.

Increasing Complexity

As currently described, this generative model has a major limitation. It cannot easily deal with cultural evolution in which features *accumulate*. Yet this is an incredibly common mode, both in technological evolution (Arthur 2009) and in the sequential acquisition of skills by developing biological individuals (Love and Wimsatt, the introduction to this book; see also Wimsatt, chapter 1). Building blocks already characterized by many features can be used to assemble an even larger invention, such as airplanes and boats combined into an aircraft carrier (Arthur 2009). If our types are defined by features at a consistent granularity, then later inventions may have more features, on average. We can capture this by allowing the average number of features to grow over time; this growth rate can be controlled by one or more parameters subject to inference. A more interesting approach would allow the data to "suggest" bundles of elemental features that should be treated as a single feature because of frequent copresence. This compression or dimensionality reduction of the feature space implements a form of parsimony; it attempts to simplify the explanation of observed facts by reducing the number of components. There are a range of approaches to so-called feature or representation learning. Matrix factorization (Bengio, Courville, and Vincent 2013) could be applied periodically or continuously to update the feature space confronting inventors; the compression schedule could be optimized so that the number of compressed features in any given invention remains relatively constant. A latent aggregation–fragmentation process provides a purely probabilistic alternative. Features can aggregate into a bundle with some small probability, and bundles can disaggregate into constituent features with another (Ghahramani 2013; Blei 2014). This would provide an explicit probabilistic model of the black-boxing process. A more radical alternative would replace surface features with latent *feature generators,* emulating topic modeling (Blei 2014). However it is implemented, such chunking (Wimsatt 2013a) is consistent with both plausible limits on working memory (Miller 1956) and the robustness of modular assembly (Simon 1969; Latour 1987; Arthur 2009). As in the exploratory analysis, the consistent chunking of several features into a bundle suggests a black-boxing event and could be used to detect such moments in the unfolding cultural-evolutionary process.[19]

Choosing Features

Once the number of features has been selected, we must choose specific features. As with parent selection, it is simplest to assume that features are selected independently. Indeed, the simplest version of feature selection would look very much like parent selection, moving through features in some order and selecting them proportional to their popularity, either over the entire past history of the system or over the set of parents.[20] Unlike parent selection, however, we allow the creative unit (the inventive individual or team) to introduce some number of new features unobserved in the parent set—and possibly never yet observed in the history of the system. This step allows for the introduction of radical novelty to the inventive system; not just the novel combination of features but the addition of entirely new features (Foster, Rzhetsky, and Evans 2015).[21] On a more mundane level, this modeling assumption allows any invention to be generated from any parent set, albeit with very small probability. This is useful computationally. It is also important substantively: it may be that inventors introduce a particular feature by plucking it from the inventive zeitgeist, rather than drawing on a particular parent. The capacity to generate new features also connects this generative model to Bayesian nonparametric processes more generally, as the number of potential features is not determined a priori in the model, although it is obviously given by the data.

Inference

Although somewhat nonobvious from the generative description, the model outlined above is remarkably close to standard phylogenetic inference in structure. Instead of a tree, the parentage assignment **P** describes a DAG that respects the time ordering of inventions **F**. Earlier inventions point toward later inventions that draw on them for features. For a given parentage assignment and values of the generative parameters Θ (i.e., the two explanatory parts of the model), we can calculate the probability of the data $\Pr(\mathbf{F} \mid \mathbf{P}, \Theta)$ directly. We have priors on the model parameters $\Pr(\Theta)$, which may be informative or uninformative. Given the parameters, the probability of any particular DAG **P** is determined. We can combine all these parts using Bayes' rule to compute the posterior distribution over the space of DAGs (i.e., explanatory histories) and parameters (i.e., explanatory processes). It is

$$\Pr(\mathbf{P}, \Theta \mid \mathbf{F}) = \frac{\Pr(\mathbf{F} \mid \mathbf{P}, \Theta)\Pr(\mathbf{P} \mid \Theta)\Pr(\Theta)}{\Pr(\mathbf{F})}. \qquad (2)$$

The denominator cannot be calculated because computing the probability of the data requires a sum over all possible DAGs **P**. We can approximate the posterior, as in phylogenetic inference, using standard methods like Markov chain Monte Carlo (Gershman and Blei 2012) to draw from or otherwise approximate the posterior distribution.

MODELING THE CULTURAL EVOLUTION OF DEVELOPING INDIVIDUALS, ORGANIZATIONS, AND INSTITUTIONS

While our approach was inspired by the challenges of modeling multiple inheritance in technological evolution, it can be applied to any cultural data for which there is dense sampling and information about the sequence of observations. One particularly exciting application concerns data in which well-defined units with temporal duration but malleable features (e.g., individual humans, organizations, genres, or states) are observed repeatedly. In this case, we can view an observation of unit j at time t as a recombination of its state at last observation with features drawn from other available "parents."

This strategy emulates the approach suggested in Boyd and Richerson (1988) for theoretical models of horizontal transmission during the life span. In other words, unit j selects its characteristics at time t by sampling from its previous state as well as from its contemporaries and predecessors. The astute reader will have noticed that this model is very close to models of social contagion; given this similarity, we must be vigilant against the possible confounding of social contagion with latent homophily (Shalizi and Thomas 2011). That said, the adopted feature(s) must come from *somewhere,* and it is possible that specific contagion versus diffuse adoption driven by latent homophilous traits can be distinguished by the presence or absence of specific influence paths in the posterior distribution over DAGs.

In practice, capturing the known features of human cultural development, such as the sequential nature of skill acquisition, would require relaxing many of the assumptions outlined above. The features retained by unit j from its past state would affect its selection of "parents" for cultural updating, as well as the features chosen from them (Foster 2018). For example, it is almost surely the case that someone who knows single-variable calculus at time t and multivariable calculus at time $t + 1$ retained his knowledge of single-variable calculus and learned the multivariable version from his teacher and/or textbook. It is also likely that this teacher is someone close in physical, social, and organizational space. Incorporating geographic or social proximity in the choice of cultural parents, including evolving markers of social identity (see chapter 12), would allow us to deal directly with cultural population structure (Wimsatt 2013a). Note that a model of "parent" selection incorporating cumulative advantage is very similar to prestige bias, an important mechanism in cultural microevolution (Boyd and Richerson 1988; Mesoudi 2011). In our running example, this imaginary student is more likely to select a popular model known for her excellent pedagogy. The student may pick up other cultural traits as a by-product of this learning relationship, such as a specific story or preference for a certain mode of investigation. This same framework would provide a powerful and precise technique for studying the evolution of organizations and institutions more broadly, as there are often repeated observations of these units.

In other words, this formal trick extends the range of our framework from cultural macroevolution to the microevolutionary dynamics of cultural change. We thereby provide an intriguing twist on Wimsatt's observation in chapter 1 of this volume that heredity and development "interchange roles in the study of biology and culture," with cultural development being more transparent to investigation and hence helping to illuminate cultural heredity. In our framework, long-term patterns of cultural heredity and short-term patterns of cultural development are treated in the same way!

TESTING MODELS OF MULTIPLE INHERITANCE

In describing our approach to the study of multiple inheritance, we have emphasized that studying cultural macroevolution requires uncertain inference of unknown processes from rich data. How might we validate these models? We briefly mentioned internal checks using model criticism, as through predictive sample reuse or posterior predictive checks (Blei 2014). Such checks are important, but they are unlikely to persuade the obdurate skeptic. Thus, we note that, just as biologists rely on paleontologists to validate the presence of particular extinct organisms at particular times, so too can students of computational cultural evolution turn to historians, sociologists, anthropologists, and archaeologists to validate particular claims about particular influence pathways and inventive events; they might also turn to cognitive scientists to validate the detailed cognitive mechanisms or processes implied by their inferences. Because such validating steps are expensive in time, labor, and expertise, validation should start with inferences that show the least uncertainty (e.g., an assembly process that shows up in 99 percent of the DAGs sampled from the posterior distribution), although weaker inferences can give provocative hypotheses as well. In this way, large-scale computational studies of cultural evolution depend on and inform a wide range of rich disciplinary perspectives and methodologies. In other words, our approach both scaffolds and is scaffolded by a much larger research agenda. It provides a way to analyze unprecedented new data sets (Evans and Foster 2011) as important model organisms for the large-scale quantitative study of cultural evolution without embracing the limiting conceptual vocabulary of a single discipline (see the introduction by Love and Wimsatt).[22]

TRIMS, TRAMS, AND THE MODE OF CULTURAL EVOLUTION

In this chapter, we described the foundations of an agnostic approach to reconstructing cultural lineages—one general enough to identify both patterns dominated by branching and patterns dominated by reticulation.[23] It is worth reflecting briefly on when and why we might expect to see these two archetypal modes of cultural inheritance. Approaches based on phylogenetic inference have leaned on the assumption that Transmission Isolating Mechanisms (TRIMs) like geopolitical boundaries, ethnocentrism, and language barriers limit the mixture and recombination of cultural components across lineages (Durham 1991; Mesoudi 2011). In the language of Wimsatt (2013a), these TRIMs mostly appeal to population structure—they are mechanisms that prevent culturally distinct populations from mixing and create distinct cultural breeding populations (see chapter 1). For example, language barriers could be viewed as institutionally induced cultural population structure. TRIMs make the pattern of cultural evolution branch-like, with a relatively slow pace—novelty is just harder to come by when new components and combinations must be produced within a cultural lineage. Hence, TRIMs create patterns of cultural evolution perfectly suited for detection by existing methods of inference that assume a single dominant inheritance pathway for each observed entity.

Although the precise TRIMs that are commonly invoked in cultural phylogenetics are much less common in the modern era of science and technology, their analogs nevertheless exist. For example, the citation of patents is slower, and radiates more slowly outward in space from the focal patent, than the citation of scientific articles (Adams, Clemmons, and Stephan 2006). Scientific communities can be largely cut off from one another by geopolitical boundaries (as in the cladogenesis that resulted in a distinctive tradition of Soviet mathematics in the mid-twentieth century) or by jargon (Vilhena et al. 2014). And population structure, whether imposed by geography, disciplines, schools of thought, or status, can substantially slow the spread of new scientific or technical knowledge, especially when it is difficult to codify (Kaiser 2009). Whenever transmissible units depend on extensive previous training or time-consuming pedagogy for reliable transmission (Kaiser 2009), their spread across populations will be slower, and cultural evolution is more likely to manifest a branching mode on some levels of analysis (Boyd et al. 1997; Wimsatt 2013a). This should be true whether the transmissible unit is crafting a stone tool or crafting an elegant proof. Thus, organizationally enabled scaffolding, while facilitating cumulative cultural evolution within a particular lineage (e.g., a discipline), promotes the development of distinct cultural breeding populations. Careers can be strongly canalized within an existing cultural population, such as when departments only hire faculty with training in their specific discipline or with degrees from a select range of similar departments (Clauset, Arbesman, and Larremore 2015). This canalization limits cumulative cultural evolution across lineages (see chapter 1).

Nevertheless, the system of modern science and technology also contains Transmission *Accelerating* Mechanisms or TRAMs, which increase the rate of horizontal transmission and recombination. These TRAMs range across the "relevant units of the cultural system" described by Wimsatt in this book. TRAMs most obviously include infrastructure such as modern transportation and communication technologies. They also include institutional conventions, like the increased dominance of English as scientific lingua franca; indeed, spoken language and writing have been powerful TRAMs and TRIMs at different scales throughout human history (see chapters 9 and 10). Classic Mertonian norms (like universalism) promote the free flow and exchange of ideas (Merton 1973), as do the explicit references, patent subclasses, and article key words associated with the publication process itself—all conventions that make information easier to find and retrieve. International conferences break down geographic population structure, while interdisciplinary meetings aim to break down the population structure created by discipline, training, and school of thought. Interdisciplinary hiring redirects careers across multiple cultural breeding populations, facilitating recombination across cultural lineages. Most intriguingly, technologies and some ideas can internalize their scaffolding so that they have easily discernible affordances. This process of black boxing allows the technology to move and recombine more easily, as described by Bruno Latour (1987) and Michel Callon (1986). In a sense, these black-boxed artifacts actually scaffold their own recombination (Wimsatt 2013a), and we hypothesize that such auto-scaffolding is the crucial TRAM driving rapid, recombinant, and cumulative technological evolution.

Our models and exploratory methods are designed precisely to allow a system of artifacts, ideas, institutions, or individuals to reveal its dominant mode of cultural evolution, whether that be branching, reticulation, or some mixture of the two. In revealing the varying tempo and mode of cultural evolution across many contexts, these methods will help us understand in detail the competition between the TRIMs and the TRAMs that together pattern the evolution of technology, ideas, and human culture more broadly. We hope that our methods, and the underlying conceptual apparatus, can accelerate the move "beyond the meme" toward the integrated, interdisciplinary, multimethod study of cultural evolution.

MATHEMATICAL APPENDIX

Here, we provide concrete mathematical details and illustrations for the methods outlined above. This appendix is best read in parallel with the main text.

Parsimony

For each invention j, let $F'_j = F_{j/novel}$ be the set of all features in invention j, once any novel features have been removed. The set of all possible parents of j is denoted $\mathcal{P}_{t_j}$. For each potential parent $p \in \mathcal{P}_{t_j}$, we compute the intersection of the set of features in j that *could have been* inherited (F'_j) and the set of features in p (F_p); call this

$$W_{jp} = F'_j \cap F_p. \tag{3}$$

We select as the first or "prime" parent the prior invention k with the maximum W_{jk}. This is the prior invention that explains the most features. If there are multiple equally explanatory inventions, we can select the invention with the smallest Δt_{jk} (recency bias), the smallest Δd_{jk} (local bias), and so forth. If all parsimony principles are exhausted, choose at random.

Now define $F'_{j/k}$ as the set of all features in invention j that remain to be explained, given that k is one of the parents. We iterate the procedure above, defining

$$W_{jp|k} = F'_{j/k} \cap F_p \tag{4}$$

and selecting as the next parent the invention m with the maximum $W_{jm|k}$ (i.e., the one that explains the most features not explained by k). We repeat this procedure until *all* heritable features of j have been explained by one or more parents. We repeat this procedure for all j to reconstruct a parsimonious history for our observed technologies. Note that this history will be a directed graph; we establish the convention that arcs run *to* invention j from parents p, k, m, and so on, so that W_{jp} counts the number of components that flow from p to j, $W_{jm|p}$ counts the number of components that flow from m to j, and so on. For notational simplicity, we will refer to $W_{jm|p}$ as W_{jm}, $W_{jq|pm}$ as W_{jq}, and so on, unless the "conditioning" is important.

Black Boxing Measures

Define $W_j = |F_j|$ —that is, the number of features in the j-th invention. Now define

$$\widehat{W}_{jp} = \frac{W_{jp}}{W_j} \tag{5}$$

as the fraction of all components in j inherited from ancestor p. The primary inheritance is just the largest $\widehat{W}_{jp}$ over all ancestors p. Call this $\widehat{W}_{jp}^{max}$. Properties of the frequency distribution of primary inheritance $\Pr(\widehat{W}^{max})$ can provide suggestive evidence for branching or reticulate evolution.

With slight abuse of notation, we can define the primary contribution

$$\widecheck{W}_{jp} = \frac{W_{jp}}{W_p} \tag{6}$$

as the fraction of components in p passed on to its descendent j in a given reconstruction, where we look at cases in which p is the primary, secondary, tertiary ancestor, and so on. The mean $\widecheck{W}_p = \frac{1}{N_j}\sum_j \widecheck{W}_{jp}$ and mode over all descendants j give us an idea of whether the components of p are typically taken as a block or whether they are separable.

The key signature for black boxing is a modal $\widecheck{W}_{jk}$ for a given k that is significantly higher than the typical mode of $\widecheck{W}_{jp}$ over the population of p's or, equivalently, a mean primary contribution $\widecheck{W}_k$ that is significantly higher than the mean $\widecheck{W}_p$ over the population of p's. When the mean primary contribution $\widecheck{W}_k$ is high but the mean fraction of components that k inherits from its ancestors $\dfrac{\sum_{p \in P_k} \hat{W}_{kp}}{|P_k|}$ is low (or, equivalently, when the primary in-

heritance is low), it is likely that invention k has sampled from several sources and black boxed the parts.

Probabilistic Models of Multiple Inheritance

As input, we again have a set of types ordered in time. Each type j is characterized by a unique set of features F_j. We can equivalently represent this as a binary feature vector $\mathbf{f}_j$ of length M, where types have M possible features. Thus, we observe a time-ordered collection of N types $\mathbf{F} = \{\mathbf{f}_1, \mathbf{f}_2, \mathbf{f}_3, \ldots, \mathbf{f}_N\}$.[24]

Number of Parents

For a given observation $\mathbf{f}_j$, the generative process begins by choosing the number of parents n_j^p. The number of parents is drawn from a distribution controlled by one or more parameters θ_p. The simplest such choice would be a Poisson distribution, Poisson(α_p), with α_p controlling the average number of parents. This is similar to the so-called Indian buffet process, where customers stop after they have sampled Poisson(α) dishes (Griffiths and Ghahramani 2011).[25] This distribution, however, would assume that there is a typical number of parents and that the distribution of n_j^p is tightly peaked around that number. This could be relaxed.

Choosing Parents

There are many ways to formalize parent choice. For simplicity, we assume that the probability of assembling a particular collection of n parents factorizes

$$\Pr(P_j = \{a_1, a_2, a_3, \ldots a_n\} \mid n_j^p = n) \sim \prod_{1}^{n} \Pr(a_1 \in P_j) \ldots \Pr(a_n \in P_j). \quad (7)$$

Number of Features

We assume that the number of features to be sampled from the parents is independent of the number or identity of the parents. As with the number of parents, the simplest choice for this distribution is Poisson(α_f), though this could be generalized to admit more complex distributions.

Increasing Complexity

Our generative model, as proposed, cannot easily deal with cultural evolution in which features accumulate. One way to deal with this is to make the parameter controlling the number of features time dependent. For example, α_f grows with t at a rate β that is also subject to inference. A more interesting approach would allow the data to "suggest" relevant bundles of elemental features that should themselves be treated as features because of frequent copresence—a compression of the feature space. This could be done in a number of ways, as described in the main text.

Choosing Features

Once the number of features has been selected, we must choose specific features. Unlike parent selection, however, we allow the creative unit to introduce some number $(0 - m)$ of new features $\sim \text{Poisson}(\alpha_{novel})$ unobserved in the parent set—and possibly never yet observed in the history of the system. This allows any observation $\mathbf{f}_j$ to be generated from any parent set while also allowing true novelty through the creation of entirely new features.

Inference

Although somewhat nonobvious from the generative description, the model defined above is remarkably close to standard phylogenetic inference in structure. The parentage assignment $\mathbf{P}$ is just a directed acyclic graph that respects the time ordering of $\mathbf{F} = \{\mathbf{f}_1, \mathbf{f}_2, \mathbf{f}_3, \ldots, \mathbf{f}_N\}$. For a given $\mathbf{P}$ and values of the generative parameters—for example, α_p and $\alpha_F \in \Theta$—we can calculate $\Pr(\mathbf{F} \mid \mathbf{P}, \Theta)$ quite directly. Then

$$\Pr(\mathbf{P}, \Theta \mid \mathbf{F}) = \frac{\Pr(\mathbf{F} \mid \mathbf{P}, \Theta)\Pr(\mathbf{P} \mid \Theta)\Pr(\Theta)}{\Pr(\mathbf{F})} \tag{8}$$

where the denominator cannot be calculated because of the required sum over all possible $\mathbf{P}$. Thus, we can approximate the posterior, as in phylogenetic inference, using standard methods such as Markov chain Monte Carlo (Gershman and Blei 2012).

NOTES

1. The frequency of horizontal gene transfer among prokaryotic taxa (Doolittle and Bapteste 2007) has created an urgent need for methods to study reticulation in biology, for example, Kunin et al. (2005). Although computational biologists have answered the call (Huson, Rupp, and Scornavacca 2010), these methods are either too generic (i.e., they are essentially

clustering) or involve too many specific biological processes (e.g., gene dele-tion or insertion) to provide a useful starting point. Until very recently, there were no Bayesian, generative model–based approaches to reticulation, though Wen, Yu, and Nakhleh (2016) may provide a way forward.

2. In some cases, horizontal transmission is reserved for trait flows within a generation (peer-to-peer), and oblique transmission is used when traits flow from nonparental individuals in an earlier generation to individuals in a later generation. We will not make this distinction here.

3. Defined by Wimsatt as "structure-like dynamical interactions with performing individuals that are means through which . . . competencies are constructed or acquired by individuals or organizations."

4. Note the epidemiological language.

5. Note here the role of several factors explored at length in this volume as TRIMs; e.g., language (see chapter 9) and identity (see chapter 12).

6. As we will argue later, the methods we propose extend unproblem-atically to some other cultural items and could even be used to model se-quential skill acquisition by developing biological individuals (see the introduction by Love and Wimsatt; see also chapter 1). For concreteness, we focus the discussion on science and technology, but the reader should keep implicit generalizations in mind throughout.

7. For many problems, a tree-based simplification could be illuminat-ing as an initial analysis of data.

8. We assume here that the features are already given, as in patent classes, PACS (physics and astronomy classification scheme) codes, or MeSH (medical subject heading) terms. In cases in which features must be con-structed by the analyst from scratch, one can draw upon a well-developed literature in feature engineering.

9. The cleaned, curated features given by patent classes, PACS codes, and MeSH terms are useful insofar as they approximate, in some fashion, the principles of vision and division that characterize the relevant communities of practice. We leave aside the very interesting question of how different com-munities of practice might break up the same invention into different ele-mentary building blocks; this would require detailed thinking about the sequentially dependent and organizationally scaffolded skill acquisition (Goodwin 2017) that would yield different ways of seeing the same inven-tion (Love and Wimsatt, the introduction to this book; Wimsatt, chapter 1), that is, different modes of "professional vision" (Goodwin 1994). Data sci-ence techniques for feature engineering may be useful for heuristic feature

construction where expert taxonomies are incomplete or nonexistent (Scott and Matwin 1999; Anderson et al. 2013).

10. While this may seem like a strong assumption, note that it has a certain plausibility in terms of search. If there are many possible histories that enrich the "primary" parent with residual features (the *three-parent* history) but only one history that pairs the right two inventions (the *two-parent* history), then we are more likely to observe someone start from the primary parent and then enrich than to observe an inventor who lands on exactly the right pair of parents.

11. Of course, this introduces a bias into our reconstruction, but absent strong evidence to the contrary, we think that the greedy assumption tends to capture more probable pathways. It is also consistent with approaches to human cognition, like case-based reasoning (Aamodt and Plaza 1994).

12. If we wish to weaken this assumption privileging significant inheritance from a single parent p, we can search over the space of pairs, triples, tetrads, etc., for the combination that contributes the most features. The computational cost for this exploration is high, however. Rather than searching through n possible parents for the single most explanatory parent (so the overall search is $\mathcal{O}(n)$), we would have to search through $\binom{n}{2}$ pairs, $\binom{n}{3}$ triples, $\binom{n}{4}$ tetrads, etc. The computational cost grows exponentially: $\mathcal{O}(n^2), \mathcal{O}(n^3), \mathcal{O}(n^4)$, etc.

13. It will be acyclic—i.e., have no loops $i \rightarrow j \rightarrow k \rightarrow i$—because the future cannot influence the past by construction.

14. This method of constructing a parsimonious evolutionary explanation is not assured to recover the actual inheritance pattern of cultural traits. Moreover, the adaptive, evolutionary significance of inheriting a particular feature may only be minimally associated with the primary inheritance on which parsimony focuses.

15. Indeed, Wimsatt notes that "black boxing is a crucial feature of most complex sequential skill acquisition" in his contribution to this book.

16. Note that, on this account invention, k (with low primary inheritance and high primary contribution) creates the black box, which persists as a packet into the next generation. In principle, persistence across multiple generations would provide stronger evidence for true black boxing.

17. The need to characterize types with discrete features or "building blocks" is an obvious limitation and a potential source of bias. These methods work best for entities that have already been characterized with features. As discussed previously, it is certainly possible to induce features when they

are not already available, but we must be especially cautious about inferences from these induced features. Independent validation of the features is a necessity. And even when features have been developed for other reasons (e.g., search or classification), these may not always correspond to the features that are relevant to the inventive process, introducing bias.

18. This will need to be relaxed in section 7 to model the sequential skill acquisition common in developing biological individuals or organizations.

19. As new, complex features are discovered—by inventors through black boxing and by analysts through feature reduction—even *old* artifacts could "acquire" new, heritable features as bundles of components are reinterpreted as coherent units. Note that the routine combination of components could take place immediately following their initial combination, could increase gradually, or could follow a discontinuous trajectory as an old combination becomes *fit* to a new environment. For example, consider the explosive rise in the use of Bayesian methods following the advent of computers, which scaffolded and explicitly catalyzed their application (see chapter 11).

20. This simplifying assumption runs roughshod over the sequential dependence of features—or even their functional interdependence.

21. The frequency of entirely new features will scale inversely with the resolution of existing features. For example, a new feature in a coarse-grained scheme might include a custom-built molecule, but this would simply be a new combination of *existing* features if atoms or molecular motifs were components (Arthur 2009).

22. Here, we are thinking especially of patents (see chapter 6).

23. While our approach allows panmixia—i.e., the selection of arbitrary parents to produce offspring (see chapter 1)—it also allows distinct cultural breeding populations to emerge from the data. It also allows researchers to encode distinct hypotheses about factors like geographic, social, or cultural distance that structure cultural breeding populations and violate panmictic assumptions.

24. We largely follow the notation and presentation of Gershman and Blei (2012) here.

25. If the feature set describing all parents is finite, we could model this by a simple Beta-Bernoulli process. There are many ways to set up a conceptually similar generative model; the important ingredients are (1) a process controlling the number of parents, (2) a process selecting the parents, and (3) a process choosing features from the set of parents.

REFERENCES

Aamodt, A., and E. Plaza. 1994. "Case-Based Reasoning: Foundational Issues, Methodological Variations, and System Approaches." *AI Communications* 7 (1): 39–59.

Adams, J. D. 2002. "Comparative Localization of Academic and Industrial Spillovers." *Journal of Economic Geography* 2 (3): 253–78.

Adams, J. D., J. R. Clemmons, and P. E. Stephan. 2006. *How Rapidly Does Science Leak Out?* Technical report. National Bureau of Economic Research.

Anderson, M. R., D. Antenucci, V. Bittorf, M. Burgess, M. J. Cafarella, A. Kumar, F. Niu, Y. Park, C. Ré, and C. Zhang. 2013. "Brainwash: A Data System for Feature Engineering." In *Proceedings of Conference on Innovative Data Systems Research* (CIDR).

Arthur, W. B. 2009. *The Nature of Technology: What It Is and How It Evolves.* New York: Simon and Schuster.

Bengio, Y., A. Courville, and P. Vincent. 2013. "Representation Learning: A Review and New Perspectives." *IEEE Transactions on Pattern Analysis and Machine Intelligence* 35 (8): 1798–828.

Bergstrom, C. T., and L. A. Dugatkin. 2012. *Evolution.* New York: W. W. Norton.

Birch, A. 1967. *The Economic History of the British Iron and Steel Industry, 1784–1879: Essays in Industrial and Economic History with Special Reference to the Development of Technology.* London: Frank Cass.

Blei, D. M. 2014. "Build, Compute, Critique, Repeat: Data Analysis with Latent Variable Models." *Annual Review of Statistics and Its Application 1,* 203–32.

Bourdieu, P. 1990. *The Logic of Practice.* Palo Alto, Calif.: Stanford University Press.

Boyd, R., M. Borgerhoff-Mulder, W. H. Durham, and P. J. Richerson. 1997. "Are Cultural Phylogenies Possible?" In *Human by Nature: Between Biology and the Social Sciences,* edited by P. Weingart et al., 355–84. Mahwah: Lawrence Erlbaum.

Boyd, R., and P. J. Richerson. 1988. *Culture and the Evolutionary Process.* Chicago: University of Chicago Press.

Bruch, E., F. Feinberg, and K. Y. Lee. 2016. "Extracting Multistage Screening Rules from Online Dating Activity Data." *Proceedings of the National Academy of Sciences* 113 (38): 10530–35.

Callon, M., 1986. "The Sociology of an Actor-Network: The Case of the Electric Vehicle." In *Mapping the Dynamics of Science and Technology,* edited by Michel Callon, John Law, and Arie Rip, 19–34. London: Palgrave Macmillan.

Claidière, N., T. C. Scott-Phillips, and D. Sperber. 2014. "How Darwinian Is Cultural Evolution?" *Philosophical Transactions of the Royal Society B* 369 (1642): 20130368.

Clauset, A., S. Arbesman, and D. B. Larremore. 2015. "Systematic Inequality and Hierarchy in Faculty Hiring Networks." *Science Advances* 1 (1): e1400005.

Cochrane, E. E., and C. P. Lipo. 2010. "Phylogenetic Analyses of Lapita Decoration Do Not Support Branching Evolution or Regional Population Structure during Colonization of Remote Oceania." *Philosophical Transactions of the Royal Society B: Biological Sciences* 365 (1559): 3889–902.

Collard, M., S. J. Shennan, and J. J. Tehrani. 2006. "Branching, Blending, and the Evolution of Cultural Similarities and Differences among Human Populations." *Evolution and Human Behavior* 27 (3): 169–84.

Doolittle, W. F., and E. Bapteste. 2007. "Pattern Pluralism and the Tree of Life Hypothesis." *Proceedings of the National Academy of Sciences* 104 (7): 2043–49.

Durham, W. H. 1991. *Coevolution: Genes, Culture, and Human Diversity.* Palo Alto, Calif.: Stanford University Press.

Evans, J. A., and J. G. Foster. 2011. "Metaknowledge." *Science* 331 (6018): 721–25.

Felsenstein, J. 2004. *Inferring Phylogenies.* Sunderland, Mass.: Sinauer Associates.

Fisher, D. C. 2008. "Stratocladistics: Integrating Temporal Data and Character Data in Phylogenetic Inference." *Annual Review of Ecology, Evolution, and Systematics* 39:365–85.

Fleming, L., and O. Sorenson. 2001. "Technology as a Complex Adaptive System: Evidence from Patent Data." *Research Policy* 30 (7): 1019–39.

Fleming, L., and O. Sorenson. 2004. "Science as a Map in Technological Search." *Strategic Management Journal* 25 (8–9): 909–28.

Foster, J. G. 2018. "Culture and Computation: Steps to a Probably Approximately Correct Theory of Culture." *Poetics* 68: 144–54.

Foster, J. G., A. Rzhetsky, and J. A. Evans. 2015. "Tradition and Innovation in Scientists' Research Strategies." *American Sociological Review* 80 (5): 875–908.

Funk, R. J., and J. Owen-Smith. 2016. "A Dynamic Network Measure of Technological Change." *Management Science* 63 (3): 791–817.

Gershman, S. J., and D. M. Blei. 2012. "A Tutorial on Bayesian Nonparametric Models." *Journal of Mathematical Psychology* 56 (1): 1–12.

Ghahramani, Z. 2013. "Bayesian Non-parametrics and the Probabilistic Approach to Modelling." *Philosophical Transactions of the Royal Society A: Mathematical, Physical, and Engineering Sciences* 371 (1984): 20110553.

Ghahramani, Z. 2015. "Probabilistic Machine Learning and Artificial Intelligence." *Nature* 521 (7553): 452–59.

Gogarten, J. P., and J. P. Townsend. 2005. "Horizontal Gene Transfer, Genome Innovation, and Evolution." *Nature Reviews Microbiology* 3 (9): 679–87.

Goodwin, C. 1994. "Professional Vision." *American Anthropologist* 96 (3): 606–33.

Goodwin, C. 2017. *Co-operative Action.* Chicago: Cambridge University Press.

Gould, S. J. 2010. *An Urchin in the Storm: Essays about Books and Ideas.* New York: W. W. Norton.

Gray, R. D., and Q. D. Atkinson. 2003. "Language-Tree Divergence Times Support the Anatolian Theory of Indo-European Origin." *Nature* 426 (6965): 435–39.

Gray, R. D., D. Bryant, and S. J. Greenhill. 2010. "On the Shape and Fabric of Human History." *Philosophical Transactions of the Royal Society B: Biological Sciences* 365 (1559): 3923–33.

Gray, R. D., A. J. Drummond, and S. J. Greenhill. 2009. "Language Phylogenies Reveal Expansion Pulses and Pauses in Pacific Settlement." *Science* 323 (5913): 479–83.

Greenhill, S. J., T. E. Currie, and R. D. Gray. 2009. "Does Horizontal Transmission Invalidate Cultural Phylogenies?" *Proceedings of the Royal Society B: Biological Sciences* 276 (1665): 2299–306.

Griffiths, T. L., and Z. Ghahramani. 2011. "The Indian Buffet Process: An Introduction and Review." *Journal of Machine Learning Research* 12:1185–224.

Huelsenbeck, J. P., and B. Rannala. 1997. "Maximum Likelihood Estimation of Phylogeny Using Stratigraphic Data." *Paleobiology* 23 (2): 174–80.

Huson, D. H., R. Rupp, and C. Scornavacca. 2010. *Phylogenetic Networks.* Cambridge: Cambridge University Press.

Huson, D. H., and C. Scornavacca. 2011. "A Survey of Combinatorial Methods for Phylogenetic Networks." *Genome Biology and Evolution* 3:23–35.

Kaiser, D. 2009. *Drawing Theories Apart: The Dispersion of Feynman Diagrams in Postwar Physics.* Chicago: University of Chicago Press.

Kunin, V., L. Goldovsky, N. Darzentas, and C. A. Ouzounis. 2005. "The Net of Life: Reconstructing the Microbial Phylogenetic Network." *Genome Research* 15 (7): 954–59.

Latour, B. 1987. *Science in Action: How to Follow Scientists and Engineers through Society.* Cambridge, Mass.: Harvard University Press.

Lipo, C. P. 2006. "The Resolution of Cultural Phylogenies Using Graphs." In *Mapping Human History: Phylogenetic Approaches in Anthropology and Prehistory,* edited by C.P. Lipo et al., 89–107. New York: Routledge.

Merton, R. K. 1973. *The Sociology of Science: Theoretical and Empirical Investigations.* Chicago: University of Chicago Press.

Mesoudi, A. 2011. *Cultural Evolution: How Darwinian Theory Can Explain Human Culture and Synthesize the Social Sciences.* Chicago: University of Chicago Press.

Miller, G. A. 1956. "The Magical Number Seven, Plus or Minus Two: Some Limits on Our Capacity for Processing Information." *Psychological Review* 63 (2): 81.

Newman, M. E. 2003. "The Structure and Function of Complex Networks." *SIAM Review* 45 (2): 167–256.

Nicholls, G. K., and R. D. Gray. 2006. "Quantifying Uncertainty in a Stochastic Model of Vocabulary Evolution." In *Phylogenetic Methods and the Prehistory of Languages,* edited by Peter Forster and Colin Renfrew, 161–71. Cambridge: McDonald Institute for Archaeological Research.

O'Brien, M. J., M. Collard, B. Buchanan, and M. T. Boulanger. 2013. "'Trees, Thickets, or Something in between?' Recent Theoretical and Empirical Work in Cultural Phylogeny." *Israel Journal of Ecology and Evolution* 59 (2): 45–61.

O'Brien, M. J., J. Darwent, and R. L. Lyman. 2001. "Cladistics Is Useful for Reconstructing Archaeological Phylogenies: Palaeoindian Points from the Southeastern United States." *Journal of Archaeological Science* 28 (10): 1115–36.

Owen-Smith, J., and W. W. Powell. 2001. "To Patent or Not: Faculty Decisions and Institutional Success at Technology Transfer." *Journal of Technology Transfer* 26 (1–2): 99–114.

Scott, S., and S. Matwin. 1999. "Feature Engineering for Text Classification." In *International Conference on Machine Learning* 99: 379–88.

Shalizi, C. R., and A. C. Thomas. 2011. "Homophily and Contagion Are Generically Confounded in Observational Social Network Studies." *Sociological Methods and Research* 40 (2): 211–39.

Silvestro, D., J. Schnitzler, L. H. Liow, A. Antonelli, and N. Salamin. 2014. "Bayesian Estimation of Speciation and Extinction from Incomplete Fossil Occurrence Data." *Systematic Biology* 63 (3): 349–67.

Simon, H. A. 1969. *The Sciences of the Artificial.* Cambridge, Mass.: MIT Press.

Swait, J., and M. Ben-Akiva. 1987. "Incorporating Random Constraints in Discrete Models of Choice Set Generation." *Transportation Research Part B: Methodological* 21 (2): 91–102.

Tehrani, J., and M. Collard. 2002. "Investigating Cultural Evolution through Biological Phylogenetic Analyses of Turkmen Textiles." *Journal of Anthropological Archaeology* 21 (4): 443–63.

Tëmkin, I., and N. Eldredge. 2007. "Phylogenetics and Material Cultural Evolution." *Current Anthropology* 48 (1): 146–54.

Uzzi, B., S. Mukherjee, M. Stringer, and B. Jones. 2013. "Atypical Combinations and Scientific Impact." *Science* 342 (6157): 468–72.

Vilhena, D. A., J. G. Foster, M. Rosvall, J. D. West, J. Evans, and C. T. Bergstrom. 2014. "Finding Cultural Holes: How Structure and Culture Diverge in Networks of Scholarly Communication." *Sociological Science* 1:221–38.

Wade, M. J. 2016. *Adaptation in Metapopulations: How Interaction Changes Evolution.* Chicago: University of Chicago Press.

Wen, D., Y. Yu, and L. Nakhleh. 2016. "Bayesian Inference of Reticulate Phylogenies under the Multispecies Network Coalescent." *PLoS Genetics* 12 (5): e1006006.

Wimsatt, W. C. 2002. "Using False Models to Elaborate Constraints on Processes: Blending Inheritance in Organic and Cultural Evolution." *Philosophy of Science* 69 (S3): S12–S24.

Wimsatt, W. C. 2013a. "Articulating Babel: An Approach to Cultural Evolution." *Studies in History and Philosophy of Science Part C: Studies in History and Philosophy of Biological and Biomedical Sciences* 44 (4): 563–71.

Wimsatt, W. C. 2013b. "Entrenchment and Scaffolding: An Architecture for a Theory of Cultural Change." In *Developing Scaffolds in Evolution, Culture, and Cognition,* edited by L. R. Caporael, J. R. Griesemer, and W.C. Wimsatt, 77–107. Cambridge: MIT Press.

6

PATENTED TECHNOLOGY AS A MODEL SYSTEM FOR CULTURAL EVOLUTION

MARK A. BEDAU

OVERVIEW AND CONTEXT

In this chapter, I argue that the study of cultural evolution would benefit from model systems that are analogous to the model organisms studied in biology and that patented technology would make an excellent model system. My argument has three main steps. First, I note an important epistemic benefit provided by model organisms in biology: knowledge about a model organism illuminates nonmodel organisms both by providing a baseline for comparison with nonmodels and by allowing us to extrapolate knowledge about the model to similar nonmodel organisms. Model organisms are often relatively easy to learn about and understand, so information about the model organisms accumulates from many perspectives, and this information becomes increasingly integrated over time. Second, I argue that the analog of model organisms, which we can refer to as *model cultural systems,* provides analogous epistemic benefits for the study of cultural evolution. Or, at least, a model cultural system *would* provide those epistemic benefits if it existed. Third, I argue that patented technology has all the hallmarks of an excellent model system for at least three important aspects of cultural evolution: the way traits flow in the *hyperparental* genealogies that are characteristic of cultural evolution, the open-ended innovation characteristic of many cultural systems, and the new automated methods and tools for mining huge digital data sets to visualize and quantify the evolution of cultural traits. Patented technology nicely illustrates these three aspects of cultural evolution and provides a relatively easy way to learn more about all three.

Although the standard methodologies for investigating cultural evolution do not include things like model organisms in biology, this is an unfortunate missed opportunity. There have been other proposals for "model organisms" for the study of cultural evolution, such as Weismann diagrams (Griesemer and Wimsatt 1989) and Punnett squares (Wimsatt 2012). When we see why model organisms are so useful in biology, we can see why the study of cultural evolution would enjoy similar epistemic benefits if it too studied appropriate model systems. The use of model systems would mark a constructive milestone in the study of cultural evolution, especially the evolution of technology.

The central aim of this chapter is to show why patented technology is an excellent candidate model system for cultural evolution. Just as biology has a number and variety of model organisms, presumably, the study of cultural evolution would benefit from a number and variety of model systems; I propose patented technology as one excellent example. Many proposed model organisms never end up being adopted by biologists as model organisms. *Lords of the Fly* (Kohler 1994) and similar studies document many unforeseen and unintended contingencies that have influenced which organisms have become successful model organisms. Analogous unforeseeable and unintended future contingencies will presumably influence which systems become successful models for cultural evolution and whether patented technology is among them.

I will use the term *cultural evolution* to refer to the change over time of any population of cultural items. Many different kinds of cultural items can exist in populations, so there can be many kinds of cultural evolution. One consists of the evolving mental states (concepts, beliefs, behaviors, fashions, designs, etc.) found in some group of humans (e.g., Cavalli-Sforza and Feldman 1981; Boyd and Richerson 1985). Another kind consists of the physical artifacts and tools that humans invent and use or the commercial products that humans buy and sell (e.g., Basalla 1988; Rogers 2003; Arthur 2009). The focus in this chapter is a third specific kind of cultural population—the new inventions that humans have created over time—for which patented technology can serve as an easily ascertainable proxy.

Many things humans invent are not patented, and many never could be. But patented technology still is a precisely delimited subset of inventions that is especially easy to study and understand. To be sure, the evolution of patented technology is not representative of all other kinds of cultural evolution; technology is one unique subset of culture, and only a small fraction of human inventions is ever patented. Of course, neither is any model biological organism representative of every aspect of the natural organisms it represents. This chapter makes the case that patented technology is a good model for at least three important characteristic phenomena in cultural evolution: the microlevel, hyperparental flow of cultural (technological) traits; macro-level, open-ended innovation in new cultural entities (inventions); and automated data-mining methods for extracting and representing the content, or meaning, of cultural entities (inventions).

Many organisms are the subjects of biological experiments, but as Ankeny and Leonelli (2011) have noted, only a few are singled out as model organisms; these include the bacterium *Escherichia coli,* the fruit fly *Drosophila,* and vertebrates like the zebra fish and mouse (a mammal). Sometimes the term *model system* is used for *populations* of organisms *(microbial model systems).* Just like model organisms and systems in biology, model cultural systems would be complex natural systems that exist in the real world, evolve over time, and are studied empirically. Conditions are ripe for students of cultural evolution to make a standard component of their methodology the focus on a few model cultural systems, such as patented technology.

Focusing on a few model systems does not ignore or deny the value of other methods for studying cultural evolution. Rather, it augments them with a powerful new method. Case studies are one traditional method in the study of actual cultural systems (e.g., Ankeny 2012). Case studies share some important features with model systems. Both empirically investigate actual cultural systems, and both focus on just one system. A narrow focus makes it easier when studying something very complex and diverse. But case studies and model systems also have an important difference. Each case study is typically unique and individual, and different scholars study different cases. *Pattern and process in cultural evolution* (Shennon 2009), for example, is full of case studies and phylogenies of material culture, and no two chapters focus on the same case. Information from different cases is collected and sometimes compared, but it is very rare for many studies to focus on the same case. For this reason case studies typically cannot support broad generalizations about other cultural systems. By contrast, model systems *are* used to support broad generalization about other systems; that is their central epistemic function. And they can perform that function because a scientific community has collectively learned a lot about a single system. Pooling the results of a great many independent studies of the same model system is one of the defining hallmarks of model systems, and it helps explain why they support generalizations about similar nonmodel systems. So, a collective focus on a few model systems would complement traditional case studies.

Amassing many unique and different case studies is valuable for the study of cultural evolution, but so is learning a lot about a few model systems if what is learned can be extrapolated to other similar cultural systems.

Most of the things that are called *models* in the study of culture are not the sort of model cultural systems I am proposing. For example, the small-scale physical model of the San Francisco Bay discussed by Weisberg (2013) was built by humans to represent a larger target system: the actual San Francisco Bay. But model cultural systems and model organisms exist naturally and independently whether anyone studies them, or not, at least initially. (It turns out that many model organisms eventually become significantly altered and reshaped by scientists, usually in order to make them easier to study in the laboratory.)

The study of model cultural systems also differs from and complements the tradition of studying cultural evolution with theoretical, mathematical, and (more recently) computational models—which I will lump together and call *formal models*. Formal models include the pioneering mathematical work of Cavalli-Sforza and Feldman (1981) and Boyd and Richerson (1985), as well as the agent-based computer model of Arthur and Polak (2006). The behavior produced by formal models can be compared with empirical observations of actual cultural systems. But though both are called "models," formal models and model cultural systems are quite different. Formal models are purely mathematical objects and not real cultural systems. By contrast, a model cultural system is a real cultural system, which is what we compare with formal models. This shows one way that model cultural systems would complement existing methods of studying cultural evolution.

Focus on a single model cultural system also complements the diverse range of work on cultural and technological evolution that concentrates on the similarities and differences between cultural and biological evolution. This includes not only verbal theories that use "memes" to describe and explain cultural evolution (Dawkins 1976) but also empirical studies of individual cases and comparison of their behavior with formal models and verbal theories (see Ziman 2000). Like the authors in *Pattern and Process in Cultural Evolution* (Shennon 2009), those in *Technological Evolution as an Evolutionary Process* (Ziman 2000) each discuss a different distinctive case.

The reasons why model organisms benefit biology suggest that model systems would also benefit the study of cultural evolution. One key reason is the vast complexity of the natural systems under investigation. Part of what makes biological organisms so hard to understand is their complexity and great variety of forms. One way in which biology copes with this complexity and variety is to identify certain organisms as "model" organisms, to learn a great detail about the model organisms, and then to leverage that knowledge to draw conclusions about nonmodel organisms. In this way model organisms help biologists "to deconstruct the complexity of nature into its constituent parts and to explore the role of each part in creating patterns in nature, first in isolation, then in combination" (Jessup et al. 2004).

Like biological evolution, cultural evolution is extraordinarily complex, "a complex beast . . . [with] multiple evolving and interdependent lineages acting on different time and size scales" (Wimsatt 2013, 564; see also Andersson, Törnberg, and Törnberg 2014). Wimsatt notes that one way to cope with this complexity is to "seek the right organism for the job" and be "opportunistic in seeking cases that are tractable and can generate relatively crisp and unambiguous data" (Wimsatt 2013, 565). If we could amass knowledge about a single model cultural system, we could then compare it with what we learn about other cultural systems and extrapolate it to other cultural systems that are similar to the model in the relevant respects. One concrete constructive way to follow Wimsatt's advice would be to adopt patented technology as a model system for cultural evolution.

The rest of this chapter elaborates the case for adopting patented technology as a model system for cultural evolution, especially for the trait flow over time in hyperparental (highly reticulate) genealogies, for the open-ended way in which cultural populations evolve, and for new methods and tools for the automated mining of big digital data sets of actual cultural population. It first reviews the epistemic hallmarks that make model organisms so useful in biology and explains why model systems would have similar benefits for the study of cultural evolution. Then it shows why patented technology excels at all the hallmarks of a model system for cultural evolution. The chapter concludes with a summary of the main argument and a glance at its larger implications.

MODEL BIOLOGICAL ORGANISMS AND MODEL SYSTEMS FOR CULTURAL EVOLUTION

There is a wealth of recent literature about model organisms (Kohler 1994; Endersby 2007; Jessup et al. 2004; Ankeny and Leonelli 2011; Ankeny 2012; Love and Travisano 2013; Levy and Currie 2014). The literature covers many issues, but this chapter focuses mainly on the hallmarks that explain the

	[A] model organisms in biology
1. The model illuminates many nonmodel systems in one or both ways:	
(1a) it is a baseline to compare with nonmodels	✓
(1b) information about it can be extrapolated to similar nonmodels	✓
2. Information about the model is amassed	✓
3. It is relatively easy to understand the model because of its excellence in:	
(3a) information quality and access	✓
(3b) scientific analysis tools	✓
(3c) empirical observability	✓
(3d) experimental manipulability	✓

Figure 6.1. The main epistemic benefits of model organisms for biology *(row 1)* and some hallmarks *(rows 2 and 3)* that explain those epistemic benefits.

important epistemic benefits of model organisms for biologists. There is general agreement in the literature about these hallmarks, and they are listed in Figure 6.1.

The central epistemic benefit provided by model organisms (Figure 6.1, *row 1*) is that what is known about a model organism illuminates many nonmodel organisms. There are at least two kinds of illumination that a model organism can provide, and they can be distinguished using the distinction between the *phenomena* (behavior) exhibited by a model or a nonmodel organism and the *mechanisms* that explain those phenomena (Love 2015). First, model organisms tend to be much better understood than nonmodel organisms, and this enables our knowledge of the characteristic phenomena involving model organisms to serve as a common baseline for comparison with the phenomena exhibited by nonmodel organisms *(row 1a)*. Comparison with a standard and well-understood baseline is informative whether or not similar mechanisms produce the phenomena in both model and nonmodel. What matters is that the behavior of model and nonmodel organisms can be compared; what matters is the similarity of their phenomena.

A second, deeper kind of illumination comes when model and nonmodel are so similar that our knowledge about the model can be extrapolated to the nonmodel organisms (Figure 6.1, *row 1b*). In practice, many model and nonmodel organisms are similar in ways that justify using the models as proxies or representatives for the nonmodels, standing in for them and sanctioning inferences about them. Ankeny and Leonelli (2011) refer to both the relatively wide "representational scope" of models organisms (nonmodels illuminated) as well as their especially wide "representational target" (questions and theories addressed).

One of the reasons we can usefully compare the phenomena and mechanisms in models and nonmodels is that information about a model organism is amassed from many perspectives (Figure 6.1, *row 2*). This information is incrementally integrated when this is thought to be relevant. A broad range of questions and theories can be addressed with a model organism because so much knowledge about one organism has been collected and annotated. Focusing on a single organism enables scientists to amass all of the gory details needed to understand the phenomena and underlying mechanisms found in even one very complex organism. The detailed knowledge accumulated about a model organism can drive the development of new scientific technologies and techniques, and it can foster productive careers for a community of professional scientists. Amassing and integrating information about model organisms are good examples of characteristic activities in what Kuhn calls "normal" science and what Lakatos calls a "progressive" research program

Row 3 of Figure 6.1 identifies another typical hallmark of model organisms: it is relatively easy to learn about the behavior of a model organism (Jessup et al. 2004). Since a model organism is easier to study, it is easier to observe and describe its behavior and therefore easier (eventually) to figure out the underlying causal mechanisms and explain its behavior. A model organism might be especially easy to understand because of the availability of abundant reliable data *(row 3a)*. For example, the quick reproductive cycle of the bacterium *E. coli* and the ability to stop, store, and restart the evolution of bacterial populations help make *E. coli* a useful model organism for experimental studies of evolution (Love and Travisano 2013). Other important practical considerations include low experimental costs and the commercial availability of standardized lines of experimental organisms (Ankeny and Leonelli 2011). In general, a key epistemic hallmark of model organisms is the availability of abundant, detailed, accurate, and inexpensive information.

The applicability of powerful scientific tools and techniques can be another reason why model organisms are easy to understand (Figure 6.1, *row 3b*). These can include laboratory practices and know-how, as well as training and mentoring practices. Shared scientific tools and techniques in a scientific community are other hallmarks of model organisms (Ankeny and Leonelli 2011) and another source of their epistemic benefits for biology.

Another hallmark of model organisms is the possibility of recording precise empirical observations about the model's behavior *(row 3c)*. Detailed and precisely controlled observations about a model's behavior in a variety of circumstances enable useful information about the model to be amassed. Similarly, practical methods for the precise experimental manipulation of model organisms *(row 3d)*, especially with microscopic and molecular techniques, are another hallmark of model organisms.

Using the hallmarks of model organisms as a guide, we can construct a table of analogous epistemic hallmarks for cultural evolution (Figure 6.2). Though the epistemic hallmarks of model cultural systems and model organisms turn out to be very similar, we will see that they are not identical.

As with model organisms, central to the epistemic benefits of a model cultural system would be the model's ability to illuminate many nonmodel systems (Figure 6.2, *row 1*). And as with model organisms, a model cultural system would illuminate nonmodels, either by serving as a common baseline for comparison *(1a)* or by knowledge about the model being extrapolated to similar nonmodels *(1b)*. The epistemic benefits of a model cultural system would also typically stem from a second hallmark shared with model organisms: amassing information about a single model from a wide variety of perspectives and sources (Figure 6.2, *row 2*). A third shared hallmark would be the relative ease with which a model cultural system can be studied and understood (Figure 6.2, *row 3*). Abundant clean data about the model would obviously help *(3a)*, as would excellent scientific tools and techniques *(3b)*. Both make it much easier to understand the model system and to share that information with a community of cooperating scientists. A third obvious epistemic benefit that model cultural systems would share with model organisms would be extensive, precise, detailed, and accurate observations of a model system in a controlled variety of circumstances *(3c)*.

Although the hallmarks of model cultural systems generally parallel those for model biological organisms, there is one important difference *(3d)*. Unlike with biological organisms, experimental manipulability of real cul-

	[A] model organisms in biology	[B] model systems for cultural evolution
1. The model illuminates many nonmodel systems in one or both ways:		
(1a) it is a baseline to compare with nonmodels	✓	✓
(1b) information about it can be extrapolated to similar nonmodels	✓	✓
2. Information about the model is amassed	✓	✓
3. It is relatively easy to understand the model because of its excellence in:		
(3a) information quality and access	✓	✓
(3b) scientific analysis tools	✓	✓
(3c) empirical observability	✓	✓
(3d) experimental manipulability	✓	×

Figure 6.2. Comparison of model organisms (column 2) and model cultural systems (column 3) with respect to their main epistemic benefits and the hallmarks that explain them. Note that the epistemic benefits are almost exactly the same. The one difference is row (3d). The check in this row indicates that excellent experimental manipulation is a hallmark of model organisms, while the × indicates that experimental difficulty or ethical constraints typically block experimental manipulation of the humans in model cultural systems.

tural populations is typically not possible with a model cultural system because it is either impractical or unethical or both. For example, the study of the patent record would benefit enormously if we could observe what would happen in counterfactual situations without certain actual technological inventions, or with certain possible inventions. But it is impossible to rewrite history, and it is very difficult to evaluate the relevant counterfactual situations with any confidence. Furthermore, even if we could experimentally manipulate the actual evolution of technology, to do so might be irresponsible or unfair or inappropriate. Because of this, a model cultural system typically lacks the experimental manipulability that is among the hallmarks of model organisms. With this one understandable exception, Figure 6.1 shows that model systems for the study of cultural evolution share most of the important epistemic benefits of model organisms. Even if actual cultural populations are not manipulated in experiments, a sufficient number of precisely controlled observations can teach us a lot about a model system, and we might be able to extrapolate what we have learned to other cultural systems.

BACKGROUND ON THE EVOLUTION
OF PATENTED TECHNOLOGY

Technology is only a small part of human culture, and only a small fraction of technologies are ever patented. Nevertheless, patents are an ideal context for investigating cultural evolution for a variety of reasons. Basalla (1988) stressed the great context for observing and describing the evolution of cultural variation provided by patented technology. More recently, Mesoudi (2011) emphasized that patent technology enables us to precisely identify individual cultural entities and to document the details of cultural phylogenies and cultural diversity. This chapter's argument for patented technology as a model cultural system follows in the same general spirit as Basalla and Mesoudi.

Part of the reason why patented technology would make a good model cultural system is the wealth of empirical analysis of patent citations collected in Jaffee and Trajtenberg (2002). Jaffee and Trajtenberg have documented the economic value of patent citations, and they have used patent citations to compare the flow of knowledge among different technology sectors, different social institutions, and different political states. The patent record includes a great deal of information about each patented invention, and human experts vet and collate the information. For example, patent examiners at the United States Patent and Trademark Office (USPTO) label each invention with a string of numerical codes that describe its important technological capacities, and these patent technology codes can be used to better understand "the nature and rate of technological change" (Strumsky, Lobo, and van der Leeuw 2012).

Dates, citations, and technology codes are just a small fraction of the information in the patent record. The information includes the unique identification number assigned to each patent, along with standardized information about when the invention was filed with the USPTO, when it was granted a patent, and the inventors of the patent. The record also contains text describing the invention's important technological features, including a title, an abstract, and a list of "claims" that describe the invention's novel technological features. Current text-analysis tools include automated methods for mining the text in the patent record and identifying the key technological features of each invention. The record also lists the USPTO technology codes that correspond to the main technological capacities contained in an invention. One can view a new invention as coming into existence (being "born") when it is granted by the USPTO and as "reproducing" whenever another patent cites it as so-called prior art" (earlier patented inventions that are relevant to a new patent's claims of originality). Then one can reconstruct the complete genealogy of every patented invention. This is one reason why patented technology is an exceptionally rich and feasible context for studying the evolution of cultural traits.

Cultural and biological evolution are often compared, so it is worth stressing that the population of patents has some properties never found in biological populations. One simple example is that existing patents never die and go out of existence. A more important example is the hyperparental genealogies that they form. On average a patent's prior art consists of roughly a dozen earlier inventions; the magnitude of this number demonstrates the hyperparental nature of patented technology. Formal tools for describing hyperparental inheritance networks have only recently been developed (Kerr and Godfrey-Smith 2009; see chapter 5 of this volume), so how well traits flow through hyperparental genealogies is an open but answerable empirical question. Hyperparentality makes a new invention's technological features a mix and combination of many earlier sources, and those features are often intentionally modified and blended by individual rational agents, so parent–offspring connections might be too degraded to enable certain kinds of natural selection to occur (Godfrey-Smith 2009, 2012).

Many complex causal webs affect cultural evolution. Patented technology is affected by things like technology inventors and designers, technology users and consumers, and economic markets and social institutions (Jaffe and Trajtenberg 2002; Arthur 2009; Caporael, Griesemer, and Wimsatt 2014). The evolution of technology involves evolution in many different populations, four of which are depicted in Figure 6.3. Each population is evolving over time, and as the arrows suggest, the populations are causally connected; entities in each interact with entities in the others, so the figure depicts only part of the story. A more exhaustive list of factors affecting cultural evolution has been compiled by Wimsatt (2013).

Population I in Figure 6.3 consists of people who design technological products to be sold in economic markets. Members of this population borrow ideas from each other, and ideas spread and diffuse through the population as people interact. Population II consists of the patented technologies

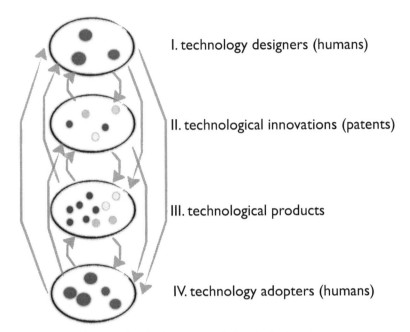

I. technology designers (humans)

II. technological innovations (patents)

III. technological products

IV. technology adopters (humans)

Figure 6.3. Four of the populations involved in the evolution of technology; the *arrows* suggest some of their interactions.

themselves; this population consists of each actual individual technological innovation, or at least those that are patented. The canonical description of a patent is simply the text in its title and abstract, as shown in the patent record. Unlike population I, the members of population II are not people but designs of specific kinds of technologies. Population III consists of commercial products in actual economic markets. These products are usually material objects, which are part of what is called *material culture.* Products compete with one another for market share and diffuse when first introduced to niches. Markets for products are affected by many kinds of factors, both endogenous and exogenous. Population IV consists of people who adopt and use technological products in their daily lives. Consumers select which technology products to buy and adopt, and preferences and fads diffuse through the population as consumers interact. Studies of cultural evolution often focus on the cultural traits of some population of *humans,* such as populations I and IV. I focus on evolution in exemplars of population II: *patented inventions.* The evolution of each of these populations and their interactions is worthy of study.

Each patented invention corresponds to a specific invention. A patented invention is not just an idea in someone's mind (e.g., those who invent or use it), for those ideas change over time and come in and out of existence. A patented invention persists even if nobody thinks of it. Patented inventions are abstract because they are *kinds* of technology with an open-ended range of instances. If a material or physical device is patented, the patent covers not some specific instance of the device but that *kind* of device.

Each patent cites some number of earlier patents. It is common to use a patent's citations of earlier patents as a proxy for a genealogical link between an invention and its technological "parents" (Jaffee and Trajtenberg 2002; Chalmers et al. 2010; Buchanan, Packard, and Bedau 2011; Bedau 2013). All of these nodes and links together comprise a genealogy that continually grows and evolves in new and unpredictable directions. We can identify the most heavily cited patents as the main drivers of the subsequent evolution of technology. Future patents build on and cite some patents more than others, and the main drivers emerge over time from this selection process. It turns out that the main drivers of technological innovation over the past forty years include bubble jet printers, polymerase chain reaction (PCR), and stents (Buchanan, Packard, and Bedau 2011).

Figure 6.4 shows part of the genealogy of U.S. utility patent number 3938459, for a certain type of minesweeper used by the U.S. Navy. Time flows from top to bottom in this genealogy, and the original minesweeper patent is the large star at the top right. All of the other patents shown in this figure are direct descendants of patent 3938459. Citations of parents by children are indicated by arrows. One can distinguish at least four large lineages, and some of them have complex internal interconnections.

Note that Figure 6.4 shows only the direct descendants of the minesweeper patent—only a small fraction of patents in the genealogy. In particular, the hyperparental structure in the genealogy is artificially downplayed. To better indicate the hyperparental structure of patent genealogies, Figure 6.5 shows all of the parents of each of the large circular nodes in Figure 6.4. While Figure 6.4 contains 36 large nodes, including the parents of those nodes reveals an order of magnitude more parents in the genealogy (Figure 6.5 contains 558 nodes). Hyperparental, indeed! Furthermore, including all of the parents of the first thirty-six descendants highlights the separate lineages in the genealogy; the four main lineages visible in Figure 6.4 are very clearly delineated in Figure 6.5, and so is the complex sublineage structure on the far left and far right. In general, Figure 6.5 shows a lot

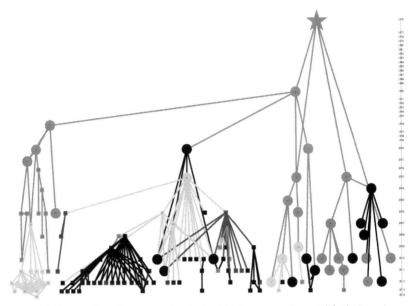

Figure 6.4. The genealogy of patent number 3938459 (*the large star at the top right*): The invention of a kind of minesweeper used by the U.S. Navy. Up to nine generations of descendant nodes are shown. Time flows from top to bottom. Nodes are individual inventions, and links indicate parents cited by their children. Only descendants that receive at least one citation are shown. Nodes and the links below them are colored six different shades of gray to reflect the patent's technological category (either Chemical, Computers and Communication, Drugs and Medical, Electrical and Electronic, Mechanical, or Other). The first three generations of descendants are shown as large circles to indicate how this is connected to the genealogy in Figure 6.5.

of cross-citation within each of the four main lineages and very little cross-citation between the main lineages.

Hyperparental network structures like those in Figure 6.5 are common in cultural populations but relatively rare in biological populations; biological genealogies are typically hypoparental rather than hyperparental. Microbes are known to experience a significant amount of horizontal gene transfer, and their resulting reticulated genealogies have some similarity to the hyperparental network structure in cultural populations. But most patents cite dozens of prior patents, so the degree and rate of hyperparental quality in the patent population is on a vastly larger scale than the horizontal gene transfer in microbial populations.

Some biological innovations are said to "open the door" to quite new and different kinds of subsequent biological innovations (Kauffman 2000; Bedau 2009); in an analogous fashion, some inventions seem to open the door to

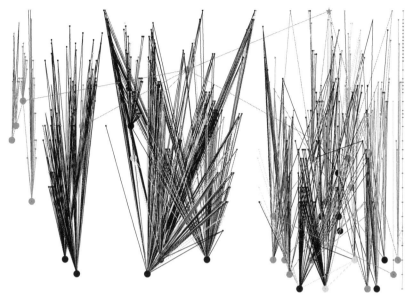

Figure 6.5. Another view of the genealogy of patent number 3938459, showing the parents of the patents in the first three generations of descendants (i.e., the large nodes in Figure 6.4); descendants in subsequent generations (and their parents) are omitted. Nodes and the links below them are colored six different shades of gray to indicate a patent's overall technological category, and only descendants that receive at least one citation are shown.

quite new and different kinds of innovations. Door-opening innovations increase the degree of innovation in the evolution of cultural populations. If one measures the degree to which an invention is door opening by the diversity of its offspring, then patent citation data show that many of the main drivers of technological innovation are highly door opening (Buchanan, Packard, and Bedau 2011).

Various tools make it easy to extract the technological *content,* or meaning, of a patent from the text in the patent record. One especially simple metric is TF-IDF *(term frequency, inverse document frequency);* the TF-IDF value of a term in a document from a corpus is the product of both the term's frequency in the document and the log of the inverse of the term's frequency in the entire corpus. This metric identifies the words in a document that most distinguish it from the other documents in a corpus. Although TF-IDF has known weaknesses and blind spots, it does a good job of automatically extracting key words that describe a patent's technological content from the text in a patent's title and abstract. Chalmers et al. (2010) extracted the most

Top patent story words

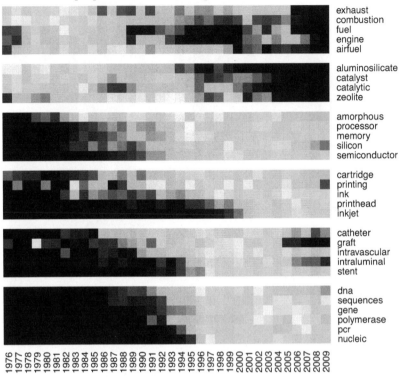

Figure 6.6. A heat map of the frequency of key words with especially high TF-IDF scores in the patents issued each year (lighter shades reflect higher values). Reprinted with permission from Chalmers et al. (2010).

important TF-IDF key words in all of the new patents issued each year and mapped their annual frequencies in a series of heat maps in which lighter colors (higher "temperatures") correspond to more frequent traits (Figure 6.6).

Heat maps like Figure 6.6 are a macrolevel description of the evolution of the content of patented inventions. By contrast, a *micro*level description might depict the content of each separate invention in a patent's genealogy, and it might show in precise and complete detail how the *content* of inventions changes over time in each lineage. The general idea would be to color the nodes in the genealogies in Figures 6.4 and 6.5 to encode the overall technological content of that invention. One easy and useful way to encode the overall technological content of each invention is to classify each into a few kinds of technologies, using the technology codes assigned to each patent by patent examiners. Figures 6.4 and 6.5 depict a familiar sixfold classification of technology categories proposed by Hall, Jaffee, and Trajtenberg (2002) and adopted by the National Bureau of Economic Research (NBER). Six gray colors encode the six general kinds of technology: Chemical, Computers and Communication, Drugs and Medical, Electrical and Electronic, Mechanical, and Other.

If we observe how traits flow in patent genealogies by close examination of the distribution of gray colors in Figures 6.4 and 6.5, we note the following conclusions. First, there are examples of all six categories among the 131 nodes shown in Figure 6.4. The minesweeping patent is classified by the USPTO as Mechanical, and all of the immediate descendants of the minesweeping patent are also classified as Mechanical. But the 36 patents in the first three generations of descendants already exhibit three other technology categories (Chemical, Computer and Communications, and Other), and the 131 patents in Figure 6.4 include examples from the two remaining categories (Drugs and Medical; Electrical and Electronic). Furthermore, the four main different lineages in Figure 6.4 contain distinctive categories of patents; the first large lineage on the far left contains mostly Mechanical and Other patents, while the second large lineage on the left contains mostly Computer and Communications and Drugs and Medical patents. The two remaining large lineages on the right both have a more complex internal citation structure, with patents from at least four different NBER categories.

Examination of the colors (NBER categories) of all the parents shown in Figure 6.5 underscores our earlier conclusions. The four different lineages each have parents from different distinctive NBER categories. The parents in the first large lineage on the far left are virtually all either Mechanical or Other patents, and the parents in the second large lineage on the left are mostly either Computers and Communications or Drugs and Medical patents. Most of the parents in the large lineage in the middle of Figure 6.5 are Electrical and Electronic parents, although parents are also in many other categories. The lineage on the right with the complex internal structure includes parents from all different categories; many are Mechanical patents, and some are Electrical and Electronic, Computer and Communications, or Chemical patents. Furthermore, we can observe that certain sublineages have different and distinctive distributions of categories of parents. Genealogies like those in Figures 6.4 and 6.5 provide detailed, empirical microlevel descriptions of what specific technological categories have flowed through the actual genealogy of any specific patent. Such genealogies can be colored to

reflect a wide variety of other kinds of traits, including those that are reconstructed from the text in a patent's title and abstract.

PATENTED TECHNOLOGY AS A MODEL CULTURAL SYSTEM

Given what we now know about patented technology, it is easy to see that it
would be an excellent model system for the study of cultural evolution, for
it has all of the hallmarks of a model cultural system (Figure 6.2), at least for
open-ended evolution, for trait flow in hyperparental cultural genealogies,
and for methods and tools for studying trait flow.

Patented technology provides an interesting form of open-ended cultural
innovation, and similar phenomena are exhibited by many other cultural systems. So, even if different mechanisms underlie innovation in different cultural systems, patented technology would still provide a standard baseline
against which other forms of open-ended cultural innovation could be compared and contrasted. Recent discussions of "revolutionary" modifications
of entrenched systems (Wimsatt and Griesemer 2007) and of "redomaining"
and its significance (Arthur 2009) suggest that there might be some significant similarity of the mechanisms behind the open-ended evolution in many
cultural systems.

There certainly is at least one important similarity in the mechanisms
that produce the trait flow that can be observed in cultural systems: their
hyperparental structure. For this reason, patented technology would be an
excellent model system for the flow, over time, of technological traits through
hyperparental patent citation networks. Since similar hyperparental mechanisms operate in many other cultural systems, lessons about hyperparental trait flow in patent populations should be applicable to many other cultural
populations. Patent citation networks can "represent" the hyperparental
mechanisms in other cultural systems and thus license inferences about the
trait flow phenomena exhibited by those systems. This is an important epistemic benefit that patented technology would bring to the study of cultural
evolution.

In addition, the same tools used to describe the flow of the content or
meaning shown in Figures 6.4 and 6.5 could be used to describe the flow over
time of the content or meaning of many other cultural systems. For this reason, patented technology would make a great laboratory for developing and
demonstrating new methods and tools for describing trait flow phenomena

and for unpacking the mechanisms behind those phenomena. Those new methods include new ways to observe cultural systems by using machine-learning and language-processing algorithms to automate the mining digital data repositories, such as patent records. Even if somewhat different mechanisms are behind the trait flow phenomena observed in different cultural systems, the mere similarity of the data produced by the different systems is enough to explain why the same methods and tools can be applied to both. The analogs of Figures 6.2, 6.3, and 6.4 can be made for many other cultural systems, given raw data over time about trait frequencies. Patented technology "represents" many other cultural systems in the respects that are relevant for applying the new scientific methods.

One reason that patented technology would make a great model system is the patent record. This public and widely accessible resource makes it relatively easy to describe and explain the diversity of existing technologies and their evolutionary origin, given a precise description of every patented invention, including when it was invented and all of its prior art. The growing mass of data about human social and cultural behavior creates confidence that lessons learned from studying patented technology will be adapted and extended to other cultural systems. In addition to citation information already available in sources like the Scientific Citation Index and LexisNexis, and in addition to textual data streaming from traditional mass media outlets like the Associated Press and the *New York Times,* a new wealth of information is being generated on the web *(Wikipedia),* including social media like Facebook and Twitter and mobile apps like texting and Tinder.

The patent record also makes it much easier to amass information about patented technology from public patent records, in no small part because of the patent record created by the USPTO. Various human experts (the patent examiners, the inventor, the inventor's lawyers, etc.) help make the patent record accurate and complete. Patented technology and citation networks have been studied in a number of scientific fields, ranging from scientometrics and bibliometrics to science and technology studies. Information about patented technology has already started to accumulate (Figure 6.2, *row 2*), and Venturini, Jensen, and Latour (2015) recently articulated the special value that these new digital repositories have for cultural studies (citations below in the original):

> The most interesting feature of digital media is that everything that they mediate becomes potentially traceable and often actually traced (Rogers 2013).

Such traceability creates data that are as rich/thick as those collected by ethnographic techniques but covering much larger populations. Everyday new public and private archives are swallowed by computer memories, economic transactions migrate online, social networks root in the Web, and the more this happens, the more traces become available on the collective dynamics that used to be hidden by the quali-quantitative divide (Latour et al. 2012).

Patented technology is also relatively easy to understand (Figure 6.2, *row 3*). Excellent information, tools, and techniques for analyzing it enable extraordinarily precise observations of this example of cultural evolution. The information in the patent record is publicly available and relatively accurate, and it is continually updated as new patents are issued. In addition, citations make it easy to reconstruct the entire genealogy of any patented invention, and powerful statistical tools and techniques make it easy to identify each patent's technological content. This makes the patent record especially fruitful for illuminating microlevel hyperparental genealogies and macrolevel open-ended evolution, as well as showing how to extract cultural content or meaning from big data.

If patented technology becomes a model cultural system, the study of patents would not be limited to mining the patent record. Quite the opposite! A system becomes like a good model for other cultural systems only after being studied by many people from many different perspectives. To amass and share detailed information about a single system takes a scientific community, and a special epistemic opportunity is created when a scientific community accumulates, curates and vets, and incrementally integrates information about a single cultural system. This accomplishment requires the investment of time and energy by a diverse cast of characters, including experts on policies and practices at the USPTO; on the psychological, social, economic, and political influences on patenting activities; and on the connections between innovation and other factors such as geography, gender, and governmental investment.

THE IDENTIFICATION of good model organisms has been a great epistemic boon for biology, helping constrain and unpack some of the complexity of life. Figure 6.2 lists the epistemic benefits of model systems for the study of cultural evolution, and patented technology exemplifies the entire list. One reason is that the patent record is full of precise and detailed information about the novel and useful features of each patented invention. Another is the existence of tools for reconstructing the genealogy of an invention and each invention's technological content (illustrated in Figures 6.2–6.4).

Patented technology could illuminate many important questions about cultural evolution. One is the impact of culture's characteristic hyperparental genealogies. Cultural evolution's hyperparental quality is one important way it differs from biological evolution. This makes cultural evolution harder to study and understand. Citations in the patent record provide a precise picture of the parent–offspring connections between patents. Patented technology also exhibits interesting macrolevel open-ended evolution that is relatively easy to describe and compare with other cultural or biological populations. Patented technology also illustrates powerful new scientific methods for extracting semantic content from textual data. Those methods could be adapted and extrapolated to describe and eventually explain the flow of semantic content in various kinds of cultural genealogies reconstructed from citations in scientific or other professional publications, in social media on the web like Facebook and Twitter, or in other digital repositories generated by texting and email.

Today, patented technology excels in all the hallmarks of model systems, and it could bring the epistemic benefits of model organisms to the study of cultural evolution. Adopting a model cultural system like patented technology could also foster a new form of interdisciplinary cooperation and collaboration in a new kind of interdisciplinary research community. It remains to be seen whether the study of cultural evolution will take advantage of this new opportunity.

NOTES

I would like to thank my many collaborators on the study of the evolution of technology, including Drew Blount, Andrew Buchanan, Devin Chalmers, Zackary Dunivin, Cooper Francis, Bobby Gadda, Alex Kosik, Alex Ledger, Jacob Menick, Norman Packard, Noah Pepper, Andre Skusa, Ricard Solé, and Sergi Valverde. Thanks also to Alan Love, Robert Meunier, Emily Parke, and Bill Wimsatt for their comments on the manuscript and to Alec Kosik for Figures 6.4 and 6.5.

REFERENCES

Andersson, C., A. Törnberg, and P. Törnberg. 2014. "Societal Systems—Complex or Worse?" *Futures,* http:// dx.doi.org/10.1016/j.futures.2014.07.003.

Ankeny, R. A. 2012. "Detecting Themes and Variations: The Use of Cases in Developmental Biology." *Philosophy of Science* 79 (5): 644–54.

Ankeny, R. A., and S. Leonelli. 2011. "What's So Special about Model Organisms?" *Studies in History and Philosophy of Science Part A* 42 (2): 313–23.

Arthur, W. B. 2009. *The Nature of Technology: What It Is and How It Evolves.* New York: Simon and Schuster.

Arthur, W. Brian, and Wolfgang Polak. 2006. "The Evolution of Technology within a Simple Computer Model." *Complexity* 11 (5): 23–31.

Basalla, George. 1988. *The Evolution of Technology.* Cambridge: Cambridge University Press.

Bedau, M. A. 2009. "The Evolution of Complexity." In *Mapping the Future of Biology: Evolving Concepts and Theories,* edited by A. Barberousse, M. Morange, and T. Pradeu, 111–30. Berlin: Springer.

Bedau, M. A. 2013. "Minimal Memetics and the Evolution of Patented Technology." *Foundations of Science* 18:791–807.

Boyd, R., and P. Richerson. 1985. *Culture and the Evolutionary Process.* Chicago: University of Chicago Press.

Buchanan, Andrew, Norman H. Packard, and Mark A. Bedau. 2011. "Measuring the Drivers of Technological Innovation in the Patent Record." *Artificial Life* 17:109–22.

Caporael, Linnda R., James R. Griesemer, and William C. Wimsatt. 2014. *Developing Scaffolds in Evolution, Culture, and Cognition.* Cambridge, Mass.: MIT Press.

Cavalli-Sforza, L. L., and M. W. Feldman. 1981. *Cultural Transmission and Evolution: A Quantitative Approach.* Princeton, N.J.: Princeton University Press.

Chalmers, D., C. C. Francis, P. Pepper, and M. A. Bedau. 2010. "High-Content Words in Patent Records Reflect Key Innovations in the Evolution of Technology." In *Proceedings of Artificial Life XII,* edited by H. Fellermann et al., 838–45. Cambridge, Mass.: MIT Press.

Dawkins, R. 1976. *The Selfish Gene.* New York: Oxford University Press.

Endersby, Jim. 2007. *A Guinea Pig's History of Biology.* Cambridge, Mass.: Harvard University Press.

Godfrey-Smith, P. 2009. *Darwinian Populations and Natural Selection.* New York: Oxford University Press.

Godfrey-Smith, P. 2012. "Darwinism and Cultural Change." *Proceedings of the Royal Society B* 367: 2160–70.

Griesemer, J. R., and W. C. Wimsatt. 1989. "Picturing Weismannism: A Case Study in Conceptual Evolution." In *What Philosophy of Biology Is,* edited by M. Ruse, 75–137. Essays for David Hull. Dordrecht, The Netherlands: Martinus-Nijhoff.

Hall, B. H., A. B. Jaffee, and M. Trajtenberg. 2002. "The NBER Patent-Citations Data File: Lessons, Insights, and Methodological Tools." Chapter 13 in Jaffe and Trajtenberg 2002, 403–54.

Jaffe, A. B., and M. Trajtenberg. 2002. *Patents, Citations, and Innovations: A Window on the Knowledge Economy.* Cambridge, Mass.: MIT Press.

Jessup, C. M., R. Kassen, S. E. Forde, B. Kerr, A. Buckling, P. B. Rainey, and B. J. M. Bohannan. 2004. "Big Questions, Small Worlds: Microbial Model Systems in Ecology." *Trends in Ecology and Evolution* 19:189–97.

Kauffman, S. 2000. *Investigations.* New York: Oxford University Press.

Kerr, B., and P. Godfrey-Smith. 2009 "Generalization of the Price Equation for Evolutionary Change." *Evolution* 63:531–36.

Kohler, Robert E. 1994. *Lords of the Fly: Drosophila Genetics and the Experimental Life.* Chicago: University of Chicago Press.

Levy, A., and A. Curry. 2014. "Model Organisms Aren't (Theoretical) Models." *British Journal for the Philosophy of Science 66 (2): 327–48.*

Love, Alan C. 2015. "Developmental Biology." In *The Stanford Encyclopedia of Philosophy,* edited by Edward N. Zalta. Fall ed. http://plato.stanford.edu/archives/fall2015/entries/biology-developmental/.

Love, Alan C., and M. Travisano. 2013. "Microbes Modeling Ontogeny." *Biology and Philosophy* 28 (2): 161–88.

Mesoudi, Alex. 2011. *Cultural Evolution: How Darwinian Theory Can Explain Human Culture and Synthesize the Social Sciences.* Chicago: University of Chicago Press.

Rogers, Everett M. 2003. *Diffusion of Innovation.* 5th ed. New York: Free Press.

Shennon, Stephen, ed. 2009. *Pattern and Process in Cultural Evolution.* Berkeley: University of California Press.

Skusa, A., and M. A. Bedau. 2002. "Towards a Comparison of Evolutionary Creativity in Biological and Cultural Evolution." In *Artificial Life VIII,* edited by R. Standish, M. A. Bedau, and H. A. Abbass, 233–42. Cambridge, Mass.: MIT Press.

Strumsky, D., J. Lobo, and S. van der Leeuw. 2012. "Using Patent Technology Codes to Study Technological Change." *Economics of Innovation and New Technology* 12:267–86.

Valverde, S., R. V. Solé, M. A. Bedau, and N. Packard. 2007. "Topology and Evolution of Technology Innovation Networks." *Physical Review E* 76:056118.1–056118.7.

Venturini, Tommaso, Pablo Jensen, and Bruno Latour. 2015. "Fill in the Gap: A New Alliance for Social and Natural Sciences." *Journal of Artificial Societies and Social Simulation* 18 (2): 11.

Weisberg, Michael. 2013. *Simulation and Similarity: Using Models to Understand the World.* Oxford: Oxford University Press.

Wimsatt, W. C. 1987. "False Model as Means to Truer Theories." In *Neutral Models in Biology,* edited by M. Nitecki and A. Hoffman, 23–55. London: Oxford University Press. Reprinted in Wimsatt 2007.

Wimsatt, W. C. 2007. *Re-engineering Philosophy for Limited Beings: Piecewise Approximations to Reality.* Cambridge, Mass.: Harvard University Press.

Wimsatt, W. C. 2012. "The Analytic Geometry of Genetics: The Structure, Function, and Early Evolution of Punnett Squares." Inaugural Biology Issue. *Archive for the History of the Exact Sciences* 37. doi:10.1007/s00407–012–0096–7.

Wimsatt, W. C. 2013. "Articulating Babel: An Approach to Cultural Evolution." *Studies in History and Philosophy of Biological and Biomedical Sciences* 44:563–71.

Wimsatt, W. C., and J. R. Griesemer. 2007. "Reproducing Entrenchments to Scaffold Culture: The Central Role of Development in Cultural Evolution." In *Integrating Evolution and Development: From Theory to Practice,* edited by R. Sansom and R. N. Brandon, 227–323. Cambridge, Mass.: MIT Press.

Ziman, John, ed. *Technological Innovation as an Evolutionary Process.* Cambridge: Cambridge University Press, 2000.

7

MODELING THE COEVOLUTION
OF RELIGION AND COORDINATION
IN BALINESE RICE FARMING

MARSHALL ABRAMS

CULTURE OFTEN SEEMS to exhibit a high degree of harmony or coherence, demonstrating various sorts of "fit" between cultural patterns in disparate domains. In this chapter, I explore one strategy for explaining some kinds of cultural coherence. I show how a gradual process by which individuals with different cultural variants influence each other could lead to such coherence. More specifically, I use computer simulations to model the spread of religious patterns specific to rice-growing regions in southern Bali. These cultural patterns show a high degree of coherence with Balinese beliefs about the natural world. I show that the religious patterns could have spread through cultural transmission biased by variation in harvest success, under the influence of local social and ecological conditions.

The simulations also highlight the potential importance of what we might call *population communication structure* in cultural transmission: in the simulations, the spread of certain religious cultural patterns was likely only when communication from distant farming regions occurred infrequently. This communication pattern allowed homogeneity in small regions to develop, creating pragmatic benefits that subsequently made religious practices in those regions attractive to individuals in other areas. This ability of partial isolation to preserve variation and to support group-beneficial effects is well known from evolutionary biology (e.g., Gillespie 1998; Godfrey-Smith 2009; W. C. Wimsatt 2002).[1] A great deal of research has been conducted on group-level effects in cultural transmission (e.g., Boyd and Richerson 2005), as well as on the effects of communication structure on cultural transmission (e.g., Alexander 2007; Atran and Medin 2008). According to some proposals, group-level effects have played an important role in the spread of certain religious practices (Norenzayan 2013; Norenzayan et al. 2016; Wilson 2002). The present research, however, illustrates a new way that explanations of harmonious cross-domain cultural patterns may be constrained by the effects of communication structure.

Shared-Culture and Gene-Culture Coevolution Traditions

Anthropologists often view a culture as something that is shared by all or most members of a society or by members of some group within it (e.g., Benedict [1934] 2005; Brown 2008; Descola 2005; C. Geertz 1973b; Lévi-Strauss [1962] 1990).[2] Although there is enormous diversity in the assumptions and methods of such "shared-culture" approaches, most anthropological research on culture probably falls within this category. Shared-culture approaches do not necessarily ignore the existence of cultural variation within a society (e.g., C. Geertz 1973b; Lienhardt 1961), but many anthropologists focus only on those variants that many individuals share. Some researchers, such as those who use *cultural consonance* methods, acknowledge a great deal of individual cultural variation but use statistical methods mainly to derive evidence of a shared core of culture that each person is able to report or embody to one degree or another (e.g., Dressler et al. 2005; Romney, Weller, and Batchelder 1986).

Research in what is known as the dual-inheritance theory or gene–culture coevolution (GCC) tradition has a different focus and often very different assumptions from shared-culture research. Rather than focusing on *a* culture as something that is common to many individuals, GCC approaches treat *cultural variants* (usually beliefs or other mental states but sometimes behavioral practices or artifacts) as attributes of individual persons. GCC approaches often focus on explaining cultural change within a population by investigating conditions that influence how cultural variants spread from individual to individual. The focus on cultural change is not unique to the GCC tradition; anthropologists working in shared-culture traditions sometimes discuss cultural change too, but when they do their focus is usually on large-scale patterns of change involving an entire society or large parts of it (e.g., Benedict [1934] 2005; Descola 1986; Lévi-Strauss [1962] 1990; Tsing 2005), rather than on ways that numerous individual interactions between many people produce this change.

Cultural Coherence

Within shared-culture traditions, culture is often presented as composed of a set of mutually interdependent, partly harmonious cultural elements. The ways in which cultural variants "cohere" varies, though. For example, Benedict ([1934] 2005) spoke of cultures as having coherent "personalities," and Lévi-Strauss ([1962] 1990) viewed many elements within a culture as conceptually complementary. Other authors implicitly describe cultures as relatively coherent by showing how certain elements of a culture are interdependent or mutually supporting (e.g., Descola 2005; C. Geertz 1973b; González 2001; Sanday 1981; Smelser 1993; Tsing 2005; W. C. Wimsatt 2002, 2014; Wimsatt and Griesemer 2007).[3]

I will talk of some sets of cultural patterns as "coherent." This idea is intentionally vague. At the most basic level, the well-known phenomenon of cognitive dissonance (e.g., Izuma et al. 2010) suggests that beliefs that are obviously contradictory, that seem unlikely to all be true, or that are simply emotionally difficult to hold at the same time will be less likely to be shared by most members of the society (which is not to say that it is impossible for contradictory sets of beliefs to be widely shared). Cultural patterns can include complex worldviews, so if cultural patterns from different societies were arbitrarily mixed together, the potential for contradiction, mere implausibility, and emotional discomfort due to conflicting values and ideas would be high. This suggests that there are at least very loose constraints on the elements of a culture that are likely to be found together. There seem to be more subtle kinds of cultural coherence as well, perhaps depending on relationships between cultural variants that involve social structure (Brown and Feldman 2009; Caporael 2014), emotional relationships involved in structured social interactions (Bourdieu 1966; Caporael 2014; C. Geertz 1973a), physical aspects of daily activities (Descola 2005; González 2001), aesthetic relationships (H. Geertz 2004; Lansing 2006), analogical and metaphorical relationships (Colby 1991; Dehghani et al. 2009; C. Geertz 1973b; Lévi-Strauss [1962] 1990; Sanday 1981; Thagard 2012; Tilley 2000), general psychological processes of association (Colby 1991), and various aspects of psychology and material culture that scaffold or facilitate learning certain cultural variants or facilitate certain behaviors (Abrams 2015a, 2015b; Kline 2015; B. H. Wimsatt 2014; W. C. Wimsatt 2014; Wimsatt and Griesemer 2007; chapter 1 of this book).

What I mean by coherence is thus quite broad, but I am particularly interested in those striking sorts of coherence that cut across disparate domains and do not immediately suggest a simple explanation. For example, Clifford Geertz (1973a) described patterns of betting behavior at Balinese cockfights. He argued that the cocks on which spectators bet and the amount of money put up reflected conflicts and alliances between various social groups whose members participated in the cockfight:

> What makes Balinese cockfighting deep is thus not money in itself, but what . . . money causes to happen: the migration of the Balinese status hierarchy into the body of the cockfight. Psychologically an Aesopian representation of the ideal/demonic, rather narcissistic, male self, sociologically it is an equally Aesopian representation of the complex fields of tension set up by the controlled, muted, ceremonial, but for all that deeply felt, interaction of those selves in the context of everyday life. The cocks may be surrogates for their owners' personalities, animal mirrors of psychic form, but . . . the cockfight is—or more exactly, deliberately is made to be—a simulation of the social matrix, the involved system of crosscutting, overlapping, highly corporate groups—villages, kingroups, irrigation societies, temple congregations, "castes"—in which its devotees live. And as prestige, the necessity to affirm it, defend it, celebrate it, justify it, and just plain bask in it . . . is perhaps the central driving force in the society, so also . . . is it of the cockfight. (C. Geertz 1973a, 436)

Geertz argues here that two very different cultural domains (relationships between social groups and betting in cockfights) exhibit structural parallels and that these parallels interact with patterns in underlying antagonisms, feelings of solidarity, and feelings about status common for men in Balinese society in the late 1950s. Other parts of Geertz's essay show how the Balinese cockfight exhibits harmony with additional dimensions of Balinese life. The essay thus provides a description of subtle relationships of coherence within a culture. Geertz's work has been both widely celebrated and widely criticized (Alexander, Smith, and Norton 2011; Brown 2008; Risjord 2007), and one might doubt whether people in any society exhibit so much psychological uniformity, but the passage provides a good illustration of what I believe has been shown by the body of qualitative ethnographic research as a whole: that subtle and complex relationships of "harmony" between cultural patterns in different domains are probably common in most societies.

GCC researchers sometimes endorse the idea that sets of cultural variants exhibit general coherence patterns (Richerson and Boyd 2005; Richerson et al. 1997). Given GCC's focus on the transmission of cultural variants between individuals, GCC researchers should take coherence to result partly from processes within and between individuals. Moreover, as suggested by the remarks above, research in cognitive science has investigated how certain thought patterns or behaviors make others more likely in an individual,[4] and it seems probable that relationships of this kind might also influence ways that different cultural variants do or do not tend to spread together within a population. However, GCC researchers have generally not made methods for investigating coherence relationships part of their focus. This is not surprising: While shared-culture researchers can easily focus on complex relationships between many cultural variants, ignoring or downplaying individual variation or focusing on only a few individuals, GCC researchers' focus on individual-level variation within populations makes the study of complex relationships between cultural patterns difficult. Some GCC-related research has looked at how adoption of some cultural variants hinders or facilitates the adoption of others,[5] but GCC researchers have rarely tried to move toward investigating the kind of subtle coherence relationships between cultural patterns that shared-culture researchers have described.

The research reported here attempts to take a small step in that direction, using simulations inspired both by the GCC tradition and by shared-culture research on subtle relationships between religious ideas, farming practices, and democratic institutions. My simulations extend those developed by Stephen Lansing and James Kremer (1993), which they used to support a hypothesis about how Balinese rice farmers came to coordinate planting and water use. Lansing's subsequent research on religious and social phenomena that interact with this coordination process motivated my simulations.

I provide a context for understanding both sets of simulations in the next section, describing Balinese farmers' sophisticated method of coordinating planting and water use and showing how Lansing and Kremer's simulations helped to explain its origin. This section also discusses Lansing's investigation of Balinese religious practices, which seem in some respects ideally suited to help maintain the institutions and practices that support the water coordination system. This is the example of cultural coherence that will be my focus in the chapter. In the following major section, I outline several hypotheses that might explain how the rice-growing regions came to have religious cultural patterns so tuned to the needs of growers. The simulation results reported following that provide support for one of these hypotheses: that success-biased cultural transmission explains the spread of religious patterns conducive to the rice grower's needs. However, the simulations also suggest that this kind of cultural transmission only makes the spread of such religious patterns likely when communication between different groups of people occurs rarely, albeit regularly. The penultimate section of the chapter notes ways that these constraints might be realized in a real society, discusses the advantages and limitations of my model, and explains why some of the alternative hypotheses mentioned below may also provide partial explanations of the spread of Balinese religious patterns. I also note that the common view that agent-based simulations should be simple in conception ought to be tempered. While simple simulations are often easier to understand, their typical focus on modeling abstract principles can keep us from discovering theoretical principles more easily noticed using somewhat complex simulations. The final section provides some summary remarks.

BACKGROUND

Balinese Rice Production

Rice growing on the southern slopes of Bali typically involves scheduling crops so that nearby fields lie fallow at the same time.[6] Otherwise, pests such as rats and insects easily move from fallow fields to those with growing plants, resulting in the unchecked growth of pests, and agricultural disaster. However, farmers depend on water flowing down the mountainside in rivers and canals, and there is usually not enough water for all farmers in the same watershed to plant simultaneously. There is thus a complex coordination problem: water management requires that planting schedules be staggered, while pest management requires that nearby fields, at least, have the same planting schedules.[7]

Balinese rice farmers have traditionally solved this problem as follows: Farmers are members of villages. Several villages and their residents are members of a single water temple, or *subak*. Groups of nearby subaks adopt the same planting schedule (see below). Each such planting schedule should, ideally, have fallow periods that differ from those of other groups of subaks sharing the same watershed. The result is a complex pattern of planting schedules, as illustrated in Figure 7.1, which shows Lansing and Kremer's (1993) map of subaks' planting schedules in two watersheds in Bali.

How did the Balinese develop this centuries-old crop-scheduling system? Schulte Nordholt ([1996] 2010, 2011) and Hauser-Schäublin (2003) argued that the system was the result of central planning by representatives of kings, beginning at least as early as the eighteenth century. We can call this the *central planning* hypothesis. Clifford Geertz (1981), followed by Lansing and his colleagues (Lansing 2006, [1991] 2007; Lansing et al. 2009; Lansing and Kremer 1993),[8] argued that the Balinese system emerged without central coordination and resulted from local decisions within democratically elected councils in subaks. According to this *distributed decisions* hypothesis, subaks do not need to confer with other subaks across an entire watershed in order to produce the kind of pattern of coordination observed.

Evidence produced in support of the central planning and distributed decisions hypotheses has come from historical documents, interviews, surveys, ethnographic work, and archaeological data. The distributed decisions position has a prima facie problem, however: Is it really plausible that such a complex system of coordination across the length of a watershed could arise from local decisions without any kind of global coordination? Using computer simulations developed by Lansing with an ecologist, James Kremer, Lansing argued that it could (Lansing 2006, [1991] 2007; Lansing and Kremer 1993).

The Lansing-Kremer Model

Lansing and Kremer's (1993) agent-based computer model treats each subak as an individual "agent" that makes a decision every year about which planting schedule to adopt. These planting schedules allow different rice varieties to be grown in various orders with different fallow months. Subaks are arranged in two watersheds based on the actual arrangement of subaks along the Oos and Petanu Rivers in Bali (cf. Figures 7.1 and 7.2), and the model tracks the effects of upstream subaks' planting schedules on downstream subaks. (Subaks use more water when their rice is growing.) The model also tracks pest growth and movement and the effect of pests on harvests. In Janssen's (2012) model, which is based on Lansing and Kremer's (1993) original model, some subaks are *pest neighbors*. Pests can only travel between pest neighbors, and thus the model isolates pests within clusters of subaks. (This idealization was relaxed in some of Lansing's later models, such as those presented in Lansing et al. [2009, 2017], producing similar results to the original Lansing–Kremer model without assuming impassable barriers to pest movement.)

Initially, each subak is assigned a random planting schedule (a sequence of rice varieties and fallow periods along with a starting month). Water flow, rice growth, pest growth, and pest movement are tracked in each timestep, which represents one month. At the end of the year, each subak reassesses its crop schedule according to the following rule:

Figure 7.1. From Lansing and Kremer (1993), Figure 10. Lines represent rivers or canals. Icons represent subak locations. Icon shapes represent planting schedules of subaks at the time of Lansing's study.

If any of our pest-neighbor subaks has a better harvest than we do, adopt their planting schedule in the coming year.

Lansing's hypothesis was that this simple, wholly local process would lead to the kind of pattern of global planting schedules that has been found to be used both currently and historically.

Indeed, though each run of a Lansing–Kremer simulation differs in details, after twenty to one hundred modeled years, the simulation always settles into a state qualitatively similar to observed arrangements of planting schedules in subaks in the Oos and Petanu watersheds. Figures 7.2 and 7.3 illustrate this relationship using a version of the Lansing–Kremer model: Figure 7.2 shows the initial state of one run of the model, and Figure 7.3 shows the configuration of planting schedules in the same run after fifty years (six hundred time steps).[9] Each gray icon represents a subak, arranged roughly as actual subaks are arranged within the two watersheds. A gray line between two subaks represents the fact that they are pest neighbors: pests are able to travel between them, and each subak will compare its harvest with the other's at the end of the year. Subaks' shapes represent their sequences of rice varieties, and the direction of black pointers represents the month in which the sequence is started.[10] Although planting schedules are initially random, after fifty years, planting schedules are identical within most clusters of connected subaks. Compare this modeled arrangement after fifty years (Figure 7.3) with the empirical arrangement (Figure 7.1).

The simple rule stated by Lansing is thus capable of generating the sort of planting schedule patterns that Balinese rice farmers actually use. The rule is also intended to choose a pattern of planting schedules that will effectively trade off water needs versus pest suppression. In the simulation, the average harvest always increases after an initial settling period, although some runs result in better average harvests than others.[11]

Lansing and Kremer's model provides a "how-possibly" explanation (Brandon 1990)[12] of the origin of water coordination: it shows that the postulated mechanism is capable of generating the phenomenon of a certain kind of planting schedule pattern, with improved harvests. Lansing has provided additional evidence showing that this mechanism is more than a merely possible explanation, though. This evidence includes interviews, surveys, historical texts (Lansing 2006, [1991] 2007; Lansing and de Vet 2012), and other data sources, including genetic evidence for a more complex hypothesis about the gradual spread of the subak system (Lansing et al. 2009).

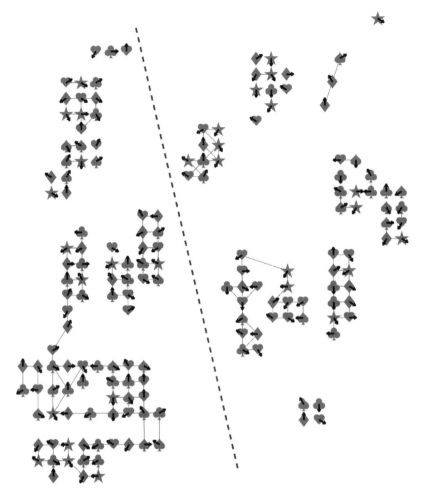

Figure 7.2. Initial random state from one run of BaliPlus, a modified version of Janssen's (2012) reimplementation of Lansing and Kremer's (1993) model, running in a configuration that reproduces the original behavior of Janssen's model. Each gray icon represents a subak, and solid gray lines connect pest neighbors. *Icon shape:* Rice variety sequence. *Pointer direction:* Month the sequence is started. *Dashed line:* Division between watersheds.

As mentioned above, critics such as Schulte Nordholt ([1996] 2010, 2011) and Hauser-Schäublin (2003) have argued for a different view, but all things considered, Lansing's view appears most plausible at this point. Tracking cultural history is always difficult, however.[13]

Note that though Lansing developed this model outside the GCC tradition, it is a cultural transmission model in which the cultural variants are

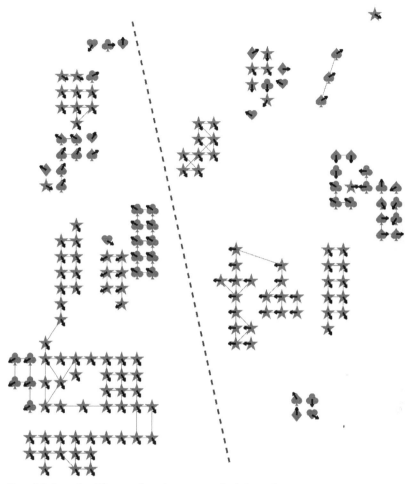

Figure 7.3. State after fifty years from the same run of BaliPlus as shown in Figure 7.2. Clusters of identical icons (*rice variety sequence*) and pointer directions (*starting month*) show that the system has evolved to a state in which pest neighbors (*connected by solid gray lines*) have the same planting schedules. See Figure 7.2 and the text for additional details.

planting schedules. Transmission occurs on a network structured by subak pest-neighbor relationships (cf. Alexander 2007; Grim et al. 2015). The model incorporates a kind of transmission bias known as *success bias* (Boyd and Richerson 1985, 2005; Richerson and Boyd 2005): specifically, individuals—subaks in this case—are more likely to copy cultural variants of other individuals that are more successful—that is, that have better harvests. It is also a niche construction model, since it incorporates feedback from the effects of human behavior on the environment (Odling-Smee, Laland, and Feldman 2003).

Struggle for Order

It seems rational, or at least reasonable, for a subak to switch its planting schedule to that of a neighboring subak with better harvests. However, Lansing's interviews and surveys, as well as historical texts, led him to suggest that subaks do not always act as rationally as his model assumed (2006, [1991] 2007). The actual behavior of subaks departs from that assumed by the model because of intra- and intersubak disputes involving power, status, greed, and so on. Lansing also suggested that religious cultural practices among rice farmers tend to suppress these disruptive tendencies toward such disputes. These practices reinforce a systematic identification of aspects of biotic and abiotic elements of the environment, psychological factors, and spiritual entities (C. Geertz 1981; Lansing 2006, [1991] 2007). For example, the Balinese world is full of demon-like entities known as *bhutakala,* which are common sources of disruption and disorder (Lansing 2006, chapters 5, 7). Some of these bhutakala are identical to rats or insects, while others are identical to aspects of the human psyche that may cause disruptive or otherwise undesirable behavior. People must work constantly to counteract the effects of bhutakala and to restore order and beauty when bhutakala succeed. This effort is simultaneously spiritual and practical, involving rituals and offerings at a hierarchical system of temples, and works to maintain harmonious relations within the democratic councils in villages and subaks that make decisions affecting water and crop management. What we call the religious aspects of Balinese culture are not just a set of policies overlaid on practical matters of water, rice, and pests but a reflection of a conception of nature as an aspect of a pervasively spiritual world. (Nevertheless, investigating cultural change requires distinguishing aspects of culture that may be inseparable from the point of view of members of the society, so I will continue to use *religion* and *religious* to refer to certain cultural patterns that seem, in the abstract, to be very distant from practical matters such as rice growth and pests.)

It may be that the relations of coherence between cultural patterns involved in the subak system include emotional and aesthetic patterns as well as conceptual and social patterns (cf. the above quotation from Geertz). The association of beauty with order is part of what gives maintenance and restoration efforts a subtle emotional dimension:

The elaborate rituals of the water temples convey a powerful message: that when individuals and subaks succeed in mastering themselves, the world (or at least the microcosm controlled by the subak) becomes more orderly. The flooded terraces resemble sparkling jewels, there are no plagues of pests, and the social life of families and communities is harmonious. On the other hand, when Reason gives way to destructive emotions, the effects are soon seen in quarreling families, disorderly fields, sickness, poverty, and pests. (Lansing 2006, 196)

These remarks suggest that the factors contributing to maintaining the orderly coordination of water management include feelings about the aesthetics of the physical systems, such as rice fields, as well as both personal and social characteristics. Moreover, descriptions of some Balinese ritual performances associated with the subak system (Eiseman Jr. 1989; C. Geertz 1981; H. Geertz 2004) suggest that such rituals might contribute to the cultivation of states of mind that reduce disruption. Eiseman mentions the need to have "a mind uncluttered with confusing or impure thoughts" (Eiseman Jr. 1989, 52) when making holy water, which plays a significant role in rituals associated with the subak system (C. Geertz 1981; Lansing 2006, [1991] 2007). If rituals associated with the subak system generally encourage similarly calm mental states, this might reduce tendencies toward disruptive states of mind.

Hypotheses about the Spread of Balinese Religion

Lansing (2006, [1991] 2007; Lansing and Fox 2011) argued that religious patterns in the rice-growing region differ from earlier royal/Brahmanic religious patterns (cf. C. Geertz 1981) and even earlier Javanese religious patterns, despite many similarities.[14] It is not clear to what extent the dimensions of the rice growers' subak system mentioned above derive from earlier sources. However, the royal system gives the king a central role in a heroic struggle to maintain spiritual purity against the forces of disorder. These will ultimately prevail, ending in the destruction of the kingdom and perhaps the world itself. By contrast, in the rice growers' religious system, every person must struggle, repeatedly, both individually and collectively, to restore social and physical order when disorder rears its head. This seems to be a change toward a system that is better for maintaining a harmonious social and agricultural system.

How does a religious system come about that seems, in some respects, as if it were tailored to the maintenance and management of the rice-growing system? Just as we cannot assume that all of an organism's traits are adaptive (Gould and Lewontin 1979), we cannot assume, as mid-twentieth-century social functionalists did, that societies must involve mutually supportive components (Kincaid 2007). The apparent harmony of some of the rice growers' religious patterns with their crop and water management methods needs a causal explanation. The fact that the values and practices of the subak system are similar to, but different from, those of earlier Balinese religious systems and their Javanese (and ultimately, Indian) predecessors suggests that the subak system arose as a modification of those systems. But why would that happen in such a way as to produce the cross-domain harmony that we see? That question is a central focus of this chapter.

Gervais et al. (2011) are surely right in suggesting that in general, biases on cultural transmission can help to explain some facts about religious cultural patterns. There are also explanations of some cross-culturally common religious patterns, such as Boyer's (2001) argument that concepts of spiritual beings are readily transmitted and retained because they are minimally counterintuitive (a cognitive property that has been shown to make concepts easier to remember). However, neither Gervais et al.'s nor Boyer's suggestions seem, by themselves, to be able to explain the way that Balinese religious patterns cohere with cultural patterns in other domains.

Elite Propaganda

One possible explanation of Balinese religious patterns among rice growers could be modeled on Schulte Nordholt's hypothesis about the water coordination system: perhaps a central royal authority constructed the rice growers' religious system and imposed it on them, even though it differed from the royal religious system. That elites may have done this in some societies seems plausible, but it would likely require a systematic propaganda campaign. I am not aware of any evidence that makes this hypothesis plausible for the present case.

Attractive Coherence

Lansing (2006, [1991] 2007) traced historical developments in Balinese religion and its predecessors but did not offer any particular hypotheses about why the rice growers' religious system came to be as it is. However, Lansing and Fox (2011, 933) suggested that the development of the complex Balinese calendar system, which plays an important role in the coordination of planting schedules, "contributed to a mental and physical landscape of pleasing harmonies and perceptible coherence." In context, I read this as a proposal that some religious patterns gradually changed *because* the newer variations helped people in the rice-growing regions feel more comfortable with their world, giving what happened in their lives a deeper explanation and meaning. This proposal implicitly involves a loosely specified psychological hypothesis that people prefer to adopt beliefs and practices that facilitate ways of thinking that are emotionally appealing. Earlier, I mentioned Lansing's (2006) suggestion that the Balinese see a well-functioning rice-growing system as harmonious and beautiful. If Lansing and Fox are right about the Balinese calendar, a similar proposal could be made about other aspects of Balinese religious ideas and practices: beliefs and practices change over time because some combinations of cultural variants feel more beautiful and harmonious together.

This proposal could be made more specific in two general ways:

1. Consciously or not, individuals gradually adjust their cultural patterns to make them more emotionally appealing.
2. When variations in cultural patterns are generated for whatever reason, those that are emotionally appealing are more likely to be copied or to be retained once copied.

The first of these provides a purely psychological explanation that could apply to a change within a single individual, without regard to interactions with others. The second hypothesis, on the other hand, postulates no internal transformations of religious patterns; it only concerns biases on what is transmitted or retained. These hypotheses are not mutually exclusive, however. Also note that if only the first hypothesis were correct, cultural transmission could still play a role in the spread of emotionally appealing patterns, but this need not involve biases toward adopting more appealing cultural patterns, as the second hypothesis requires. Other biases, toward patterns that are commonplace or that are held by high-status individuals, for example, might still play a role. The kind of individual adjustment postulated by the first hypothesis could also play a substantial role in cultural transmission in another way. Sperber and his colleagues (Claidière, Scott-Phillips, and Sperber 2014; Claidière and Sperber 2007; Sperber 1996) have argued that the spread of cultural patterns has more to do with ways in which what is learned is transformed by internal psychological processes. When a cultural pattern is transmitted, it may be *transformed* into something that is more emotionally appealing. I include this as one possible consequence of the first hypothesis.

Lansing and Fox seem to reject the second hypothesis as an explanation of the development of the Balinese calendar; they say that the development of the calendar "is not well captured by a Darwinian perspective" on cultural transmission (Lansing and Fox 2011, 933). However, this seems unwarranted. If pleasing harmonies are appealing to people, why could that not generate biases toward copying some beliefs and practices rather than others?

Applied to religious patterns, the first hypothesis that certain patterns are more likely to be adopted because they feel good looks like ones that Boyer (2001, chapter 1) has critiqued. However, Boyer's argument was that the emotional appeal of some forms of religion cannot answer the question "Why is there religion?" Here the question is, rather, "Why did certain religious patterns change into particular other patterns?" Moreover, the first hypothesis fits loosely with research on cognitive dissonance (e.g., Izuma et al. 2010), according to which people sometimes modify their beliefs to avoid distressing thoughts. There is also research suggests that people have a slight preference for accepting or constructing analogies to what they already believe (Gentner, Holyoak, and Kokinov 2001; Holyoak and Thagard 1995; Thibodeau and Boroditsky 2013). (Note that most accounts of analogy processing treat analogies as involving higher-order coherence relations—a kind of quasi-isomorphism—between sets of beliefs.)

It is worth mentioning here a variation of the second hypothesis that does not depend on emotional effects. There is evidence suggesting that certain kinds of coherence between cultural variants make them easier to retain once learned from other individuals (Bransford and National Research Council 2000; Kline 2015; Mesoudi and Whiten 2004; cf. Caporael, Griesemer, and Wimsatt 2014; Wimsatt and Griesemer 2007). For example, in Upal's (2011) experiments, stories in which all elements are easy to make sense of together were more easily remembered. Thus, as suggested above, it could turn out that religious patterns that fit with what is believed about people and pests and rice paddies are simply more likely to be retained and hence passed on than alternatives, regardless of emotional appeal.

Group Selection

Wilson (2002), commenting on Lansing ([1991] 2007), seemed to suggest that religious dimensions of the subak system evolved by a selection process of some kind, but he did not explain what sort. Wilson's book focuses mostly on forms of group selection, however. Was there competition between subaks, with successful subaks creating new ones? Lansing and colleagues (2009) argued that in some parts of Bali, there is evidence that the subak system spread as some individuals left old subaks and originated new subaks downstream. In the Petanu River watershed, Lansing et al. (2009) argued that this subak "budding" hypothesis is supported by genetic data. This is consistent with a group selectional explanation of change in religious patterns: It may have been that there were chance variations that arose from earlier religious patterns and that some of these variations turned out to help rice farming and water management. Those subaks with the beneficial variants had more food, on average, and as a result their populations increased more rapidly than those of other subaks. Eventually, some members of the high-growth subaks left to form new subaks. This is a form of what Boyd and Richerson call *cultural group selection* (Bell, Richerson, and McElreath 2009; Richerson and Boyd 2005; Soltis, Richerson, and Boyd 1995), whereby some groups grow and create new groups more rapidly than others because of cultural differences between groups.

Despite the evidence for a budding process in the Petanu watershed, Lansing et al. (2009) found no genetic evidence for a similar process in the Sungi River watershed. It is possible that a budding process occurred there at such an early date that migration and intermarriage between subaks, or other factors, have erased genetic evidence of the budding process. Studies of this area were in fact the basis of Schulte Nordholt's ([1996] 2010, 2011) arguments that the water coordination system originated in central planning by representatives of the king. It may be that in some parts of Bali, such as the Petanu watershed, cultural group selection explains the spread of religious patterns that support the water coordination system but that in others, such as the Sungi watershed, a different explanation would be needed.[15]

My Lansing-style budding process explanation of religious change is somewhat similar to an explanation by Norenzayan (2013) and his collaborators (Norenzayan et al. 2016) of the spread of religions that postulate one or more "Big Gods"—all-powerful, omniscient, moralizing, supernatural beings. These researchers argue that Big Gods religions have spread through cultural group selection and success-biased transmission (see below) because the religions promote within-group cooperation. Balinese gods and spiritual beings share some of the qualities of Big Gods, and the subak-local religion can be viewed as a borderline case of what Norenzayan et al. (2016) describe. However, the cultural variants that support rice farming go well beyond the

general Big Gods pattern—for example, identifying rats with disruptive spiritual forces (Lansing 2006, chapter 7). Norenzayan et al.'s (2016) proposals
can at best provide a partial explanation of the spread of subak-local Balinese religious patterns.

Success-Biased Transmission

The subak religious system may have spread through imitation rather than
cultural group selection. If religious patterns prevalent in some subaks resulted in greater food production, people in other subaks may have been more
likely to copy these patterns. This is what Richerson and Boyd (2005) call
cultural transmission with success bias, a form of model-based bias (since it
depends on properties of the individual copied, i.e., the "model"). It is not
necessary that members of the society have a clear understanding of the
mechanisms by which religious patterns result in better harvests in order for
the patterns to be preferentially copied.

The simulation model that I will describe below is designed to show that
a success-biased transmission hypothesis is plausible given a Lansing-style
emergence model of water coordination. More specifically, my version of this
hypothesis is the following:

1. Initially, it was fairly common for disruptive individuals to cause subaks
   to choose a planting schedule other than one that was most successful in
   neighboring subaks. Rice-growing peasants used religious patterns that
   did not mitigate this effect particularly well. Religious patterns at that
   point may have been closely analogous to Brahmanic patterns and/or
   may have reflected older Javanese or native Balinese religious patterns.
2. New ideas involving religious patterns similar to those described above
   developed in one or more individuals—somewhat randomly, though
   probably partly due to analogies with existing religious patterns and possibly also partly due to analogies with existing practices directly related
   to water management.[16] A Lansing/Fox style cognitive process favoring
   emotionally appealing cultural patterns might have played a role in
   generating new patterns too.
3. Some individuals in some subaks adopted these initially rare concepts,
   beliefs, and practices, perhaps solely due to chance effects, partly because
   people are somewhat drawn to analogies of sets of preexisting beliefs
   (Dehghani et al. 2009; Hofstadter and Sander 2013; Holyoak and

Thagard 1995; Thagard 2000, 2012; Thibodeau and Boroditsky 2011, 2013), or because of other appealing properties of the new patterns.

4. At some point, when *most* members of some local group of subaks had adopted religious patterns like those described above, the result was that people in those subaks had better harvests, on average, because these patterns help to suppress attitudes or behaviors that interfere with the functioning of the water/pest coordination system. (Among other things, the religious patterns might have involved analogies or metaphors that help motivate ways of thinking and behaving that help harvests. For example, an analogy between rats and demons helps to think of rats as agents of disorder. If analogies turn into identifications—rats are themselves demons—then that strengthens the effects of religious variants, since practical matters are then identical to spiritual matters.)

5. Success-biased transmission between local groups then resulted in other subaks adopting the same religious patterns. The idea is that individuals in one subak noticed that another subak had better harvests and guessed that part of the reason was because their religious beliefs and practices were correct or beneficial due to spiritually mediated effects. (This idea is analogous to one dimension of Norenzayan's [2013; Norenzayan et al. 2016] proposal that "Big Gods" religions such as the Abrahamic religions spread because of features that facilitated within-group cooperation. Those adopting a religion will often attribute the success of those they are copying to spiritually mediated help, even if the same success can be explained by the religion's effects on practical behavior. In the Balinese case, too, it seems that the benefits of certain religious patterns can be explained without recourse to spiritual hypotheses.)

An Extended Lansing Kremer–Style Model

In order to investigate the preceding hypothesis, I developed an extended Lansing-Kremer model (BaliPlus). The results show that under some conditions, processes like the one just described can indeed lead to the spread of new religious patterns because of their effects on harvests. The model represents the cultural forces and patterns discussed in earlier sections with a considerable degree of abstraction. The preceding discussion is valuable, I feel, because it allows an understanding of what kind of real-world phenomena motivate the model and what it is that is being abstracted away.

Overview

In order to explain how elements of the hypothesis described above are represented in the model, I will begin with a general description of the BaliPlus model and then describe how the preceding hypothesis can be tested using this model.

GENERAL CHARACTERISTICS OF THE MODEL

To model success-biased transmission of religious variants in an abstract fashion, I began with Janssen's (2012) version of Lansing and Kremer's model and added additional variables and functions:

- The model has a global "capriciousness" variable to represent the overall effects of greed, jealousy, and so forth on individual subaks' planting-schedule choices. More specifically, I treat capriciousness as a probability of an individual subak randomly choosing a different planting schedule than that of the neighboring subak that has the best harvest.[17]
- Each subak has a variable representing its "religious" cultural variant, represented as a number between 0 and 1. Religious variants near 1, which represent a degree of closeness to recent Balinese patterns, tend to suppress the effects of capriciousness. I investigated several mappings of values of this variable to effects on capriciousness, including a linear relationship and three functions that gave greater weight to higher-valued religious variants.
- Once per year, subaks copy religious variants from other subaks that have better harvests. This copying process is not perfect: The new religious value received by a subak is roughly normally distributed around the transmitting subak's value. (It would be precisely normally distributed, except that I restrict religious values to the interval [0,1]. If the sum of the transmitter's religious value and a normally distributed random number with mean 0 lies outside this interval, the subak is assigned, as its religious variant, the nearest extreme value.)
- Subaks may copy religious variants from either pest neighbors or from members of a randomly selected set of subaks from the global population. The rationale for this rule is that there is no good reason to restrict religious copying to pest neighbors, but pest neighbors are near each other, so copying from them would probably be more likely. I investigated various ways to implement this idea.

In summary, subaks copy both planting schedules and religious variants from subaks that have more successful harvests, but they copy religious variants from a larger set of subaks, and religious values influence the tendency of subaks to choose merely random planting schedules instead of the best pest neighbor's schedule.

The model operates as follows:

- Subaks are initially assigned randomly chosen planting schedules and uniformly distributed random religious variants between 0 and 1.
- After a six thousand-month = five hundred-year "burn-in" period,
- the model runs for twenty-four thousand months = two thousand years.[18]

Since the model is stochastic, we ran one hundred simulation runs for each set of fourteen parameter variants described below,[19] under five different pest and rainfall parameter combinations—a total of seven thousand simulation runs. An appendix summarizes all parameter settings.

OVERVIEW OF HYPOTHESES TESTED

Earlier, I described a hypothesis about how success-biased transmission could lead to the spread of religious patterns that facilitate the coordination of water and crop management. How can we translate this hypothesis about cultural processes in the world into the framework of the BaliPlus model sketched above? Two things need to be shown.

1. Capriciousness should reduce harvests: A background assumption of the success-biased transmission hypothesis was that some members of the community engage in disruptive behaviors that interfere with the coordination of planting and water use, and this results in a reduction of harvests relative to what would happen otherwise.[20] In BaliPlus, random planting schedule choices controlled by the capriciousness variable represent this kind of disruption. We need to make sure that capriciousness in BaliPlus does in fact lead to a decrease in harvest success compared to the original Lansing-Kremer model. This must be done without allowing religious patterns to suppress effects of capriciousness so that we can understand the impact of religious effects later.

2. After adding the transmission of religious variants that are capable of suppressing capricious effects:

    (a) The population average harvest level should be higher with religious transmission than with capriciousness and no effects of religious variants because

    (b) capriciousness-suppressing religious variants—those with values near 1—should become widespread within the population (and thus suppress the effects of capriciousness).

THE PROBLEM WITH GLOBAL TRANSMISSION

I initially believed that part 2 of the preceding hypothesis would be satisfied when

- each subak chose another subak from which to copy religious values by examining the harvests of all 172 subaks in the population and then copying from the subak with the best harvest.

Preliminary experiments suggested that this would almost never lead to the spread of religious variants near 1. When subaks use success-biased copying from the entire population, all subaks quickly converge to an apparently randomly chosen, narrow range of religious variants. The narrow cluster of religious values of the population then shifts to higher or lower values in what looks much like a random walk.

    The explanation for this behavior is this: In order for harvest success bias to lead to the spread of high religious values, there has to be a positive correlation of harvest success and high-value religious variants. The model begins without any such correlation. All subaks then copy the religious variant of the one subak (or few subaks) that happens to have the highest harvest value. This value is roughly as likely to lie in any one region of [0,1] as in any other. All subaks then have approximately the same religious variant, and there is no variation in religious variants on which success bias can operate. The correlation between religious values and harvest success remains low, not because both values are randomly distributed but because of the lack of variation in religious variants. Success-biased copying from the global population never gets off the ground because there is no variation in religious behavioral patterns. To summarize:

- At the beginning of the simulation, when subaks first compare harvests across the entire population, there will be no association between religious values and harvest success.

- Since all subaks copy the best harvests, variation in religious variants disappears.
- After that, narrowly clustered religious variants random-walk (roughly speaking) due to transmission noise.

So religious variants with values near 1—those that suppress capriciousness—are unlikely to be spread due to success bias when success-biased copying considers the entire population. Thus, in the simulations described below, members of various smaller, partially random subsets of the global population serve as possible sources of religious variants.

Details of Simulations

CULTURAL TRANSMISSION NETWORK STRUCTURE
Subaks examine possible sources for religious variants according to the following rules:

1. Each subak always considers imitating the religious variants of its pest neighbors.
2. Each subak also considers imitating the religious variants of a Poisson-distributed number of subaks from the entire population, with the mean number of subaks equal to one of the following three values: .025, 1, 50. This (randomly chosen) number of subaks is then randomly chosen from the other subaks, without replacement. If the number of subaks that results is greater than 171 (the total number of other subaks), all 171 subaks are examined. Note that a pest neighbor can be chosen, in which case the choice of this "additional" subak from the global population has no effect.

CAPRICIOUSNESS AND THE EFFECT OF RELIGIOUS VARIANTS
In simulation runs that include capriciousness, after every subak has acquired a new planting schedule from a pest neighbor or has retained its previous schedule, a probability of acquiring a new randomly chosen planting schedule is calculated. This probability is set to 0.3 if there is no religious transmission. If there is also religious transmission, then the probability of acquiring a random planting schedule is 0.3 times the distance from 1 of the subak's *religious effect* (see below), reduced by a factor of two-thirds.[21] That is, when there is religious transmission, the probability of choosing a new, random planting schedule is:

$$0.3 \times 2/3 \times (1-\text{religious-effect}) = 0.2 \times (1-\text{religious-effect})$$

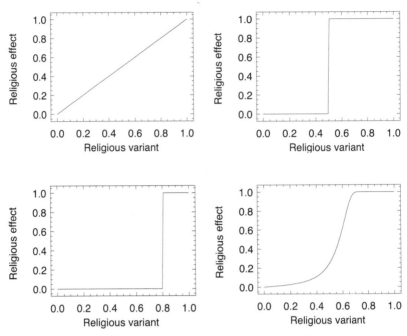

Figure 7.4. Four religious effect curves, each used in a different set of simulation runs.

What is *religious effect*? This number is set as a function of the subak's religious variant. The simplest way to do this would be to set the religious effect for a subak (in a given year) equal to the value of the subak's religious variant. This might not accurately represent what goes on in the world, however. It may be that as one acquires more components of a complex cultural pattern, its cumulative effect increases nonlinearly because of the way that its components reinforce each other. Thus, I ran simulations using four different *religious-effect functions*, each of which maps a subak's religious value in [0,1] to a degree of capriciousness suppression. The first religious-effect function simply treats the value of a religious variant as the strength of the religious effect; the other three produce different kinds of threshold effects, with a sharp increase in the intensity of the religious effect once the religious variant reaches a certain level (Figure 7.4):

1. **Linear:** religious effect = religious variant.
2. **Step at 0.5:** A function that maps religious variants below 0.5 to 0 and religious variants greater than or equal to 0.5 to 1.

3. **Step at 0.8:** A similar step function with a step at 0.8 rather than 0.5.

4. **Sigmoidey:** This is similar to a step function but is designed to allow a gradual increase in the suppression of capriciousness as the value of the religious variant increases. For religious variant v, this function is

$$\tanh \frac{v}{e^{2.25}(1-v^2)e^{1.7}}.$$

I chose the form of this "sigmoidey" function partly by trial and error. It is not important that one understand the details of this function; it is simply a function that allows a wide variety of monotonically increasing curves between 0 and 1 to be generated by substituting other numbers for 2.25 and 1.7. I chose this particular curve with parameters 2.25 and 1.7 because it was step-like yet gradual and similar to the two pure step functions.

PESTS AND RAINFALL

Janssen's (2012) NetLogo (Wilensky 1999) version of the Lansing-Kremer model allows rainfall to be set at three levels, *low, middle,* and *high.* There are also variables that control pests' growth rate (with values ranging from 2.0 to 2.4) and the rate of pests' dispersal to pest-neighbor subaks (values ranging from 0.6 to 1.5).[22] We ran fourteen sets of one hundred simulations, each with the following five rainfall and pest parameter combinations:[23]

- High pest, high rain
- High pest, low rain
- Low pest, high rain
- Low pest, low rain
- Middle pest, middle rain

High pest means that pests' growth and dispersal values were set to the highest values allowed by the NetLogo model; *low pest* means that these values were set to the lowest allowed values. *Middle pest* means that the pest growth rate was set to 2.2 and that the pest dispersal rate was set to 1.0. These are the values that Janssen (2007) used as intermediate values for his analysis.[24]

To summarize, for each of the five configurations just mentioned, we ran one hundred simulations in each of the following fourteen conditions (seven thousand simulations in all):

1: The Lansing-Kremer model: no capriciousness, no effects of religious variants.

2: The same model with the addition of the effects of capriciousness but no effects from religious variants.

3-14: Twelve different configurations in which both capriciousness and religious variants have effects. These twelve hundred simulations cover each combination of the three communication network structure parameters and the four religious-effect functions described above.

The first two configurations were intended to test the first hypothesis described in the section titled "Overview of Hypotheses Tested." The other configurations were intended to test the second hypothesis described there.

Results

Since all five pest/rainfall configurations gave qualitatively similar results, it will be easiest to present results only from the fourteen hundred simulations in the high pest, low rain configuration; this is one of the two configurations that produced the least striking confirmation of my main claims.[25] (Plots from the other four pest and rainfall configurations will be available online or by request from the author.)

GENERAL REMARKS

In the pure Lansing–Kremer configuration and in the runs with both capriciousness and the effects of religion, the population average harvest usually settles down to a value around which there are small fluctuations, with occasional long-term shifts, also small. There is a bit more variation in harvest values in the configuration with capriciousness and no religious effects, but these changes usually remain close to a central value. For each combination of conditions, the average harvest value around which there are small fluctuations varies from run to run.

Populations' average religious variants often vary quite a bit over the two thousand years in each simulation run. Close examination of the data makes it clear that even in runs in which the average religious variant has a high value during most timesteps, it sometimes takes a long time to arrive at that value, and in some cases the average subsequently drops down to a much lower value. As the figures below show, meaningful differences between the fourteen parameter settings concern distributions of *average* population-level effects over post-burn-in years in the one hundred runs with the same parameters. It is perhaps realistic that no condition in the model guarantees a particular result, especially in the case of religious variants.

HYPOTHESIS 1: EFFECT OF CAPRICIOUSNESS ALONE ON HARVESTS

Figure 7.5 shows that adding capriciousness to the pure Lansing–Kremer model does indeed reduce harvests, on average (the first hypothesis described in the "Overview of Hypotheses Tested" section). Since in any given year not all subaks have the same harvest, I used a measure of the per-year-population-average harvest *(avgharvestha)*. This value fluctuates from year to year in each model run, so I averaged it over two thousand years after the five hundred-year initial settling period. What Figure 7.5 shows, then, is the distributions of the resulting value in one hundred runs without capriciousness and in one hundred runs with capriciousness. Note that although capriciousness reduces harvests on average through reduction in the coordination of planting schedules, the overlap in the two curves in

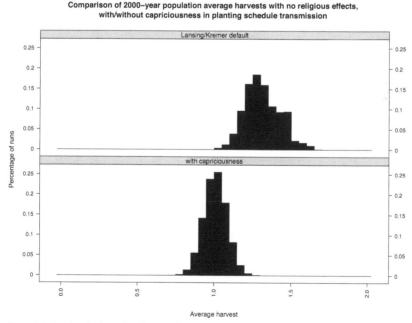

Figure 7.5. One hundred-run distribution of averages over two thousand years and all subaks of the per-year population-level average harvest values *(avgharvestha)* under two conditions. *Horizontal axis:* harvest value. *Vertical:* relative frequency in one hundred runs. *Top:* pure Lansing–Kremer model without added capriciousness. *Bottom:* Lansing–Kremer model with added capriciousness.

Figure 7.5 (between about 1.0 and 1.25) shows some capriciousness runs *(bottom panel)* with better average harvests than in runs under the pure Lansing–Kremer condition *(top panel)*. Capriciousness does not guarantee a poorer harvest.

HYPOTHESIS 2A: EFFECT OF RELIGIOUS VARIANTS ON HARVESTS
As noted above, in addition to the two sets of one hundred simulation runs just described, we ran twelve sets of one hundred simulations in which religious variants spread (three transmission schemes) and suppressed the effect of capriciousness on subaks when their religious variants were near 1 (four religious-effect functions). The results varied between the twelve conditions, but in each of the twelve conditions, harvests were better on average than in the pure capriciousness condition. Figure 7.6 shows this. In each of the twelve plots, an outline histogram shows the same one hundred-run distribution of two thousand-year average harvest values with capriciousness

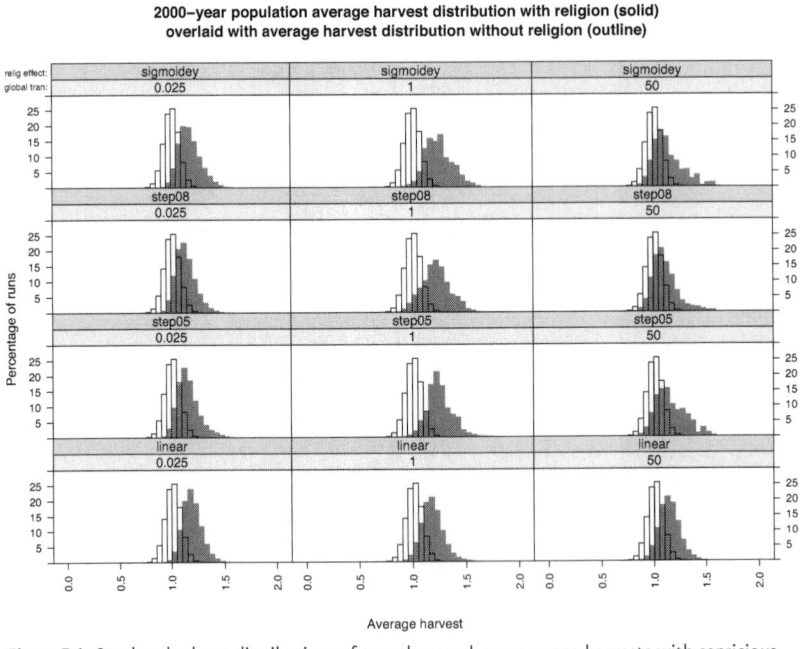

Figure 7.6. One hundred-run distributions of two thousand-year average harvests with capriciousness and spread of religious variants *(solid bars)* under twelve parameter combinations, compared to the same distribution with capriciousness alone *(outline bars)*. Upper row of labels *(sigmoidey, step08, step05, linear)*: religious-effect function. Lower row of labels *(0.025, 1, 50)*: approximate mean number of subaks from global population examined for religious transmission.

but no suppression by religion (corresponding to the lower plot in Figure 7.5), with a different solid-color plot of average harvests, one for each of the twelve religious-effect conditions. It is worth remarking that though the combination of the effects of religion with capriciousness produces greater harvests on average than with capriciousness alone, the effects of religion do not completely undo the effects of capriciousness. Under most of the parameter combinations, the mean of the religious-effects-plus-capriciousness curve is intermediate between the means for capriciousness alone and for the pure Lansing–Kremer model (not shown in Figure 7.6).

HYPOTHESIS 2B: SPREAD OF RELIGIOUS VARIANTS

Recall that in BaliPlus, in order to determine whether to copy another subak's religious variant, each subak looks at each of its pest neighbors as well as members of a (possibly empty) randomly chosen set of subaks from the entire population. The *size* of this set of randomly chosen subaks is itself random, with three different possible mean *global transmission* values for the probability of sizes. That is, the number of subaks examined is chosen randomly for each subak in each year, but the mean of the random distribution over these numbers is set once for each simulation run. This global transmission mean thus represents an average tendency for communication about religion across the entire population of subaks.

Figure 7.7 shows that capriciousness-suppressing religious values tend to spread when communication between nonneighbors exists but is rare (global transmission mean = 0.025 or 1). The effect is more pronounced with some religious-effect functions *(displayed top to bottom)*. By contrast, when the number of nonneighbors considered for comparison is large (global transmission mean = 50), there is no pronounced tendency for religious variants with high values to spread. Informal exploratory simulations using BaliPlus suggest that allowing communication with even higher numbers of subaks would decrease the likelihood of the spread of high-value religious variants further.

Close examination of the ways in which average religious variants change over time shows that even in those conditions that tend to spread high-valued religious variants, the average religious variant sometimes dips down to intermediate or low values for extended periods of time *(not shown)*.[26] Some global communication parameter combinations do better at "capturing" high-valued average religious variants for extended periods of time. For example, in the simulation runs in which communication across the entire

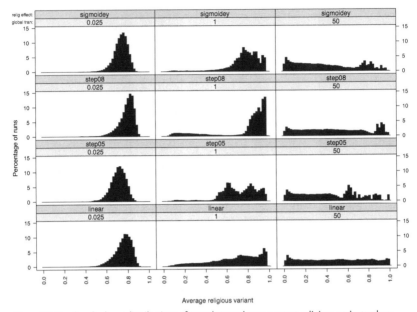

Figure 7.7. One hundred-run distributions of two thousand-year average religious variant values with capriciousness and spread of religious variants under twelve parameter combinations. Upper row of labels *(sigmoidey, step08, step05, linear)*: religious-effect function. *Lower row of labels (0.025, 1, 50)*: approximate mean number of subaks from the global population examined for religious transmission.

population is common—summarized in the right column of Figure 7.7— population averages for religious variants usually wander across much of the range of possible values, although religious values spend more time at high values in a few simulation runs. In the other two communication conditions *(left, center columns),* population averages wander until they happen to reach higher values and then stay there—usually. It appears that the manner in which the average religious variant does or does not wander in various situations is what accounts for the distributions represented in the histograms in Figure 7.7.

DISCUSSION

General Remarks

The results described above show that under certain conditions religious patterns can spread because they have, as one of their effects, suppression of behavior that interferes with mechanisms that otherwise produce widely desirable results such as larger harvests. The simulations support a "how-possibly" (Brandon 1990) explanation of the spread of these religious patterns in Bali. Specifically, the model lends support to the hypothesis that religious patterns involved in the Balinese planting/water-coordination system spread through success-biased cultural transmission between members of different subaks, with stronger and more regular influences between neighboring subaks than between more distant subaks. According to this hypothesis, those religious patterns that facilitated the subak-based crop and water management system by suppressing capricious behavior were those that managed to spread after partially random changes allowed some clusters of subaks to develop new, beneficial patterns. People in some subaks decided that others' religious patterns that seemed to lead to better harvests were worth copying. By running the simulations with a variety of parameter combinations, the simulations suggest that religious patterns that reduce capriciousness can spread by this kind of mechanism under a broad range of conditions.

My model illustrates how different domains of a culture can come to exhibit coherence in two senses. First, that religious patterns have beneficial effects on practices that support successful farming is a kind of coherence between religion and farming practices; the simulations show how this kind of coherence might come about. Second, although my model does not represent details of Balinese religious patterns, it is inspired by them and represents them in an abstract way. It can thus be viewed as a model of the spread of these more detailed religious patterns. These patterns seem to allow Balinese people to treat threats to harvest success—such as pests and greed—as threats to a spiritual order that is seen as emotionally and aesthetically attractive. Restoration of order is supposed to be sought both through religious practices implemented by individual farmers—offerings at local shrines, for example—and by religious practices of groups, which in turn are linked to democratic institutions at the levels of villages, subaks, and groups of subaks. Balinese religious patterns in the subak system thus exhibit various detailed coherence relations between religious and pragmatic practices of various kinds. The simulations show how such patterns might have spread.

This project is unusual in trying to explain certain kinds of coherence in a particular culture in terms of a specific mechanism of cultural transmission and in using computer simulations to do so. Explanations of general kinds of cross-cultural change, such as Norenzayan (2013), Norenzayan et al. (2016), or Sanday (1981), can be important, but much of what is interesting about culture is specific to particular societies. Illustrating new strategies for investigating local cultural change, as I do here, is valuable.

Intermittent Copying

In order for the success-biased transmission hypothesis to explain the cultural patterns that are my focus here, it appears to be necessary that subaks' tendencies to copy others' religious patterns involve only intermittent copying from more distant members of the global population. This is because capriciousness-suppressing religious patterns help harvests only when subaks connected by pest-neighbor relations adopt these same cultural patterns, allowing them to come to have the same planting schedules. If the harvests of all or many subaks were examined in order to determine which religious patterns were to be copied, then—assuming religious variation is initially random with respect to harvest success—many subaks would copy the religious patterns of a small number of subaks that happen to have the best harvests. This results in very little religious variation across the population, and without sufficient variation, it is unlikely that any cluster of subaks would come to have religious patterns that suppress capriciousness. Yet without concentration in clusters, capriciousness-suppressing religious patterns would have no particular advantage, so other subaks would not preferentially copy them. Thus, capriciousness-suppressing patterns would not spread.

On the other hand, if—as in some simulations described above—subaks always examine neighbors' harvests but occasionally also examine more distant subaks in order to decide whether to copy religious patterns, it is possible—again, by chance—for the members of one local cluster of subaks to adopt capriciousness-suppressing religious patterns from each other. This cluster will be likely to maintain its religious patterns over time; the members' harvests will usually be better than those of other subaks, so there will be no reason for them to copy religious patterns from outside the cluster. Then, when subaks from elsewhere eventually examine the harvests of members of this cluster, they will see that their harvests are better and will copy their religious patterns. At some point this process will result in a second cluster in which capriciousness-suppressing religious patterns are the norm, increasing the speed of the spread of these patterns. Over time, this process will lead to capriciousness-suppressing religious patterns spreading throughout the population. Subsequently, random factors can occasionally result in periods of time in which capriciousness-suppressing patterns are not widespread, but those periods will usually be relatively short-lived.

The way in which this model explains religious patterns raises a question for empirical research: Is it likely that such intermittent communication did occur in Bali? There are a variety of ways in which it might have occurred. First, note that subaks are composed of villages, which in turn are composed of many individuals. Second, note that real Balinese religious patterns are enormously more complex than the simple numeric values that BaliPlus uses to summarize variation. These two points allow for a variety of network effects that could produce the kind of intermittent influence modeled in BaliPlus:

1. It may simply be that contact between members of different subaks is itself intermittent. Various factors might interact here: distance, trade, kinship, friendship, and so on.

2. Within any given social group, those who communicate more often may be more likely to influence common cultural patterns, thus making it more difficult for cultural patterns held by others to spread within that group (cf. Abrams 2014; Alexander 2007; Caporael 2014; Morris 2000; Young 1998; chapters 1 and 12 of this book). The idea is that communication between people who are all in the same subak can reinforce others' cultural patterns. This could make it difficult for other cultural patterns from distant subaks to be taken seriously, even if communication with those subaks was not uncommon, and the distant subaks' religious practices were well known and appealing due to success bias. Other sorts of social identity might constrain communication as well (see chapter 12).

3. Some individuals within a group may have more influence than others due to having power of various sorts or being successful in ways not reflected in a simple model—perhaps due to likeability, charisma, or a reputation for wisdom or knowledge (cf. Durham 1991; Henrich and Broesch 2011; Richerson and Boyd 2005; Smaldino 2014). If particular individuals of this kind are not receptive to new cultural patterns, that fact can make it less likely that new patterns will spread in the social group.

4. As discussed in the first section of this chapter, some combinations of religious beliefs or practices may be infelicitous with others, create cognitive dissonance together, or even be logically contradictory, while other combinations may be more acceptable to many individuals, given

other prevalent cultural patterns. These relationships between cultural patterns generate a kind of interpersonal network structure, in that some cultural variants strengthen or resist influence from others: even if each member of subak S_1 is bombarded by influences from members of all other subaks S_k, it may be that because of relative incompatibility between cultural variants, certain religious patterns new to S_1 have a low probability of influencing anyone in S_1 and do so only occasionally (cf. Abrams 2013; Atran and Medin 2008; Axelrod 1997; Hegselmann and Krause 2002; Mueller, Simpkins, and Rasmussen 2010; Zollman 2013). One way in which this kind of phenomenon can occur is when some cultural variants scaffold or otherwise facilitate the learning of others; if an individual has not yet adopted the former patterns, the adoption of new religious beliefs or practices may be difficult or unlikely (cf. Abrams 2015a, 2015b; Kline 2015; B. H. Wimsatt 2014; W. C. Wimsatt 2014; Wimsatt and Griesemer 2007; chapter 1 of this book). However, the ways in which adopting cultural variants facilitate or hinder the adoption of others need not always exhibit the kind of typical linear sequence suggested by the concept of scaffolding or the analogy with biological development, as discussed by Wimsatt and Griesemer (2007; W. C. Wimsatt 2014; Wimsatt, chapter 1 of this volume). In Abrams (2015b) I suggested that all such cases could be conceptualized in terms of *transition probability interaction*: that probabilities of adoption of some cultural variants are conditional on what other cultural variants have been adopted (cf. Abrams 2015a).

What Is It a Model Of?

Some aspects of the BaliPlus model are clearly unrealistic. The original Lansing–Kremer model represented months and years in timesteps in order to organize modeled water flow, pest behavior, and planting schedules. It also made the simplifying assumption that subaks only consider changing planting schedules at the end of each year. In BaliPlus, subaks consider copying religious variants on the same annual schedule, but I have no empirical justification for the assumption that religious transmission should happen on the same timescale as agricultural decisions. I chose various other parameters (see the appendix) somewhat arbitrarily. For example, there is no strong reason for choosing 0.3 as the base probability of randomly choosing a planting schedule. However, the point of the model is to explore the possibility that certain kinds of religious patterns might spread, probabilistically, under the influence of factors specified by the parameters described above (and in an appendix). The rationale for choosing basic parameters such as this one is that the parameters allowed the possibility of generating the kinds of effects I was interested in investigating. Still, because of these relatively arbitrary assumptions, two thousand "years" of *communication as represented in the model* is not necessarily realistic, despite the fact that two thousand years is a period over which, realistically, there might have been rice farming in Bali (Lansing et al. 2009; see note 18). Nevertheless, the point of the BaliPlus model is to show how a particular *kind* of process might explain the spread of religious patterns conducive to planting and water management. We can view these simulations as illustrating certain kinds of processes by which cultural patterns can spread because of the very indirect influence on outcomes that are clearly valued (rice production, in this case).

There is, moreover, a more general point that has emerged from the simulations reported above. By embedding the transmission of religious patterns capable of influencing decisions about planting into simulations that had already modeled interactions involving rice growing, water flow, and the effects of pests, we learn the following: At least in cases sufficiently analogous to those modeled here, the practical effects of religious patterns can explain their spread *under the condition that this transmission is usually local and intermittently global.* Of course, what counts as sufficiently analogous to the Balinese case as modeled here is not clear. (One reason for this has to do with the complexity of the ecological processes modeled in Lansing and Kremer's and Janssen's simulations and in BaliPlus.)

Other Hypotheses

Even if success-biased transmission does explain the spread of certain Balinese religious patterns, that does not rule out some of the other explanations in the section on hypotheses about the spread of Balinese religion. Humans are complex, so there may be complementary explanations of cultural change that depend on different, potentially interacting processes. As noted above, Wilson (2002) seemed to suggest that religious patterns among Balinese rice farmers could be explained by group selection. While Lansing et al.'s (2009) "budding" model of the spread of the subak system fits Wilson's group selection hypothesis, it is not entirely clear whether this model fits all regions of Bali in which the subak system is common. Group selection may be part of the explanation for Balinese religious patterns, as might Lansing and Fox's (2011) hypothesis that certain cultural patterns arose in individuals because these patterns were psychologically more satisfying. I suggested that such psychological effects might also bias cultural transmission to favor the transmission and the retention of certain patterns. Perhaps religious patterns in Balinese rice-growing regions arose and spread due to a combination of individual psychological transformations, biases due to psychological attractiveness and harvest success associated with certain religious patterns, and group selection resulting from better harvests.

Moderate-Complexity Model Benefits

I developed another set of simulations (Intermittran, https://github.com /mars0i/intermittran) that were inspired by, but not directly based on, Lansing and Kremer's model. These simulations are not my focus here, so I will not go into detail about them, but it is worth mentioning some differences with BaliPlus. In Intermittran, I simplified BaliPlus's complex ecological feedback effects on harvest success to a simple function of nearby subaks' religious values combined with random noise. With this model, it is challenging, though not impossible, to produce results that are qualitatively similar to those in BaliPlus. The problem is that Intermittran makes it too easy to cause high-value religious variants to spread to all subaks and too easy to subsequently maintain a high average value. By contrast, in BaliPlus, even in those runs with parameter values that tend to make a population spend large amounts of time with high average religious variant values (Figure 7.7, *left and middle columns*), quite a few runs spend significant amounts of time with lower average religious variant levels—even after many years at high values *(not shown)*.[27] I was able to produce qualitatively similar behavior in Intermittran only through somewhat careful tuning of the random distribution that affects harvest success.[28] Because of this, I am skeptical that the kind of noisiness produced by the real-world ecological relationships modeled in BaliPlus can easily be approximated by reducing them to random distributions of the kind typically chosen by modelers.[29]

There is a common view (e.g., Epstein 2006) that it is best if agent-based simulations are *simple-agent models* (Abrams 2013)—that is, models in which the behaviors of agents are governed by a few simple rules. Otherwise, given a large number of interacting agents, it can be difficult to understand what is significant in the production of the model's behavior. The simple-agent strategy is a good heuristic, but it is not clear that the insights gotten from BaliPlus could have been gotten using only simple-agent models such as Intermittran. In Abrams (2013) I argued that there is value in developing agent-based models that involve somewhat more complex processes. The fact that it is difficult for Intermittran, a simple-agent model, to reproduce behavior like that in BaliPlus—apparently because of the way in which BaliPlus models complex ecological processes—provides some additional support for this point.

RELIGION IS an important topic because of its important role as a distinct cultural domain in modern industrialized societies. This chapter was not motivated by an interest in religion, however. My use of *religion* and *religious* to describe the cultural patterns that were the focus here merely provided a convenient shorthand for certain patterns within the Balinese rice-growers' culture. Since for the rice growers the physical world is spiritual and what is spiritual is continuous with the physical world, Balinese culture is one of those in which it is misleading to conceptualize religion as a distinct cultural domain (e.g., Descola 2005; C. Geertz 1973b; H. Geertz 2004; González 2001; Lansing 2006; Lienhardt 1961; Tilley 2000).[30] This is not to say that "religious" change must also involve changing all dimensions of culture (farming, eating, hunting, dress, etc.). Otherwise, all cultural change would require radical cultural saltations. I think that the evidence from historical ethnographic research such as Lansing's suggests otherwise.

What I find fascinating is that cultural patterns I have labeled *religious* seem, at first glance, to have no direct impact on practical needs, such as the provision of food or shelter. For example, the Balinese may have thought that religious practices intended to mitigate the effects of demons are relevant to farming because rats are in fact demons, but an outsider may find it mysterious why those practices should improve rice growing. By contrast, some cultural patterns can readily be understood as direct responses to subsistence needs, given environmental conditions and prior cultural traditions. The fact that Balinese farmers grow rice in paddies is probably a response to ecological facts about Bali and how rice can profitably be grown, in combination with the existence of rice growing in the societies from which theirs descended.

Part of what sometimes makes those cultural patterns classified as religious puzzling is that they seem distant from such pragmatic concerns, providing little obvious material benefit and costing a great deal. When religious patterns seem to cohere, in some clear sense, with patterns in other more pragmatic cultural domains, there is an additional puzzle: What explains the relationship between patterns in a pragmatic domain and those in one that could have been, one would think, completely independent of it?

What I have illustrated here is one strategy for understanding how such patterns could spread and come to "fit" with those that have more immediate practical consequences: cultural patterns without direct pragmatic consequences can have the effect of adjusting behaviors in subtle or complex ways so that the behaviors end up having improved practical consequences, perhaps for reasons that are not apparent to the participants. This can lead to success-biased preferences for copying those cultural patterns and thus spreading the "impractical" patterns, creating and maintaining a harmony between apparently disparate cultural domains.[31]

APPENDIX: SIMULATION PARAMETERS

NetLogo 5.2.1
Source file: src/LKJplus/BaliPlus.nlogo
(Versions of November 2015; some with trivial modifications from January and February 2016)
burn-in-months = 6,000 (500 years)
months per run: 30,000 (2,000 years plus 500 years burn-in)

Pest and rainfall configurations:

	pestgrowth-rate	*pestdispersal-rate*	*rainfall-scenario*
high/high	2.4	1.5	"high"
high/low	2.4	1.5	"low"
low/high	2.0	0.6	"high"
low/low	2.0	0.6	"low"
mid/mid	2.2	1.0	"middle"

Notes: The high/low configuration is the one from which data were reported in the text.

relig-tran-stddev = 0.02
relig-influence = 1.5
Used only the five crop plans that include only traditional rice varieties (1 and 2) (I.e., rice variety 3 was not used in any model.)
For runs with capriciousness, *ignore-neighbors-prob* = 0.3
Religious effect functions:
  - step at 0.5
  - step at 0.8
  - linear (i.e., suppression effect = value of religious cultural variant)

- "sigmoidey" with *relig-effect-center* = 2.25; *relig-effect-endpt* = 1.7 (see text for function definition)
Poisson means for the addition of subaks from the global population to those neighboring subaks who are candidates for transmission: *subaks-mean-global* = 0.025, 1, 50

NOTES

1. See also arguments, such as Page's (2007), that drawing upon diverse cultural backgrounds or ways of thinking can be valuable, for example, in problem solving.

2. This is not the place to discuss views that hypostatize culture as something that lies beyond the mental states, behaviors, and artifacts of a society (Clark 1999; Risjord 2014).

3. Kuhn (1996) describes similar patterns in scientific communities that share a paradigm; these might be called scientific cultures. Much of the evidence for cultural coherence comes from qualitative research, but Dressler, Balieiro, and dos Santos (2017) provide statistical evidence for coherence relations between different domains of life among urban Brazilians.

4. For example, Banaji and Greenwald (2013); Gentner, Holyoak, and Kokinov (2001); Hofstadter and Sander (2013); Holyoak and Thagard (1995); Izuma et al. (2010); Thibodeau and Boroditsky (2013).

5. See Abrams (2013, 2015a, 2015b); Alam et al. (2010); Boyd and Richerson (1985, 1987); Castro and Toro (2014); Cavalli-Sforza and Feldman (1981); Claidière, Scott-Phillips, and Sperber (2014); Claidière and Sperber (2007); Fogarty, Strimling, and Laland (2011); Henrich and McElreath (2003); Kashima (2000); Mesoudi and Whiten (2004); Sperber (1996).

6. The material in this section is based primarily on C. Geertz (1981); Janssen (2007); Lansing (2006, [1991] 2007); Lansing et al. (2009); Lansing and de Vet (2012); Lansing and Kremer (1993); Lansing, Kremer, and Smuts (1998), except where noted.

7. An alternative strategy was tried at the recommendation of Green Revolution planners in the 1960s and 1970s: Many farmers planted continuously, using new rice varieties and pesticides. After numerous attempts to fine-tune this strategy to avoid extremely poor results, the strategy was dropped, and Balinese rice farmers returned to the traditional methods sketched here.

8. See also Janssen (2007); Lansing and Fox (2011); Lansing, Kremer, and Smuts (1998).

9. This model was written primarily by Marco Janssen (2012), who was kind enough to make his model publicly available. I subsequently made modifications. More precisely, these figures show results of the BaliPlus model described below—which is based on Janssen's model—but with all of my extensions to Janssen's model disabled and with graphics tailored for the present display. My students Blake Helms and Jackson Hyde also modified the code, helping to develop the graphics for Figures 7.2 and 7.3, among other things.

10. Like the simulations described below, this simulation run used only those five planting sequences in the model that used only traditional, pre–Green Revolution rice varieties.

11. Also, in some runs, the average harvest drops a little when the large cluster of subaks in the lower-left corner of the display settles on a single planting schedule. Apparently, it is locally better for each subak to choose the same planting schedule as its neighbors in this cluster because that reduces pest growth. The result is that too many subaks are planting at the same time, so water use in the watershed is not optimal.

12. Richerson and Boyd (2005) call such explanations "why-maybe" explanations. Huneman (2014) calls them "candidate" explanations.

13. In recent work (Lansing et al. 2014; Lansing and Fox 2011; Lansing and Miller 2005), Lansing and his colleagues have argued that there are important cultural differences between upstream and downstream subaks. These are differences in attitudes and values that have to do with ways in which subaks interact in the water coordination system. I do not address these differences.

14. There appears to have been significant influence from Java, which is adjacent to Bali, by at least the ninth century c.e. (Lansing 2006), although there is evidence of much earlier contact with Indians or other Asians (Lansing et al. 2004).

15. In later papers Lansing and his colleagues gave further arguments against the hypothesis that planting schedules were centrally managed, even in the Sungi watershed (Lansing and de Vet 2012; Lansing and Fox 2011).

16. In simulations that I do not describe here, I have used methods introduced in Abrams (2013) to investigate processes by which analogies might have played a role in the spread of new religious patterns in Bali.

17. A new model of Balinese water and crop management by Lansing et al. (2017) also incorporates randomness in planting schedules due to disruptive behavior. This randomness has a different purpose in Lansing's model,

so his model does not include a factor that plays the capriciousness-suppressing role that religion does here. Though direct comparison of parameters in Lansing's model and the one described here is difficult, I believe that the amount of random disruption in his model is effectively less than in mine and may be analogous to the capriciousness remaining in the present model after new religious patterns have become widespread.

18. Lansing et al. (2009) report that the earliest evidence of possible rice cultivation in Bali is from about 2660 years ago and argued that it is more realistic to think that rice cultivation began 2000 years ago. I actually ran the models for 5000 years after the 500-year burn-in but only report the first 2000 years here. Results for the full 5000-year runs are qualitatively identical to what I present here. As discussed below, the amount of communication about religious variants in the model is not calibrated to data about actual communication, so it may be appropriate to think of years in the model as simply abstract markers of time. Nevertheless, I prefer to focus on a realistic number of years because the stochastic dimensions of the model that come from the vagaries of pest and water distribution are calibrated to actual years. I discuss this further in the penultimate section of the paper.

19. My students Blake Helms and Jackson Hyde ran most of the simulations.

20. This was also an assumption implicit in the group selection hypothesis.

21. I chose 0.3 as the base probability of choosing a random planting schedule and reduced the impact of the religious variant by two-thirds because these values allowed variation in other parameters to produce a wide range of interesting behavior.

22. For further details on the meaning of these parameters, see Janssen (2007), Lansing and Kremer (1993), and the documentation that comes with Janssen's (2012) model.

23. Most of these simulations were performed by two of my students, Christopher Blake Helms and Jackson Hyde.

24. Janssen (2007) used a different way of quantifying the pest dispersal rate. *pestdispersal-rate* = 1.0 in the NetLogo model corresponds to $d = 0.3$.

25. The other one is the high pest, high rain configuration.

26. A color plot illustrating this point is available from the author.

27. Since it may be common for there to be a great deal of fluctuation in religious patterns in some societies, the fact that BaliPlus illustrates this possibility is interesting. According to Hildred Geertz's (2004) description of

Balinese villages in the 1980s, a wide variety of nominally inconsistent variations on traditional (Hindu-based) Balinese religious patterns coexisted and interacted.

28. I initially experimented with various Gaussian distributions but had more success with beta distributions, which allow greater control over distributions' shapes.

29. In Abrams (2017) and Abrams (unpublished manuscript), I argue that some biological mechanisms and evolutionary processes may involve what is called *imprecise probability* (e.g., Fierens, Rêgo, and Fine 2009), a generalization of probability. Every process in BaliPlus is either deterministic or probabilistic in the usual sense, which would imply that there are no imprecise probabilities in BaliPlus. Nevertheless, I think it may be possible to argue that patterns of harvest success in BaliPlus have certain properties that would also be common in processes involving imprecise probabilities but uncommon in those involving probabilities. This is an issue for future investigation.

30. Some of Howe's (2001) remarks about religion in Bali suggest that for many Balinese, religion has in recent years become a distinct cultural domain.

31. I am grateful to Bill Wimsatt and Alan Love for detailed, helpful comments on an earlier draft of this paper and to Bill for earlier feedback; to Blake Helms and Jackson Hyde for work on the BaliPlus source code and for running the simulations; to Stephen Lansing for answering questions about his work and sharing unpublished material; to Marco Janssen for making available his NetLogo version of Lansing and Kremer's model and for answering questions. Others who provided helpful feedback include Adrian Currie, Barbara Wimsatt, Bill Dressler, Bret Beheim, Brett Calcott, Byron Kaldis, Cailin O'Connor, Christopher Lynn, Colin Garvey, Dan Grunman, Daniel Singer, David Henderson, Heidi Calloran, Jason DeCaro, Jim Bindon, Kathryn Oths, Lesley Weaver, Margaret Schabas, Mark Risjord, Melissa Brown, Michael Weisberg, Michiru Nagatsu, Murray Leaf, Paul Smaldino, Pete Richerson, Tyler Curtin, Yoichi Ishida, as well as others at several presentations. I owe my interest in Lansing's work to Emily Schultz's recommendation; conversations with Emily have influenced my thinking in ways that are reflected in this chapter but that may not be apparent. Finally, I am sincerely grateful to the University of Alabama at Birmingham IT Research Computing unit for making time available on the Cheaha computing cluster. Here is an official statement of acknowledgment:

This work was supported in part by the research computing resources acquired and managed by UAB IT Research Computing. Any opinions, findings, and conclusions or recommendations expressed in this material are those of the author and do not necessarily reflect the views of the University of Alabama at Birmingham.

REFERENCES

Abrams, M. Unpublished manuscript. "Imprecise Probability and Biological Fitness."

Abrams, M. 2013. "A Moderate Role for Cognitive Models in Agent-Based Modeling of Cultural Change." *Complex Adaptive Systems Modeling* 1 (16): 1–33. doi:10.1186/2194-3206-1-16.

Abrams, M. 2014. "Maintenance of Cultural Diversity: Social Roles, Social Networks, and Cognitive Networks." *Behavioral and Brain Sciences* 37:254–55.

Abrams, M. 2015a. "Coherence, Muller's Ratchet, and the Maintenance of Culture." *Philosophy of Science* 82 (5): 983–96.

Abrams, M. 2015b. "Cultural Variant Interaction in Teaching and Transmission." *Behavioral and Brain Sciences* 38:e32.

Abrams, M. 2017. "Probability and Chance in Mechanisms." In *The Routledge Handbook of Mechanisms and Mechanical Philosophy,* edited by S. Glennan and P. Illari, 169–83. New York: Routledge.

Alam, S. J., A. Geller, R. Meyer, and B. Werth. 2010. "Modelling Contextualized Reasoning in Complex Societies with 'Endorsements.'" *Journal of Artificial Societies and Social Simulation* 13 (4): 6. http://jasss.soc.surrey.ac.uk/13/4/6.html.

Alexander, J. C., P. Smith, and M. Norton, eds. 2011. *Interpreting Clifford Geertz: Cultural Investigation in the Social Sciences.* Basingstoke, UK: Palgrave MacMillan.

Alexander, J. M. 2007. *The Structural Evolution of Morality.* Cambridge: Cambridge University Press.

Atran, S., and D. Medin. 2008. *The Native Mind and the Cultural Construction of Nature.* Cambridge, Mass.: MIT Press.

Axelrod, R. 1997. "The Dissemination of Culture: A Model with Local Convergence and Global Polarization." *Journal of Conflict Resolution* 41 (2): 203–26.

Banaji, M. R., and A. G. Greenwald. 2013. *Blindspot: Hidden Biases of Good People.* New York: Delacorte Press.

Bell, A. V., P. J. Richerson, and R. McElreath. 2009. "Culture Rather than Genes Provides Greater Scope for the Evolution of Large-Scale Human Prosociality." *Proceedings of the National Academy of Sciences* 106 (42): 17671–74.

Benedict, R. (1934) 2005. *Patterns of Culture.* Boston: Mariner.

Bourdieu, P. 1966. "The Sentiment of Honour in Kabyle Society." In *Honour and Shame,* edited by J. G. Peristiany and translated by Philip Sherrard, 192–241. Chicago: University of Chicago Press.

Boyd, R., and P. J. Richerson. 1985. *Culture and the Evolutionary Process.* Chicago: University of Chicago Press.

Boyd, R., and P. J. Richerson. 1987. "The Evolution of Ethnic Markers." *Cultural Anthropology* 2 (1): 65–79. Reprinted in Boyd and Richerson 2005.

Boyd, R., and P. J. Richerson. 2005. *The Origin and Evolution of Cultures.* Oxford: Oxford University Press.

Boyer, P. 2001. *Religion Explained.* New York: Basic Books.

Brandon, R. N. 1990. *Adaptation and Environment.* Princeton, N.J.: Princeton University Press.

Bransford, J., and National Research Council. 2000. *How People Learn: Brain, Mind, Experience, and School.* Washington, D.C.: National Academy Press.

Brown, M. J. 2008. "Introduction: Developing a Scientific Paradigm for Understanding Culture." In *Explaining Culture Scientifically,* 3–16. Seattle: University of Washington Press.

Brown, M. J., and M. W. Feldman. 2009. "Sociocultural Epistasis and Cultural Exaptation in Footbinding, Marriage Form, and Religious Practices in Early 20th-Century Taiwan." *Proceedings of the National Academy of Sciences* 106 (52): 22139–44.

Caporael, L. R. 2014. "Evolution, Groups, and Scaffolded Minds." In Caporael, Griesemer, and Wimsatt 2014, chapter 2.

Caporael, L. R., J. R. Griesemer, and W. C. Wimsatt, eds. 2014. *Developing Scaffolds in Evolution, Culture, and Cognition.* Cambridge, Mass.: MIT Press.

Castro, L., and M. A. Toro. 2014. "Cumulative Cultural Evolution: The Role of Teaching." *Journal of Theoretical Biology* 347:74–83.

Cavalli-Sforza, L. L., and M. W. Feldman. 1981. *Cultural Transmission and Evolution: A Quantitative Approach.* Princeton, N.J.: Princeton University Press.

Claidière, N., T. C. Scott-Phillips, and D. Sperber. 2014. "How Darwinian Is Cultural Evolution?" *Philosophical Transactions of the Royal Society of London B: Biological Sciences* 369 (1642): 1–8.

Claidière, N., and D. Sperber. 2007. "The Role of Attraction in Cultural Evolution." *Journal of Cognition and Culture* 7 (1): 89–111. Reply to Henrich, J., and

R. Boyd. "On Modeling Cognition and Culture." *Journal of Cognition and Culture* 2 (2): 2002.

Clark, S. 1999. *Thinking with Demons: The Idea of Witchcraft in Early Modern Europe.* Oxford: Oxford University Press.

Colby, B. N. 1991. "The Japanese Tea Ceremony: Coherence Theory and Metaphor in Social Adaptation." In *Beyond Metaphor: The Theory of Tropes in Anthropology,* edited by J. W. Fernandez, 244–60. Palo Alto, Calif.: Stanford University Press.

Dehghani, M., S. Sachdeva, H. Ekhtiari, D. Gentner, and K. Forbus. 2009. "The Role of Cultural Narratives in Moral Decision Making." In *Proceedings of the 31st Annual Conference of the Cognitive Science Society,* edited by N. Taatgen and H. van Rijn, 1912–17. Austin, Tex.: Cognitive Science Society.

Descola, P. 1986. *La nature domestique: Symbolisme et praxis dans l'écologie des Achuar.* Paris: Editions de la Maison des Sciences de L'Homme. English translation in Descola 1994.

Descola, P. 1994. *In the Society of Nature: A Native Ecology in Amazonia.* Translated by Nora Scott. Paris: Editions de la Maison des Sciences de L'Homme; Cambridge: Cambridge University Press.

Descola, P. 2005. *Par-delà nature et culture.* Translated by Janet Lloyd. Paris: Editions Gallimard. English translation in Descola (2005) 2013.

Descola, P. (2005) 2013. *Beyond Nature and Culture.* Translated by Janet Lloyd. Chicago: University of Chicago Press.

Dressler, W. W., M. C. Balieiro, and J. E. dos Santos. 2017. "Cultural Consonance in Life Goals and Depressive Symptoms in Urban Brazil." *Journal of Anthropological Research* 73 (1): 43–65.

Dressler, W. W., C. D. Borges, M. C. Balieiro, and J. E. dos Santos. 2005. "Measuring Cultural Consonance: Examples with Special Reference to Measurement Theory in Anthropology." *Field Methods* 17 (4): 331–55.

Durham, W. H. 1991. *Coevolution: Genes, Culture, and Human Diversity.* Palo Alto, Calif.: Stanford University Press.

Eiseman Jr., F. B. 1989. *Bali: Sekala and Niskala: Essays on Religion, Ritual, and Art.* Vol. 1. North Clarendon, Vt.: Periplus Editions.

Epstein, J. M. 2006. *Generative Social Science: Studies in Agent-Based Computational Modeling.* Princeton, N.J.: Princeton University Press.

Fierens, P. I., L. C. Rêgo, and T. L. Fine. 2009. "A Frequentist Understanding of Sets of Measures." *Journal of Statistical Planning and Inference* 139:1879–92.

Fogarty, L., P. Strimling, and K. N. Laland. 2011. "The Evolution of Teaching." *Evolution* 65 (10): 2760–70.

Geertz, C. 1973a. "Deep Play: Notes on the Balinese Cockfight." In chapter 15 of *The Interpretation of Cultures: Selected Essays*. Oxford: Oxford University Press.

Geertz, C. 1973b. *The Interpretation of Cultures: Selected Essays*. Oxford: Oxford University Press.

Geertz, C. 1981. *Negara: The Theatre State in 19th Century Bali*. Princeton, N.J.: Princeton University Press.

Geertz, H. 2004. *The Life of a Balinese Temple*. Honolulu: University of Hawai'i Press.

Gentner, D., K. J. Holyoak, and B. N. Kokinov, eds. 2001. *The Analogical Mind: Perspectives from Cognitive Science*. Cambridge, Mass.: MIT Press.

Gervais, W. M., A. K. Willard, A. Norenzayan, and J. Henrich. 2011. "The Cultural Transmission of Faith: Why Innate Intuitions Are Necessary, but Insufficient, to Explain Religious Belief." *Religion* 41 (3): 389–410.

Gillespie, J. H. 1998. *Population Genetics: A Concise Guide*. Baltimore: Johns Hopkins University Press.

Godfrey-Smith, P. 2009. *Darwinian Populations and Natural Selection*. Oxford: Oxford University Press.

González, R. J. 2001. *Zapotec Science*. Austin: University of Texas Press.

Gould, S. J., and R. C. Lewontin. 1979. "The Spandrels of San Marco and the Panglossian Paradigm: A Critique of the Adaptationist Programme." *Proceedings of the Royal Society of London Series B: Biological Sciences* 205 (1161): 581–98.

Grim, P., D. J. Singer, C. Reade, and S. Fisher. 2015. "Germs, Genes, and Memes: Function and Fitness Dynamics on Information Networks." *Philosophy of Science* 82 (2): 219–43.

Hauser-Schäublin, B. 2003. "The Precolonial Balinese State Reconsidered: A Critical Evaluation of Theory Construction on the Relationship between Irrigation, the State, and Ritual." *Current Anthropology* 44 (2):153–81.

Hegselmann, R., and U. Krause. 2002. "Opinion Dynamics and Bounded Confidence: Models, Analysis, and Simulation." *Journal of Artificial Societies and Social Simulation* 5 (3): 1–33. http://jasss.soc.surrey.ac.uk/5/3/2.html.

Henrich, J., and J. Broesch. 2011. "On the Nature of Cultural Transmission Networks: Evidence from Fijian Villages for Adaptive Learning Biases." *Philosophical Transactions of the Royal Society B* 336:1139–48.

Henrich, J., and R. McElreath. 2003. "The Evolution of Cultural Evolution." *Evolutionary Anthropology: Issues, News, and Reviews* 12 (3): 123–35.

Hofstadter, D. R., and E. Sander. 2013. *Surfaces and Essences: Analogy as the Fuel and Fire of Thinking*. New York: Basic Books.

Holyoak, K. J., and P. Thagard. 1995. *Mental Leaps: Analogy in Creative Thought*. Cambridge, Mass.: MIT Press.

Howe, L. 2001. *Hinduism and Hierarchy in Bali*. Oxford: James Currey/School of American Research Press.

Huneman, P. 2014. "Mapping an Expanding Territory: Computer Simulations in Evolutionary Biology." *History and Philosophy of the Life Sciences* 36 (1): 60–81.

Izuma, K., M. Matsumoto, K. Murayama, K. Samejima, N. Sadato, and K. Matsumoto. 2010. "Neural Correlates of Cognitive Dissonance and Choice-Induced Preference Change." *Proceedings of the National Academy of Sciences* 107 (51): 22014–19.

Janssen, M. A. 2007. "Coordination in Irrigation Systems: An Analysis of the Lansing Kremer Model of Bali." *Agricultural Systems* 93:170–90.

Janssen, M. A. 2012. *Lansing-Kremer Model of the Balinese Irrigation System (version 2)*, October 18. CoMSES Computational Model Library.

Kashima, Y. 2000. "Maintaining Cultural Stereotypes in the Serial Reproduction of Narratives." *Personality and Social Psychology Bulletin* 26 (5): 594–604.

Kincaid, H. 2007. "Functional Explanation and Evolutionary Social Science." In Turner and Risjord 2007, 213–49.

Kline, M. A. 2015. "How to Learn about Teaching: An Evolutionary Framework for the Study of Teaching Behavior in Humans and Other Animals." *Behavioral and Brain Sciences* 38:e31.

Kuhn, T. S. 1996. *The Structure of Scientific Revolutions*. 3rd ed. Chicago: University of Chicago Press.

Lansing, J. S. (1991) 2007. *Priests and Programmers: Technologies of Power in the Engineered Landscape of Bali*. Princeton, N.J.: Princeton University Press. Reprinted with a new foreword and preface in 2007.

Lansing, J. S. 2006. *Perfect Order: Recognizing Complexity in Bali*. Princeton, N.J.: Princeton University Press.

Lansing, J. S., S. A. Cheong, L. Y. Chew, M. P. Cox, M.-H. R. Ho, and W. A. Arthawiguna. 2014. "Regime Shifts in Balinese Subaks." *Current Anthropology* 55, no. 2 (April): 232–39.

Lansing, J. S., M. P. Cox, S. S. Downey, M. A. Jannsen, and J. W. Schoenfelder. 2009. "A Robust Budding Model of Balinese Water Temple Networks." *World Archaeology* 41 (1): 110–31.

Lansing, J. S., and T. A. de Vet. 2012. "The Functional Role of Balinese Water Temples: A Response to Critics." *Human Ecology* 40 (3): 453–78.

Lansing, J. S., and K. M. Fox. 2011. "Niche Construction on Bali: The Gods of the Countryside." *Philosophical Transactions of the Royal Society B: Biological Sciences* 366 (1566): 927–34.

Lansing, J. S., and J. N. Kremer. 1993. "Emergent Properties of Balinese Water Temple Networks: Coadaptation on a Rugged Fitness Landscape." *American Anthropologist* 95 (1): 97–114.

Lansing, J. S., J. N. Kremer, and B. B. Smuts. 1998. "System-Dependent Selection, Ecological Feedback, and the Emergence of Functional Structure in Ecosystems." *Journal of Theoretical Biology* 192 (3): 377–91.

Lansing, J. S., and J. H. Miller. 2005. "Cooperation, Games, and Ecological Feedback: Some Insights from Bali." *Current Anthropology* 46 (2): 328–34.

Lansing, J. S., A. Redd, T. M. Karafet, J. Watkins, I. Ardika, S. Surata, and M. Hammer et al. 2004. "An Indian Trader in Ancient Bali?" *Antiquity* 78, no. 300 (June): 287–93.

Lansing, J. S., S. Thurner, N. N. Chung, A. Coudurier-Curveur, C. Karakas, K. A. Fesenmyer, and L. Y. Chew. 2017. "Adaptive Self-Organization of Bali's Ancient Rice Terraces." *Proceedings of the National Academy of Sciences* 114 (25): 6504–9.

Lévi-Strauss, C. 1966. *The Savage Mind*. Chicago: University of Chicago Press.

Lévi-Strauss, C. [1962] 1990. *La pensée sauvage*. New York: Pocket. English translation in Lévi-Strauss 1966.

Lienhardt, G. 1961. *Divinity and Experience: The Religion of the Dinka*. Oxford: Oxford University Press.

Mesoudi, A., and A. Whiten. 2004. "The Hierarchical Transformation of Event Knowledge in Human Cultural Transmission." *Journal of Cognition and Culture* 4 (1): 1–24.

Morris, S. 2000. "Contagion." *Review of Economic Studies* 67:57–78.

Mueller, S. T., B. Simpkins, and L. Rasmussen. 2010. "Incorporating Representation When Modeling Cultural Dynamics: Analysis of the Bounded Influence Conjecture." In *Proceedings of the Workshop on Cognitive Social Sciences: Grounding the Social Sciences in the Cognitive Sciences?*, edited by R. Sun, 29–34. Technical Report 2010-RS-0001. Troy, N.Y.: Rensselaer Polytechnic Institute.

Norenzayan, A. 2013. *Big Gods: How Religion Transformed Cooperation and Conflict*. Princeton, N.J.: Princeton University Press.

Norenzayan, A., A. F. Shariff, A. K. Willard, E. Slingerland, W. M. Gervais, R. A. McNamara, and J. Henrich. 2016. "The Cultural Evolution of Prosocial Religions." *Behavioral and Brain Sciences* 39, e1. doi: 10.1017/ S0140525X14001356.

Odling-Smee, F. J., K. N. Laland, and M. W. Feldman. 2003. *Niche Construction: The Neglected Process in Evolution.* Princeton, N.J.: Princeton University Press.

Page, S. E. 2007. *The Difference: How the Power of Diversity Creates Better Groups, Firms, Schools, and Societies.* Princeton, N.J.: Princeton University Press.

Richerson, P. J., and R. Boyd. 2005. *Not by Genes Alone.* Oxford: Oxford University Press.

Richerson, P. J., R. Boyd, M. Borgerhoff-Mulder, and W. H. Durham. 1997. "Are Cultural Phylogenies Possible?" In *Human Nature, between Biology and the Social Sciences,* edited by P. Weingart, P. J. Richerson, S. D. Mitchell, and S. Maasen, 355–86. Mahwah, N.J.: Lawrence Erlbaum. Reprinted in Boyd and Richerson 2005, 310–36.

Risjord, M. 2007. "Ethnography and Culture." In Turner and Risjord 2007, 99–128.

Risjord, M. 2014. *Philosophy of Social Science: A Contemporary Introduction.* Abingdon, UK: Routledge.

Romney, A. K., S. C. Weller, and W. H. Batchelder. 1986. "Culture as Consensus: A Theory of Culture and Informant Accuracy." *American Anthropologist,* n.s., 88, no. 2 (June): 313–38.

Sanday, P. R. 1981. *Female Power and Male Dominance.* Cambridge: Cambridge University Press.

Schulte Nordholt, H. (1996) 2010. *The Spell of Power: A History of Balinese Politics, 1650–1940.* Leiden, The Netherlands: KITLV Press; reprint, Leiden: Brill Academic.

Schulte Nordholt, H. 2011. "Dams and Dynasty, and the Colonial Transformation of Balinese Irrigation Management." *Human Ecology* 39 (1): 21–27.

Smaldino, P. 2014. "The Cultural Evolution of Emergent Group-Level Traits." *Behavioral and Brain Sciences* 37:243–54.

Smelser, N. J. 1993. "Culture: Coherent or Incoherent." In *Theory of Culture,* edited by R. Munch and N. J. Smelser, 3–28. Berkeley: University of California Press. Accessed April 14, 2014.

Soltis, J., P. J. Richerson, and R. Boyd. 1995. "Can Group-Functional Behaviors Evolve by Cultural Group Selection?" *Current Anthropology* 36, no. 3 (June): 473–94.

Sperber, D. 1996. *Explaining Culture: A Naturalistic Approach.* Oxford: Blackwell.

Thagard, P. 2000. *Coherence in Thought and Action.* Cambridge, Mass.: MIT Press.

Thagard, P. 2012. "Mapping Minds across Cultures." In *Grounding Social Sciences in Cognitive Sciences,* edited by R. Sun, 35–62. Cambridge, Mass.: MIT Press.

Thibodeau, P. H., and L. Boroditsky. 2011. "Metaphors We Think With: The Role of Metaphor in Reasoning." *PLoS ONE* 6 (2): e16782. http://dx.doi.org /10.1371%2Fjournal.pone.0016782. doi:10.1371/journal.pone.0016782.

Thibodeau, P. H., and L. Boroditsky. 2013. "Natural Language Metaphors Covertly Influence Reasoning." *PLoS ONE* 8 (1): e52961. http://dx.doi.org/10.13 71%2Fjournal.pone.0052961. doi:10.1371/journal.pone.0052961.

Tilley, C. 2000. *Metaphor and Material Culture.* Oxford: Blackwell.

Tsing, A. L. 2005. *Friction: An Ethnography of Global Connection.* Princeton, N.J.: Princeton University Press.

Turner, S. P., and M. W. Risjord, eds. 2007. *Philosophy of Anthropology and Sociology.* New York: Elsevier.

Upal, M. A. 2011. "Memory, Mystery, and Coherence: Does the Presence of 2–3 Counterintuitive Concepts Predict Cultural Success of a Narrative?" *Journal of Cognition and Culture* 11 (1/2): 23–48.

Wilensky, U. 1999. *NetLogo.* Center for Connected Learning and Computer-Based Modeling, Northwestern University, Evanston, Illinois. http://ccl.northwest ern.edu/netlogo/.

Wilson, D. S. 2002. *Darwin's Cathedral: Evolution, Religion, and the Nature of Society.* Chicago: University of Chicago Press.

Wimsatt, B. H. 2014. "Footholds and Handholds: Scaffolding Cognition and Career." In Caporael, Griesemer, and Wimsatt 2014, chapter 16.

Wimsatt, W. C. 2002. "Using False Models to Elaborate Constraints on Processes: Blending Inheritance in Organic and Cultural Evolution." *Philosophy of Science* 69 (S3): S12–24.

Wimsatt, W. C. 2014. "Entrenchment and Scaffolding: An Architecture for a Theory of Cultural Change." In Caporael, Griesemer, and Wimsatt 2007, 77–105.

Wimsatt, W. C., and J. R. Griesemer. 2007. "Reproducing Entrenchments to Scaffold Culture: The Central Role of Development in Cultural Evolution." In *Integrating Evolution and Development,* edited by R. Sansom and R. N. Brandon, 227–323. Cambridge, Mass.: MIT Press.

Young, H. P. 1998. *Individual Strategy and Social Structure: An Evolutionary Theory of Institutions.* Princeton, N.J.: Princeton University Press.

Zollman, K. J. 2013. "Network Epistemology: Communication in Epistemic Communities." *Philosophy Compass* 8 (1): 15–27.

8

CONTENT MATTERS
The Materiality of Cultural Transmission and
the Intersection of Paleolithic Archaeology with
Cultural Evolutionary Theory

GILBERT B. TOSTEVIN

THE PROBLEM

With respect to the cultural behavior of other primates, the derived capacity for the complex dynamics of cumulative culture has evolved in our lineage over the last three million years. As Robert Boyd aptly states, this derived capacity "is an essential part of the human adaptation, and as much a part of human biology as bipedal locomotion or thick enamel on our molars."[1] Yet despite the existence of numerous case studies from the Pleistocene *fossil* record on the gradual evolution of bipedality and enamel thickness, the *archaeological* record of the Pleistocene has not provided complementary case studies regarding how this derived cultural capacity *itself* evolved through time. While archaeology has been able to point to changes throughout the Pliocene and Pleistocene in artifact morphologies and the technical complexity of the methods by which those morphologies were achieved (Perreault et al. 2013), it has not provided particularly useful behavioral case studies with quantitative support of the gradual evolution of specific cultural transmission (CT) processes, structures, and scaffolds (sensu Wimsatt and Griesemer 2007).

Paleolithic archaeology should—but is not currently able to—provide data that clarify and characterize early hominin CT processes at different points in time in the past in order to compare and contrast them with the complexity of institutionally diversified cultures found in the present. We should be striving to contribute data to the resolution of questions such as: At different points in human evolution, did the material culture require stimulus enhancement (Charman and Huang 2002; Franz and Matthews 2010; Matthews, Paukner, and Suomi 2010), emulation learning (Tomasello 1996), or imitation learning via triadic attention (Whiten et al. 2009; Tomasello et al. 2005)? When did experienced performers of artifactual skill begin to actively correct mistakes by more novice individuals? Under what adaptive contexts did gestural instruction become significant? In what contexts and when did linguistically assisted instruction play a more important role than observation? When did skill levels in artifact production become diversified enough that particular individuals assumed achieved status as institutionalized role models because of their skill rather than based on other aspects of age, kin selection, or social ranking? What were the structural ramifications of specific behavioral innovations becoming exapted scaffolds for CT, such as the use of fire for storytelling (Wiessner 2014)? What was the population size of the group that could sustain a given level of technological innovation in a specific artifactual medium in a given environment? What can we learn from comparing trends in hominin encephalization with an archaeologically measurable ratchet effect on cumulative culture during human evolution (Donald 1998; Tennie, Call, and Tomasello 2009)?

While we may never be able to answer these questions completely or to our satisfaction, cultural evolutionary theory can only advance if we struggle to engage these central questions, all of which reside at the intersection of many fields, including but not limited to primatology, cognitive science, developmental biology, and population genetics. Yet archaeology is the one field that has access to the physical results of the intergenerational loop between CT and cultural replication that is material culture. And material culture is implicated, if not central, to all of these questions. Archaeologists have engaged with these questions (Pigeot 1990; Karlin et al. 1993; Ploux and Karlin 1993; Grimm 2000; Wynn 2002; Roche 2005; Shipton 2010; Kuhn 2012; Schillinger, Mesoudi, and Lycett 2014; Hiscock 2014), and many have offered carefully argued answers. However, due to the historical rather than quantitative nature of the data traditionally produced in archaeology and the difficulty of connecting our data with bodies of theory from different disciplines (Garofoli and Haidle 2014), it is still possible for two archaeologists to start from basic principles and end up concluding opposed answers. One salient example is the diametrically opposed interpretations of the minimal pedagogical requirements for the most studied artifact in the Paleolithic record, the Acheulean handaxe. Some archaeologists conclude that simple rules of production, acquired without abundant instruction, can produce the variability seen among Acheulean handaxes (e.g., McPherron 2000; Davidson 2010), while others conclude that complex forms of instruction and apprenticeship are necessary for their production (e.g., Wynn 2002; Shipton 2010; Hiscock 2014). Even further outliers, such as Corbey et al. (2016), argue that these artifact forms are as genetically controlled as birds' nests. Without a quantitative and anthropologically sound body of archaeological theory to disprove some of these diametrically opposed hypotheses, there is not much scientific progress to be had within Paleolithic archaeology around the subject of artifactual learning (see additional commentary within Tennie et al. [2017]). We need to develop an archaeology that explicitly addresses the evolution of learning processes, an *archaeology of pedagogy*, to borrow Tehrani and Riede's (2008) term.

The failure to use the quantitative strength of the archaeological record to contribute to answering these questions is a missed opportunity. Paleolithic CT processes were likely to have been simple systems, and studying simple systems in detail can provide an enormously improved understanding of how such processes work in more complex contexts. Studies of Darwin's finches on Daphne Major in the Galápagos Islands demonstrate how the examination of a simple context through time can reveal the workings of a complex process such as natural selection (Grant and Grant 2011, 2014). Evolutionary biologists are in a better position to understand how natural selection works in more complex contexts because of these studies. Paleolithic archaeology should be serving the same role for the development of a comprehensive approach to cultural evolution; nowhere else but in the Pleistocene archaeological record will we find data pertaining to a simple CT context close to the evolutionary appearance of the cultural capacity itself. Studying modern human foragers (Hewlett et al. 2011; Hewlett 2013) and living primates (Whiten, Schick, and Toth 2009; Tomasello et al. 2012) is extremely useful but also limits us to reasoning by analogy and restricts our understanding of how culture *actually* evolved since our last common ancestor with the genus *Pan*.

This chapter describes two obstacles that have caused this unfortunate state of affairs and outlines ongoing research that can move us forward toward solutions. One problem, which I do not count among the two obstacles, is the indirect nature of archaeology as a historical science. Archaeologists excavate data that is indirect when compared with the fossil record; behavior preserves even more ephemerally than bone. Unlike the Grants on Daphne Major, we cannot watch our subjects in real time, and it is debatable whether the significance of these studies would have been realized if the inferences were dependent on the fossil record of finches on the island. Neontology (sensu S. Gould 2002, 778) does have benefits over paleontology. Archaeologists can practice experimental archaeology and ethnoarchaeology, the study of how living humans' behavior forms the archaeological record (Yellen 1977; Binford 1978; R. Gould 1980). However, we are limited to studying only modern humans with their fully developed, institutionalized CT structures, such as fictive kinship (Read 2011), reciprocal altruism via exchange systems (Wiessner 1982, 2002), and scaffolding via storytelling (Wiessner 2014). The hominins responsible for the earliest transmission of material culture, the Oldowan (2.6–1.7 million years ago) or even the slightly older but newly discovered Lomekwian (3.3 million years ago; Harmand et al. 2015), likely did not have any of these CT scaffolds.

OBSTACLES TO A MORE MEANINGFUL CONTRIBUTION OF PALEOLITHIC ARCHAEOLOGY TO CULTURAL EVOLUTIONARY THEORY

Two obstacles are currently making it difficult to utilize the study of the Pleistocene behavioral record for the development of a robust cultural evolutionary theory. The first is the absence of a connection between the types of data produced by most lithic (i.e., stone tool) analysts in the Old World and the cultural learning sets that operate as units of change in CT theory. The second is the overly abstract, nonmaterial nature of how the transmission process is most frequently modeled by the CT community. As a consequence, the transmission process appears far less structured than ethnoarchaeologists and behavioral archaeologists know it to be. I will take each of these obstacles in turn and then explore possible means to overcome them.

Units of Analysis in Paleolithic Systematics

Paleolithic archaeologists tend to structure their data in ways that are inappropriate for studying CT transmission processes, not to mention the developmental complexities of cultural evolutionary theory. From the point of view of the stone tool record, which made up 98 percent of the archaeological record until a few thousand years ago, there are two dominant forms of stone tool data produced by most Paleolithic archaeologists. Within the history of the discipline, the older method focuses on the presence or absence of rarer artifacts and the variations in their morphology, which are interpreted as being highly functional or symbolic, such as large shaped cutting tools (e.g., the Acheulean bifacial handaxes), nodules of rock from which

sharp flakes were struck in specific sequences (e.g., cores of particular exploitation strategies such as the Levallois method), and small projectile points
(e.g., spearpoints and arrowheads). These "pretty" pieces constituted the (almost fetishistic) focus of research during the youth of Paleolithic archaeology, despite their actual rarity in the record (Monnier 2006).

Ironically, it is these types of artifacts that current lithic analysts most
interested in advancing CT research in archaeology have concentrated on
over the last twenty years. If one examines the CT archaeology programs begun by Bettinger, Boyd, and Richerson (1996), Bettinger and Eerkens (1999),
and summarized nicely in Eerkens and Lipo (2007) and Lycett (2015), it is
the study of the variation in these rare shaped objects that has been used to
argue for different modes of CT being active at given times and places in the
record. Similarly, the phylogenetic and cladistic approaches espoused by
O'Brien, Darwent, and Lyman (2001), O'Brien and Lyman (2003), Lycett and
von Cramon-Taubadel (2008), Buchanan and Collard (2008), Lycett (2010),
Riede (2011), and others have focused exclusively on morphological analysis
of such rare shaped objects, rather than utilizing the entirety of stone artifact assemblages to discover the physical evidence of the cultural learning
sets that should be the units of analysis in CT research. While I cannot over-
state the enormous advances made to date by CT archaeologists through
their introduction of new quantitative methods and new theoretical perspectives, their approach is still handicapped by their contentment to studying
only the "finished" pretty pieces, the cultural phenotype represented by the
final shape of these objects of long use-life. Having drawn their method and
theory from paleontology, they seem content to treat the variation in that
raw morphology as unproblematic reflections of cultural inheritance, opening them to substantial critique by archaeologists specialized in studying artifactual manufacturing techniques (Bamforth and Finlay 2008). In contrast,
I would argue, a cultural genotype exists in the physical behaviors observed
and internalized by learners during CT. As these learned behaviors were
later physically reenacted in the creation of new objects and so preserved in
the resulting manufacturing debris, it is the more ubiquitous manufacturing debris that we should target as better proxies for what the observers
learned.

The second dominant form of stone tool data created by Paleolithic archaeologists dates to the 1960s instead of the 1860s and utilizes the entirety
of a collection of artifacts, including manufacturing debris, from one geological layer of an archaeological site. Each collection is studied to reconstruct

the flintknapping process (i.e., how the stone tools were made) during the period captured within that geological stratum. These reconstructions typically take the form of an assemblage-wide operational sequence, which is the sequence of steps used to reduce raw nodules of stone into cores from which usable flakes with suitable cutting edges were removed and then reshaped for use. Whether produced via the Continental European approach of the *chaîne opératoire* school (for a useful review of this approach, see Soressi and Geneste [2011]) or the Anglo-American approach of core reduction sequence analysis (Shott 2003), these operational sequences have the potential to more closely approximate the units of cultural learning in CT theory. This is because they include the artisan's choices, which must be made at specific points within the sequence of steps in the production of the assemblage of tools (Riede 2006). Unfortunately, however, even if Paleolithic archaeologists use quantitative and transparent methods for constructing these sequences, which is not always the case (see Bar-Yosef and Van Peer 2009), most tend to assign the detailed sequence from a given assemblage into one of several immutable, essentialized "types" of reduction methods. Alternatively, they invent a new label to add to the long list of existing categorical entities, variously called reduction methods, industrial types, technocomplexes, and technological types (inter alia). It is these categorical entities that are then used as units of analysis for positing a historical narrative of which cultural entities existed, what behaviors they pursued, and where and when they were found in the Pleistocene.

This approach to Paleolithic research continues to obscure the exact behavioral variation we should be studying (Monnier and Missal 2014). The epistemological problem of incomparability between "technological types," much like the proverbial comparison of apples and oranges, eliminates the power of a CT approach when applied to such data (Tostevin 2009, 2011b). John J. Shea (2014) has recently emphasized this same point in his critique of *named archaeological stone tool industries,* or NASTIES, as being obstacles to studying behavioral evolution in this period. I have argued at length that evaluating hypotheses of CT between populations in time and space requires the deconstruction of these generalized categorical types through the recognition within individual artifact assemblages of behavioral units that can be evaluated as potential instances of learning between entities (Tostevin 2007, 2009, 2011b). *If Paleolithic archaeologists are to study culture change in an evolutionarily informed, nonessentialist paradigm,* as is required for studying evolutionary processes and CT (Tschauner 1994), *archaeologists need to study this change through time* within *the abstractions we call tech-nocomplexes rather than* between *these typologically defined categories* (sensu Adams and Adams 1991; see also Read 2007).

Yet, even as new categorical types are added to the recognized list of NASTIES, older labels are rarely eliminated from the literature, as Shea (2014) points out. Monnier (2006) has shown how archaeologists' existing views of Paleolithic cultural evolution through time have been influenced more by the inherited history of their research traditions than by newly excavated data (a situation that Wimsatt would recognize as entrenchment; sensu Wimsatt and Griesemer 2007). Indeed, Shea once overheard a senior Paleolithic archaeologist complain (Shea, personal communication, 2014), "We are all prisoners of de Mortillet," referring to Gabriel de Mortillet (1821–1898), the archaeologist who published the first widely used classification of the Paleolithic in 1869. If this applied to the study of the fossil record, current paleontologists would be constrained to use the same immutable units of analysis as those of Georges Cuvier (1769–1832), the founder of comparative anatomy. Instead of being able to utilize the pattern of Retzius lines in the microstructure of dental enamel to understand different developmental growth rates between taxa (Smith et al. 2007), modern paleontologists would be constrained to discussing taxa only in terms of their pointy versus flat canines.

Materiality and Structure in Current CT Literature

The second obstacle confronted by those trying to unite Paleolithic archaeology with cultural evolutionary theory is that current CT models tend to ignore the materiality of the process, such that many Paleolithic archaeologists find the models unsuitable to the material culture they study. Specifically, archaeologists who study artifactual manufacturing sequences, particularly behavioral archaeologists who specialize "in the concrete interactions that take place in the activities constituting the life histories of artifacts and people" (Schiffer and Skibo 1997, 28), have long recognized that to learn how to make an item of material culture is to learn two different and highly structured bodies of knowledge: (1) knowing what you *should* do in the conceptual sense, the *connaissance* of the behavioral gesture in the parlance of the French *chaîne opératoire* school (Pelegrin 1990); and (2) knowing *how* to do it as a bodily action, through the development of the patterned neural connections that enable the correct choice of bodily gesture to be enacted in the correct way—that is, the savoir faire. This is a specific type of developmental structure in the CT process that is lacking in the current literature. Thus, it is not a general lack of attention to how structured content or structured populations affect the results of CT that makes current research less attractive to archaeologists. In fact, compared to the origins of CT research (Cavalli-Sforza and Feldman 1981; Boyd and Richerson 1985), recent studies have contributed significantly to exposing how structure plays out during CT. For instance, CT studies have recently incorporated structural elements such as changes in skillfulness through time (Andersson 2013; Andersson, Törnberg, and Törnberg. 2014; Andersson and Read 2016), the effect of prerequisites within sequentially structured knowledge (Mesoudi and O'Brien 2008; Madsen and Lipo 2015), the costs of acquisition of new knowledge (Mesoudi 2011), and the ramifications on cultural variability resulting from how transmission occurs on a spatial scale (Premo and Kuhn 2010; Perreault and Brantingham 2010; Premo and Scholnick 2011; Premo 2012b; Premo 2015; Premo and Tostevin 2016). Instead, what is lacking in the current approach to CT research is a focus on how the differences in learning these two bodies of knowledge would make the structure of the CT process itself dependent upon the physical realities of each material culture medium. In other words, an archaeologically applicable CT approach needs to model how the results of the transmission can be altered by differences in the material requirements of learning one content versus another—that is, learning an idea versus learning the bodily performance involved in the manufacturing techniques for a specific artifact. This is where closer collaboration with archaeologists can help.

An illustration will help clarify this issue. Boyd and Richerson (2000) artfully point out how the inherent variability in CT units is one of several factors that make cultural evolutionary processes so distinct from biological evolutionary processes:

> Unlike genes, ideas usually are not passed intact from one person to another. Information in one person's brain generates a behavior, and then someone else tries to infer the information required to do the same thing. Breakdowns in the accurate transmission of ideas can occur because differences in the genes, culture or personal background of two individuals can cause one person to make a wrong assumption about what motivated the other's behavior. (54)

Boyd and Richerson's article pictorially captured this variability in the CT unit in a sketch by Dušan Petričić and serves as an excellent critique of meme theory (Dawkins 1976; Blackmore 2000), which posits that the unit of CT is

Figure 8.1. My update of Boyd and Richerson's (2000, Figure 1) classic portrayal of the CT process: "IDEAS often mutate as they pass from one person to another." Inspired by Dušan Petričić's original sketch for *Scientific American* of a letter A morphing as it is handed off between three individuals, I have turned the letter into a word to indicate more clearly how the unit of transmission changes as the context of its material expression changes. Here the word *At* represents the preposition, indicating the location where an individual wants to be found for future correspondence. Thus, moving from left to right, an individual hands a large handwritten word, *At* (as one would sign a personal letter for hand delivery: Miss Jane Marple, At the Vicerage, St. Mary Mead), to another figure, who then hands a Courier-font *At* (as on a typewriter-addressed envelope) to the next figure, who passes on an @ sign (as part of an address for computer-based email, jmarple @AgathaChristie.com) to another figure with a Facebook icon, an indication of a one-to-many communication, where one individual can post multimedia to many people via smartphone and social media apps. Illustration by G. B. Tostevin.

in fact gene-like. I have redrawn their concept in an updated form in the present Figure 8.1. The transmitted unit is reshaped, both cognitively and behaviorally, by the process through which it is learned. While this makes the necessary point against any straightforward view of memes, it does not go far enough. How physical requirements for the ideas being transmitted influence the possible variation in how far the ideas can morph between role model and learner is still relegated to the background. In the metaphor of Figure 8.1, CT research needs to start to explore how the shape and slipperiness of the pliable letter (the content) changes how each individual in the process needs to grip, squeeze, and manhandle it between hand offs, with subtle changes in how the letter is distorted in each case. Some letter shapes are easier to hand off without distorting their lines; others require a harder grip that more substantially changes the shape.

Consider the behavioral choices within the operational sequence for how to make a stone tool. These learned choices are not simply susceptible to conceptual misunderstanding in the mind of the learner, akin to simple "cognitive mutation." As hinted at above, these choices have to be learned at *two* levels: the connaissance of the behavioral gesture and the savoir faire to successfully execute the gesture. The savoir faire of flintknapping is extremely specific. It requires the control of thousands of timed muscular contractions to deliver a successful blow of the stone hammer to strike a flake off a core. The motion of the arm delivering the blow occurs in less than a second and can rarely be altered after it has begun. Once the hammer stone touches the core (at a rate of approximately 2.4 meters per second), the rate of fracture propagation separates the flake from the core at a speed of 630–1100 meters per second, depending on the hardness of the stone (Cotterell and Kamminga 1987, 680). In neither the delivery of the blow nor the physics of its result is there time for a knapper to think about the delivery or the consequences of the action. Depending on the physical requirements of learning both the connaissance and savoir faire of each unit of transmission (i.e., a combination of the appropriate choice and appropriate enactment of the choice), there could be *more or less* fidelity in transmission between what is demonstrated and what is learned. Tostevin (2012, chapter 4) provides a conceptual model for how the variables known to control the flake-by-flake knapping process can be altered (i.e., can experience cultural mutation) between the demonstrator and the learner in a simple observation of a flintknapping event. In addition to the variation caused by the two-part learning process, we also know that perception errors resulting from limits on human visual acuity relative to the size of the material being copied contribute to variation in transmission (Eerkens and Bettinger 2001; Eerkens and Lipo 2005; Kempe, Lycett, and Mesoudi 2012). Whether a technology is additive (as in adding clay to a pot during its production) or reductive (removing stone flakes from a core or wood from a carving) also affects the transmission process (Skibo and Feinman 1998; Schillinger, Mesoudi, and Lycett 2014). *The materiality of the content matters and fundamentally changes the process.* This requires us to pay attention to where and how fidelity variation is created in the learning of even the earliest action of material culture creation in the archaeological record, the striking off of one flake from a core.

Lithic technology is not the only material culture whose transmission structure is affected by the physicality of its content. Mark Bedau's (see chapter 6) analysis of inheritance and adaptive radiations in U.S. patents is a perfect and far more recent example of how the physicality of the transmission event changes the pattern of the cultural evolution of technology. The U.S. Patent and Trademark Office (USPTO) requires inventors to cite in their applications all applicable prior patents as the basis for its evaluation of the sufficient novelty in a given application to warrant approval by the USPTO. This physical requirement in the application process makes multiparental inheritance explicit and helps to define the shape of a new technology. In the context of designing an innovation with a mind to patenting it, innovators must not only contextualize what elements they are inheriting but also distinguish their creations from these antecedent patents more than they might otherwise have done. "An important scaffold for the evolution of technology is the inheritance (citation) network among inventions, and content flow in the network is strongly affected by the network's multiparental structure" (see chapter 6). This context contributes a structure of descent with modification to the process that is more explicit than in most processes of cultural transmission and makes patents the best example of a transmitted unit akin to memes yet demonstrated. Patents, and their design elements, are excellent examples of *transmissible elements* (see chapter 1), and this is because of the physicality of the application requirements in the approval process. Because of the more obvious transmissible element, Bedau has been able to demonstrate fascinating cultural evolutionary patterns, including pivotal "door-opening" innovations, within this data set. For Paleolithic archaeologists, to recognize similar cultural evolutionary patterns (or at least to construct data that articulates with cultural evolutionary questions), we need to recognize units of analysis that are equivalent to transmissible elements within the process of learning how to flintknap. This is where material culture began and where we must start if we are to understand what CT processes were utilized by the first hominin populations exploiting cumulative CT.

BRIDGING THE OBSTACLES TO A MORE MEANINGFUL CONTRIBUTION OF PALEOLITHIC ARCHAEOLOGY TO CULTURAL EVOLUTIONARY THEORY

The Need for Comparability within Paleolithic Data for Contributing to CT Theory

The first obstacle—the absence of a connection between the structure of Paleolithic data and the cultural learning sets needed in CT theory—is surmountable if Paleolithic archaeologists choose to analyze the record with more attention to how those analyses will be used for answering specific questions. Specifically, methodological approaches that do not produce an analytical structure that allows the evaluation of predictions from high-level theory should be rejected despite being sanctified by long historical use in the discipline (Tostevin 2011a). Here I am relying on a typology of archaeological method and theory as manifested at three levels of operation: low-level, middle-level, and high-level theory (Thomas 1998, 66–94). Low-level theories include observations obtained in archaeological fieldwork, such as the products of measurement techniques, inferences from the qualitative examination of artifacts, statistical representations of counts and attributes, and published artifact illustrations. Low-level theory is thus "data," the beginning of the archaeological method. Philosophers of science might call this the theory-conditioning of data (William Wimsatt, personal communication). Middle-level theories (or middle range; sensu Binford 1977) connect these observations of the archaeological record to patterns of human behavior. Connections are established through experimental archaeology, ethnoarchaeology, and other types of research designed to recognize causal relationships between the processes of human behavior and their resultant effect on the formation of the archaeological record. High-level theories provide the context for what archaeologists are interested in examining as a research target. They provide the intellectual goals related to asking certain questions of the archaeological record, usually from a specific orientation to explaining the past. Thomas's three-level distinction in method and theory forces us to consider each step of argumentation between the data and the research question. "The three-level distinction allows one to understand how low- and middle-level theories need to be shaped in a particular way in order to achieve the goals of high-level theory" (Tostevin 2011a, 294).

The present high-level theory goals of most Paleolithic archaeologists working on the reconstruction of operational sequences from lithic data are not *inappropriate* goals, but they tend not to produce low-level theory (data) commensurate with other desirable high-level theory goals. This is because most Paleolithic archaeologists strive for *more detail and more richness* in their reconstructions of operational sequences. This leads them to produce "data" that is so specific as not to be comparable in any fashion between contexts, such as different sites.

> In contrast to most other disciplines, archaeology does not aim to reduce a wealth of data to a few essentials. It does the reverse, putting flesh and clothing on "bare bones." Its logic is therefore very different from the logic of the natural sciences, but also from that of the social sciences. (Van der Leeuw 2004, 118)

Paleolithic archaeologists' logic is not unscientific, however, despite Van der Leeuw's observation, but rather aimed at maximizing what can be learned from each specific case of "putting flesh and clothing on the 'bare bones.'"

In particular circumstances, this approach can produce remarkable results. For instance, in Pigeot's (1987, 1990) reconstructions of the flintknapping that took place at Etiolles, a Magdalenian campsite in the Paris Basin, she was able to reconstruct where a master knapper sat demonstrating her/his knapping while surrounded by knappers early in their learning process. This is a rare but convincing argument for the presence of apprenticeship eighteen thousand years ago; an astounding demonstration of an actual CT scaffold in a Stone Age site. Yet it was not data analysis beyond the artifacts of this one site that allowed this result but the astonishing preservation of artifact contexts within the site, particularly the intra-site comparison of reduction sequences that showed execution errors with those that were flawless. In fact, while there is value in the astounding detail of the best reconstructions of operational sequences of lithic technology (Pigeot 1987; Cattin 2002; Bullinger, Leesch, and Plumettaz 2006), pyrotechnology (Plumettaz 2007), and organic technology (Knecht 1993) produced by my Paleolithic colleagues, these studies do not do enough to advance the collaborations that are needed to answer questions about the evolution of CT structures. Because such Pompeii-premise sites are so rare, we cannot move forward with an archaeology of pedagogy without comparable data beyond these well-fleshed-out snapshots.

Endeavoring to articulate low- and middle-level theory with my high-level theory goals of studying CT through Pleistocene archaeological data, I have developed an analytical method for replacing the categorical entities (technocomplexes, reduction methods, industrial types, and other NASTIES) in lithic research with quantitative, behavior-by-behavior reconstructions of assemblage-wide lithic operational sequences that allow comparisons of similarity and dissimilarity between assemblages (Tostevin 2000, 2003a, 2003b; Tostevin and Škrdla 2006). This approach goes a long way to solving both the deceptive emphasis on rare artifacts and the apples-versus-oranges problem of NASTIES in Paleolithic research.

The second obstacle—the need to recognize the creation of structure in the CT process as a result of the materiality of the unit being transmitted—requires an even more drastic reconfiguration of traditional Paleolithic analytical methods. In response to this need, I have proposed an ethnographic-based middle-range theory for predicting which behaviors within a lithic operational sequence are learnable in different contexts of contact between foragers of different social intimacy (Tostevin 2007, 2012). The strategy is to let the CT process itself determine the units of analysis. This is equivalent to taking an evolutionary developmental approach to CT archaeology. How does the observer learn the behavioral details of a lithic operational sequence by watching the performance of a knapper? The physicality of the observational context by which the connaissance is learned, as well as the subsequent repetitions/practice on the part of the learner by which the savoir faire is mastered, determine the structure of the CT process. It is this level of the materiality of the process, which is currently lacking in the literature within CT theory, that makes the unification of Paleolithic archaeology with CT research difficult.

Solving this second obstacle involves archaeologists examining the variables that we know control the shape of each flake as it is removed from the core, the same variables that the observer saw and learned through his own replication within the social intimacy of the group's enculturating environment. This approach goes a long way toward overcoming the obstacle of incorporating CT research into the archaeology of human evolution, regardless of whether it is focused on lithic technology or another material culture. With the help of John Shea's talent for acronyms, I have dubbed this the *behavioral approach to cultural transmission* (BACT).

The Behavioral Approach to Cultural Transmission

BACT considers two sets of questions as a means to structuring lithic analysis to articulate with cultural evolutionary theory. First, *How does dual inheritance occur on the landscape in foraging societies?* This question can be decomposed into more detailed questions: *Where and when are foragers enculturated? Where, how, and when do they witness technological performances that affect their adoption of technological choices, and how do their observations and the feedback they receive in training affect their own performances?* I have endeavored to answer these questions through the construction of a middle-range theory built on ethnographic data (Wiessner 1982, 1983, 1984; Lee and DeVore 1976; Kelly 1995) and anthropological theory (Carr 1995; Wobst 1977; Sackett 1990) directed at understanding how, where, and when individual foragers learn and transmit their cultural behavior (Tostevin 2007, 2009, 2012). Tostevin (2007) presents the kernel of the middle-range theory for predicting which aspects of a lithic operational sequence reflect behaviors that are learned and learnable only in contexts of social intimacy among foragers. Tostevin (2012) develops these ideas in greater detail within the context of an evolutionary approach to Pleistocene CT, building off of dual-inheritance modeling within CT theory (Boyd and Richerson 1985, 1987, 1996; Cavalli-Sforza and Feldman 1981; Richerson and Boyd 1978, 2002, 2005).

Taking a behavioral approach (Schiffer 1975, 1976, 1996) to flintknapping, an artifact assemblage is recognized as the central tendencies and dispersions in flake attributes reflecting specific decisions a knapper must make during the reduction of a core for flake blanks, which are subsequently made into tools to be used on the landscape. These decision nodes, which must be learned over the years of the enculturation of the individual as a skilled knapper, must be taken regardless of the option used at a given node in a given assemblage, making them consistently comparable units of analysis across space and time. Thus, the decision nodes can be treated as cultural instruction sets that would have been visible and thus learnable by foragers present at the different site localities being compared. The exposure of socially intimate individuals to flintknapping performances at base camps and raw-material procurement sites, where enculturation occurs, would have allowed these individuals to witness and learn the body techniques and behavioral details involved in flake production. The social intimacy between the observer and the performer would have afforded the observer the chance not only to learn the connaissance of the behavioral details but, given enough time, to develop the savoir faire of the body techniques. This exposure differs from that of socially distant individuals who would be exposed to the mobile tool kit only, the products of the end of the operational sequence. Because the artifacts of the mobile tool kit are carried onto the pathways of the landscape (Gamble 1999, 68–71), these tools become more visible to socially distant individuals but visible only from "bow-shot" range, the likely range for contact between strange foragers (Wiessner 1983). Given the equifinality in lithic reduction, exposure to mobile tool kits on pathways of the landscape or from discarded tools at retooling camps would not be sufficient for a stranger to produce the same debitage-wide central tendencies for all of the behaviors in the process, even if a few of the options were intuited from a curated tool. Independent innovation or convergence of behaviors within flake production, representing homoplasy, is thus always a possibility but not a high probability. This is the basis of the *taskscape visibility* concept, defined as the relationship between where, when, and with whom a cultural trait, such as a flintknapping behavior, is performed and the possible CT modes (sensu Boyd and Richerson 1985) available for promulgating the trait into the next generation.

Derived as it was from archaeological and ethnographic method and theory alone, Premo and Tostevin (2016) set out to evaluate the taskscape visibility concept using a formal, spatially explicit, agent-based model. Using an established model for the transmission of cultural traits among central-place foragers (Premo 2012a, 2012b), the simulation evaluated the equilibrium diversity of two selectively neutral traits that differed only in their taskscape visibility—that is, where they were learnable on the landscape. The simulation showed that the trait with the lower visibility, which was learnable only at residential base camps, had higher equilibrium diversity levels than the trait with the higher visibility, which was learnable at both base camps and logistical foray camps. Without the recognition of the role of taskscape visibility, which was the only difference between the traits, the difference in the observed equilibrium diversity levels of the two traits might have been incorrectly interpreted as resulting from qualitatively different forms of biased cultural transmission. These results suggest that the theoretical principles derived by archaeologists such as Sackett (1990), Carr (1995), and Wobst (1997) should be incorporated more closely into future CT research.

While the first set of questions addressed by BACT revolves around where interactions of different levels of social intimacy occur on the taskscape, the second set of questions focuses on the microscale, the observational learning of artisan choices: *Which emic choices of the artisan are visible as etic observations by the learner? Which observations of the learner also are etically observable by the archaeologist?* Anthropology's distinction between emic and etic perspectives may be one of the most important contributions to the development of cultural evolutionary theory. The distinction is most often associated with Marvin Harris (1976) and his cultural materialism agenda in cultural anthropology over the last quarter of the twentieth century. Harris, however, did not invent the terms but co-opted them from Kenneth Pike (1967). Pike coined the term *emic* to refer to the internal rules or logic of a behavior from the perspective of a member of the society that practices that behavior. *Etic,* on the other hand, refers to the external perspective of an anthropologist trying to understand a culture-specific behavior in light of participant observation, as well as comparison with other cultures. Pike constructed these terms from a similar distinction in linguistic anthropology: etic comes from *phonetic* (the possible sounds made by different parts of the human vocal anatomy across all humans) and emic from *phonemic* (the subset of etic sounds that a given culture recognizes as making a difference in meaning or semantics).

Recognizing the emic/etic distinction helps illustrate how the process of learning a physical skill such as flintknapping by visual and auditory observation can structure both the forms of the artifacts produced and the means by which archaeologists reconstruct the behaviors that were both learned and performed in a given society. From controlled experiments in fracture mechanics (Dibble and Pelcin 1995; Pelcin 1997, 1998; Dibble and Rezek 2009; Rezek et al. 2011; Lin et al. 2013), we know that the knapper needs to choose particular physical variables on a core to remove a flake with a set of physical properties.[2] To remove a flake, she needs to decide how much of the convexity on the face of the core she wants to remove for the flake to have the desired shape, such as being pointed or round, long or broad, and so on. She chooses these aspects of the dorsal surface of the core by identifying where on a platform opposite this convexity she will strike. She decides how far into the platform from the edge of the core's dorsal surface to strike (the platform thickness or depth) to determine where the fracture plane will intersect the core. She can choose to remove more volume with a deeper platform thickness or alter the exterior platform angle between the platform and the dorsal surface to achieve the same result, since multiple controlled experiments have shown that external platform angle and platform thickness together predict the mass of the removal. The dorsal convexity, on the other hand, contributes most significantly to giving specific shape to that mass. All of these "choices" can be made consciously before the delivery of the blow but are executed together with the split-second delivery of the strike, a movement that cannot be altered after the brain sends the message for the movement of the arm to begin. In a profound way, these "choices" are determined by the unconscious training of motor–neural pathways developed over years of practice.

From the point of view of the observer learning the process, he can tell roughly where the knapper is gazing but not exactly what platform variables she is choosing *emically*. He can estimate the speed (and thus the force) of the delivery of the strike from the position and gesture of the percussing arm. He also can estimate the angle of attack controlled by the arm and leg supporting the core. However, the knapper has the full-body experience of precisely controlling all of these variables in the split second it takes to deliver the blow and remove the flake. The observer has an etic viewpoint, whereas the knapper has a fuller, emic viewpoint, experiencing the blow from the alpha to the omega of the performance. At best, by watching the knapper and even examining the knapper's products as they are removed from the core, the observer has only an etic appreciation of what the knapper actually did rather than what she may have intended to do (i.e., her emic choice). Thus, he can learn the position of that blow within the sequence of removals he has just witnessed (the strategic knowledge inherent in the *connaissance* of the blow) but not the *tactical, savoir faire* know-how to make those removals himself. He must practice for weeks and months, if not years, to develop a full emic level of skill. Thus, in my use of the emic/etic distinction (Tostevin 2012), archaeologists and prehistoric novice flintknappers have parallel relationships. Archaeologists are by necessity relegated to the etic perspective; we cannot access the minds of prehistoric artisans who have their own emic perspective. *But the prehistoric observer at the beginning, if not the end, of the CT process is also limited to the etic perspective, at least for the savoir faire of the content, even though the observer is part of the enculturating environment of that culture.*

Recognizing the Need for the Connaissance/Savoir Faire and Etic/Emic Perspectives in CT Research

The distinction between the parts of the learning process implied by connaissance and savoir faire requires further elaboration. This dichotomous view of *knowing* in the French language has long played a significant role in the understanding of technological performance, including flintknapping in the Old World (Mauss 1935; Chamoux 1978; Pelegrin 1990; Karlin 1991). Apel (2008, 98) provides a helpful discussion of the topic and unpacks the concepts using the English word *knowledge* for *connaissance* and *know-how* for *savoir faire.*

> Knowledge is an integral part of a recipe for action, it is a form of declarative memory and thus consists of theoretical information only, while know-how is an important part of the teaching framework, especially self-teaching by trial and error, since it is a form of muscle memory that can be acquired only through practice (Apel 2001; Roux and Brill 2006). Pelegrin's terms [*connaissance* and *savoir faire*] have the advantage that they make a sharp distinction between information acquired from a source outside the body and the type of know-how that can only be achieved by coordinating the muscles involved in a gesture.

Connaissance/knowledge is thus learnable to a far greater degree by observation alone (possibly aided by verbal communication), whereas savoir faire/

know-how must be learned by an individual through extensive bodily repetition.

Wynn and Coolidge (2004) also provide a useful discussion of this distinction in relation to the working memory concepts from cognitive science (Baddeley and Logie 1999; Baddeley 2001) and cognitive anthropology literature on the phenomenological acquisition of skill (Keller and Keller 1996). For Wynn and Coolidge, both knowledge and know-how are part of Keller and Keller's blacksmith's "stock of knowledge," as well as part of Ericson and Kintsch's (1995) "long-term working memory" from cognitive psychology, which allows the enactment of complicated tasks with little loss of attention to other behaviors. Wynn and Coolidge's synthesis of these perspectives points to ten years of practice for the acquisition of expert know-how.

Figure 8.2 presents an unpacking of these concepts according to different authors. To these oppositions, I add that savoir faire in flintknapping constitutes the tactical know-how or skill to successfully execute a blow to

Connaissance/Knowledge	Savoir faire/Know-how	Source
Explaining	Acting	Apel (2008, Table II)
Explicit memory	Unconscious memory	Apel (2008, Table II)
Communicative	Intuitive	Apel (2008, Table II)
Theoretic memory	Muscle memory	Apel (2008, Table II)
Lost in case of conscious memory loss	Not lost in case of conscious memory loss	Apel (2008, Table II)
Semantic	Nonsemantic visual, tactile, and aural imagery	Wynn and Coolidge (2004)
Declarative knowledge	Skill/ability to replay motor behaviors	Wynn and Coolidge (2004)
Concept	Experience	Apel (2008, Table II), modified by Tostevin
Strategic knowledge: the plan for a sequence of removals within core reduction, including contingency plans for error corrections.	*Tactical know-how:* the skill to successfully execute a blow dictated by the strategic plan.	Tostevin (present paper)

Figure 8.2. The unpacking of *connaissance* versus *savoir faire* according to Apel (2008, Table II), Wynn and Coolidge (2004), and the present author.

remove a desired flake. The flake-by-flake variables I described above thus equate to the tactical know-how of *savoir faire*. *Connaissance* for flintknapping, on the other hand, constitutes the strategic knowledge or plan for exploiting the core volume down to exhaustion through the removal of a long sequence of flakes. The strategic plan includes the creation of the relationship between core surface convexities, as well as the subsequent rotation of the core for the exploitation of different platforms. Strategic knowledge also includes contingency plans for correcting errors in the ever-changing morphology of the core that could cause its premature discard. Thus, while tactical decisions are enacted with each flake in a reduction, strategic decisions are made at the level of each core reduction.

Given how strategic knowledge must be observed etically to be learned but how tactical know-how must be observed etically and then practiced emically to be learned, the physicality of the transmission process puts the observer and the archaeologist in the same etic perspectives to the transmission event. Thus, for the archaeologist, tactical decisions within an assemblage of stone tools from a given site are characterizable through the central tendencies and dispersions in etically observable variables across the population of flakes in the assemblage, just as they were to the observer as she or he continuously practiced to get products to approximate the morphology of the products of the original performer. The strategic decisions are etically characterizable, on the other hand, at the level of the entire assemblage (or the smallest level of meaningful geoarchaeological association, such as raw material units, e.g., Turq et al. [2013]; Machado et al. [2013, 2016]). The fact that these choices are as observable to the archaeologist through a quantitative attribute analysis (Figure 8.3) as they were to the observer allows archaeologists to avoid the epistemologically dangerous task of guessing the emic logic of the prehistoric knapper, as often happens with teleological reconstructions of operational sequences (Dibble et al. 2017). Instead, BACT for lithic technology allows one to characterize an assemblage in terms of the quantitative choices enacted at different parts of the knapping process that had to be learned etically and then practiced emically in socially intimate contexts. These behavioral choices are thus suitable as transmissible elements for the investigation of cultural evolutionary processes in the archaeological record.

For operational sequences of sufficient complexity, the separation of the material content of the learning process into two levels (tactical know-how vs. strategic knowledge) creates a distinct transmission isolating mechanism

Flintknapping domain	Decision node characterized by archaeological observations	Type of knowledge
Core modification	Core orientation: extant core morphologies	Strategic knowledge
	Core convexity management: refits, diagnostic reparations	Strategic knowledge
Pattern of core rotation during reduction	Early exploitation: dorsal scar patterns of blanks vs. blank length	Strategic knowledge
	Late exploitation: dorsal scar patterns of blanks vs. blank length	Strategic knowledge
Platform maintenance	Platform Treatment	Tactical know-how
	Exterior platform angle	Tactical know-how
	Platform thickness	Tactical know-how
Dorsal surface convexity	Longitudinal extent of the surface removed: length/width ratio	Tactical know-how
	Vertical convexity of the mass removed: width/thickness ratio	Tactical know-how
	Longitudinal shape of the surface: lateral edge type	Tactical know-how
	Dorsal ridge system: number of ridges defining the convexity: cross-section type	Tactical know-how
	Curvature of the core surface removed: profile type	Tactical know-how

Figure 8.3. Archaeologically observable decision nodes in a flintknapping operational sequence according to the type of knowledge implied by the distinction between strategic knowledge (learnable by etic observation of the process) and tactical know-how (learnable to an emic level only through bodily practice). Methods for the measurement and characterization of each decision node are provided in Tostevin (2012, chapter 4).

(TRIM) (Durham 1991; Mesoudi 2011). As Foster and Evans (see chapter 5) emphasize:

> Whenever transmissible units depend on extensive previous training or time-consuming pedagogy for reliable transmission, their spread across populations will be slower and cultural evolution more likely to manifest a branching

mode on some level of analysis (Boyd et al. 1997; Wimsatt 2013). This should be true whether the transmissible unit is crafting a stone tool or crafting an elegant proof.

I also would add that for Pleistocene hunter–gatherers the cultural evolutionary branching pattern is likely to be symmetric with the branching pattern of biological inheritance for the individuals involved, since emic-level training in foragers does not happen unless the individuals involved are socially intimate enough to be members of the same gene pool (Tostevin 2007).

MODELING THE INTERACTION BETWEEN SCAFFOLDS AND THE CT PROCESS FOR ACQUIRING FLINTKNAPPING SKILL

If the operation of the etic/emic and connaissance/savoir faire structural oppositions in the process of CT for flintknapping creates a TRIM, to what extent can the process vary depending on the support of transmission accelerating mechanisms (TRAMs)? To ask the question another way, how does the support of scaffolds (sensu Wimsatt and Griesemer 2007) affect the acquisition of both types of knowledge? Attempting to answer this question is critical to the development of a robust cultural evolutionary theory, since it will determine how CT content that differs in its material requirements vis à vis strategic versus tactical knowledge affects the role of scaffolds and other evolutionary forces. As scaffolds and other CT structures likely played significant roles in the evolution of human society from the Pleistocene to the Holocene, understanding their roles in even simple technological systems should be useful. For the final section of this chapter, I offer a comparison of a series of conceptual models of the role of scaffolds in the acquisition of flintknapping skill.

Wimsatt and Griesemer (2007) recognize three types of scaffolds, building off of developmental psychology's artifactual metaphor for the role of teachers' and others' behaviors that facilitate a child's development (Greenfield 1984; Bickhard 1992; Lave and Wenger 1991).

1. *Artifact Scaffolding:* "Artifacts can scaffold acts when they make acts possible, feasible, or easier than they otherwise would have been" (Wimsatt and Griesemer 2007, 60).
2. *Infrastructure Scaffolding:* "The most important mode[s] of infrastructural scaffolding are forms without which culture and society would not

be here at all. Going backwards in time: written language, settlements and agriculture, and animal husbandry and trade practices (developing into economic systems) were major infrastructural innovations central to all that followed. Spoken language with oral traditions and tools use antedate all of these by many tens to hundreds of thousand years. All are generatively entrenched so deeply as to be virtually constitutive of all of our forms of life, limiting the kinds of presence-and-absence comparisons we would like to have to assess their effects" (65).

3. *Developmental Agent Scaffolding*: "Scaffolding skills in agents where the scaffold is (or includes) another agent are particularly interesting: the scaffold is or involves another person, social group, or organization, often in spatial and temporally organized dynamical arrangements with artifacts" (66).

In the present case, I take the cognitive capacities of prehistoric actors to be elements of infrastructure scaffolding. Are these scaffolds or prerequisites? It is difficult to say, and thus the distinction between infrastructure and artifact scaffolding is useful. As each of the questions asked in the introduction to this chapter concerning the development of CT structures during the course of human evolution includes one or more of these types of scaffolds, how can we conceive of these scaffolds affecting the fidelity of learning knapping skills?

Figures 8.4–8.9 present scenarios that diagram the gradual development of knapping skills across the duration of the transmission process *(moving from the top of the figure to the bottom)* due to the influence of a "knowledgeable knapper (K)" on a "naïve observer (O)." Scenarios differ based on the action of the different types of scaffolding structures that have been proposed as significant in the evolution of the cumulative capacity for culture (see, e.g., Sterelny 2012). The three types of scaffolds serve as column headings running across the top of the figure and the gradual development of a naïve individual's etic and emic perspectives on the observed/transmitted content runs down the right-hand side of the figure.

Scenario A (Figure 8.4) has the most minimal of scaffolding possible while still giving K some influence on the learning of O. Here, the infrastructure scaffolding consists of O's cognitive capacity for emulative learning—that is, learning the goal but not the step-by-step procedure for an operation (Tomasello 1996). K only serves as a developmental agent scaffold in that her social tolerance of O's presence allows O to learn from K's activities with

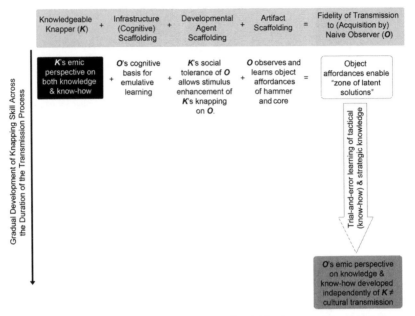

Figure 8.4. Scenario A, emulation learning: A conceptual model for the most minimal of roles for scaffolds in the gradual development of knapping skills by *Naïve Observer (O)* (top right), due to the influence of a *Knowledgeable Knapper (K)* (top left). Beginning with K's complete emic perspective on both strategic knowledge and tactical know-how, the column headings at the top of the diagram represent the summation of the infrastructure scaffolding, developmental (agent) scaffolding, and artifactual scaffolding that can contribute to O's learning of K's knowledge. The gradual development of the naïve individual's etic and emic perspectives on the observed content is represented from the beginning to the end of the process as the movement from the top of the figure to the bottom along the arrow in the rightmost column. The fidelity of O's acquisition of K's knowledge is represented by the degree of saturation of the grayscale coloration of the text boxes and arrow, with a fully saturated black coloration equating to complete fidelity between O and K, while a white text box indicates no similarity between O's and K's knowledge. Rows within the middle of the diagram represent that action of scaffolds available in the given scenario. The vertical placement of the rows of scaffolding actions does not imply any sequential order or absolute timing to their actions within the developmental process of O's acquisition of knowledge. It is assumed that the actions of these scaffolds are cumulative both vertically and horizontally across the diagram.

hammer and core, a process known as stimulus enhancement (Charman and Huang 2002; Franz and Matthews 2010; Matthews, Paukner, and Suomi 2010). As a result, O learns the object affordances of the artifacts (artifact scaffolds) and by her own trial-and-error experimental learning acquires strategic knowledge of the utility of making a cutting edge by conchoidal fracture. In this scenario, there is no other feedback between the learning activities of O on the part of K. Thus, Scenario A represents scaffolding that facilitates the *zone of latent solutions* (sensu Tennie, Call, and Tomasello 2009; Tennie et al. 2017) that we see in chimpanzee societies. Whether this scenario applies to the australopiths or early *Homo* remains to be seen. But this scenario serves as the absolute base from which we can enrich the process with more and more scaffolds. In many CT theories, the independent discovery of both knowledge and know-how in this scenario indicates that there was *no* cumulative CT, depending on whether one considers low-fidelity social learning, such as stimulus enhancement, as a mechanism that would lead to cumulative culture (see Tennie et al. [2017] for a diversity of opinions on this question).

Scenario B (Figure 8.5) differs from A because O's cognitive capacity now privileges her focus on sequential behaviors as meaningful to her own behavior. In other words, her infrastructure scaffolding includes imitative learning (Whiten et al. 2009), the learning of not only the goal but the means to achieve it. This change from Scenario A allows O to learn more from K's proximity in that she can learn K's sequence of blows—the strategic knowledge of the process accessible via an etic perspective. The artifact scaffolds also take on a different role in that O's examination of K's core and flakes can serve as models for her own practice knapping, which is still vital because she begins with no know-how. This scenario thus produces a gradual increase in the emic-level learning of O to that of moderate fidelity to that of K and is diagrammed in the scenario through the increase in gradient from white to gray in the arrow on the right of the figure.

Scenario C (Figure 8.6) has both K and O possessing joint attention toward O's learning to knap, another increase in infrastructure scaffolding. Tomasello et al. (2005) refer to this as triadic attention. The joint attention produces a greater involvement of K in O's learning through the social intimacy of K to O and K's active pointing and gestures of direction to O. These interventions of K might include actively taking O's core from her hands to correct an error of platform management by reparation removals before returning the core for O to continue the pursuit of her strategic plan. Ferguson's (2008) experimental work has demonstrated that this is a successful scaffold in increasing the speed of modern humans learning to knap. The social intimacy afforded O now allows her to repeatedly practice in company with K and thus have continuous opportunities to compare body motions, core-holding configurations, and the resultant artifacts between her and K's reductions. This produces a faster acquisition of tactical know-how and

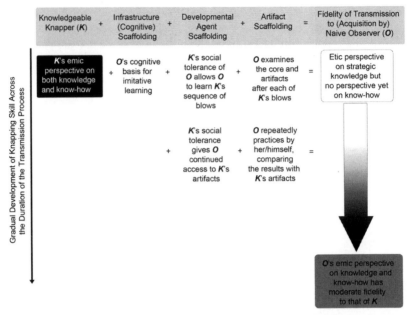

Figure 8.5. Scenario B, imitation learning: A conceptual model for the role of scaffolds in the gradual development of knapping skills by *Naïve Observer (O)* (top right), due to the influence of a *Knowledgeable Knapper (K)* (top left). Beginning with K's complete emic perspective on both strategic knowledge and tactical know-how, the column headings at the top of the diagram represent the summation of the infrastructure scaffolding, developmental (agent) scaffolding, and artifactual scaffolding that can contribute to O's learning of K's knowledge. The gradual development of the naïve individual's etic and emic perspectives on the observed content is represented from the beginning to the end of the process as the movement from the top of the figure to the bottom along the arrow in the rightmost column. The fidelity of O's acquisition of K's knowledge is represented by the degree of saturation of the grayscale coloration of the text boxes and arrow, with a fully saturated black coloration equating to complete fidelity between O and K, while a white text box indicates no similarity between O's and K's knowledge. Rows within the middle of the diagram represent that action of scaffolds available in the given scenario. The vertical placement of the rows of scaffolding actions does not imply any sequential order or absolute timing to their actions within the developmental process of O's acquisition of knowledge. It is assumed that the actions of these scaffolds are cumulative both vertically and horizontally across the diagram.

strategic knowledge from the beginning of the process, which results in a high degree of fidelity in transmission. Even when O engages in purely trial-and-error learning on her own, the social intimacy of O and K would produce a feedback loop between K and O based on K's evaluation of O's products.

Scenario D (Figure 8.7) shows K and O sharing linguistic abilities and a common language as the infrastructure scaffolding. K can now actively teach

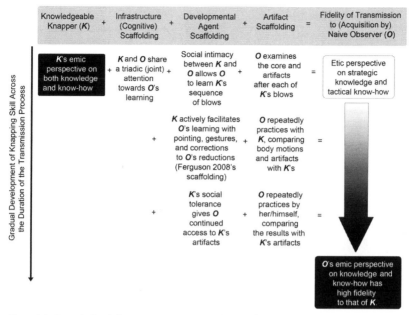

Figure 8.6. Scenario C, triadic attention: A conceptual model for the role of scaffolds in the gradual development of knapping skills by *Naïve Observer (O)* (top right), due to the influence of a *Knowledgeable Knapper (K)* (top left). Beginning with K's complete emic perspective on both strategic knowledge and tactical know-how, the column headings at the top of the diagram represent the summation of the infrastructure scaffolding, developmental (agent) scaffolding, and artifactual scaffolding that can contribute to O's learning of K's knowledge. The gradual development of the naïve individual's etic and emic perspectives on the observed content is represented from the beginning to the end of the process as the movement from the top of the figure to the bottom along the arrow in the rightmost column. The fidelity of O's acquisition of K's knowledge is represented by the degree of saturation of the grayscale coloration of the text boxes and arrow, with a fully saturated black coloration equating to complete fidelity between O and K, while a white text box indicates no similarity between O's and K's knowledge. Rows within the middle of the diagram represent that action of scaffolds available in the given scenario. The vertical placement of the rows of scaffolding actions does not imply any sequential order or absolute timing to their actions within the developmental process of O's acquisition of knowledge. It is assumed that the actions of these scaffolds are cumulative both vertically and horizontally across the diagram.

O the emic logic behind the strategy of removals, which might include ritual and superfluous steps to buffer the fidelity of the transmission through *overimitation* (Mace and Jordan 2011; McGuigan 2012). Compared to Scenario C, O can now achieve an emic perspective on strategic knowledge far earlier, and the ability of K to communicate with verbal cues during O's reductions may accelerate the development of tactical know-how, although

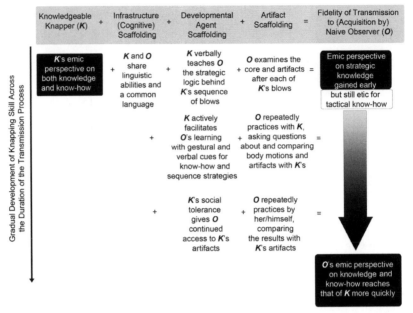

Figure 8.7. Scenario D, linguistic instruction: A conceptual model for the role of scaffolds in the gradual development of knapping skills by *Naïve Observer (O) (top right)*, due to the influence of a *Knowledgeable Knapper (K) (top left)*. Beginning with K's complete emic perspective on both strategic knowledge and tactical know-how, the column headings at the top of the diagram represent the summation of the infrastructure scaffolding, developmental (agent) scaffolding, and artifactual scaffolding that can contribute to O's learning of K's knowledge. The gradual development of the naïve individual's etic and emic perspectives on the observed content is represented from the beginning to the end of the process as the movement from the top of the figure to the bottom along the arrow in the rightmost column. The fidelity of O's acquisition of K's knowledge is represented by the degree of saturation of the grayscale coloration of the text boxes and arrow, with a fully saturated black coloration equating to complete fidelity between O and K, while a white text box indicates no similarity between O's and K's knowledge. Rows within the middle of the diagram represent that action of scaffolds available in the given scenario. The vertical placement of the rows of scaffolding actions does not imply any sequential order or absolute timing to their actions within the developmental process of O's acquisition of knowledge.

verbal communication can only do so much to aid this step. Yet O's emic perspectives on both strategic knowledge and tactical know-how are acquired even faster with this level of scaffolding compared to Scenario C, as is reflected in the darker gradient in the arrow on the right of the figure.

In moving from Scenario A to D, we can see how the different scaffolds can actually change the mode of transmission (sensu Boyd and Richerson 1985) of flintknapping skill. Scenario A can be characterized as predominantly guided variation, with the inheritance of the kernel of a concept, in this case the affordance of hammer and core to make a sharp flake, supplemented by O's trial-and-error learning. With Scenario D, the mode has become a form of biased transmission with less reliance on trial-and-error learning. While the observer still needs repetitive practice to approach K's emic skill level, the movement from etic to emic skill is faster, and thus the source of the biased transmission is more favored than the individual's trial-and-error learning. This would have the effect of increasing during transmission the coherence of design recipes, the fidelity of elements within behavioral packages, and the resultant covariation of variables measurable by archaeologists. *Recognizing the role of scaffolds in changing the mode of transmission thus has repercussions for how archaeologists and cultural evolution modelers think about what "modes" mean.* Bettinger and Eerkens's (1999) influential analysis of the adoption of bow-and-arrow projectile technology over that of spear-thrower technology in the American Great Basin at 1,350 years before the present used the absence of covariation in measurements associated with arrowhead design as being a result of guided variation in the transmission of Eastern Californian arrowhead knowledge compared to the strong covariation between these elements in Central Nevada, which was argued to be the result of indirect bias transmission. Rethinking Bettinger and Eerkens's argument, we can understand this difference in terms of the action of different developmental agent scaffolds related to social intimacy during transmission in each regional context. This observation removes much of the sting in Bamforth and Finlay's (2008) strong critique of Bettinger and Eerkens's assumptions about the meaning of variance in stone tool attributes. Citing the experimental work of Ferguson (2008), Bamforth and Finlay point out that large versus small variance in a given measurement can indicate different degrees of skill, not mode of transmission. But recognizing that the different modes of transmission in fact represent the effects of *different scaffolds for learning skill,* we can see that Bettinger and Eerkens and Bamforth and Finlay are arguing from two sides of the same coin.

This approach to modeling the role of scaffolds of different types in the fidelity (and even ability, given Scenario A) of CT can also be used to diagram other complex scenarios of "learning." For instance, Figure 8.8 presents a diagram depicting the scaffolds available during an episode of stimulus diffusion (Kroeber 1940)—that is, the transmission process in which the context of contact limits the transmission between individuals to only the idea of an object but not its techniques of production. Under the taskscape visibility concept for lithic technology, a stimulus diffusion scenario depicts the transmission of the idea of a tool, such as its morphology, to a Stranger (S) without the transmission of the detailed, specific knowledge to produce the morphology within the original enculturating environment exemplified by K's knapping. This process would occur when a socially distant individual gains access only to the limited results of K's flintknapping, either when S encounters K's discarded mobile tool kit when K is not present or when S encounters K with her tool kit at a logistical foray camp where the core reduction that produced the tool kit is not pursued (Tostevin 2007). In this scenario, however, S is an expert flintknapper, with both the strategic knowledge and tactical know-how of her own group, but she is unfamiliar with the material culture of K's enculturating group. S's task in this scenario is thus not to develop tactical know-how but to acquire K's strategic knowledge.

Recently, there has been some theoretical discussion of the role of lithic artifacts from preceding periods being, for all intents and purposes, artifact scaffolds for the reinvention of lost methods by artisans in later periods. Hiscock (2014) raises this possibility, even going so far as to describe the persistence of lithic artifacts on the landscape surface for millennia as a "library of stone" from which later knappers will learn. Such unburied artifacts on the landscape would certainly be sources of stimulus diffusion, and there is artifactual evidence that much older artifacts served as blanks for subsequent reshaping into new tools in later periods, such as Middle Paleolithic artifacts serving as blanks for tools made during the Upper Paleolithic (Belfer-Cohen and Bar-Yosef 2015). Yet beyond artifact reuse, the results of transmission under Hiscock's hypothesis would be limited to stimulus diffusion by the effect of equifinality, the archaeological observation that there are multiple ways to reduce a core (by the application of different bodies of strategic knowledge) that will produce similar flake morphologies (e.g., Boëda 1995, Figure 4.13). When S encounters K's object without K's performance, there is no guarantee that S will be able to reverse engineer K's original strategic knowledge from the old object's morphology. Being a skilled knapper herself, however, S would likely be able to reengineer the strategic technology to a generic level, perhaps equivalent to the largest recognized units of global variability in stone tools, Shea's (2013) modes A–I. Upon encountering a blade of particular dimensions, for instance, S would be able to recreate a blade technology of those dimensions but likely not the specific variety of blade technology (pyramidal vs. *semitournant*, bidirectional vs. unidirectional, etc.). Given equifinality, the exact reverse engineering of the details

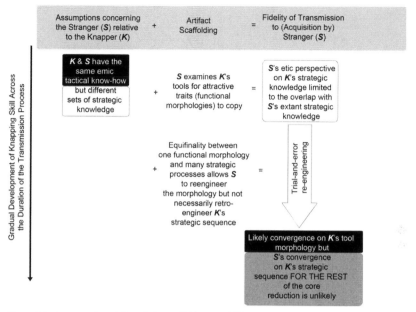

Figure 8.8. Conceptual modeling of the scaffolding available to an expert knapper during the stimulus diffusion (sensu Kroeber 1940) of the morphology of a lithic tool transported to a location where the early part of the operational sequence is not pursued. The diagram depicts the accurate transmission of the idea of the tool to *Stranger* (S), with S's independent trial-and-error learning to reengineer the core reduction details to produce the tool morphology. The likelihood of the reengineering as retro-engineering that would match K's original strategic knowledge depends on the breadth of equifinality between the multiple pathways of strategic knowledge that lead to that tool form and S's ingenuity and willingness to invest time in accuracy beyond simply achieving the morphology. The point of this conceptual model is to diagram the limited number of scaffolds available to the expert prehistoric knapper beyond her own emic levels of strategic and tactical knowledge.

of platform thickness, exterior platform angle, core rotation patterns, reparation techniques, and so on of K's knowledge seems unlikely. It is possible but the lack of scaffolds for S's reverse engineering, compared to the scaffolds afforded O in Scenarios B–D, makes stimulus diffusion more likely than diffusion of both the strategic plan and the final tool morphology.

My argument for the likelihood of stimulus diffusion over full transmission in such a case is quantitative only in comparing the count of available scaffolds between Figures 8.4–8.7 and Figure 8.8. In representing the strength of each scaffold, I can only be qualitative, as this subject has simply not been the focus of adequate quantitative experimental research. Therefore, I speak only of likelihoods and not exact probabilities. Further, I am extremely

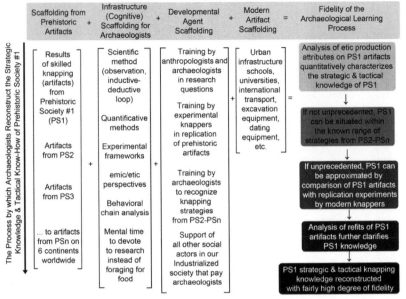

Figure 8.9. Conceptual modeling of the scaffolding available to professional archaeologists for learning (reconstructing) the strategic knowledge and tactical know-how of stone tool artisans from a specific prehistoric society. The types of scaffolding available are listed at the top of the diagram as column headings, starting at the top left with prehistoric artifacts of many societies, beginning with prehistoric society #1, the target for the present reconstruction, but also including reference training sets of artifacts from prehistoric society #2, #3, and so on, through to prehistoric society #n. As the number of scaffolds within each type are too numerous and interdependent to link from left to right, the effects of the scaffolds are depicted within each scaffold type as a matrix set. The matrix sets are summed to contribute to the archaeological learning processes in the rightmost column.

conscious of the wisdom of John Shea's observation that "time and again, the stone tool evidence shows that the surest way to be wrong in human origins research is to under-estimate Pleistocene hominins' behavioral variability" (Shea 2017, 191). Yet one of the reasons for my willingness to risk running afoul of his warning is the difficulty we archaeologists ourselves face in attempting to accomplish a similar task. This point can be illustrated by examining what is required in terms of scaffolds of all three types for the reverse engineering of a prehistoric technology within the contexts of an industrially supported archaeological community. In Figure 8.9, the left-hand axis now depicts the process by which archaeologists reconstruct the strategic knowledge and tactical know-how of a specific prehistoric society (in this case, prehistoric society #1, or PS1). The required scaffolds are now so numerous within each type that the individual scaffolds are presented within matrix sets so that the vertical axis only applies to the rightmost column in which the fidelity of the archaeological learning process is given. As with the stimulus diffusion scenario, note the absence of prehistoric artisans themselves in the diagram.

In constructing Figure 8.9, I was forced to recognize more developmental agent scaffolds in my own training than I am used to, which was a humbling process. Even so, the diagram is perhaps overly optimistic about the accuracy of archaeological reconstructions. The archaeological learning processes begin at the top of the rightmost column and gradually darken as they approach complete accuracy in reconstructing the prehistoric methods by the bottom of the rightmost column. This is certainly an idealized situation if not a pipe dream, for, apart from a dozen truly expert knappers worldwide, most lithic analysts do not possess within themselves the tactical know-how to match the best prehistoric artisans, particularly from periods in which lithic craft specialization was an occupation. Where archaeologists have an edge is their greater breadth of strategic knowledge gained from the examination of artifact collections from the Pliocene to the modern period from all six prehistorically occupied continents. In other words, while most archaeologists are not as tactically skilled as the knappers in the past, their purview on strategic knowledge is far greater. We are trained to recognize all of Shea's modes A–I, whereas most prehistoric societies practiced only a subset of these modes. Thus, an archaeologist has the advantage in retro-engineering a specific strategic knowledge set, whereas an expert knapper engaged in a prehistoric stimulus diffusion scenario would have an advantage, and thus a tendency, to prioritize effective reengineering given her extant and more limited strategic knowledge.

RECENT EXPERIMENTAL INVESTIGATION OF SCAFFOLDING AND MUTATION RATES IN THE LEARNING OF LITHIC TECHNOLOGY

I have endeavored to convey the complexity of the scaffolds necessary even for an archaeologist, with all of our industrial support, to replicate a technology represented by the complex interplay between tactical knowledge held in body memory and strategic knowledge held in conscious memory. This complexity puts the lie to Gould's (1987, 70) often quoted characterization of CT as "five minutes with a wheel, a snowshoe, a bobbin, or a bow and arrow may allow an artisan of one culture to capture a major achievement of another." Even giving Gould the benefit of the doubt in assuming that his scenario represents stimulus diffusion rather than the transmission of both true strategic knowledge and tactical know-how, I would not bet on the success of his bow and arrow. With only five minutes of learning, he would likely starve.

Unfortunately, Gould is not alone in misrepresenting the prehistoric learning process. Other scientists, including archaeologists, at times ignore the importance of the dual nature of learning bodily performances when conducting controlled experiments on learning. Morgan et al. (2015) have recently published an experimental study in which 184 individuals were trained in flintknapping under five varying parameters of transmission. The parameters included *reverse engineering* in which the naïve observer was given a hammer stone and a core and shown stone tools but never a knapper in action (akin to the stimulus diffusion scenario above, save that their knappers were completely naïve); *imitation/emulation* (equivalent to Scenario B above); *basic teaching* in which the demonstrator could alter the grip of the learner on the core and slow his own demonstrations but not use gestures beyond these (more limited than the scaffolding presented in Scenario C above); *gestural teaching* (fully equivalent to Scenario C above); and *verbal teaching* (equivalent to Scenario D above). What is striking in the experimental structure of Morgan et al.'s study is first, the large sample size of learners, articulated into transmission chains of learners teaching learners in iterations of five to ten "generations." For the first time, a knapping experiment has achieved a sufficient sample size of learners to produce statistically analyzable results. Second, the similarities between Morgan et al.'s five *transmission mechanisms* and the scaffolding scenarios above (B, C, D, and stimulus diffusion) show a clear convergence in how scholars are conceiving of the additive nature of learning mechanisms since my conceptual modeling above was independently created before the publication of Morgan et al.'s study. Admittedly, Morgan et al. do not mention "scaffolding" or similar structural relationships within the physical constraints within lithic cultural transmission, but the overall intent is similar. Further, Morgan et al. usefully compare their results to the question of how hard different prehistoric technologies were to learn with different mechanisms. Overall, they concluded that for most of their measures only verbal teaching consistently produced a positive effect on learning, and thus lithic technology more complex than

Oldowan reduction, which their experiment replicated, likely required verbal instruction in the past.

What is markedly different between our views of the learning process, however, is Morgan et al.'s ignorance, given how they structured their experiment, of the importance of bodily practice in the acquisition of both strategic knowledge and tactical know-how. *Ironically, like Gould, they fixed on a five-minute educational window.* Each learner was only exposed to the learning environment for five minutes before being required to teach the next generation. As a result, the learners situated later in the transmission chains performed more and more poorly, until by generation five the learners in the verbal teaching cohort were performing as poorly as those in the reverse engineering and imitation/emulation groups, which possessed the fewest scaffolds. *The results in fact demonstrated that the quality and efficacy of CT declined through time within the experiment. There was in fact little to no preservation of learned behaviors between generations beyond what observation alone could accomplish.* Contrary to the authors' conclusions, the only evidence for maintenance of skill and possibly a ratchet effect—that is, improvement in knapping skill between generations one and five—were the reverse engineering and imitation/emulation groups that showed a very slight increase in the proportion of viable flakes produced, although below the level of statistical significance (Morgan et al. 2015, Figure 2h). My hypothesis in this case is that learners in these latter two groups were allowed to focus more on their own trial-and-error learning of tactical know-how without the interruption of less than accurate scaffolding attempts by the increasingly inept teachers found in the later generations of the cohorts of the "more complex" mechanisms. It is laudable that Morgan et al. documented the decline in both the frequency and accuracy of the verbal communications by instructors across the transmission chains, although they did not contextualize the breadth of knowledge or length of teaching experience of the initial trained experimenters at the beginning of each transmission chain, issues that are relevant to teaching lithic technology (Bamforth and Finlay 2008; Shea 2015). In summary, their experiment only modeled the transmission of strategic knowledge, as it did not allow enough time for the development of any tactical know-how, although the measures used to assess the success of the transmission were directly related to tactical skill, not strategic skill.

Despite these limitations, however, the scope of the Morgan et al. study across these mechanisms and across such numbers of learners sets a new bar for experimental work in this area. If the study were repeated with sufficiently long periods of instruction to achieve a stable CT environment, much could be learned from this type of experiment. The decision on what constitutes a sufficient length of instructional time, regardless of which mechanism is being used, could be informed by flintknappers who have taught these technologies in practice. Shea (2015), who has taught flintknapping for over three decades, argues for at least one hour of instructor time to teach Acheulean handaxes and three to six hours for hierarchical cores such as Levallois or blade technology. From my experience teaching a flintknapping course of twelve students once a year for fifteen years, the necessary time for student practice *after* this initial exposure can be as much as double the instructional time.

Morgan et al. (2015) are not alone in ignoring the role of savoir faire learning in a CT experiment related to flintknapping. As Lycett et al. (2015) summarize, their research team conducted two experiments designed to evaluate the effects of size mutation (Kempe, Lycett, and Mesoudi 2012) and shape mutation (Schillinger, Mesoudi, and Lycett 2014) in the copying of an Acheulean handaxe. In the first study, naïve participants were asked to use a tablet computer's touch screen to resize an image of an Acheulean handaxe to that of an example image. In the latter experiment, participants were asked to use a stainless steel table knife to carve the shape of a model Acheulean handaxe out of a standardized plasticine block. While each experiment was well executed and in its way ingenious, as was their overall purpose in creating a "model organism" context to stimulate CT research (Lycett et al. 2015), neither experiment made any effort to approximate the material reality of the process involved in acquiring or utilizing the savoir faire of flintknapping. Studying the effects of size mutation (Kempe, Lycett, and Mesoudi 2012) might arguably be a question of connaissance, but surely the rate of shape mutation (Schillinger, Mesoudi, and Lycett 2014) relies upon the fidelity of the transmission of savoir faire far more than connaissance and so should not be removed from the experiment. The results of both studies are thus highly suspect if they are to be applicable to the knapping of stone.

The Schillinger, Mesoudi, and Lycett (2014) study, however, represents another interesting case of convergence. I myself have used closed-cell foam in many of my flintknapping classes, starting in 2003, to serve as proxy stone cores for teaching students the strategic differences between core reduction methods, such as bifacial, Levallois, and blade technology. I gave each student a foam core and a little saw and asked them to saw off flakes in the appropriate series of removals. I did this precisely because I wanted to test their knowledge of the connaissance of each technology (the sequence and direction of each removal) *without* the interference of their poor savoir faire since they had not yet had time to develop adequate bodily gestures to execute each technology.[3] Thus, one can see a similar solution to removing the constraints of how hard it is to learn how to knap stone. The question is, how should we use such solutions?

A more recent experiment attempted to do the opposite of Schillinger, Mesoudi, and Lycett (2014) and my foam-core teaching method—that is, to learn what stone technology looks like when the connaissance/strategy of the core reduction is removed from the equation, leaving only the savoir faire/tactical know-how of individual flake removal. In a unique and rather brilliant experiment, Moore and Perston (2016) endeavored to eliminate strategic-level cognition as much as possible from the knapping procedure by randomizing platform selection, where an expert knapper is to strike on the core, between flake removals. The goal was to simulate what an assemblage of stone tools might look like under a rule of "least effort" flake production such as might characterize the earliest of lithic technologies in which only one flake was desired at a time. Further experiments along this line, which *utilize* rather than *ignore* the difference between tactical and strategic knapping skill, will move us much closer to an archaeology of pedagogy.

CONCLUSION

In this chapter I have endeavored to illustrate the importance of considering the dual structural oppositions between etic/emic perspectives and strategic (connaissance)/tactical (savoir faire) bodies of knowledge for making CT research materially explicit enough to accommodate Paleolithic data on lithic artifact production sequences. The differences between etic and emic perspectives are already recognized to some degree within CT research, given the indirect learning exemplified by Boyd and Richerson (2000) and my Figure 8.1. Many disciplines, however, have converged on a similar recognition of the indirect nature of cognition and the learning process. As Salikoko Mufwene has observed (see chapter 9), the indirect nature of social learning would make *cultural replication* a more accurate descriptor than *cultural transmission* for the process that interests us. The term *cultural replication* would avoid what Michael Reddy, a linguistic anthropologist, calls the *conduit metaphor,* a concept in spoken and written English that frames our language about language.[4] The framing ignores the construction of meaning in the mind of the listener by privileging the perspective of the speaker, both for the creation of meaning and as the responsible party for moving the meaning between interactors (Reddy 1979). Phrases like "get your thoughts *across* better" and "you still haven't *given me* any idea of what you mean" exemplify the conduit metaphor. Reddy's critique of the conduit metaphor demonstrates how frequently our language about language creates the fictitious idea that "ideas" are passing through the ether between interactors. The cultural "transmission" seen in Figure 8.1 as a series of hand-off events relies on the conduit metaphor and so perpetuates the conflict between the (false) sender-oriented metaphor in CT theory's title and what we know about the receiver-oriented nature of cognition itself. The archaeologist Michael Schiffer (1999) has made this point explicit in his emphasis on how humans learn from the environment via an exclusively receiver-oriented perspective, including the communicative acts of other humans. Arguing that language is the most obvious but least omnipresent medium of communication, Schiffer proposes a three-interactor model for human inferences, with material culture—rather than language—acting as a vehicle for the majority of information humans glean from their environment. By replacing the typical linguistic two-body model of sender and receiver critiqued by Reddy, Schiffer advocates a model with a "sender" that alters the physical properties of an "emitter," which then cues cognitive responses *(correlons)* in the mind of the "receiver" that inform the receiver about its environment. Each of the three roles (sender, emitter, and receiver) can be played by a person, an artifact, or even a natural phenomenon *(extron)*. As material culture plays the role of emitter in most of Schiffer's basic communication processes, the receiver-oriented approach should take on added significance for CT theory.

In building the Behavioral Approach to Cultural Transmission (BACT), I have endeavored to keep a receiver-oriented approach for how novice flintknappers learn their skill sets through their enculturating environment. This approach, as a result, relies heavily upon the strategic knowledge/tactical know-how distinction inherent in lithic technology. Given Mufwene's question, therefore, should we not rename *cultural transmission theory* to *cultural replication theory* in order to avoid the demonstrable perils of Reddy's conduit metaphor? It is too early to say. Boyd and Richerson (1985) already tried to move the discipline away from Cavalli-Sforza and Feldman's (1981) original *cultural transmission* descriptor by emphasizing the name *dual inheritance modeling,* and that did not stick. But I do believe that the greater recognition and utilization of the perspectival difference between a sender and a receiver would go a long way toward making CT research more realistic for cases of craft production because it forces one to remember the physical process during a "transmission event." Thus, with certain technologies, adding the strategic knowledge versus tactical know-how distinction would encourage us to see strategic knowledge as a product of something short, an *event of transmission,* whereas tactical know-how is something that takes longer to acquire, as a product of a *process of development.* Rather than advocating another title, we can instead embrace the need for both the developmental *and* phylogenetic perspective when conducting research on how individuals acquire knowledge (Wolcott 1991).

In this chapter I have also endeavored to highlight the efficacy of considering in detail how the CT of a given technology can be affected by scaffolding of many types. From the six conceptual models I provide, the reader may ask, What is the utility of the scaffolding scenarios? Am I not going to peg specific archaeological NASTIES to each scenario? No, I will not engage in what can only be seen as guesswork at the moment. However, I will present a challenge. If more archaeologists can apply the quantitative and behavioral methods discussed here widely enough to produce sufficient assemblage data sets, we will begin to be able to test behavioral hypotheses derived from CT modeling against real Paleolithic data. To date, there are a limited number of archaeologists besides myself using these methods (Nigst 2012; Nigst et al. 2014; Scerri 2013; Scerri et al. 2014; Scerri et al. 2016). But if we can begin to construct larger difference matrices from the comparison of assemblages (as Tostevin and Škrdla [2006, Table 4] began for the Early Upper Paleolithic in the Middle Danube basin), we will improve our ability to make scientific progress, at least for the quantitative comparison of change through time in instructional learning sets. Having quantitative measures of what was or was not learned in different times and places is a first step toward making Paleolithic archaeology useful for testing hypotheses concerning CT. Imagine if Paleolithic archaeologists were able to calculate diversity values, such as F_{ST} values, from Paleolithic data to compare with F_{ST} predictions from agent-based models (e.g., Premo 2012b) and other population genetics–inspired modeling. Each scenario above can be modeled, even if the task is difficult. *If Paleolithic archaeologists can meet the modelers halfway, imagine what a developmental agent scaffold that would be for the growth of a theory of cultural evolution.*

In this chapter, I have provided only the briefest illustration of how Paleolithic archaeology should redesign its analytical approach in order to be more useful in the larger endeavor of building cultural evolutionary theory. My intent has been to point out the potential of Paleolithic data should the needed analytical changes be adopted and highlight conceptual areas where CT theorists and Paleolithic archaeologists can come together as closer collaborators, as has been done in other contexts (Premo and Kuhn 2010; Premo and Tostevin 2016). There is a productive traction for abundant research between the processual thinking of such modelers and the archaeologists who excavate and measure artifacts. This has been my experience with the exploration of the concept of generative entrenchment as applied to blade technology and compound tools in the Late Pleistocene (Tostevin 2013), a direct result of interacting with William Wimsatt at the University of Minnesota. And it is even true when the theorist (Wimsatt) did not have enough time to complete his flintknapping training in my lab. He certainly got the connaissance, if not sufficient practice in order to internalize the procedure as emic, savoir faire body knowledge.

NOTES

Many thanks to William Wimsatt for inviting me to the original workshop and for all of our productive discussions over the last few years. I would also like to thank Genevieve Tostevin for her help with the figures in this paper.

1. Boyd's Arizona State University biographical statement, https://webapp4.asu.edu/directory/person/1952328, accessed September 1, 2014.

2. Tostevin (2012, chapter 4) provides an illustrated discussion of the flintknapping process and these variables, showing their visibility from the point of view of the observer versus that of the performer.

3. My thanks to Liliane Meignen, one of the founders of the French *chaîne opératoire* school of lithic analysis, for the idea of using a substitute core when teaching students who have not yet achieved sufficient savoir faire skills to flintknap stone themselves. Dr. Meignen advised me in 1993 to use large raw potatoes as cores. I tried this approach for a few years when I first began teaching myself but found that the potato "flakes" were too messy, whereas the foam cores and flakes could be taken home by students as teaching kits.

4. My thanks to David Valentine for introducing me to Reddy's article many years ago.

REFERENCES

Adams, W. Y., and E. W. Adams. 1991. *Archaeological Typology and Practical Reality: A Dialectical Approach to Artifact Classification and Sorting.* Cambridge, Mass.: Cambridge University Press.

Andersson, Claes. 2013. "Fidelity and the Emergence of Stable and Cumulative Sociotechnical Systems." *PaleoAnthropology* 2013:88–103.

Andersson, Claes, and Dwight Read. 2016. "The Evolution of Cultural Complexity: Not by the Treadmill Alone." *Current Anthropology* 57 (3): 261–86.

Andersson, Claes, Anton Törnberg, and Petter Törnberg. 2014. "An Evolutionary Developmental Approach to Cultural Evolution." *Current Anthropology* 55 (2): 154–74.

Apel, Jan. 2001. *Daggers, Knowledge, and Power.* Uppsala: Coast to Coast Books 3.

Apel, Jan. 2008. "Knowledge, Know-How, and Raw Material—The Production of Late Neolithic Flint Daggers in Scandinavia." *Journal of Archaeological Method and Theory* 15:91–111.

Baddeley, A. D. 2001. "Is Working Memory Still Working?" *American Psychologist* 11:851–64.

Baddeley, A. D., and R. H. Logie. 1999. "Working Memory: The Multiple-Component Model." In *Models of Working Memory: Mechanisms of Active Maintenance and Executive Control,* edited by A. Miyake and P. Shah, 28–61. New York: Cambridge University Press.

Bamforth, Douglas B., and Nyree Finlay. 2008. "Introduction: Archaeological Approaches to Lithic Production Skill and Craft Learning." *Journal of Archaeological Method and Theory* 15:1–27.

Bar-Yosef, O., and P. Van Peer. 2009. "The *Chaîne Opératoire* Approach in Middle Paleolithic Archaeology." *Current Anthropology* 50 (1): 103–31.

Bedau, Mark A. 2013. "Minimal Memetics and the Evolution of Patented Technology." *Foundations of Science* 18:791–807.

Belfer-Cohen, A., and O. Bar-Yosef. 2015. "Paleolithic Recycling: The Example of Aurignacian Artifacts from Kebara and Hayonim Caves." *Quaternary International* 361:256–59.

Bettinger, R. L., R. Boyd, and P. J. Richerson. 1996. "Style, Function, and Cultural Evolutionary Processes." In *Darwinian Archaeologies,* edited by H. D. G. Maschner, 133–64. New York: Plenum Press.

Bettinger, R. L., and J. Eerkens. 1999. "Point Typologies, Cultural Transmission, and the Spread of Bow and Arrow Technology in the Prehistoric Great Basin." *American Antiquity* 64:231–42.

Bickhard, M. H. 1992. "Scaffolding and Self-Scaffolding: Central Aspects of Development." In *Children's Development within Social Contexts: Research and Methodology,* edited by L. T. Winegar and J. Valsiner, 33–52. Mahway, N.J.: Lawrence Erlbaum.

Binford, Lewis R. 1977. In the introduction to *For Theory Building in Archaeology: Essays on Faunal Remains, Aquatic Resources, Spatial Analysis, and Systematic Modeling,* edited by L. Binford, 1–10. New York: Academic Press.

Binford, Lewis R. 1978. *Nunamiut Ethnoarchaeology.* New York: Academic Press.

Blackmore, Susan. 2000. *The Meme Machine.* Oxford: Oxford University Press.

Boëda, Eric. 1995. "Levallois: A Volumetric Construction, Methods, a Technique." In *The Definition and Interpretation of Levallois Technology,* edited by H. L. Dibble and O. Bar-Yosef, 41–68. Madison: Prehistory Press.

Boyd, R., M. Borgerhoff-Mulder, W. H. Durham, and P. J. Richerson. 1997. "Are Cultural Phylogenies Possible?" In *Human by Nature: Between Biology and the Social Sciences,* edited by Peter Weingart, Sandra D. Mitchell, Peter J. Richerson, Sabine Maasen, 355–84. New York: Psychology Press.

Boyd, Robert, and P. J. Richerson. 1985. *Culture and the Evolutionary Process.* Chicago: University of Chicago Press.

Boyd, Robert, and P. J. Richerson. 1987. "Evolution of Ethnic Markers." *Cultural Anthropology* 2 (1): 65–79.

Boyd, Robert, and P. J. Richerson. 1996. "Why Culture Is Common but Cultural Evolution Is Rare." *Proceedings of the British Academy* 88:77–93.

Boyd, Robert, and P. J. Richerson. 2000. "Meme Theory Oversimplifies Cultural Change." *Scientific American,* October, 54–55.

Boyd, Robert, and P. J. Richerson. 2005. *The Origin and Evolution of Cultures.* New York: Oxford University Press.

Buchanan, Briggs, and Mark Collard. 2008. "Phenetics, Cladistics, and the Search for the Alaskan Ancestors of the Paleoindians: A Reassessment of Relationships among the Clovis, Nenana, and Denali Archaeological Complexes." *Journal of Archaeological Science* 35:1683–94.

Bullinger, Jérôme, Denise Leesch, and Nicole Plumettaz. 2006. "Le site magdalénien de Monruz, 1: Premiers éléments pour l'analyse d'un habitat de plein air." *Archéologie neuchâteloise* 33. Neuchâtel: Service et musée contonal d'archéologie.

Caporael, Linnda R., James R. Greisemer, and William C. Wimsatt, eds. 2014. *Developing Scaffolds in Evolution, Culture, and Cognition.* Vienna Series in Theoretical Biology. Cambridge, Mass.: MIT Press.

Carr, Chris. 1995. "A Unified Middle-Range Theory of Artifact Design. In *Style, Society, and Person: Archaeological and Ethnological Perspectives,* edited by C. Carr and J. Neitzel, 171–258. New York: Plenum Press.

Cattin, Marie-Isabelle. 2002. "Hauterive-Champréveyres 13: Un campement magdalénien au bord du lac de Neuchâtel, exploitation du silex (secteur 1)." *Archéologie neuchâteloise* 26. 2 vols. Neuchâtel: Service et musée contonal d'archéologie.

Cavalli-Sforza, L. L., and M. W. Feldman. 1981. *Cultural Transmission and Evolution: A Quantitative Approach.* Princeton, N.J.: Princeton University Press.

Chamoux, M. N. 1978. "La transmission des savoir-faire: Un objet pour l'ethnologie des techniques." *Techniques et Culture. Bulletin de l'équipe de recherche* 191 (3): 46–83.

Charman, T., and C. T. Huang. 2002. "Delineating the Role of Stimulus Enhancement and Emulation Learning in the Behavioral Re-enactment Paradigm." *Developmental Science* 5:25–27.

Corbey, Raymond, Adam Jagich, Krist Vaesen, and Mark Collard. 2016. "The Acheulean Handaxe: More Like a Bird's Song than a Beatles' Tune?" *Evolutionary Anthropology* 25:6–19.

Cotterell, Brian, and Johan Kamminga. 1987. "The Formation of Flakes." *American Antiquity* 52 (4): 675–708.

Davidson, Ian. 2010. "Stone Tools and the Evolution of Hominin and Human Cognition." In *Stone Tools and the Evolution of Human Cognition,* edited by A. Nowell and I. Davidson, 185–206. Boulder: University Press of Colorado.

Dawkins, Richard. 1976. *The Selfish Gene.* Oxford: Oxford University Press.

Dibble, H. L., S. Holdaway, S. Lin, D. Braun, M. Douglass, R. Iovita, S. P. McPherron, D. Olszewski, and D. Sandgathe. 2017. "Major Fallacies Surrounding Stone Tool Artifacts and Assemblages." *Journal of Archaeological Method and Theory* 24:813–51.

Dibble, H. L., and A. Pelcin. 1995. "The Effect of Hammer Mass and Velocity on Flake Mass." *Journal of Archaeological Science* 22:429–39.

Dibble, H. L., and Z. Rezek. 2009. "Introducing a New Experimental Design for Controlled Studies of Flake Formation: Results for Exterior Platform Angle, Platform Depth, Angle of Blow, Velocity, and Force." *Journal of Archaeological Science* 36:1945–54.

Donald, Merlin. 1998. "Hominid Enculturation and Cognitive Evolution." In *Cognition and Material Culture: The Archaeology of Symbolic Storage,* edited by Colin Renfrew and Chris Scarre, 7–18. McDonald Institute Monographs. Cambridge: McDonald Institute for Archaeological Research.

Durham, W. H. 1991. *Coevolution: Genes, Culture, and Human Diversity.* Stanford, Calif.: Stanford University Press.

Eerkens, J. W., and R. L. Bettinger. 2001. "Techniques for Assessing Standardization in Artifact Assemblages: Can We Scale Material Variability?" *American Antiquity* 66:493–504.

Eerkens, Jelmer W., and Carl P. Lipo. 2005. "Cultural Transmission, Copying Errors, and the Generation of Variation in Material Culture in the Archaeological Record." *Journal of Anthropological Archaeology* 24:316–34.

Eerkens, Jelmer W., and Carl P. Lipo. 2007. "Cultural Transmission Theory and the Archaeological Record: Providing Context to Understanding Variation and Temporal Changes in Material Culture." *Journal of Archaeological Research* 15:239–74.

Ericsson, K. A., and W. Kintsch. 1995. "Long-Term Working Memory." *Psychological Review* 102:211–45.

Ferguson, Jeffrey. 2008. "The When, Where, and How of Novices in Craft Production." *Journal of Archaeological Method and Theory* 15:51–67.

Franz, Mathias, and Luke J. Matthews. 2010. "Social Enhancement Can Create Adaptive, Arbitrary, and Maladaptive Cultural Traditions." *Proceedings of the Royal Society B* 277:3363–72.

Gamble, Clive. 1999. *The Palaeolithic Societies of Europe.* Cambridge: Cambridge University Press.

Garofoli, D., and M. N. Haidle. 2014. "Epistemological Problems in Cognitive Archaeology: An Anti-relativistic Proposal towards Methodological Uniformity." *Journal of Anthropological Sciences* 92:7–41.

Gould, Richard A. 1980. *Living Archaeology.* Cambridge: Cambridge University Press.

Gould, Stephen Jay. 1987. *An Urchin in the Storm: Essays about Books and Ideas.* New York: W. W. Norton.

Gould, Stephen Jay. 2002. *The Structure of Evolutionary Theory.* Cambridge, Mass.: Harvard University Press.

Grant, Peter R., and B. Rosemary Grant. 2011. *How and Why Species Multiply: The Radiation of Darwin's Finches.* Princeton Series in Evolutionary Biology. Princeton, N.J.: Princeton University Press.

Grant, Peter R., and B. Rosemary Grant. 2014. *40 Years of Evolution: Darwin's Finches on Daphne Major Island.* Princeton, N.J.: Princeton University Press.

Greenfield, P. M. 1984. "A Theory of the Teacher in the Learning Activities of Everyday Life." In *Everyday Cognition: Its Development in Social Context,* edited by B. Rogoff and J. Lave, 117–38. Cambridge, Mass.: Harvard University Press.

Grimm, Linda. 2000. "Apprentice Flintknapping: Relating Material Culture and Social Practice in the Upper Palaeolithic." In *Children and Material Culture,* edited by J. S. Derevenski, 53–71. London: Routledge.

Harmand, Sonia, Jason E. Lewis, Craig S. Feibel, Christopher J. Lepre, Sandrine Prat, Arnaud Lenoble, and Xavier Boës. 2015. "3.3-Million-Year-Old Stone Tools from Lomekwi 3, West Turkana, Kenya." *Nature* 521 (7552): 310–15.

Harris, Marvin. 1976. "History and Significance of the Emic/Etic Distinction." *Annual Review of Anthropology* 5:329–50.

Hewlett, Bonnie. 2013. "'*Ekeloko*' The Spirit to Create: Innovation and Social Learning among Aka Adolescents of the Central African Rainforest." In *Dynamics of Learning in Neanderthals and Modern Humans Volume 1: Cultural Perspectives,* edited by T. Akazawa, Y. Nishiaki, and K. Aoki, 187–95. Tokyo: Springer Japan.

Hewlett, B. S., H. N. Fouts, A. H. Boyette, and B. L. Hewlett. 2011. "Social Learning among Congo Basin Hunter–Gatherers." *Philosophical Transactions of the Royal Society B* 366:1168–78.

Hiscock, Peter. 2014. "Learning in Lithic Landscapes: A Reconsideration of the Hominid 'Toolmaking' Niche." *Biology Theory* 9:27–41.

Karlin, Claudine. 1991. "Connaissances et savoir-faire: Comment analyser un processus technique en préhistoire, introduction." In *Tecnología y Cadenas Operativaí Líticai, Treballi d'Arqueologia* 1:99–124.

Karlin, C., S. Ploux, P. Bodu, and N. Pigeot. 1993. "Some Socio-economic Aspects of the Knapping Process among Groups of Hunter-Gatherers in the Paris Basin Area." In *The Use of Tools by Human and Non-human Primates,* edited by A. Berthelet and J. Chavaillon, 318–40. Oxford: Clarendon Press.

Keller, C., and J. Keller. 1996. *Cognition and Tool Use: The Blacksmith at Work.* Cambridge: Cambridge University Press.

Kelly, Robert. 1995. *The Foraging Spectrum: Diversity in Hunter-Gatherer Lifeways.* Washington, D.C.: Smithsonian Institution Press.

Kempe, Marius, Stephen Lycett, and Alex Mesoudi. 2012. "An Experimental Test of the Accumulated Copying Error Model of Cultural Mutation for Acheulean Handaxe Size." *PLoS ONE* 7 (11): e48333.

Knecht, Heidi. 1993. "Early Upper Paleolithic Approaches to Bone and Antler Projectile Technology." In *Hunting and Animal Exploitation in the Later Palaeolithic and Mesolithic of Eurasia.* Archeological Papers of the American

Anthropological Association 4 (1): 33–47. Washington, D.C.: American Anthropological Association.

Kroeber, Alfred. 1940. "Stimulus Diffusion." *American Anthropologist* 42 (1): 1–20.

Kuhn, Steven L. 2012. "Emergent Patterns of Creativity and Innovation in Early Technologies." In *Creativity, Innovation, and Human Evolution,* edited by S. Elias, 69–87. Developments in Quaternary Science Volume 16. Amsterdam: Elsevier.

Lave, J., and E. Wenger. 1991. *Situated Learning: Legitimate Peripheral Participation.* New York: Cambridge University Press.

Lee, R., and I. DeVore, eds. 1976. *Kalahari Hunter-Gatherers: Regional Studies of the !Kung-San and Their Neighbors.* Cambridge, Mass.: Harvard University Press.

Lin, Sam C., Zeljko Rezek, David Braun, and Harold L. Dibble. 2013. "On the Utility and Economization of Unretouched Flakes: The Effects of Exterior Platform Angle and Platform Depth." *American Antiquity* 78 (4): 724–45.

Lycett, Stephen J. 2010. "Cultural Transmission, Genetic Models, and Palaeolithic Variability: Integrative Analytical Approaches." In *New Perspectives on Old Stones: Analytical Approaches to Paleolithic Technologies,* edited by S. J. Lycett and P. R. Chauhan, 207–34. New York: Springer Science+Business Media.

Lycett, Stephen J. 2015. "Cultural Evolutionary Approaches to Artifact Variation over Time and Space: Basis, Progress, and Prospects." *Journal of Archaeological Science* 56:21–31. doi:10.1016/j.jas.2015.01.004.

Lycett, Stephen J., Kerstin Schillinger, Marius Kempe, and Alex Mesoudi. 2015. "Learning in the Acheulean: Experimental Insights Using Handaxe Form as a 'Model Organism.'" In *Learning Strategies and Cultural Evolution during the Palaeolithic,* edited by A. Mesoudi and K. Aoki, 155–66. Replacement of Neanderthals by Modern Humans Series. Tokyo: Springer Japan.

Lycett, Stephen J., and N. von Cramon-Taubadel. 2008. "Acheulean Variability and Hominin Dispersals: A Model-Bound Approach." *Journal of Archaeological Science* 35:553–62.

Mace, R., and F. M. Jordan. 2011. "Macro-evolutionary Studies of Cultural Diversity: A Review of Empirical Studies of Cultural Transmission and Cultural Adaptation." *Philosophical Transactions of the Royal Society B* 366:402–11.

Machado, Jorge, Cristo M. Hernández, Carolina Mallol, and Bertila Galván. 2013. "Lithic Production, Site Formation, and Middle Palaeolithic Palimpsest Analysis: In Search of Human Occupation Episodes at Abric del Pastor Stratigraphic Unit IV (Alicante, Spain)." *Journal of Archaeological Science* 40 (2013): 2254–73.

Machado, Jorge, Francisco J. Molina, Cristo M. Hernández, Antonio Tarriño, and Bertila Galván. 2016. "Using Lithic Assemblage Formation to Approach Middle Palaeolithic Settlement Dynamics: El Salt Stratigraphic Unit X (Alicante, Spain)." *Archaeological and Anthropological Sciences.* doi:10.1007/s12520-016-0318-z.

Madsen, M., and C. Lipo. 2015. "Behavioral Modernity and the Cultural Transmission of Structured Knowledge: The Semantic Axelrod Model." In *Learning Strategies and Cultural Evolution during the Palaeolithic,* edited by A. Mesoudi and K. Aoki, 67–83. Tokyo: Springer Press.

Matthews, Luke J., Annika Paukner, and Stephen J. Suomi. 2010. "Can Traditions Emerge from the Interaction of Stimulus Enhancement and Reinforcement Learning? An Experimental Model." *American Anthropologist* 112 (2): 257–69.

Mauss, Marcel. 1935. "Body Techniques." In *Sociology and Psychology: Essays,* edited by M. Mauss, 97–135. London: Routledge and Kegan Paul. A 1979 English translation by B. Brewster of *Les Techniques du Corps, Journal de Psychologie* 32, no. 3–4 (1935): 271–93.

McGuigan, N. 2012. "The Role of Transmission Biases in the Cultural Diffusion of Irrelevant Actions." *Journal of Comparative Psychology* 126 (2): 150–60.

McPherron, Shannon P. 2000. "Handaxes as a Measure of the Mental Capabilities of Early Hominids." *Journal of Archaeological Science* 27:655–63.

Mesoudi, Alex. 2011. Variable Cultural Acquisition Costs Constrain Cumulative Cultural Evolution. *PLoS ONE* 6:e18239. doi:10.1371/journal.-pone.0018239.

Mesoudi Alex, and Michael J. O'Brien. 2008. "The Learning and Transmission of Hierarchical Cultural Recipes." *Biological Theory* 3:63–72.

Monnier, Gilliane F. 2006. "The Lower/Middle Paleolithic Periodization in Western Europe: An Evaluation." *Current Anthropology* 47 (5): 709–44.

Monnier, Gilliane F., and Kele Missal. 2014. "Another Mousterian Debate? Bordian Facies, chaîne opératoire Technocomplexes, and Patterns of Lithic Variability in the Western European Middle and Upper Pleistocene." *Quaternary International* 350:59–83.

Moore, Mark W., and Y. Perston. 2016. "Experimental Insights into the Cognitive Significance of Early Stone Tools." *PLoS ONE* 11 (7): e0158803. doi:10.1371/journal.pone.0158803.

Morgan, T. J. H., N. T. Uomini, L. E. Rendell, L. Chouinard-Thuly, S. E. Street, H. M. Lewis, C. P. Cross, C. Evans, R. Kearney, I. de la Torre, A. Whiten, and

K. N. Laland. 2015. "Experimental Evidence for the Co-evolution of Hominin Tool-Making Teaching and Language." *Nature Communications*. doi:10.1038/ncomms7029.

Nigst, Philip R. 2012. *The Early Upper Paleolithic of the Middle Danube Region.* Studies in Human Evolution Series. Leiden, The Netherlands: Leiden University Press.

Nigst, Philip R., Paul Haesaerts, Freddy Damblon, Christa Frank-Fellner, Carolina Mallol, Bence Viola, Michael Götzinger, Laura Niven, Gerhard Trnka, and Jean-Jacques Hublin. 2014. "Early Modern Human Settlement of Europe North of the Alps Occurred 43,500 Years Ago in a Cold Steppe-Type Environment." *Proceedings of the National Academy of Sciences of the United States 2014.* Early Ed. www.pnas.org/cgi/doi/10.1073/pnas.1412201111.

O'Brien, M. J., J. Darwent, and R. L. Lyman. 2001. "Cladistics Is Useful for Reconstructing Archaeological Phylogenies: Palaeoindian Points from the Southeastern United States." *Journal of Archaeological Science* 28:1115–36.

O'Brien, M. J., and R. L. Lyman. 2003. *Cladistics and Archaeology.* Salt Lake City: University of Utah Press.

Pelcin, Andrew. 1997. "The Formation of Flakes: The Role of Platform Thickness and Exterior Platform Angle in the Production of Flake Initiations and Terminations." *Journal of Archaeological Science* 24:1107–13.

Pelcin, Andrew. 1998. "The Threshold Effect of Platform Width: A Reply to Davis and Shea." *Journal of Archaeological Science* 25:615–20.

Pelegrin, Jacques. 1990. "Prehistoric Lithic Technology: Some Aspects of Research." *Archaeological Review from Cambridge* 9 (1): 116–25.

Perreault, Charles, and P. J. Brantingham. 2010. "Mobility-Driven Cultural Transmission along the Forager–Collector Continuum." *Journal of Anthropological Archaeology* 30:62–68.

Perreault, Charles, P. J. Brantingham, Steven L. Kuhn, Sarah Wurz, and Xing Gao. 2013. "Measuring the Complexity of Lithic Technology." *Current Anthropology* 54 (S8): S397–S406.

Pigeot, Nicole. 1987. *Magdaléniens d' Étiolles: Économie de Débitage et Organization Sociale.* XXVe Supplément à Gallia Préhistoire. Paris: Éditions du Centre National de la Recherche Scientifique.

Pigeot, Nicole. 1990. "Techniques and Social Actors: Flintknapping Specialists and Apprentices at Magdalenian Etiolles." *Archaeological Review from Cambridge* 9 (1): 126–41.

Pike, Kenneth L. 1967. *Language in Relation to a Unified Theory of the Structures of Human Behavior.* 2nd ed. The Hague: Mouton.

Ploux, S., and C. Karlin. 1993. "Fait technique et degré de sens dans l'analyse d'un processus de débitage magdalénien." *Techniques et culture* 21:61–78.

Plumettaz, Nicole. 2007. *Le site magdalénien de Monruz, 2: Étude des foyers à partir de l'analyse des pierres et de leurs remontages.* Neuchâtel : Service et musée contonal d'archéologie (Archéologie neuchâteloise 38).

Premo, Luke S. 2012a. "Local Extinctions, Connectedness, and Cultural Evolution in Structured Populations." *Advances in Complex Systems* 15 (1/2): 1150002-1-1150002-18.

Premo, Luke S. 2012b. "The Shift to a Predominantly Logistical Mobility Strategy Can Inhibit Rather than Enhance Forager Interaction." *Human Ecology* 40:647–49.

Premo, Luke S. 2015. "Mobility and Cultural Diversity in Central-Place Foragers: Implications for the Emergence of Modern Human Behavior." In *Learning Strategies and Cultural Evolution during the Palaeolithic,* edited by A. Mesoudi and K. Aoki, 45–65. Replacement of Neanderthals by Modern Humans Series. Tokyo, Japan: Springer.

Premo, Luke S., and Steven Kuhn. 2010. "Modeling Effects of Local Extinctions on Culture Change and Diversity in the Paleolithic." *PLoS ONE* 5 (12): e15582.

Premo, Luke S., and Jonathan B. Scholnick. 2011. "The Spatial Scale of Social Learning Affects Cultural Diversity." *American Antiquity* 76 (1):163–76.

Premo, Luke S., and Gilbert B. Tostevin. 2016. "Cultural Transmission on the Taskscape: Exploring the Effects of Taskscape Visibility on Cultural Diversity." *PLoS ONE* 11 (9): e0161766. doi:10.1371/journal.pone.0161766.

Read, Dwight W. 2007. *Artifact Classification: A Conceptual and Methodological Approach.* Walnut Creek, Calif.: Left Coast Press.

Read, Dwight W. 2011. *How Culture Makes Us Human: Primate Social Evolution and the Formation of Human Societies.* Walnut Creek, Calif.: Left Coast Press.

Reddy, Michael J. 1979. "The Conduit Metaphor: A Case of Frame Conflict in Our Language about Language." *Metaphor and Thought* 2:164–201.

Rezek, Z., S. Lin, R. Iovita, and H. L. Dibble. 2011. "The Relative Effects of Core Surface Morphology on Flake Shape and Other Attributes." *Journal of Archaeological Science* 38:1346–59.

Richerson, P. J., and R. Boyd. 1978. "A Dual Inheritance Model of the Human Evolutionary Process: I. Basic Postulates and a Simple Model." *Journal of Social and Biological Structures* 1:127–54.

Richerson, P. J., and R. Boyd. 2002. "Culture Is Part of Human Biology: Why the Superorganic Concept Serves the Human Sciences Badly." In *Probing Human*

Origins, edited by P. J. Richerson, 59–85. Cambridge, Mass.: American Academy of Arts and Sciences.

Richerson, P. J., and R. Boyd. 2005. *Not by Genes Alone: How Culture Transformed Human Evolution.* Chicago: University of Chicago Press.

Riede, Felix. 2006. "*Chaîne Opératoire, Chaîne Evolutionaire?* Putting Technological Sequences into an Evolutionary Perspective." *Archaeological Review from Cambridge* 21 (1): 50–75.

Riede, Felix. 2011. "Steps towards Operationalising an Evolutionary Archaeological Definition of Culture." In *Investigating Archaeological Cultures: Material Culture, Variability, and Transmission,* edited by B. W. Roberts and M. Vander Linden, 245–70. New York: Springer Science+Business Media.

Roche, H. 2005. "From Simple Flaking to Shaping: Stoneknapping Evolution among Early Hominins." In *Stone Knapping: The Necessary Conditions for a Uniquely Hominin Behaviour,* edited by V. Roux and B. Bril, 35–48. McDonald Institute Monographs. Cambridge: McDonald Institute for Archaeological Research.

Roux, V., and B. Brill, eds. 2006. *Stone Knapping: The Necessary Condition for a Uniquely Hominid Behaviour.* Cambridge: McDonald Institute for Archaeological Research.

Sackett, James R. 1990. "Style and Ethnicity in Archaeology: The Case for Isochrestism." In *The Uses of Style in Archaeology,* edited by M. Conkey and C. Hastorf, 32–43. Cambridge: Cambridge University Press.

Scerri, Eleanor M. L. 2013. "The Aterian and Its Place in the North African Middle Stone Age." *Quaternary International* 300:111–30.

Scerri, Eleanor M. L., Nick A. Drake, Richard Jennings, and Huw S. Groucutt. 2014. "Earliest Evidence for the Structure of *Homo Sapiens* Populations in Africa." *Quaternary Science Reviews* 101:207–16.

Scerri, Eleanor M. L., Brad Gravina, James Blinkhorn, and Anne Delagnes. 2016. "Can Lithic Attribute Analyses Identify Discrete Reduction Trajectories? A Quantitative Study Using Refitted Lithic Sets." *Journal of Archaeological Method and Theory* 23:669–91.

Schiffer, M. B. 1975. "Behavioral Chain Analysis: Activities, Organization, and the Use of Space." *Chapters in the Prehistory of Eastern Arizona IV, Fieldiana: Anthropology* 65:103–19. Chicago: Field Museum of Natural History.

Schiffer, M. B. 1976. *Behavioral Archaeology.* New York: Academic Press.

Schiffer, M. B. 1996. "Some Relationships between Behavioral and Evolutionary Archaeologies." *American Antiquity* 61:643–62.

Schiffer, M. B. 1999. *The Material Life of Human Beings: Artifacts, Behavior, and Communication.* With Andrea R. Miller. New York: Routledge.

Schiffer, M. B., and J. Skibo. 1997. "The Explanation of Artifact Variability." *American Antiquity* 62:27–50.

Schillinger, Kerstin, Alex Mesoudi, and Stephen J. Lycett. 2014. "Copying Error and the Cultural Evolution of 'Additive' vs. 'Reductive' Material Traditions: An Experimental Assessment." *American Antiquity* 79 (1): 128–43.

Shea, John J. 2013. "Lithic Modes A–I: A New Framework for Describing Global-Scale Variation in Stone Tool Technology Illustrated with Evidence from the East Mediterranean Levant." *Journal of Archaeological Method and Theory* 20:151–86.

Shea, John J. 2014. "Sink the Mousterian? Named Stone Tool Industries (NASTIES) as Obstacles to Investigating Hominin Evolutionary Relationships in the Later Middle Paleolithic Levant." *Quaternary International* 350:169–79.

Shea, John J. 2015. "Making and Using Stone Tools: Advice for Learners and Teachers and Insights for Archaeologists." *Lithic Technology* 40 (3): 231–48.

Shea, John J. 2017. *Stone Tools in Human Evolution: The Archaeology of Behavioral Differences among Technological Primates.* Cambridge: Cambridge University Press.

Shipton, Ceri. 2010. "Imitation and Shared Intentionality in the Acheulean." *Cambridge Archaeological Journal* 20:197–210.

Shott, Michael. 2003. "*Chaîne Opératoire* and Reduction Sequence." *Lithic Technology* 28 (2): 95–105.

Skibo, James M., and Gary M. Feinman, eds. 1998. *Pottery and People: A Dynamic Interaction.* Salt Lake City: University of Utah Press.

Smith, Tanya, Paul Tafforeau, Donald J. Reid, Rainer Grün, Stephen Eggins, Mohamed Boutakiout, and Jean-Jacques Hublin. 2007. "Earliest Evidence of Modern Human Life History in North African Early Homo Sapiens." *Proceedings of the National Academy of Sciences* 104 (15): 6128–33.

Soressi, Marie, and Jean-Michel Geneste. 2011. "The History and Efficacy of the *Chaîne Opératoire* Approach to Lithic Analysis: Studying Techniques to Reveal Past Societies in an Evolutionary Perspective." *PaleoAnthropology* 2011:334–50.

Sterelny, Kim. 2012. *The Evolved Apprentice: How Evolution Made Humans Unique.* Cambridge, Mass.: MIT Press.

Tehrani, Jamshid J., and Felix Riede. 2008. "Towards an Archaeology of Pedagogy: Learning, Teaching, and the Generation of Material Culture Traditions." *World Archaeology* 40 (3): 316–31.

Tennie, Claudio, Josep Call, and Michael Tomasello. 2009. "Ratcheting up the Ratchet: On the Evolution of Cumulative Culture." *Philosophical Transactions of the Royal Society B* 364:2405–15.

Tennie, Claudio, L. S. Premo, David R. Braun, and Shannon P. McPherron. 2017. "Early Stone Tools and Cultural Transmission: Resetting the Null Hypothesis." *Current Anthropology* 58 (5): 652–72.

Thomas, David Hurst. 1998. *Archaeology*. 3rd ed. Fort Worth: Harcourt College.

Tomasello, Michael. 1996. "Do Apes Ape?" In *Social Learning in Animals: The Roots of Culture,* edited by C. M. Heyes and B. G. Galef Jr., 313–46. New York: Academic Press.

Tomasello, Michael, Malinda Carpenter, Josep Call, Tanya Behne, and Henrike Moll. 2005. "Understanding and Sharing Intentions: The Origins of Cultural Cognition." *Behavioral and Brain Sciences* 28:675–735.

Tomasello, Michael, Alicia P. Melis, Claudio Tennie, Emily Wyman, and Esther Herrmann. 2012. "Two Key Steps in the Evolution of Human Cooperation: The Interdependence Hypothesis." *Current Anthropology* 53 (6):673–92.

Tostevin, Gilbert B. 2000. "Behavioral Change and Regional Variation across the Middle to Upper Paleolithic Transition in Central Europe, Eastern Europe, and the Levant." PhD diss., Harvard University, Cambridge, Mass.

Tostevin, Gilbert B. 2003a. "Attribute Analysis of the Lithic Technologies of Stránská Skála II–III in Their Regional and Inter-regional Context." In *Stránská Skála: Origins of the Upper Paleolithic in the Brno Basin,* edited by J. Svoboda and O. Bar-Yosef, 77–118. American School of Prehistoric Research Bulletin 47. Dolní Věstonice Studies 10. Cambridge, Mass.: Peabody Museum of Archaeology and Ethnology, Harvard University.

Tostevin, Gilbert B. 2003b. "A Quest for Antecedents: A Comparison of the Terminal Middle Paleolithic and Early Upper Paleolithic of the Levant." In *More than Meets the Eye: Studies on Upper Palaeolithic Diversity in the Near East,* edited by A. N. Goring-Morris and A. Belfer-Cohen, 54–67. Oxford: Oxbow Press.

Tostevin, Gilbert B. 2007. "Social Intimacy, Artefact Visibility, and Acculturation Models of Neanderthal-Modern Human Interaction." In *Rethinking the Human Revolution: New Behavioural and Biological Perspectives on the Origins and Dispersal of Modern Humans,* edited by P. Mellars, K. Boyle, O. Bar-Yosef, and C. Stringer, 341–57. McDonald Institute Research Monograph. Cambridge: McDonald Institute for Archaeological Research.

Tostevin, Gilbert B. 2009. "The Importance of Process and Historical Event in the Study of the Middle to Upper Paleolithic Transition." In *Transitions in*

Prehistory: Papers in Honor of Ofer Bar-Yosef, edited by J. Shea and D. Lieberman, 166–83. American School of Prehistoric Research Monograph. Oxford: Oxbow Books.

Tostevin, Gilbert B. 2011a. "Introduction to the Special Issue: Reduction Sequence, *Chaîne Opératoire,* and Other Methods: The Epistemologies of Different Approaches to Lithic Analysis." *PaleoAnthropology* 2011:293–96.

Tostevin, Gilbert B. 2011b. "Levels of Theory and Social Practice in the Reduction Sequence and Chaîne Opératoire Methods of Lithic Analysis." *PaleoAnthropology* 2011:351–75.

Tostevin, Gilbert B. 2012. *Seeing Lithics: A Middle-Range Theory for Testing for Cultural Transmission in the Pleistocene.* American School of Prehistoric Research Monograph Series. Oxford and Oakville, Conn.: Peabody Museum, Harvard University, and Oxbow Books.

Tostevin, Gilbert B. 2013. "Applying Evolutionary Development to Lithic Technology: The Generative Entrenchment of Blade Core Technology." Poster presented at the Paleoanthropology Society Meeting, Honolulu, April 2–3.

Tostevin, Gilbert B., and Petr Škrdla. 2006. "New Excavations at Bohunice and the Question of the Uniqueness of the Type-Site for the Bohunician Industrial Type." *Anthropologie* (Brno) 44 (1): 31–48.

Tschauner, H. 1994. "Archaeological Systematics and Cultural Evolution: Retrieving the Honour of Culture History." *Man* 29 (1): 77–93.

Turq, Alain, Wil Roebroeks, Laurence Bourguignon, and Jean-Philippe Faivre. 2013. "The Fragmented Character of Middle Palaeolithic Stone Tool Technology." *Journal of Human Evolution* 65:641–55.

Van der Leeuw, Sander E. 2004. "Why Model?" *Cybernetics and Systems: An International Journal* 35:117–28. doi:10.1080/01969720490426803.

Whiten, Andrew, Nicola McGuigan, Sarah Marshall-Pescin, and Lydia M. Hopper. 2009. "Emulation, Imitation, Over-Imitation, and the Scope of Culture for Child and Chimpanzee." *Philosophical Transactions of the Royal Society B* 364:2417–28.

Whiten, Andrew, Kathy Schick, and Nicholas Toth. 2009. "The Evolution and Cultural Transmission of Percussive Technology: Integrating Evidence from Palaeoanthropology and Primatology." *Journal of Human Evolution* 57:420–35.

Wiessner, Polly W. 1982. "Risk, Reciprocity, and Social Influences on !Kung San Economics." In *Politics and History in Band Societies,* edited by E. Leacock and R. B. Lee, 61–84. Cambridge: Cambridge University Press.

Wiessner, Polly W. 1983. "Style and Social Information in Kalahari San Projectile Points." *American Antiquity* 48:253–76.

Wiessner, Polly W. 1984. "Reconsidering the Behavioral Basis for Style: A Case Study among the Kalahari San." *Journal of Anthropological Archaeology* 3:190–234.

Wiessner, Polly W. 2002. "Hunting, Healing, and Hxaro Exchange: A Long-Term Perspective on !Kung (Ju/'hoansi) Large-Game Hunting." *Evolution and Human Behavior* 23:407–36.

Wiessner, Polly W. 2014. "Embers of Society: Firelight Talk among the Ju/'hoansi Bushmen." *Proceedings of the National Academy of Sciences of the United States* 111(39): 14027–35. www.pnas.org/cgi/doi/10.1073/pnas.1404212111.

Wimsatt, W. C. 2013. "Articulating Babel: An Approach to Cultural Evolution." *Studies in History and Philosophy of Science Part C: Studies in History and Philosophy of Biological and Biomedical Sciences* 44 (4): 563–71.

Wimsatt, William C., and James R. Griesemer. 2007. "Re-producing Entrenchments to Scaffold Culture: The Central Role of Development in Cultural Evolution." In *Integrating Evolution and Development: From Theory to Practice,* edited by R. Sansome and R. Brandon, 228–323. Cambridge, Mass.: MIT Press.

Wobst, H. M. 1977. "Stylistic Behavior and Information Exchange." In *For the Director: Essays in Honor of James B. Griffin,* edited by C. Cleland, 317–42. Anthropological Papers of the University of Michigan 61. Ann Arbor: Regents of the University of Michigan.

Wolcott, Harry. 1991. "Propriospect and the Acquisition of Culture." *Anthropology and Education Quarterly* 22:251–73.

Wynn, Thomas. 2002. "Archaeology and Cognitive Evolution." *Behavioral and Brain Sciences* 25:389–438.

Wynn, Thomas, and Frederick L. Coolidge. 2004. "The Expert Neandertal Mind." *Journal of Human Evolution* 46:467–87.

Yellen, John. 1977. *Archaeological Approaches to the Present: Models for Reconstructing the Past.* Vol. 1. New York: Academic Press.

9

THE EVOLUTION OF LANGUAGE
AS TECHNOLOGY
The Cultural Dimension

SALIKOKO S. MUFWENE

DEBATES ABOUT THE phylogenetic emergence of language have generally included the question of whether this protracted process was driven by biology or by culture, as if the processes associated with one or the other were mutually exclusive. For some, such as Bickerton (1990, 2010) and Chomsky (2010), hominines could not have developed language without first acquiring a "biological endowment for language"—also called *Universal Grammar* (UG) or the "language organ," which was originally identified as the *language acquisition device* (LAD). It is presumed to have facilitated the emergence of language, which, for them, was a saltatory event. For others, such as Evans and Levinson (2009) and Everett (2012), this evolution is primarily cultural, as it depends on learning by inference and proceeds faster than biological evolution.[1]

I submit that both biological and cultural evolutions are equally predicated on the conditions articulated by Lewontin (1970): (1) variation; (2) heredity/inheritance, which presupposes multiple generations, with the later ones inheriting genes or learning techniques (or construction materials) from earlier ones; and (3) differential reproduction, with the later generations exhibiting different genetic recombinations and thus producing different organisms or reproducing different variants of their culture. There are indeed differences in the specific ways that materials and information are "transmitted" or "inherited." In biology it is literally through the transmission of units, whereas in culture transmission is through learning, by inference in many cases (Atran and Sperber 1991; Mufwene 2001). Additionally, there are interspecific differences within both biology and culture that reflect differences in the ontogenetic properties of particular species or cultural domains (see also Wimsatt, chapter 1). For instance, practices in material culture, such as weaving or face painting, are not learned in exactly the same ways as those in nonmaterial culture, such as religion or governance. However, it is not evident that one must posit a theory of cultural evolution that is so different from that of biological evolution that one would have to claim either that language is only the outcome of biological evolution or that it is exclusively a product of cultural evolution.

I argue below that the phylogenetic emergence of language presupposed a particular stage of biological evolution. It occurred after hominines were endowed with a particular mental capacity that generated (more) complex thinking, greater need to domesticate their natural ecology, and larger and more social organizations. The same mental capacity also exerted more pressure to exchange rich and diversified information explicitly and to expedite the growth of knowledge. However, languages are cultural phenomena on a par with others such as religion, hunting practices, farming, and folk music. The basic and nonspecialized aspects are typically learned by inference and thus with modification. The essence of vernacular linguistic systems (used for day-to-day communication and learned before one is taught the standard variety of their language in school) is learned the same way, by inference, piecemeal, incrementally, and from interacting with others. Consequently, languages exhibit characteristics associated with "cultural evolution," particularly horizontal transmission, imperfect replication, and fast rate of change (Mufwene 2001, 2017).

Below, I approach the subject matter in the following order: In part 2, I introduce the conception of language as communication technology and therefore as a cultural artifact. I use it to show that the debate over whether language evolution is biological or cultural has been framed inadequately. I argue that cultural evolution itself presupposes a particular stage of biological evolution, which sets humans apart from other primates. Biological evolution produced a brain that was not only language-ready but also culture-ready. That is, after reaching a certain evolutionary stage, the hominine brain was capable of mental activities that produced not only language but other cultural phenomena not observable among nonhuman primates and other animals.

In part 3, I argue against positing UG or a language organ as the prerequisite for the emergence of language. One would otherwise have to posit similar constructs for the emergence of other cultural phenomena, such as music and social organization. The mind (interpreted here as the state of the brain in activity) appears to operate in a more economical way than suggested by phrenology, with some mechanisms, such as syntax and recursion, applying also outside language. I also elaborate on the idea of language as technology, which enables me to further flesh out the cultural aspects of language. In addition, I show how the specific materials used, such as sounds or manual signs, impose specific constraints on how the technology can be developed.

In part 4, I explain the particular role that naming must have played in the phylogenetic emergence of language, as it facilitates communication also about the past and the future. It actually drove the expansion of phonetic inventories in different languages. I show again how culture is a consequence of the particular way in which a population does things and should not be the explanation of how languages evolved. In part 5, I articulate the role played by Generative Entrenchment (Wimsatt 2000) and by successive scaffolding (Wimsatt and Griesemer 2007) in the gradual emergence of language as communication technology. I conclude the chapter in part 6.

BIOLOGY AND CULTURE ARE NOT MUTUALLY EXCLUSIVE IN THE EVOLUTION OF LANGUAGE

Along with scholars such as Jackendoff (2010) and Sperber and Origi (2010), I argue that biology and culture are not mutually exclusive in the phylogenetic emergence of language. While biological evolution generated the hominine "language-ready" brain (Arbib 2012), the latter produced languages that, because they vary from one population to another, are also characterized as culture-specific. Note that culture, as explained below, is not antecedent to language if we interpret it, roughly, as the particular ways in which members of a population behave and do things conventionally.[2]

The cultural fold of a language lies in the specific way that the particular population that has developed it as its communication technology has shaped it (viz., its phonemes, morphemes, words, and the relevant norms of usage) at variance with the ways other populations have done theirs (Mufwene 2013a). This is indeed comparable to, for example, two populations that have developed knowledge to protect themselves from elements in nature but have not used exactly the same materials nor produced the same styles for their clothing and shelters. Such differences occur not only due to alternative ways of solving the same problems but also because the challenges to which they respond are not identical. Typological variation among languages reflects this state of affairs. Conceiving of languages as communication technologies helps to address the question of the role of biology and the significance of culture in the phylogenetic emergence of languages, without suggesting that there is a cultural evolution that is fundamentally different from biological evolution.

A first step in connecting the biological and cultural aspects of the phylogenetic emergence of language consists of addressing the fallacy of the phrase *language and/in culture*. It is certainly not the same as *language and/in society*. We must ask what *culture* is and whether it has some existence prior to how members of a population behave and do things. This question is related to whether or not cultures are static or dynamic. In my view, populations shape their cultures as they behave and do things; as they develop or borrow new ways of growing food, or cooking and eating meals, or dressing and protecting themselves from the elements. These are the kinds of changes that encourage us to say that a population has changed its culture or that the culture has evolved.

However, does a population change its culture deliberately? Or do changes often occur undetected, with its members noticing them in hindsight? Both kinds of changes occur, but the latter is probably more pervasive. The reason the changes are detected in hindsight appears to be because of untutored social learning, which is the typical pattern in folk culture. As noted above, the learning proceeds by inference, based on observing other members of the population that have experience in what they do, and its outcome is typically imperfect replication. Changes are the outcome of the cumulation of (often minor) details that are modified during the learning and/or execution process.

Culture is dynamic; it is constantly reshaped by its practitioners as they do things, express their beliefs, and behave with or act toward one another under current ecological pressures. Culture is not knowledge, which is precisely why we can speak of *knowing a culture*. It is practice, and practices are shaped in part by learning from other members of the population. Knowledge consists of representations or schemas about how to behave on particular occasions, how to do things, or how to interpret the universe and life (frequently formulated unconsciously) and thus how to practice a culture. On the other hand, how did the initial patterns we find in a culture emerge? We can address this question by singling out particular cultural practices, such as building dwellings, clothing ourselves, organizing ourselves socially (e.g., into nuclear or extended families), and communicating with each other in a particular language.

As cultural phenomena, languages also fall into the category of practice and behavior, consistent with the new wave of quantitative sociolinguistics and with linguistic anthropology. There, it has become customary to speak of *communities of practice* shaped by actual interactions (Eckert and McConnell-Ginet 1992). This is different from the traditional terms *language community* or *speech community,* which are defined by the potential that members of a population have to interact with other members. In a community of practice, the members shape their norms through their interactions and are not assumed to have simply inherited them from previous speakers. Their interactions also define their communities.

Communication as transfer or exchange of information remains a constant in this approach to culture. The pressure to communicate more information and in the most satisficing way is part of what, from the evolutionary perspective, triggered the expansion of the vocabulary and of linguistic structures. This still happens today in the ontogenetic development of language, from child-like to adult-like communication. The structural expansion may also involve exaptations of current structures, as can still be observed now in grammaticization processes, such as when the motion verb *go* is co-opted to also function as a marker of *future* in *be going to* + VERB.[3] All these changes cumulate into evolution, assuming the phylogenetic emergence of language was incremental (Mufwene 2013a), as I show below in part 4.

I conceive of languages as technologies for transmitting information (McArthur 1987; Koster 2009; Lee et al. 2009; Everett 2012; Mufwene 2013a; Dor 2015). Like computers, they are technologies of a mixed kind, consisting of physical units (vocal or signed) and nonphysical elements (semantic units and principles called *rules* or *constraints* on many levels: phonology, morphology, and syntax). Hominines developed them to solve a problem: how to convey even complex information or knowledge explicitly from one mind to another (Arthur 2009) and with high fidelity in transmission (Morgan et al. 2015).[4] In so doing, cooperation was enhanced; knowledge grew more rapidly at both the individual and the communal levels, as innovations could be shared and spread, thanks to the world-creating capacity of language in narratives (Mufwene 2015). According to Morgan et al. (2015), the emergence of symbolic communication, then certainly still distant from the earliest phonetic forms of communication (assumed to have started only around two hundred thousand years ago),[5] helped hominines evolve from the seven hundred thousand-year stasis of Oldowan toolmaking technology to the more complex Acheulean technology (about 1.7 mya). The transition must have required teaching the relevant knowledge and making more explicit to learners the different steps involved in manufacture.

Languages are also like other emergent, collective, and cumulative folk technologies in the sense that they have not been produced by elite groups of thinkers, in a laboratory, and then taught to others after testing how well they work. Languages have evolved piecemeal and incrementally, especially when one focuses on the principles and constraints followed by speakers in combining sounds into words, and words into phrases and sentences, in developing their linguistic systems. Anybody that has the capacity to innovate and produce an utterance successfully has the potential of contributing to the emergent system of their group. As the population thrives (if it is not overtaken by another), their language evolves in response to pressures to also meet its novel communicative needs, which keep arising from changes in their universe of experience or imagination.

Thus, languages are adaptive technologies in ways comparable to expansions in social organizations or the growing complexity of material technologies, such as computers or airplanes. Although a great amount of explicit thinking was engaged in the production of computers and airplanes, these technologies—like languages—grow organically akin to folk technology (i.e., by additive collective actions and cumulatively). Thus, it is for a good reason that some emergentists characterize languages as complex adaptive systems (e.g., Beckner et al. 2009; Cornish, Tamariz, and Kirby 2009; Steels 2000; Lee et al. 2009; Kretzschmar 2015; Massip-Bonet 2012; Mufwene, Coupé, and Pellegrino 2017).

Focusing on the relation of languages to cognition, many linguists have preferred to characterize them as representation systems, thus as sorts of structured snapshots of their speakers' knowledge of the world (e.g., Bickerton 1990). To be sure, this characterization is not false, as languages *do* have a multifaceted architecture and convey information about diverse cognitive domains. However, the representation-system facet appears to be a consequence of the particular ways in which chunks of information are packaged for transmission in a language. The packages vary from one population to another, for instance, whether speakers use one or two separate words for siblings, differentiated by gender (as in European languages, viz., *brother* vs. *sister* in English) or by age (as in Bantu languages, viz., *yaya* 'older sibling' vs. *leke* 'younger sibling' in Kikoko-Kituba). They may also vary depending on whether siblings must be distinguished from cousins (as in European languages, viz., *brother/sister* vs. *cousin* in English) or lumped together in the same kinship category (as in Bantu languages, availing typically the same term as for siblings). Similar cross-linguistic variation applies to whether all nouns must combine with a classifier when they co-occur with a demonstrative or a quantifier (as in Chinese, e.g., 5/THIS + CLASSIFIER + BOOK). Sometimes there is variation even among those claiming to speak the same language—for instance, whether or not one should speak of *heads of children* in the same way they speak of *heads of cattle*.

The semantics of languages on both lexical and sentence levels vary cross-culturally, just like the physical components of their architectures (viz., their phonologies, morphologies, and syntaxes). In this respect, they are also like other cultural artifacts, as different populations do not cook in identical ways, build their dwellings in identical fashions, or clothe themselves in identical styles. Although the materials used and the purposes of their practices may be the same, their implementations vary, just like the ways that, for instance, cars and computers are made, not according to exactly the same design from one manufacturer to another. Languages are thus cultural phenomena, like cooking, dwellings, clothing, religions, and a host of other cultural products, although there are systemic and complexity differences that are consequences of differences in the ontological properties of cultural phenomena. Together, they are constructed as cultures, which distinguish us from other animals, including the great apes, which are assumed to be anatomically and mentally the closest to mankind.

Species-wise, the minds that produced human languages reflect a specific and common stage of hominine biological evolution, especially that of the brain. Nonetheless, the languages they produced are cultural artifacts because these also reflect particular ways of behaving and doing things that vary from one population to another. Consequently, it is inaccurate to speak of *language and/in culture* because the phrase implies that culture is separate from language. If anything, a language as technology contributes to defining the culture of a particular population. If it holds a distinctive status in society, it is simply because it enables the production of more knowledge and of other cultural phenomena that presuppose communication.

Morgan et al. (2015) provide a good example of this in their discussion of the transition from Oldowan to Acheulean stone technology. Although still in its preprimordial stages (if we focus on phonetic communication), symbolic language appears to have facilitated both the innovation of Acheulean technology and how fast it apparently spread within *Homo erectus*. In modern times, language has been critical to the transmission of complex cultural knowledge through teaching (see also Wimsatt, chapter 1; Tostevin, chapter 8), especially in the case of specialized professional skills. Examples include weaving and knitting, some aspects of gardening, building animal traps, manufacturing hunting tools, and witchcraft, although one can argue that teaching has not involved all the details of the skills; some are still learned by inference, leaving room for both variation and innovation.

This line of reasoning prompts us to question an oft-repeated claim that language is what makes humans unique in the animal kingdom, in part because it enables us to express complex and abstract thoughts. There are many cultural phenomena besides language that distinguish us from other animals (Mufwene 2013a, 2015). For instance, we cook or process food items (e.g., by seasoning, marinating, drying, or smoking them); we clothe ourselves (although among some humans it is just a matter of covering the genitals); we hold religious beliefs (including atheism); we build dwellings that are adapted to our residential and mobility patterns; we have various levels of social organization beyond the nuclear family and stricter norms against incest; we have political organizations and trade practices; and we resort to a wide range of material technologies (however primitive) to solve practical problems.[6]

All these peculiarities suggest that something more fundamental than language distinguishes us from all other animals—namely, the human mind. If the human brain is anatomically still very similar to that of nonhuman primates (as made evident by, for instance, the behaviors of mirror neurons [see, e.g., Gallese and Goldman 1998; Arbib 2012], our cognitive capacity appears to be exponentially different from theirs. As part of the mind, it drove the emergence of language and other cultural phenomena in our species, although language may stand out simply because it has enabled innovations of complex technologies (Morgan et al. 2015), has prevented most of us from having to reinvent the wheel, and has enabled the rapid spread and growth of knowledge in mankind.

As I am reminded by Alan Love (personal communication, November 23, 2015), the relationship between language and other cultural phenomena is quite complex. For instance, while it is true that language has played an important role in the emergence of political and administrative organizations, it has also been pointed out that changes in social organization, such as extended-family and larger hunter–gatherer groups, must have exerted important ecological pressures in the phylogenetic emergence of language. Nonetheless, these social organizations can hardly be sustained without efficient communication much more explicit and informative than, for instance, the nonlinguistic communication means used by humans themselves and other primates. To date, our social structures still scaffold the ontogenetic development of language as practice and as system. From an emergentist perspective, linguistic systems can be claimed to have emerged as consequences of repeated instances of phonetic or manual communication.[7]

A feedback loop appears to have emerged too (consistent with Odling-Smee, Laland, and Feldman's [2003] idea of "niche construction"), as language appears to have expedited the expansion of the other cultural domains, and these in turn exerted pressures on language to expand accordingly. The common producer of all these cultural phenomena is the human mind—that is, the state of activity in which the brain is engaged. One may invoke the human mental capacity, too, though it is not clear to me what difference or improvement the alternative wording makes. In any case, what is important is the observation that it is not language that makes us uniquely human. Collectively, human cultures do. From a reductionist perspective, the mind, which produced them, distinguishes us from other animals. Language is only one of the many relevant cultural phenomena.

It is at this juncture that we must discuss the role of biology or, more specifically, that of a brain architecture capable of cognitive capacities achieved only by hominines at some specific stages of their evolution since, probably, *Homo erectus,* in the emergence of human cultures. Assuming polygenesis (Mufwene 2013a, 2013c), it appears that all the cultural phenomena mentioned above emerged at more or less the same phylogenetic time in different hominine colonies, during or after the emergence of the 1–2 percent of genetic materials that distinguish us from chimpanzees. The significance of the biological infrastructure lies in producing those critical peculiarities of the hominine brain circuitry, apparently located in the cortex (Lieberman 2012), that generate a mental capacity able to situate events in the past, present, or future (Corballis 2011), thus capable of foresight and planning.

This capacity is reflected in our narratives, in which we can navigate in the present, past, and future (Mufwene 2015). Language is the kind of technology that hominines produced to share knowledge, feelings, attitudes, dispositions, and plans. Note that, as explained by Arthur (2009), technology need not be material, monolithic, or planned; it can become complex by the accumulation of contributions from different members of a population. In this sense, religions and myths, too, may be considered technologies, just like scientific hypotheses, as they both help make sense of the world.

In more or less the same way as computers, which deserve this name only when both their hardware and software are taken into account, languages are hybrid technologies. They have been useful in helping hominines evolve more explicit and more reliable communication, from the point of view of transmission fidelity, not only about the present and the observable but also about the absent and the imaginable. Languages have a world-creating capacity—evident in narratives—that has generated both myths and scientific discourse (Mufwene 2015); they have also evolved complex architectures that meet the communicative needs of the communicators. The ability of these technologies to convey complex knowledge about the past, present, or future, or to express feelings and sensations, or to make requests or impart orders or instructions, is commensurate with the level of cognitive development in the communicators, both phylogenetically and ontogenetically. Note that child language is less complicated than adult language largely because the child has less complex information to convey.

UNIVERSAL GRAMMAR (UG) AND THE EMERGENCE OF LANGUAGE

The position that UG guided or drove the phylogenetic emergence of languages has been disputed by some linguists, including those cited at the outset of this chapter and linguists closer to Chomsky in spirit, such as Jackendoff (2010). The strongest evidence may come from those modeling language evolution (e.g., Steels 2011, 2012), who can get their models to produce some aspects of language, including syntax, without a counterpart of the putative UG. We can thus safely conclude that what is captured by UG is a consequence of the relative uniformity in the way that the similar brains generating similar minds at various stages of hominine evolution have produced the same fundamental basic architecture in the mechanics of languages. This occurred despite the variation in the ways that different populations selected their phonetic inventories, developed constraints on how to combine them into words (phonology) and into sentences (syntax), and so on.

The cross-community variation evident in all modules of the architecture of language (viz., phonology and morphology, which work in the lexicon; syntax, which regulates the structure of sentences; and semantics, which applies to both the lexicon and sentences) is comparable to what is observable in the development of several other technologies. Consider, for instance, the algorithms that run the operating systems of Apple and Microsoft computers. They are divergent but do similar jobs for the consumer. In all such technologies, the fundamental principles are nonetheless similar, not because there is a special UG-like mechanism that generated them but simply because the material used imposes its own constraints on how a particular technology can be developed (Mufwene 2013a).

Working with sounds imposes strict linearity, as one cannot produce two phonetic sounds simultaneously. For this reason, different populations have developed conventions on acceptable combinations thereof in the words of their languages and on how the words can be combined into larger phrases, including sentences. Thus, a German word can start with the consonant clusters /ps/ or /ts/, as in *Psychologie* and *Zug* 'train,' which are not allowed in English at the beginning of English words. The first syllable of *psychology* in English is pronounced /saiᵢ/, not */psaiᵢ/. German requires that the auxiliary verb be extrapolated to the end of the subordinate clause, whereas such a construction would be ungrammatical in English: for instance, *den Mann den Ich gesehen habe* 'the man whom I have seen' but not *the man whom I seen have*. There are languages that, unlike English, start their sentences with a verb, whereas some others end them with a verb. These kinds of typological variation reveal the cultural dimension of language evolution, amounting simply to how particular populations chose to develop their communication technologies in their own ways. They vary without violating the fundamental principle of linearity (viz., sounds follow each other, and words follow each other) or that of combining the sounds into larger and larger units (words and phrases), called *duality of patterning* (Hockett 1959; see below).

The rigid linearity attested in spoken languages is a consequence of the fact that the mouth can produce only one sound at a time. Thus, syntax, which starts at the level of combinations of sounds into words, is attested in the phonology, morphology, and "syntax" modules of the architecture of language. It is a consequence of the linearity imposed by the material. So are the other aspects of syntax, in the traditional sense of the term in linguistics, which have to do with identifying constituents and dependency relations between constituents, as captured by phrase structures or agreement in case marking.[8] Communal norms emerge because members of the relevant population of speakers converge on which particular combinations of sounds, morphemes, and words yield acceptable utterances (of various lengths) and which ones do not. Thus, grammars reflect these norms, also characterized as *conventions* in linguistics, which are cultural peculiarities.

Recursion, which, since Hauser, Chomsky, and Fitch (2007), has generated so much controversy about whether it is a peculiarity of languages alone, is a practice that reduces the number of different kinds of units and structures that speakers use in communication. Implementing economy in the system, it enables usage of the same structure or kind of construction several times over at different levels, just like some formulae in algebra. It makes it possible to produce longer and more complex utterances without increasing the number of grammatical rules. Although several examples of recursion outside language or computer algorithms can be cited, the practice underscores again the role of the same mental capacity in solving problems in various human productions that constitute cognition and culture. Thus, recursion is far from being an exclusive peculiarity of language or UG (Lieberman 2012). As a matter of fact, it reflects how the mind works. Christiansen and Chater (2015) argue that this strategy must have "piggyback[ed] on domain-general sequence learning abilities" (11) that hominines evolved before the emergence of language; nonhuman primates are apparently not capable of it, at least not to the same extent as humans. Language "is subserved by the same neural mechanisms as used for sequence processing" (5).

That the material used in a particular technology acts as a constraint finds evidence in some differences between spoken and sign languages. Since sign languages use hands as articulators, which are larger than speech organs and are used in a much larger space, communication would be much slower if they were structured in a strictly linear way. Signers take advantage of the tridimensional space in which the hands move and can incorporate several kinds of information into one signed word. Thus, multiword English expressions, such as *rapidly slither up/down* or *slowly wiggle one's way,* can be signed in single words. This peculiarity of sign languages, known as *incorporation,* enables signing to be as fast as speech. As a matter of fact, I contend in Mufwene (2013a) that phonology and morphology are conflated into one module in signed languages, without losing the particularity of duality of patterning, which Hockett (1959) invoked as typical of human languages.

Duality of patterning, which Martinet (1960) identified as "double articulation," is a misnomer for the fact that in a spoken language words can be broken down into meaningless sounds, composite words into morphemes, phrases into words, sentences into phrases, and so on. *Duality* is a misnomer for what turns out to be several levels, not just two, in a hierarchical structure. On each level, the smaller, lower-level units make it possible to differentiate two sequences that could otherwise be confused—for instance, the contrast between /t/ and /p/ in the words *tear* versus *pear/pair* or that enabled by the suffixes {-d} and {-z} in the pair *legged* versus *legs* or the opposition permitted by *top* and *leg* in *tabletop* versus *table leg*. Sign languages still exhibit similar contrasts, on different levels, even if one does not posit a phonological module that is distinct from a morphological one. The smallest unanalyzable units in sign languages are not meaningless and are indeed fewer in number than in the phonetic inventory of a spoken language. Along with the more numerous complex units, they correspond to morphemes in spoken languages.

The material-related constraints invoked here regarding how one can package information in language are indeed reminiscent of those one can observe in other cultural domains. For instance, how one can produce a chair varies depending on whether the material used is natural wood, wicker, plastic, or metal. Likewise, how one plays the American national anthem varies depending on the specific instrument used, such as the flute as opposed to the saxophone or the piano, just like its vocal production varies depending on whether the singer is an alto or soprano (and in this respect interindividual variation in the specifics of the buccopharyngeal structure of the singers is a relevant factor). The shapes of dwellings have changed significantly from the primordial constructions with tree branches and leaves, through mud-wall and thatched-roof houses, all the way to brick constructions and steel and glass skyscrapers. Even the choice of logs, bricks, or stones alone as materials for walls imposes different constraints on the latitude that the builder has regarding the shape of a house. These examples all provide evidence for arguing that languages are technologies, and their grammars are in some ways consequences of the specific materials used to package information, viz., sounds or manual signs.

The study of the emergence of language as communication technology entails focusing on how the technology evolved, through successive exaptations of the anatomy and of current structures, driven by increasing ecological pressures for more and more complex communication permitting the higher-fidelity transmission of information.

SOME CONJECTURES ON THE PHYLOGENETIC EMERGENCE OF LANGUAGE AS COMMUNICATION TECHNOLOGY

What I present below is very much inspired by how children learn language, though I do not subscribe to the position that ontogeny recapitulates phylogeny. I hypothesized in Mufwene (2013a) that the initial steps in the phylogenetic emergence of language must have involved naming individuals (persons and animals), entities, activities, and states/conditions. This behavior is the closest to pointing, which, according to Tomasello (2008), distinguishes us from other animals, including nonhuman primates.

Naming may actually have scaffolded on pointing, which is linked to joint attention. It certainly constituted an important advance over pointing because it enabled our hominine ancestors to refer to individuals, entities, and so on that are not present, including those in the past, if the situation prompted memories. Later, it enabled modern humans to identify imaginary entities and activities, such as in myths. It played a central role in the development of narratives.

The naming of common objects, activities, events, and states/conditions (e.g., *ax, motion, dance, strong,* and *strength*), which differ from proper names in that they do not single out individuals that are unique in our universe of knowledge or socialization, is also associated with another milestone. This is the ability to lump in the same category instances of kinds of objects, events, activities, states/conditions, or behaviors that are similar (though not necessarily identical). This capacity to categorize and structure the universe of experience or knowledge definitely goes beyond the ability to individuate entities singled out by pointing. It marks the emergence of common nouns, with which, as one can imagine, a speaker could not specify reference efficiently without, for instance, demonstratives, articles, quantifiers, or grammatical number markers (e.g., *this/that boy, a boy, the boy, boys, the boys, those boys,* and *boy* as in *Boy meets girl*). Attention to referents could thus be directed less vaguely. The addressee would therefore know whether the speaker was speaking about one token or a plurality, whether the referents were supposed known (thus definite) or unknown (viz., indefinite) to the interlocutor, whether reference was being made generically (as in *boy meets girl*), or to a whole denotational class (e.g., *The lion is king of the jungle*), and so forth. Some languages even use noun class markers (e.g., in Bantu) or numeral classifiers (e.g., in Chinese) to do part of the job. The development of all these strategies improved hominines' capacity to communicate about their universes of experience or knowledge in various ways that are more informative, especially when the referents are not present.

Cross-linguistic differences between strategies of establishing reference (not only between languages such as English and French, which both use articles, but also between languages that use articles, those that use noun class markers, and those that use numeral classifiers) highlight again the cultural dimension of the emergence and evolution of languages. Different populations did not solve the communication problem in identical ways any more than they behave identically or assume the same beliefs about the world.

Although naming did not displace pointing, which can still disambiguate reference or establish reference in the present (in case one does not remember the name), it started what Hockett (1959) called *displacement*, which is the ability to refer to or communicate about what is not present. Contrary to the way linguists explain the architecture of spoken language, with phonetic sounds as the basic physical units, it is apparently naming, thus words, that drove the evolution of phonetic systems (Mufwene 2013a). As the hominines' cognitive capacity and their need to communicate grew, pressure increased to expand the vocabulary needed to name various things, activities, events, conditions, attitudes, dispositions, and more.

The need to clearly distinguish one word from another (by the relation that Saussure [2016] identified as "opposition") and to avoid repeating the same syllable over and over in long words exerted internal ecological pressure to expand the phonetic inventory by producing more sounds. One can definitely expand one's vocabulary more significantly, say, with fifteen different sounds than with just five sounds. It also appears that Rousseau (1755) was not mistaken in speculating that consonants were produced to make speech more fluid than if we spoke with vowels only. They apparently make it easier to transition from one syllabic peak (typically a vowel) to another and to distinguish words from one another perceptually.

I submit that naming fostered the emergence of digital vocalizations, which hominines share only with songbirds, though the latter have capitalized on what corresponds to prosody (tones or melody) in human speech. Digital vocalization is indeed what speech is, whereas other animals have only continuous and holistic ones. (See, e.g., Fitch 2010.) The digitization of vocalizations made it possible to accomplish syllabic variegation (MacNeilage 2008) and thus to produce words even longer than two syllables without repeating the same syllables. Human digital vocal communication is more productive than birdsong. It makes it possible for populations to produce as much vocabulary as they need with only a limited inventory of phonetic sounds (15–85), which they combine in some conventional, culture-specific ways in sequences of variable lengths.

This is also when it becomes obvious that it takes more than the anatomical capacity for digital vocalization to speak. As long pointed out by

Darwin (1871), parrots can mimic speech but do not speak in the sense of providing original information to others, because they are not endowed with the mind or mental capacity that drove the phylogenetic emergence of spoken language. If one factors in the fact that parrots do not even use the same organs as humans in mimicking speech, it becomes more evident that if humans had the same organs as parrots, they would still be able to produce speech and spoken language, provided they were endowed with a mind that enabled the emergence of the latter. Part of the evidence for the critical role of the mind in the emergence and evolution of language also lies in the fact that humans who cannot produce speech have developed signing, which can communicate information as richly and explicitly as speech. Additional, though negative, evidence comes from the fact that parrots that mimic speech do not use it to communicate with each other. It does not serve their natural communicative needs. Though one might suspect it would endow them with the capacity to communicate as richly as humans do, their minds do not appear to perceive the benefit of adopting a language made by others. Humans naturally become multilingual and use additional language(s) with people of the same ethnolinguistic background often because of some communicative advantages they derive from the process. For instance, some scholars feel more comfortable discussing academic subject matters in their language of scholarship when it is different from their native vernacular.

As the hominine cognition and representation of the universe became more complex, involving several relations, it exerted more ecological pressure for the emergence of predication. The reason is that one directs the attention of one's interlocutors to individuals, entities, events, and more not just for their own sake but to convey information about them, about oneself, about them and oneself, or about them and others. I submit that predication was the next step in the emergence of language as technology to communicate about one's universe of experience or knowledge. From there on, most utterances other than imperatives would consist of arguments interpreted as *agents* or *patients* and of predicates.

The cultural dimension in this case lies in whether the syntax of a language imposes a strict Noun/Verb distinction, with only verbs allowed to head predicate phrases (as in English), or has a more permissive system (as in Mandarin or the Kwa languages of West Africa), in which even adjectives and prepositions can also head a predicate phrase. Thus, one does not have to say 'John is mad' or *Jean est fou* (in French) but *Jan mad* (in an English

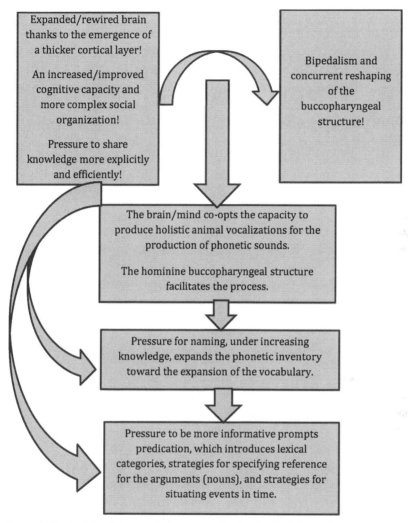

Figure 9.1. Sketch of the different evolutionary stages involved in the phylogenetic emergence of language.

Creole). Culture bears on every mechanical and structural aspect of language because, as noted above, the mind availed different populations differing options in the ways they could solve their communication problems. Assuming polygenesis, cultural differences subsume typological variation among the world's languages, though it is another story to demonstrate whether polygenesis is the fundamental reason for linguistic diversity.

Predication brought with it pressure for more informative communication, such as specifying reference for the arguments and situating activities/events and states/conditions in time. As suggested above, within certain ranges of variation, different populations developed culture-specific strategies for specifying reference (through markers of number, gender, definiteness, etc.), specifying time (through markers of tense and aspect), and for establishing a degree of responsibility regarding the veracity of the information communicated (through mood distinctions).

The need to collaborate with one's cohorts also exerted pressure to distinguish statements from commands and from requests for information (i.e., questions). Appropriate strategies have been developed in all human languages to meet all these communication needs, although the details of their implementation vary from one population to another, which underscores the cultural aspects of languages as technologies. That is, while the mind that drove the evolution of language reflects particular stages and trajectories in biological evolution that distinguish us from other animals, it leaves plenty of room for variation from one population to another, just like between individuals, in the way they solve problems.

While all modern specimens of *Homo sapiens* represent apparently the same biological evolutionary stage, they have often followed separate evolutionary trajectories. From the point of view of culture, they have developed different ways of responding to their natural ecologies, different patterns of behaving with one another, different social organizations, different belief systems, and of course different communication conventions. A noteworthy consequence of this evolution that exhibits both commonalities and divergences is the clear distinction in the grammatical behaviors of nouns (as prototypical arguments) and verbs (as prototypical predicates) in virtually all human languages. As noted above, there is some variation regarding prepositions and adjectives, which can also function predicatively in some languages (though they are distinguished from verbs) when they exist as grammatical categories distinct from nouns and verbs. However, it is less clear when and how these categories emerged.[9]

As the hominine cognitive capacity increased and improved and social organization became more complex (always in ways that vary from one population/culture to another), therefore increasing communication needs, ecological pressures also increased for languages that are more and more complex, with larger vocabularies (as noted above) and with longer and more complex utterances.[10] Practicality would have dictated working economically, resorting to, for instance, recursion, made possible by the duality of patterning, to generate longer and more complex utterances. Beyond basic words, where recursion is limited to concatenation, the strategy works in conjunction with constituent structure (what others may call "construction"), which facilitates processing in a vocal medium that is strictly linear.

As explained above, recursion is not so much a unique characteristic of human languages as it is a reflection of the way human minds work in solving problems economically. At the level of clauses embedded within each other, this observation corroborates Corballis's (2011) position that recursion reflects hominines' capacity to travel mentally, such as being able to connect events that did not take place at the same time or to embed one event in another event. This is all consistent with the capacity for displacement (Hockett 1959), which every modern language satisfies. As usual, there are cultural differences in the ways the characteristic is implemented. Thus, serializing languages do not operate exactly like languages that resort to subordination in the way they expand sentences to express more complicated ideas. While in the former languages one would say something corresponding to *John swim cross the river,* the French alternative is *Jean a traversé la rivière en nageant* (literally, 'John crossed the river by swimming') for the English *John swam across the river.* The same meaning is expressed through different syntactic strategies. So far there is no explanation other than cultural arbitrariness for why different populations do not settle on one strategy for the same function, although they use minds at comparably the same stage of evolution.

GENERATIVE ENTRENCHMENT AND SCAFFOLDING IN THE EMERGENCE OF LANGUAGE

There is much more to explain regarding the emergence of the architecture of languages, although I admit to having no clues yet about some aspects of this. Progress in the scholarship on language typology may better inform our speculations about the incremental evolutionary trajectories of modern languages. The evidence for my speculations is both indirect (as also confirmed recently by Hillert [2015]) and language-internal. We learn languages incrementally, starting with the most fundamental things (typically, naming and then very simple sentences), and there are structures that appear to have evolved from others (Mufwene 2013a). What I would like to underscore below is the significance of Wimsatt's (2000) Generative Entrenchment (GE) and Wimsatt and Griesemer's (2007) (Self-) Scaffolding throughout this protracted evolution. The former notion has to do with the incremental way complex structures emerge, by building new structures upon older ones, in such a way that the later ones could in fact collapse or cease to operate if the older ones were removed. Although Wimsatt explains GE best with material technology (as do, for instance, Michel Janssen and Gilbert B. Tostevin in this book), one can actually invoke the way humans develop knowledge in different domains, with the later additions grafted onto what is already known and becomes more entrenched and necessary. For instance, in algebra, one learns the more complex equations based on an understanding of the simpler ones. One must understand $(a \times b)^2$ before understanding $((a \times b)^2 \times c)^2$. Scaffolding has to do with the support that earlier structures provide in the development of new structures. I explain below how both notions applied in the phylogenetic emergence of language.

Starting with GE, vocalizations had already been in use among all mammals and other animals for communication. Hominines just made them more generative and productive by digitizing them (during the initial naming practices), introducing more functional variegation (MacNeilage 2008), and resorting to some syntax (which starts indeed at the level of phonology, the syntax of sounds) to produce exponentially larger vocabularies and longer utterances from limited inventories of sounds. Typological variation among languages around the world shows that, past a critical mass of consonants and vowels, what matters is really what combinatorial conventions different populations develop to generate various words and utterances. Recursion appears to be an initial implementation of GE in that speakers reuse structures already in place to produce larger ones. Various ways of expanding structures in language seem to illustrate this, such as in preposition phrases (e.g., *the book on the coffee table in my house in Hyde Park in Chicago*) and in relative clauses (e.g., *the dog that chased the cat that ran after the mouse that ate the cheese*).

What is particularly noteworthy in cases of structural innovations, as is evident from the scholarship on grammaticalization, is the extent to which the novel creations are constrained by extant structures. For instance, using *go* as a future auxiliary in English is constrained by how it is used as a motion verb—namely, in the progressive to express a process, in combination with an auxiliary *be,* which is required by the less verby nature of the progressive form. Because it has its own auxiliary and only one auxiliary can be

The verb *go* is co-opted in the progressive construction to express the FUTURE tense when its DIRECTION complement is a verb, as in *He is going to write*.

However, unlike other auxiliary verbs, it still behaves in some ways as the MOTION verb, thus as a main verb, inflected in the progressive, although *to* does not function as a preposition any more. *To* has been reanalyzed as a complementizer. Also the form *going* combines with the copula/auxiliary *be*, which can combine with the negation marker *not* and can be inverted with the subject in question. Thus, *he is not going to write* but not **he is going not to write*.

On the other hand, like other auxiliary verbs and unlike the MOTION verb in the progressive, the semi-auxiliary *going* can be coalesced with *to* into *gonna*, which can be reduced to *gon* in some dialects, such as African American English. In this respect, it behaves like other auxiliary verbs, for instance, *will not → won't*.

Throughout these processes, the basic syntax of the semi-auxiliary *go* remains that of the MOTION verb *go*, which constrains its overall evolution, as predicted by scaffolding!

Figure 9.2. Different evolutionary stages in the grammaticalization of the verb *go* as a semi-auxiliary verb for *future* tense.

inverted in questions or precede a negation marker (e.g., *not* or *never*), only *be* can participate in these syntactic rules but not the present participle *going* (thus, *Is he going to write?* but not **Is going he to write?*). It is only after satisfying these constraints that *going* as a marker of future can develop the peculiarities that distinguish it from the motion *going to*—for instance, the fact that this construction can contract into *gonna* or even *gon* in some dialects (*He's gonna write* but not **He's gonna town*). These developments are made possible by the fact that, as a semantic modifier/auxiliary, *going* bears weak stress. Its grammaticalization into an auxiliary verb also prevents it from combining with the preposition *from*, which also suggests that the *to* it combines with is a complementizer but not a preposition anymore. Throughout this evolution, GE has imposed on the semi-auxiliary *going* a syntactic frame that restricts the modifications that its grammaticalized usage can undergo (thus *Is he going to/gonna/gon write?* but not **Going he to write?* or **Gon(na) he write?*).

Particularly noteworthy in the evolution of language is the fact that the addition of new elements is supported by extant structures, though it proceeds in an ad hoc fashion, depending on the particular needs that arise at particular points in time. For instance, although they need a technical metalanguage (which may include complex, nontransparent formulae), scholars still have to write their prose according to the syntax of the schooled layperson's language. Things happen this way simply because it is less costly to make ad hoc adjustments to a system than to redesign it from scratch.

The above discussion also illustrates self-scaffolding. From a physical point of view, speech is scaffolded on hominines' innate capacity for vocalizations. We co-opted our masticatory organs to diversify our vocalizations and to introduce syllabic variegation. We also domesticated our breathing patterns in the process. This makes speech a very inexpensive technology, which proceeded by exaptation without having to resort to any anatomical organ that hominines did not already use for some other vital function. Once we were able to produce words and increase the vocabulary, the foundation of syntax as combinations of words into longer utterances were laid. That is, syntax as a consequence of using a technology that can be produced only linearly started within the vocabulary. This consists of words, which are formed from constrained combinations of sounds and can be distinguished from other recombinations of sounds even if exactly the same sounds and numbers thereof are used. For instance, it is because the same sounds are combined differently that one can tell *pit* from *tip* or *dog* from *god*. These contrasts instantiate the same principle used in syntax, at the level of combinations of words and of phrases, between *Paul loves Mary* and *Mary loves Paul* and between *the dog chased the cat* and *the cat chased the dog*. The duality of patterning is thus a consequence of the self-scaffolding of the possibility of combining units into larger ones. Recursion, illustrated above, is a special case of this.

It does not look like a dedicated language organ, rather than a general-purpose mind, was needed for this particular evolution of communication systems in the hominine species. What distinguishes us from other animals is a mind capable of solving problems at a low cost, by exaptation, drawing on available resources that could be adapted for new functions quite different from their original ones. The structural complexity of languages (such as in the strategies for specifying reference, using a complement clause where a noun would function as an object and using a relative clause to modify a noun) appears to have emerged incrementally, thanks to new adaptations that were not anticipated at the earliest stages but were needed later on to match hominines' increasing mental and cognitive capacities. I have, of course, not articulated all the details of the relevant evolutionary processes. These remain part of the research program I am engaged in.

CONCLUSION

The emergence of language is undoubtedly the outcome of a particular biological evolutionary trajectory that hominines do not share with other animals. Hominines evolved anatomical peculiarities that their more powerful mental capacity could co-opt conveniently for linguistic communication. These include the particular shape of their buccopharyngeal structure with a permanently descended larynx (pushed down by a descended tongue root) and bipedalism, which freed their more agile hands for doing other things with them even while in motion. The range of things that the hands could be exapted for (that is, for functions other than grabbing objects) includes signing.[11]

However, variation in the ways that different populations have structured their respective languages highlights the cultural dimension of this particular technological evolution. Although the basic communication pressures were presumably very similar across hominine populations, their language-ready minds left plenty of room for variation in the details of the architecture of the languages they developed. Thus, it does not matter whether a population uses a verb at the beginning or at the end of a sentence, or in some position inside; whether it relies on case markers (as in Latin) or postpositions (as in Korean) to make explicit the syntactic functions of nouns or relies only on word order (as in English); or whether the modifying phrase (e.g., an adjective phrase or relative clause) precedes or follows the head noun, and so on. Such variation is comparable to some populations keeping left on the road, whereas others keep right, or different populations using differing keyboards then on their typewriters and now on their computers.

It is not so much that there was a cultural evolution that differed from biological evolution; it is that, like mutations, alternative innovations by the mind and reproductions of these by untutored learning (driven by observation and inference) account for why the "transmission" of cultural phenomena proceeds differently from that of biological materials.

To be sure, one can argue that biological reproduction is not entirely faithful because the genes from the same pool are recombined variably in ways that make every offspring unique. However, in the case of language, as in other cultural domains, it is difficult to identify an elemental unit comparable to a gene, despite common invocations of memes in cultural evolution. Memes are of different sizes, depending on what one learns. Some of them are really complex. They are not reproduced faithfully from one communicator to another, especially in the case of sounds and meanings.

Institutional attempts to standardize the grammars and vocabularies of languages, thus to reduce if not eliminate variation, are unnatural, although they are helpful efforts to minimize instances of miscommunication. They also reflect how much political power some segments of a population (wish to) exert on others. Standardization in other cultural domains, such as in construction materials, facilitates cooperation, even though this takes additional economic and political dimensions.

Languages appear to be cultural phenomena like many others that distinguish humans from other animals, but their evolution need not be seen as independent of or excluding biological evolution. A question that arises at this point is the following: Does "cultural evolution" mean 'an evolution that proceeds differently from biological evolution' or 'evolution as it applies to cultural phenomena'? Could there not be just one notion of evolution (interpreted as 'change in heritable traits' or 'gradual directional change') whose specifics vary depending on what it applies to? I favor the latter interpretation.

NOTES

I am grateful to Alan Love and Bill Wimsatt for very constructive and detailed feedback on the first draft of this chapter. I alone am responsible for all the remaining shortcomings.

1. Note that this literature has capitalized on animal biology, for which transmission is vertical and unidirectional and a consequence of mating between two partners. It has generally not considered virology, in which horizontal transmission and polyploidy are typical. This is actually what language evolution should be compared with (Mufwene 2001, 2008). Thus, speed of evolution is not an issue anymore; nor is the question of whether the concept of *transmission* really applies to culture too (e.g., Fracchia and Lewontin 1999). The answer is that, while transmission is enabled by mating in

animal biology, it is made possible by interactions and learning in the case of culture.

2. Throughout this chapter I will refer alternately to language and to languages. In the former case, I direct attention to the essence or common properties of languages but not necessarily to a disputable common primordial language or protolanguage, whereas in the latter case I intend to conjure up the diversity that occurs among them, which deserves just as much attention. I discuss the relevant issues in the body of the chapter.

3. I elaborate on this process in part 5.

4. In this chapter I will often invoke the *mind* where others may invoke the *brain* or *cognitive capacity/structure* to account for the emergence of language in mankind. I think of it as the condition of the brain in activity, when the neurons are interacting with each other and enable the bearer to be aware of his/her surroundings, to respond to stimuli and challenges, and to think and solve problems, among other things. This is close enough to Searle's (2013) interpretation of it, from the point of view of consciousness as a central feature of the mind. It is not enough to invoke the brain because this may be dead and useless, as in corpses. I assume that cognitive capacity is a feature of the mind, while cognitive structure conjures up some organization, in which something can be integrated. I focus on the activity part of the brain, which can generate something new, such as concepts, systems, and of course language.

5. In their own words: "This need not imply that Acheulean hominins were capable of manipulating a large number of symbols or generating complex grammars. Our findings imply that simple forms of positive or negative reinforcement, or directing the attention of a learner to specific points (as was common in the gestural teaching condition), are considerably more successful in transmitting stone knapping than observation alone" (6).

6. To be sure, some animals, including birds, resort to some technology (in Brian Arthur's sense) to solve problems, but not with as much diversity, or with the same level of complexity, as folk technology among humans.

7. Attempts by academies and political institutions to prescribe how particular populations should speak their languages or which particular languages or dialects they should speak (in particular situations) are very late developments in hominine evolution. They tell us more about how efforts to control language evolution politically usually fail than about the ecological factors that influenced the phylogenetic emergence of language in mankind.

8. Identifying constituents correctly helps parse strings of words meaningfully. For instance, in English, the conjunction *and* in *the boy and the girl* goes with *the girl* but does not form a constituent with *the boy*. The combination **the boy and* is ill-formed and harder to interpret without the following conjunct. In Latin, the suffix *que* 'and' combines with the second conjunct but not with the first: *Maria Petrusque* 'Mary and Peter' but not **Mariaque Petrus*. In processing an English sentence, one must first distinguish the subject noun phrase from the verb phrase, regardless of how complex either constituent is, before getting into their details. For instance, [*The tall woman*] [*stood in the doorway*] or [*The tall woman in the red dress*] [*stood in the doorway*] or [*The tall woman*] [*stood with a defying look in the doorway and summoned Paul*], etc. Sentences are not always structured or parsed this way in all languages, consistent with the cultural dimension of language evolution.

9. It is debatable whether some languages have adjectives at all. For instance, Bantu languages use verbs or nominal modifying phrases connected to the head noun by a connective, where English uses an adjective. When an adjective-like item (of which there are very few) is used predicatively, its status is as indeterminate as that of *fun* in such a function in English; one cannot tell for sure whether it is an adjective or a noun in *It was fun/a lot of fun/more fun/?funner/?very fun*. Also, according to some students of grammaticalization (e.g., Heine and Kuteva 2007), prepositions have evolved from erstwhile nouns or verbs, but it is not evident that this is the case for all of them, let alone in all languages.

10. I will dodge here the elusive issue of how to conceptualize complexity in language, as it does not boil down to a system with more units (e.g., a larger phonetic inventory and vocabulary) and more rules. There is also the kind of complexity, more significant perhaps, that arises from the interactions of the different units, rules, and modules of the architecture of a language with one another (Mufwene, Coupé, and Pellegrino 2017).

11. One can of course add clenching the hands into fists in aggression, or using the palm or back of the hand for the same purpose, and a host of other things that other animals cannot do (equally well) with their forelimbs.

REFERENCES

Arbib, Michael. 2012. *How the Brain Got Language: The Mirror System Hypothesis*. Oxford: Oxford University Press.

Arthur, Brian W. 2009. *The Nature of Technology: What It Is and How It Evolves.* New York: Free Press.

Atran, Scott, and Dan Sperber. 1991. "Learning without Teaching: Its Place in Culture." In *Culture, Schooling, and Psychological Development,* edited by Liliana Tolchinsky Landsmann, 39–55. Norwood, N.J.: Ablex.

Beckner, C., R. Blythe, J. Bybee, M. H. Christiansen, W. Croft, N. C. Ellis, J. Holland, J. Ke, D. Larsen-Freeman, and Tom Schoenemann. 2009. "Language Is a Complex Adaptive System: A Position Paper." *Language Learning* 59 (1): 1–26.

Bickerton, Derek. 1990. *Language and Species.* Chicago: University of Chicago Press.

Bickerton, Derek. 2010. *Adam's Tongue: How Humans Made Language, How Language Made Humans.* New York: Hill and Wang.

Chomsky, Noam. 2010. "Some Simple Evo Devo Theses: How True Might They Be for Language." In Larson, Déprez, and Yamadiko 2010, 45–62.

Christiansen, M. H., and N. Chater. 2015. "The Language Faculty That Wasn't: A Usage-Based Account of Natural Language Recursion." *Frontiers in Psychology* 6:1182. doi:10.3389/fpsyg.2015.01182.

Corballis, Michael C. 2011. *The Recursive Mind: The Origins of Human Language, Thought, and Civilization.* Princeton, N.J.: Princeton University Press.

Cornish, H., M. Tamariz, and S. Kirby. 2009. "Complex Adaptive Systems and the Origins of Adaptive Structure: What Experiments can Tell Us. *Language and Learning* 59 (1): 187–205.

Darwin, Charles. 1871. *The Descent of Man.* Amherst, N.Y.: Prometheus Books.

Dor, Daniel. 2015. *The Instruction of Imagination: Language as a Social Communication Technology.* Oxford: Oxford University Press.

Eckert, Penelope, and Sally McConnell-Ginet. 1992. "Think Practically and Look Locally: Language and Gender as Community-Based Practice." *Annual Review of Anthropology* 21:461–90.

Evans, Nicholas, and Stephen C. Levinson. 2009. "The Myth of Language Universals: Language Diversity and Its Importance for Cognitive Science." *Behavioral and Brain Sciences* 32:429–92.

Everett, Daniel. 2012. *Language: The Cultural Tool.* New York: Pantheon Books.

Fitch, W. Tecumseh. 2010. *The Evolution of Language.* Cambridge: Cambridge University Press.

Fracchia, Joseph, and R. C. Lewontin. 1999. "Does Culture Evolve?" *History and Theory: Studies in the Philosophy of History* 38 (4): 52–78.

Gallese, Vittorio, and Alvin Goldman. 1998. "Mirror-Neurons and the Simulation Theory of Mind-Reading." *Trends in Cognitive Sciences* 2:493–501.

Hauser, Mark, Noam Chomsky, and W. Tecumseh Fitch. 2007. "The Language Faculty, What It Is, Who Has It, and How Did It Evolve?" *Science* 298:1569–79.

Heine, Bernd, and Tania Kuteva. 2007. *The Genesis of Grammar: A Reconstruction.* Oxford: Oxford University Press.

Hillert, Dieter G. 2015. "On the Evolving Biology of Language." *Frontiers in Psychology* 6, article 1796.

Hockett, Charles F. 1959. "Animal 'Languages' and Human Language." *Human Biology* 31:32–39.

Jackendoff, Ray. 2010. "Your Theory of Language Evolution Depends on Your Theory of Language." In Larson, Déprez, and Yamadiko 2010, 63–72.

Janssen, Michel. 2019. "Arches and Scaffolds: Bridging Continuity and Discontinuity in Theory Change." In Love and Wimsatt 2019.

Koster, Jan. 2009. "Ceaseless, Unpredictable, Creativity." *Biolinguistics* 3:61–92.

Kretzschmar, William A., Jr. 2015. *Language and Complex Systems.* Cambridge: Cambridge University Press.

Larson, Richard K., Viviane Déprez, and Hiroko Yamadiko, eds. 2010. *The Evolution of Human Language: Biolinguistic Perspectives.* Cambridge: Cambridge University Press.

Lee, Namhee, Lisa Mikesell, Anna Dina L. Joacquin, Andrea W. Mates, and John H. Schumann. 2009. *The Interactional Instinct: The Evolution and Acquisition of Language.* Oxford: Oxford University Press.

Lewontin, Richard C. 1970. "The Units of Selection." *Annual Review of Ecology and Systematics* 1:1–18.

Lieberman, Philip. 2012. *The Unpredictable Species: What Makes Humans Unique.* Princeton, N.J.: Princeton University Press.

Love, A. C., and W. C. Wimsatt, eds. 2019. *Beyond the Meme: The Role of Structure in Cultural Evolution.* Minneapolis: University of Minnesota Press.

MacNeilage, Peter. 2008. *The Origin of Speech.* Oxford: Oxford University Press.

Martinet, André. 1960. *Elements de linguistique générale.* Paris: Armand Colin.

Massip-Bonet, Àngels. 2012. "Language as a Complex Adaptive System: Towards an Integrative Linguistics." In *Complexity Perspectives on Language, Communication, and Society,* edited by Àngels Massip-Bonet and Albert Bastardas-Boada, 35–60. Berlin: Springer Verlag.

McArthur, D. 1987. Le langage considéré comme une technologie. *Cahiers de Lexicologie* 50:157–64.

Morgan, T. J. H., N. T. Uomini, L. E., L. Chouinard-Thuly, S. E. Street, H. M. Lewis, C.P. Cross, C. Evans, R. Kearney, I. de la Torre, A. Whiten, and K. N.

Laland. 2015. "Experimental Evidence for the Co-evolution of Hominin Tool-Making Teaching and Language." *Nature Communications* 6 (6029): 1–8.

Mufwene, Salikoko S. 2001. *The Ecology of Language Evolution.* Cambridge: Cambridge University Press.

Mufwene, Salikoko S. 2008. *Language Evolution: Contact, Competition and Change.* London: Continuum Press.

Mufwene, Salikoko S. 2013a. "Language as Technology: Some Questions That Evolutionary Linguistics Should Address." In *In Search of Universal Grammar: From Norse to Zoque,* edited by Terje Lohndal, 327–58. Amsterdam: John Benjamins.

Mufwene, Salikoko S. 2013b. "The Origins and the Evolution of Language." In *The Oxford Handbook of the History of Linguistics,* edited by Keith Allan, 13–52. Oxford: Oxford University Press.

Mufwene, Salikoko S. 2013c. "What African Linguistics Can Contribute to Evolutionary Linguistics." In *Selected Proceedings of the 43rd Annual Conference on African Linguistics: Linguistic Interfaces in African Languages,* edited by Olanike Ola Orie and Karen Wu, 52–67. Somerville, Mass.: Casadilla Press.

Mufwene, Salikoko S. 2015. Des langues et des récits dans l'espèce humaine: Une perspective évolutive. In *Corps en scenes,* edited by Catherine Courtet, Mireille Besson, Françoise Lavocat, and Alain Viala, 127–37. Paris: CRNS Editions.

Mufwene, Salikoko S. 2017. "Language Evolution, by Exaptation, with the Mind Leading." In *New Horizons in Evolutionary Linguistics,* guest edited by Gang Peng and F. Wang, 27:158–89. *Journal of Chinese Linguistics* Monograph Series.

Mufwene, Salikoko S., Christophe Coupé, and François Pellegrino, eds. 2017. *Complexity in Language: Developmental and Evolutionary Perspectives.* Cambridge: Cambridge University Press.

Odling-Smee, F. John, Kevin N. Laland, and Marcus W. Feldman. 2003. *Niche Construction: The Neglected Process in Evolution.* Princeton, N.J.: Princeton University Press.

Rousseau, Jean-Jacques. 1755. *Discours sur l'origine et les fondements de l'inégalité parmi les hommes.* Amsterdam: M. M. Rey. Translation of *Essai sur l'origine des langues,* in J. H. Moran and A. Gode, 1966, *On the Origin of Language: Two Essays by Jean-Jacques Rousseau and Gottfried Herder.* New York: F. Ungar Pub. Co.

Saussure, Ferdinand de. 2016. *Cours de linguistique générale,* edited by Charles Bally and Albert Sechehaye, in collaboration with Albert Riedlienger. Paris: Payot.

Searle, John. 2013. "Theory of Mind and Darwin's Legacy." *Proceedings of the National Academy of Science of the United States of America* 110:10343–48.

Sperber, Dan, and Gloria Origi. 2010. "A Pragmatic Perspective on the Evolution of Language." In Larson, Déprez, and Yamadiko 2010, 124–31.

Steels, Luc. 2000. "Language as a Complex Adaptive System." In Vol. 1917 of *Parallel Problem Solving from Nature—PPSN VI ; 6th International Conference,* edited by M. Schoenauer, K. Deb, G. Rudolph, X. Yao, E. Lutton, J. J. Merelo, and H-P Schwefel, 17–26. LNCS. Heidelberg: Springer Verlag.

Steels, Luc. 2011. "Modeling the Cultural Evolution of Language." *Physics of Life Review* 8:339–56.

Steels, Luc, ed. 2012. *Experiments in Cultural Language Evolution.* Amsterdam: John Benjamins.

Tomasello, Michael. 2008. *Origins of Human Communication.* Cambridge, Mass.: MIT Press.

Tostevin, Gilbert B. 2019. "Content Matters: The Materiality of Cultural Transmission and the Intersection of Paleolithic Archaeology with Cultural Evolutionary Theory." In Love and Wimsatt 2019.

Wimsatt, William C. 2000. "Generativity, Entrenchment, Evolution, and Innateness." In *Biology Meets Psychology: Constraints, Connections, Conjectures,* edited by Valerie Gray Hardcastle, 139–79. Cambridge, Mass.: MIT Press.

Wimsatt, William C., and James R. Griesemer. 2007. "Reproducing Entrenchments to Scaffold Culture: The Central Role of Development in Cultural Evolution." In *Integrating Evolution and Development: From Theory to Practice,* edited by R. Sansom and R. Brandon, 227–323. Cambridge, Mass.: MIT Press.

10

WRITING IN EARLY MESOPOTAMIA
The Historical Interplay of Technology, Cognition, and Environment

MASSIMO MAIOCCHI

DESPITE THE FACT that language is just one among the many manifestations of a given culture, it is usually considered one of its core features. In particular, writing (i.e., the visual representation of language) is regarded as one of the crucial inventions in the history of humanity because it dramatically enhances communication potential and promotes persistent cultural memory. Even today, the advent of writing is commonly accepted as the dividing line between history and prehistory.[1] The impact of writing is evident not only from the amount of data that philologists can recover from ancient inscriptions but also from the profound changes in cognition, society, and environment that it has brought about. In this sense, writing has been defined as a *Kulturtechnik,* which stresses the bond between its material representation, operative aspects, and transmission within a given cultural environment.[2] Writing effectively extends cognitive facilities by allowing the externalization of previously embodied meaningful information clusters in the form of linguistic symbols, which in turn can then be easily compared at a glance. For instance, the creation of indexes, catalogs, glosses in margins, or simple indentations may produce a superimposed hierarchy of sections, suggesting associations between chunks of text that would otherwise have no obvious relation to one another (see section 3). This quick nonlinear access to information is otherwise impossible in spoken language.[3] In this way, writing assists in identifying associations, shaping thought, and intensifying the cognitive apparatus in a reciprocal feedback process, which can produce cascade effects on other techniques and fields of knowledge. For instance, records of empirical observations may lead to the creation of a formalized institutional calendar, which in turn allows for a more precise management of environmental resources, as aptly demonstrated by the Mayas in Mesoamerica. This, in turn, can maximize production and thereby create a surplus, which encourages the development of structures for its management. Simultaneously, a more formally structured religious ideology seems to arise in response to the need for social stability generated by the productive system, which in Mesopotamia became increasingly asymmetrical in terms of labor and access to resources.[4] Going back to the impact of writing on cultural evolution, it is important to stress that this technology makes access to information possible regardless of whether the encoder of the information is physically present. Contrary to what typically happens in modern societies, this information access has usually been restricted to the social elite in antiquity. Nevertheless, under certain circumstances, writing could be displayed to a broadly illiterate audience to reaffirm the rank of those individuals within the social hierarchy who are able to access that message.

Because writing can be perennial, or at least stable, over long periods of time in particular forms, it is perceived as magic, sacred, or even taboo in many ancient societies, and the people associated with it inherit these qualities. As a consequence, writing is frequently invoked as a prime determinant for cultural change in modern theories of cultural evolution. In their earliest formulation,[5] these theories framed writing systems as originating and changing historically in a linear evolutionary sequence, beginning with a primitive stage based on the massive use of logography (i.e., word-signs) and progressing to a more "advanced" logosyllabic stage, followed by a fully developed alphabetic system, which is celebrated as the incarnation of Western democracy and scientific advancement. This alphabetocentric (ethnocentric) paradigm is nowadays obsolete due to advancements in several disciplines that intermingle with the study of writing (archaeology, history, philology, linguistics, semiotics, etc.), as well as clear counterexamples. (Written Japanese, one of the most sophisticated writing systems presently used, did not hamper technological advancements or social achievements, despite the difficulty in learning the system). Nevertheless, even today, there is a certain bias in evaluating the potential expressed by the invention and adoption of writing in ancient societies. This is no doubt due to the fact that people who study writing are (inevitably) literate; they are embedded in a deeply entrenched paradigm—the literate paradigm. After years of training, we take pride in using writing technology, just as ancient scribes certainly did. Hence, we tend to associate civilization with the use of writing—and barbarism with its absence—regardless of stubborn facts to the contrary. The Incas, who did not read or write, were as civilized as the Mayas, who did read and write. For reasons possibly linked to overspecialization and lack of imagination, certain aspects and functions of writing are therefore overemphasized, whereas other aspects or functions are neglected. Most of the duties performed by written records can be carried out by means of nonlinguistic symbolic systems and mnemonic devices. In addition, common misconceptions about the nature of writing, as well as the lack of a commonly accepted definition of it, blur the overall picture. In this chapter I address some of the issues concerned with early writing, especially its connection with culture and environment. In my view, one of the most overlooked factors in the analysis of early writing is how writing systems emerged in particular environments. To encompass this, I shall treat writing both as a material and immaterial technology, beginning with a detailed account of the historical evidence and then proceeding to evaluate the environmental, technological, and conceptual dependencies of the writing technique.

SCAFFOLDINGS FOR WRITING

In contrast to diffusion theories popular in the middle of the last century that assumed a single origin for writing, it appears that writing was independently invented several times in history. Grammatologists (those involved in the study of writing systems) use the evocative term *grammatogenesis* (or *grammatogeny*) to label this process. Modern scholars recognize four pristine (i.e., independently generated) grammatogenetical events that occurred in different cultures at different time periods and in distinct geographical contexts as the result of long incubations involving deep transformations in society and environment: (1) cuneiform script in southwestern Asia (~middle of the fourth millennium B.C.E.), (2) Egyptian hieroglyphic in northern Africa (~middle of the fourth millennium B.C.E.),[6] early Chinese script in Central Asia (late thirteenth century B.C.E.), and (4) Mayan hieroglyphics in Mesoamerica (~fourth century B.C.E.). Other scripts, interesting in their own right, can be seen as derived products of cultural contact between literate and illiterate societies. For reasons of space and personal competence, these scripts are referred to only marginally in this chapter.[7] The grammatogenesis of these pristine writing systems is not creation ex nihilo; they invariably relate to other visual systems for storing information, such as calculi, numerical tags, and calendrical systems.

Of the four writing systems, cuneiform represents the best case for exploring the emergence of writing and its implications for cultural development, cognitive enhancement, social diversification, and environmental change. The available evidence is not only abundant but also covers the long period prior to the establishment of writing as the main technology for record keeping. Nevertheless, generalizations based on the cuneiform scenario are inherently risky. Writing is the product of a complex society and therefore is bound to a variety of intertwined factors standing in multilevel, superimposed, and asymmetrical relationships with one another. The hunt for universals in writing systems, while feasible to some extent, especially in terms of structural features shared by any representational system, is subject to considerations of language, culture, technological changes, sociopolitical developments, and environmental context. These diverse considerations hamper the creation of models of writing as scaffolding for the establishment of complex urban societies.[8] In this regard, it is worth noticing that writing is only one of several technological innovations that contributed to the emergence of the so-called urban revolution in Mesopotamia.

The label *urban revolution* is clearly a misnomer because the "revolution" lasted for roughly one millennium, but it is partly justified by its profound impact on the subsequent modes of human interaction. By the beginning of the third millennium B.C.E., Uruk (modern Warka, in southern Iraq) was a metropolis of 2.5 to 5.5 square kilometers in size (including the lower town), approximately twice the size of classical Athens (fifth century B.C.E.) and only half the size of imperial Rome (first century A.D.).[9] The city probably hosted forty thousand to fifty thousand people, an astonishing number, especially if compared with other settlements of this and subsequent periods. These figures, significant as they may be, only hint at the complexity that characterized urban life in the late fourth millennium B.C.E. The multiplicity of social niches attested to in both archaic written sources (especially the list of professions; see section 5) and archaeological data (monumental complexes, residential quarters, iconographic motives, etc.) is an outcome of prolonged anthropic contact with a variegated landscape, whose cyclical fluctuations in terms of water regimes made possible the development of different strategies for the exploitation of natural resources. The southern alluvium is an area rich in ecological diversity. The Tigris and Euphrates Rivers create a diverse landscape, alternating wetlands, marshes, steppe, plains, lagoons, seas, and wadis, as well as sandy and rocky desert.[10] It is within this landscape that the early urban society first made language visible.

Tokens and Clay Envelopes

As for Uruk's technological background, we should consider the prehistoric developments that took place over a period of roughly five thousand years, from 8,500 to 3,500 B.C.E.[11] Small clay objects, shaped in a number of different ways, were found in several sites *scattered* over the whole area of southwestern Asia, from Iran to Turkey (Figure 10.1). These objects are referred to in modern literature as *tokens*. According to shape and other features, they may be classified into a threefold typology: simple tokens (shaped as disks, cones, spheres, and other basic geometric forms); derived simple tokens (simple geometric shapes but bearing one or two incisions);[12] and complex tokens (shaped in elaborate ways, bearing several incisions, perforations, painting, or other modifications). Their archaeological context is mostly unclear, but they can be dated with some confidence to this long phase before the advent of writing.

Examples from the first category (i.e., simple tokens) appear roughly at the same time as the domestication of plants and animals by early settlers (roughly 8,000 B.C.E.). Derived simple tokens and complex tokens are found only much later (4,500 B.C.E. in Uruk and 3,500 B.C.E. in Susa and Syria). The interpretation of these objects is still debated. On the basis of later evidence (e.g., the system of the bullae, described below), it seems reasonable—albeit not provable—that simple tokens were used to count some sort of goods.[13] Assemblages of simple tokens might have been put in leather bags or some other sort of perishable container that leaves no trace in the archaeological record. In this case, besides being used as calculi, they might have served the same function of later bullae, which are hollow balls of clay used as envelopes to enclose tokens. (They vary in size from a golf ball to a baseball.) Several details concerning the function of these enigmatic objects are still debated. Apparently, we lack several pieces of the puzzle because strings could have been added to these artifacts to hang the bulla or attach other perishable additional parts carrying information.[14]

Clay envelopes first appear in Uruk and the surrounding region around 3,500 B.C.E. and possibly slightly later in Susa (Iran).[15] The surface of these artifacts is usually covered in its entirety with impressions of cylinder seals, which are administrative devices usually made of stone (a rare material in Uruk, as well as in the whole southern Mesopotamian area), carved with iconographical motifs. These artifacts were rolled on fresh lumps of clay used to seal bullae, jars, rooms, and (later) cuneiform tablets.[16] Up to three different seals might be present on a single bulla, each functioning as a sort of signature of an official or an individual in contact with the local administration. The aim was to guard against the falsification of an envelope's content by covering the entire surface with impressions that are not easy to replicate, making any infraction evident.

The study of bullae is complicated by two factors. First, the total number of archaic bullae presently known is rather limited (roughly around 130 exemplars). Second, the content of most of these objects is unknown, since museums are rightfully unwilling to damage these precious documents to look for tokens contained inside. Recently, noninvasive techniques (such as axial tomography) have reached the necessary resolution to allow for the noninvasive study of bullae, but the data are still unpublished. Nevertheless, it seems clear that only simple tokens, or in a few cases derived simple tokens, feature inside the ancient clay envelopes. Derived simple tokens are few in number compared to the abundance of complex tokens. It seems therefore reasonable to conclude that complex tokens served a different function than simple tokens (e.g., complex tokens are possibly unrelated to accounting).[17] The situation is complicated by the fact that some complex tokens show striking similarities with protocuneiform signs found on actual tablets. Nevertheless, there is no correlation between the frequency of complex tokens and the frequency of the alleged corresponding signs. For instance, the sign for *sheep* (a circle with a cross in it) is exceedingly common on tablets, but the *corresponding* token is very rare. One may conclude that there was a shared set of symbols used by early accountants, but the system was fluid. More than one code was probably in use by different people involved (at various levels) in the early urban system.[18]

Returning to the features of bullae, it is remarkable that some of them show what appear to be numerical impressions on the surface, produced in a number of different ways, such as with tokens, fingers, or a reed stylus. The impressions on the outside may or may not correspond to the number of tokens contained inside the bulla. However, it is clear that these marks represent numbers, which are consonant with the metrological systems attested to on another type of document—namely, the numerical tablets found in Uruk, Susa, and Godin Tepe (western Iran on the Zagros Mountains, north of Susa), as well as with protocuneiform tablets of Uruk and Susa.[19] These metrological impressions are not randomly placed on the bulla's surface but instead respect some ordering principle, grouping numerical signs of the same kind in columns or lines. Keeping in mind that more than one code could have been in use at a given time, it seems that the ancient accountants impressed numbers representing larger units first and then added impressions for smaller units last. This implies an "advanced numerical syntax,"[20] which is impossible to express using simple tokens alone. One might even speculate that the need for clarifying this syntax in an environment where multiple codes operated brought about the practice of impressing numerical signs on the bulla's surface.[21] Alternatively, the impressions on the outside may serve to prevent (as much as possible) the necessity of breaking the artifact for inspection, at the same time obliterating its future validity. In other words, a quick look at the impressions on a bulla's surface may have been sufficient to retrieve the information concerning its content, thus making the bulla a "double document."[22] The existence of this syntax implies metrological standardization (still an ongoing process at this stage), as well as shared conventions that are transmitted within the frame of an incipient bureaucratic system.

From this short survey, we have seen that bullae are complex artifacts. Besides a clay envelope and tokens, they typically have seal impressions, numerical impressions, and (in some cases) strings passing through them. Only the first two features (envelopes and tokens) in this list are necessary, but in most cases all of them are present in a given bulla. The proper relationship between these elements is hard to ascertain, but later evidence (e.g., sealed contracts from the late third millennium onward) suggests that at least some sealed bullae served as legal documents, binding two parties in a mutual agreement, or as receipts for goods (typically grain or cattle). In case of litigation, or simply when the contract expired, one of the parties could break the bulla and inspect the contents. This situation happened in antiquity because a number of bullae have been found broken in situ.

Insight into the functions of these complex artifacts can be gleaned by examining a much later bulla, dated to 1,400 B.C.E., from the site of Nuzi (Yorghan Tepe, northern Iraq).[23] The document was found in a private house, together with a cuneiform tablet that sheds light on the use of this specific bulla, which contained forty-nine tokens. Both the bulla and the cuneiform tablet bear impressions from the same seal, as well as a cuneiform inscription in Akkadian. The inscription on the bulla appears to be a shortened and perhaps complementary version of the more elaborate inscription on the tablet. The text mentions "49 sheep and goats belonging to Puhishenni, the son of Musapu, which were given over to the care of Ziqarru, the son of Shalliya, the shepherd." It seems therefore that this clay envelope was part of a contract between the literate owner of the flock and an illiterate shepherd, who was to pasture the animals in the surroundings of Nuzi for a period of several months. The tablet was meant to protect the owner (e.g., against loss or the substitution of animals with less valuable ones) and remained with him, whereas the bulla was meant to protect the shepherd (e.g., against possible accusation of theft). The envelope possibly traveled with him, but in this exceptional case was returned together with the flock.

Generalizations on the basis of this unique example are not possible, especially since the bulla itself is inscribed with a cuneiform inscription. Nonetheless, it is reasonable to assume that the prehistoric bullae from Uruk and other sites share at least the character of accounting documents, and some of them also potentially have a legal character. This may explain the practice of sealing the entire surface of these objects.[24]

Numerical Tablets: A Space for Counting, a Space for Thought

As noted, numerical impressions are a feature shared both by bullae and "numerical tablets" that are mostly found in Uruk and Susa but also in other sites in Iran (Chogha Mish, Godin Tepe, Tepe Sialk), Iraq (Jemdet Nasr, Nineveh), and Syria (Mari, Nagar, Habuba Kabira, Jebel Aruda). Their proper dating is unclear, but the general consensus is that they probably appeared at the same time as the earliest bullae (3,500 B.C.E.) or possibly slightly later (3400 B.C.E.). The later dating is primarily dictated by reasons of convenience, as it is tempting to place bullae and numerical tablets in a linear "evolution" sequence, but the later dating finds some partial support in the stratigraphic evidence as well.[25]

It must be kept in mind, however, that bullae do not disappear with the advent of numerical tablets or with the rise of cuneiform documents, as proved by a bulla from Tepe Yahya roughly datable to 2,700 B.C.E.,[26] an unprovenanced Old Akkadian bulla (2,300–2,200 B.C.E.),[27] and the much later Nuzi bulla, dated to 1,400 B.C.E. (described above). The same holds true for simple tokens. A very recent find in Tushan (Ziyaret Tepe, southeastern Turkey) proves that these devices, first introduced in the eighth millennium B.C.E., were still in use in a provincial capital of the neo-Assyrian Empire toward the middle of the first millennium B.C.E.[28] Remarkably, several cuneiform tablets were also unearthed there, proving that full-fledged writing and archaic accountability systems coexisted over millennia due to different levels of literacy and bureaucratic demands, which implies that a variety of social niches were extant within the urban ecosystem.

The label *numerical tablets* is motivated by the fact that only numerical impressions are found on these artifacts (i.e., they do not feature cuneiform signs). Similar to clay envelopes, their surface is usually covered with seal impressions. Contrary to bullae, tablets are not spherical and therefore cannot contain tokens. Contrary to proper cuneiform tablets (from 3,300 B.C.E. on), numerical tablets imply only a limited literacy. This consideration may explain the use of abnormal repetitions of numerical signs otherwise usually bundled together on a bunch of numerical tablets from Jebel Aruda (northern Syria) and other sites in the North. But this explanatory approach may derive from our tendency to rigidly systematize the available data.[29] It has been suggested that numerical tablets may serve the same function as bullae (albeit this is difficult to determine with certainty) but were more practical to produce in comparison to clay envelopes since they do not require shaping tokens or producing a spherical artifact. At Susa, some numerical tablets and bullae were found in the same room and even in the same container.[30] In addition, the same seal impressions are occasionally found on both tablets and bullae at both Uruk and Susa.[31] It is therefore tempting to consider the idea that, at least in some cases, numerical tablets served the same function as the much later Nuzi tablet, found together with the bulla and having the same seal impression. Rather than being an "evolution" of bullae, numerical tablets may therefore have been part of a complementary system of accounting. In any case, one should allow that the system was flexible: the absence of seal impressions on some documents suggests that they may have served a variety of different functions.

Although numerical tablets are not yet connected to spoken language, they represent an important step in the history of writing and, more generally, in the history of human cognition. Having two flat surfaces, they inherently arrange space into distinct parts: obverse, reverse, and edges. The information these tablets provide is thus embedded not only in the numerical signs per se but also in the position within the tablet where the signs occur. In other words, the writing space is *semanticized*. Interestingly, some numerical tablets also feature column division and arrangement into boxes (cases), which involved impressing lines on the surface.[32] This feature appears in protocuneiform and later tablets where there is a fully developed writing system. The advantage of flat "rectangular" tablets over bullae is also one of storage space and ease of filing. We know little about the original archival context of Uruk documents. Early tags (i.e., small perforated tablets) may have been attached with a rope to baskets containing tablets or other items, as is the case in later periods. We may also compare the practice of filing tablets on shelves, attested to in Ebla (Tell Mardikh, Syria, ~2,400 B.C.E.). There, some tablets bear inscriptions on the edges and mention the period of time covered by the individual accounts. These indicators apparently served as labels to quickly find documents within the archive, much like book titles in a modern library. Similar conventions had already manifested in a few numerical tablets, bearing numerical impressions on the edge, with or without other impressions on the obverse and reverse, but with seal impressions on the tablet's surface.

The interpretation of the numeric impressions is unclear, but it seems that the numbers on the edges are not sums of the numbers on the obverse or reverse. This tentative explanation may be true since in the earliest stages of protowriting the information space extends beyond the physical limits of the individual documents, allowing a mapping of textual groups and navigation within a possible archive. Whether or not this is operative already in Uruk is hard to say, but at least from the middle of the third millennium B.C.E., this appears to be the case.

These different forms of structuring information open up important cognitive possibilities. Access to data is not only effective but allows for quick comparison between various meaningful segments of information. The indexing in this archaic period ultimately resided in the decoder's mind.[33] For instance, parallel textual sections belonging to multiple texts can be easily compared, evaluating similarities and differences. Even though this process can be replicated in an illiterate society by committing the storage of information to mnemonic systems and nonlinguistic devices, the effort and time required make the task difficult to the point of being exceedingly impractical. In this regard, writing (even protowriting) works as a catalyzing agent for the cognitive process, scaffolding the organization of ideas in ways that are otherwise impossible for spoken language.

THE FIRST SIGNS

Shortly after the middle of the fourth millennium, more complex tablets appeared in Uruk (southern Iraq), Susa (eastern Iran), and Nagar (Tell Brak, northern Syria). These artifacts have been labeled *logonumeric* (or *numeroideographic*) tablets to stress the fact that they bear only one or two signs associated with numbers.[34] The signs presumably denote the items counted or perhaps refer to the individuals (or institutions) involved in the movement of goods. Some logonumeric tablets bear cylinder seal impressions. Whereas the signs in both Uruk and Susa tablets make use of the same formal conventions found on later tablets (e.g., signs depicting quadrupeds render the animal head only), the two Nagar tablets seem to use a different convention, representing whole animals.[35] This practice is most probably explained as a local deviation from southern standards; it reflects the fluid situation that characterizes early writing in the vast area of the ancient Near East.[36]

If we focus on differences in representational convention, one notes that numerical signs in Uruk are placed in front of the signs they refer to, whereas the opposite happens in Susa. Yet in both cases, the sign repertoire provided by the logonumeric tablets is rather limited. Nevertheless, it appears that already in this early stage of script development, nonpictographic (or noniconic, but possibly indexical) signs were introduced side by side with pictographic ones (e.g., signs representing jars, plants, birds, or body parts). One must be careful in this respect, though, because we may be unable to identify the right referents of ancient items due to cultural distance. Despite this caution, there is a clear tendency that suggests an intellectual effort to create conventions and possibly to borrow or adapt preexisting elements of the fluid symbolic systems discussed above.

Besides logonumeric tablets, roughly eighteen hundred protocuneiform tablets (including fragments) have been found in Uruk. Each is inscribed with several signs and usually framed in a set of several boxes (cases). These are the most archaic tablets presently known. They are labeled Uruk IV, from the name of the archaeological level of the site, and can be dated to approximately 3,300 B.C.E. A second, larger group (roughly forty-five hundred texts and fragments), labeled Uruk III, is dated to around 3,100–2,900 B.C.E. Unlike Uruk IV tablets, Uruk III tablets stem from many sites in southern Mesopotamia (Eshnunna, Kish, Larsa, Umma, Jemdet Nasr, Tell Uqair, Ur) and show a rather quick diffusion of the writing technology. The signs on protocuneiform tablets are arranged in rectangular cases (or boxes). Each case contains an administrative entry, composed by numerals, logograms (word-signs), or both. Within each case, numerals are grouped together in formalized sequences, respecting older conventions already found in bullae, whereas logograms (word-signs) are freely placed. That means that there is no "grammatical" order if two or more signs are present within the same case; this must be supplied by the decoder. The textual cases vary in size, but their mutual position is not random. Depending on tablet format, the division into columns and subcolumns allows the decoder to retrieve information on the relationship between the content of a given box and those that it surrounds. The writing space is thus semanticized into units that may relate to one another, such as in the case of balanced accounts or in texts showing rather elaborate summations.

This spatial syntax was already present in the numeric tablets, but it lacked the systematization found on Uruk IV and Uruk III tablets. Intriguingly, the dissemination of this writing technology corresponds to the end of the Uruk phenomenon. Contemporary to Uruk III in Susiana, an entirely different system was conceived (proto-Elamite). It is mostly undeciphered because it died out shortly after its introduction. The proper archaeological context of most of the tablets from Uruk is both unclear—because the original excavators were not yet aware of stratigraphic methods—and disturbed—because they were found in dump areas. In addition, the content of protocuneiform texts is partly opaque to us due to their archaism. Regardless, it is clear that the documents fall into two distinct categories: administrative texts and lexical lists. Most documents belong to the former category,[37] which includes records of various kinds of commodities in relation to individuals and institutions, whereas lexical lists are documents listing thousands of words, mostly thematically organized: animals, cities, fish, food items, professions, metals, plants, vessels, garments, and wood objects (inter alia). On the basis of this evidence, there is little doubt that writing in Mesopotamia emerged in response to practical needs—namely, to keep track of the goods produced and moved within the early state.

SOME REFLECTIONS ON THE USE OF WRITING AND ITS COGNITIVE IMPLICATIONS

Lexical lists have often been described as the prime tool for the transmission of scribal knowledge.[38] Their great authority is evident from the fact that these documents were copied over and over again for centuries, with only minimal deviations from the original. The archaic list of professions is already attested to in Uruk IV (3,300 B.C.E.) and spread from there to most of the Mesopotamian world. It was still copied in Nippur, the most prestigious center for scribal education at that time, in the very heart of Babylonia, around 1,800 B.C.E. Despite the fact that lexical lists do not contain the totality of cuneiform signs observable in the documents of a given period, it is clear that they contributed to the stability and perpetuation of the writing system. This facilitated scribes in learning how to produce well-formed and meaningful signs, but it also had cognitive implications. Lexical lists promoted framing concepts within a visual, symbolic representation system. To some extent, cuneiform writing inherently generates a taxonomy, which in turn stimulates intellectual reflection on the world as perceived through the prism of written language. Lists are thus a new way of looking at the world, ordering reality into fixed architectures upon which scribal knowledge is structured.

This willingness to classify and order the cosmos is also evident in the effort to establish standards for weights and measures—a crucial concern of any administration. This segmentation of reality into discrete units (an early "digitalization process") led to the artificial division of a day into twenty-four hours and a month into thirty days. Administrative time was born. This was essential to calculate things such as the workforce needed and the grain rations to be disbursed for construction work or some similar task, as possibly recorded on several Uruk tablets. The metrology of these archaic texts is indeed rather intricate. Several systems were in use at the same time, depending on what was being counted. Besides area and time measures, one finds two different sexagesimal systems used to count dairy products and textiles, on the one hand, and dead animals and jars, on the other. Two bisexagesimal systems were used to count different grain products, cheese, and fish disbursed as rations. As for the cognitive implication of this standardization, it seems relevant here that "slaves" appear to be treated according to the same metrological conventions that apply to animals. They are also represented on cylinder seal motifs.

Despite the fact that this practice is primarily bureaucratic in nature, there is little doubt that it contributed to the mental process of self-identification within the literate part of society in terms of a contrast with its subordinate. The emergence of writing is of little impact in promoting empathy as a structural feature of cooperative behavior, except perhaps among those who share the technology.[39] Instead, the application of this technology seems to stimulate social stratification, especially in terms of the enslavement of foreign people, which Lévi-Strauss had already concluded for cultures in South America.[40] In his words:

> Si mon hypothèse est exacte, il faut admettre que la fonction primaire de la communication écrite est de faciliter l'asservissement. L'emploi de l'écriture à des fins désintéressées, en vue de tirer des satisfactions intellectuelles et esthétiques, est un résultat secondaire, si même il se réduit pas le plus sovent à un moyen pour renforcer, justifier ou dissimuler l'autre.[41]

This provocative position seems too extreme when applied to the origin of writing in Mesopotamia. Writing technology emerged there as a consequence of accountability needs that were not immediately related to enslavement. It is difficult to state with certainty a "primary function" for this writing technology, but it is possibly most tightly connected to the prediction of future events within the productive system on the basis of past accounts. This seems to be a contrastive element of writing as opposed to other mnemonic devices used in early city administration. The difference lies not just in the fact that a full-fledged writing system is capable of expressing any message, whereas other systems, such as the bullae, do not. In more practical terms, the difference is that writing enormously facilitates the quantification and statistical prediction of future recurrent events based on recorded history. In conformation with this view, it is worth noticing that many of the protocuneiform texts have recently been interpreted as contingency tables, such as a means for estimating the amount of grain to be harvested in the forthcoming season based on data recorded in previous seasons.[42] In addition, writing surpasses other solutions for retaining information when reporting to a higher authority. A well-structured bureaucratic apparatus necessitates the rigid verbalization of written records.[43]

It is worth stressing that our modern perception of writing as a pervasive phenomenon within contemporary society has little to do with ancient evidence. Scribal knowledge was limited to a few individuals belonging to the urban elite. Additionally, it took roughly seven hundred years for Mesopotamian scribes to conceive and create a document that was not either administrative or lexical in nature. Thus, the domain of writing remained restricted to city administration for a very long period of time, proving that there is no obvious evolutionary progression in the history of writing. It is only close to the twenty-sixth century B.C.E. that literary texts appear as a different genre, possibly as a result of the prolonged contact between Sumerian and Akkadian cultures in the South.

ENVIRONMENT AND WRITING:
THE CASE OF MESOPOTAMIA

When dealing with the invention of cuneiform writing, most authors hold that clay was chosen as a medium because it was cheap and abundant in Mesopotamia. This explanation is rather simplistic because clay was also abundant in the environments where other pristine writing systems emerged but other media were preferred.[44] In Egypt, the earliest writing is attested to on bone and ivory tags; only later is it found on stone and papyrus.[45] In China, turtle shell or bone was used in addition to bamboo strips.[46] In Mesoamerica, Mayan scribes wrote on animal skin, bark paper, vessels, and stone.[47] In light of these facts, clay is not so obvious a choice in Mesopotamia despite its wide availability. Instead, the choice of clay is better explained in terms of the existence of the bullae system, which in turn makes sense only in the variegated environment hinted at above (see section 2).

The alternating wet and dry areas in southern Mesopotamia promoted the emergence of what has been labeled a *dimorphic* society, where seminomadic human groups coexisted with permanent settlers in a mutually dependent relationship that was established over a long period of progressive climatic drying.[48] The dynamics of social interactions between these two groups are not always easy to ascertain because of biases in the available documentation. Seminomad pastoralists leave few traces in the archaeological record and are seldom mentioned in the written sources concerned with urban bureaucracy.[49] This is especially true for fourth millennium Uruk because the site was only partly excavated, and the relatively few cuneiform texts unearthed there are not completely understood due to their very archaic nature. It is risky to use data from much later periods (e.g., the end of third and the beginning of second millennia B.C.E. Mesopotamia) as a basis for projecting back to the situation toward the end of the first urbanization phase. What can be observed from later sources is that permanent settlers progressively developed a production system based heavily on cereal monocultures, which were mass produced thanks to technological innovations first introduced in the period of incipient urbanization (e.g., seeder plow, threshing sledge, water canalization, short-field irrigation), as well as social stratification (e.g., organization of labor to work and maintain the fields or dig canals). The seasonal contact between settlers and seminomads occurred right after harvesting, when the flocks were taken to graze fields in a mutually beneficial situation: the animals fertilized the soil while using up the remainder of plant stalks as fodder.

The social boundaries between the "movable" and "immovable" parts of this society were not rigid. "Settlers" could certainly transit in and out of "seminomadic" clans and vice versa.[50] How much this applies to Uruk everyday life is difficult to state, in part because the city's economy seems to have relied not only on agriculture and animal husbandry but also on the exploitation of marsh resources, such as fish and reed, which were abundant in this period. However, it seems reasonably certain that animal husbandry was mostly performed outside the cities, regardless of the proper social connotation of the local human groups involved in seasonal large-scale movements. According to both epigraphic and iconographic evidence (e.g., cylinder seal motifs), domesticated animals (mostly sheep, goats, oxen, and pigs) were exploited in Uruk as alimentary resources (meat and dairy products), for the production of goods (wool, sinew, etc.), and possibly as draft animals (though this practice is rarely attested to in the earliest documents).[51] When settlements grew in size and complexity, institutions in charge of the management of surplus cattle and grain emerged within the newly established urban society. This created a need for an accounting system, whose development over millennia can be traced as described above.

The bullae system was likely invented within this scenario and perhaps subsequently adapted to account not only for animals and grain but also other goods and labor, but this remains speculative. Clay envelopes operated as points of contact between an increasingly literate social group belonging to the city administration and an illiterate one deeply embedded in and circulating around the rural landscape. The ecological factor, intertwined with cultural development and environmental exploitation, was crucial for the development of an accountability system that propelled writing into the *Kulturtechniken* expressed in ancient Near Eastern societies while freely borrowing a number of features from other preexisting solutions. Clay was chosen as a medium not only because of its availability but also because of the habit of producing sealed documents and the need to continue doing so. Clay is well suited as a sealing medium, providing a continuous surface that can also bear identifying marks (e.g., seal impressions), and is much more durable and less expensive than textiles, leather, or other containers. Additionally, the materiality of bullae might depend on much older practices, such as the production of pottery and bricks. From this point of view, the choice of writing medium was one of the most deeply entrenched features in the process of knowledge transmission, which depended on ecological circumstances that nurtured processes of cultural evolution and facilitated the origin of writing and subsequent transformations in human cognition.

IN THIS CHAPTER I intentionally avoided applying the term *evolution* to writing systems.[52] This is partly due to my expertise, which is limited to the field of ancient Near Eastern studies, but also reflects a common practice within

Figure 10.1. Distribution through time of accounting devices and written documents. Modified after Woods (2010) to include new data published by MacGinnis et al. (2014) and Monaco (2014).

this field that prefers more neutral terms, such as *change, transformation, development,* or *adaptation* of the system. This practice originated as a reaction to the *Ex oriente lux* paradigm, a reformulation of the once fashionable diffusion theory, according to which civilization first appeared in Mesopotamia and spread from there. Even taking the term *evolution* as a metaphor, few modern scholars are willing to consider protocuneiform tablets as evolving from the bullae system since the former is glottographic (i.e., it conveys meaning and words, as expressed in a given language), whereas the latter is semasiographic (i.e., it conveys meaning without expressing a specific language). This distinction is functional, but it masks an important shared feature in the fluid development of accountability systems, which constitutes a fundamental step in the construction of both material and immaterial structures that are manifestations of the cultural evolution process.

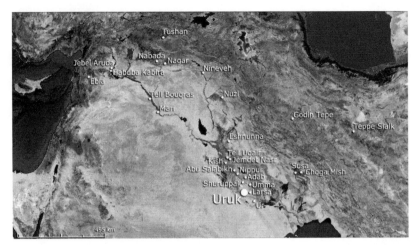

Figure 10.2. Map of the ancient Near East showing the sites mentioned in this chapter. Map data: Google; DigitalGlobe.

Without arguing explicitly for a linear evolutionary path from bullae to cuneiform tablets, it is worth keeping in mind that the numerical systems expressed in bullae are consonant with those appearing in protocuneiform tablets. In addition, bullae may share some of the functions expressed by cuneiform documents, such as their possible legal nature. There is no clear-cut boundary between writing and nonwriting within the cuneiform evidence; for a long period of time, tablets expressed extremely limited linguistic information (i.e., they were *mostly* semasiographic in nature). Most notably, clay as a medium for both systems stands out as a deeply entrenched feature in the development of the early Mesopotamian writing system. In light of these considerations, it seems useful to reconsider parallel attempts in the creation of tools for the maintenance of early city bureaucracy (sealing practices, tokens, bullae) and interpret them as scaffolds for the emergence of protocuneiform, rooted in a varied environment that exhibits the primary factors accounting for the existence of the writing technology. The need for managing a surplus created within the first urban societies stimulated the advent of writing as a technology for the more efficient exploitation of natural and human resources. In turn, this promoted cognitive developments, intellectual achievements, social diversification, craft specialization, and the possibility of more effective preservation, transmission, and intensification of knowledge, which we now perceive as one of the most important features of our own civilization.

NOTES

1. To a contemporary historian, this is clearly an exaggeration. First, there is obviously a "history before history"; modern archaeological techniques make available large amounts of data that rival in size, and often complement, what is known from written records. Second, depending on the definition adopted, writing may be considered either as an invented technique or as a slow development that emerged over several centuries. In addition, important ancient civilizations, such as the Inca in Mesoamerica, may or may not be acknowledged as literate, which invites us to reflect on how fragile modern definitions can be when applied to complex systems. Finally, we should consider that several undeciphered scripts, no matter how sophisticated they may appear, may or may not turn out to be actual writing (e.g., Rongorongo on Easter Island or the Indus script).

2. Cancik-Kirschbaum (2012, 131–32).

3. It is worth noticing here that the exploitation of associative capabilities is a built-in feature of all pristine writing systems, which are nonalphabetic in nature. An in-depth treatment of the typologies of writing systems and of their structural features is not possible here. It suffices to say that logographic systems, which are based on logograms—i.e., word-signs, invariably combine basic graphemes in order to be able to express large amounts of words with a limited repertoire of signs. For instance, in cuneiform the sign for *female worker, female slave,* read /geme/ in Sumerian (probably the underlying language of early cuneiform), is obtained by juxtaposition of the signs for *woman* and for *mountain, foreign land,* read *munus* and *kur,* respectively. The resulting sign thus suggests the para-etymology *foreign woman, woman from the mountains* for female slave, which is just not there in spoken language (cf. also the discussion on *creative etymology* in Glassner [2003, 54]). Again, for reasons of economy, in order to contain the total number of signs to be learned by the encoder, logographic systems exploit the so-called rebus principle: words that sound similar are written down with the same sign. For instance, the sign for *garden,* read *sar* in Sumerian, is also used to express the word *to write,* again pronounced *sar.* Not surprisingly, the goddess of writing is primarily connected with vegetation in the Sumerian pantheon. The application of the rebus principle thus connects otherwise semantically unrelated words. Thus, the visual nature of written language promotes indexicality and associations of otherwise poorly connected ideas.

4. It is not possible to explore here the details of the development of religious thought in early Mesopotamia. It suffices to say that the joint efforts of the workforce under the supervision of a central authority, combined with technological innovations and favorable environmental conditions, produced a large surplus. However, the producers were required to deliver such surplus to organizations embedded in a system primarily devoted to accumulation and redistribution—a painful process for the producers, which requires an ideological explanation ultimately residing in religious thought via divine legitimization of the elite. Cf. Liverani (2006, 33).

5. Gelb (1952).

6. Whether Egyptian hieroglyphics resulted from a stimulus-diffusion process with cuneiform is still debated. The absolute chronology of the earliest Uruk evidence is not well established, so it may or may not turn out to be older than Egyptian hieroglyphics. However, the inclusion of the latter as distinct is granted by the fact that the system is radically different from cuneiform, even if cultural contact with Mesopotamia promoted its invention.

7. The grammatogenesis of nonpristine systems is complex and includes nuances such as the basic idea of making language visible and that a certain set of systemic features may pass from one side to the other in the interaction process. In most cases, a certain linguistic competence and proficiency in reading and writing by the inventor of the new system is implied, as well as ideological motivations. The phenomenon is therefore labeled *sophisticated* grammatogenesis (Daniels 1996a, 579–85), as opposed to writing systems produced by individuals with no previous training in reading and writing. An example of an unsophisticated grammatogenesis is provided by the alphabet (eighth century B.C.E., eastern Mediterranean coast). It is best conceived as a case of imperfect transmission of knowledge between a literate Phoenician and an illiterate Greek (Gnanadesikan 2009, 208–28). This is different from the internal development of a script, which is a much slower process that happens within a given literate entity, such as a scribal school, and thus subject to conservative rules.

8. The term *complexity* is sometimes abused in modern literature (see chapter 13; Verhoeven 2010). The complexity of the ancient Uruk urban system does not derive merely from the increase in the total number of people settling this site, which can be conceived as nodes in a network diagram, or by counting the number of possible interactions within the extended group of individuals (edges connecting the nodes). Social life in Uruk is complex

in the sense that the interactions belong to different systemic elements, which include administrative entities that are hierarchically organized in addition to social groups and nuclear and extended families. The existence of such a complex network of material and immaterial relationships (e.g., the exchange of goods, services, knowledge, ideas, and ideology) is an identifying mark of cities as opposed to villages and towns (cf. Liverani 2006, 20–22).

9. Estimates vary on the actual size of Uruk at the very end of the fourth millennium B.C.E. due to the fact that the site is only partly excavated (cf. Nissen 1988, 71–72; Finkbeiner 1991, 193–94).

10. For a more detailed description of the water regimes and ecology of southern Iraq, see Pournelle (2013, 13–23, 28–29).

11. See Schmandt-Besserat (2010) and Michalowski (1993). The discussion here is limited to possible direct antecedents of writing, but several other crucial technical developments, such as the domestication of plants and animals, techniques for storing alimentary items (pottery), maximizing production (clay sickles, seeder plows, or threshing sledges), and processing food (grindstones and ovens) appear over this long period. These important innovations underpinned the possibility of accounting and writing and therefore may be regarded as scaffolds for scaffolds.

12. Englund (2006, 17); Monaco (2014).

13. What kind of goods exactly remains unclear. Due to the vast geographical extension of the token system, it seems improbable that only one code was in use. Tokens of a given typology were probably used to count different items in different areas, or possibly even among different human groups in the same area. The interpretation of tokens as calculi was put forward by Amiet (1966) and further developed by Schmandt-Besserat (1992, 1995, 2012; for a critical review, see Zimansky 1993 and Michalowski 1993). The practice of tallying (and possibly of basic arithmetic operations) is rooted in a much more distant past when other products of the human symbolic mind first emerged, including the practice of inhumation, jewelry making and wearing, and painting. The earliest tally sticks, such as the Lebombo bone (43,000 to 41,000 B.C.E.) and the Ishango bone (18,000 to 20,000 B.C.E.), stem from Africa and Western Asia. The interpretation of the latter is controversial. It is possible that this artifact was used not just for counting (e.g., keeping track of time elapsed from a certain event, such as the last new moon) but to perform simple mathematical calculations (the addition of numbers up to sixty and division by two). For an overview of the development of a symbolic repertoire in ancient Near Eastern art and material culture, see

Stordeur (2010). All these and similar objects deserve more attention than can be given here.

14. Woods (2015).

15. Cf. Englund (2004, 28n7) for a possible attribution of early unprovenanced bullae to the sites of Umma and Adab, respectively some 40 and 120 kilometers north/northeast of Uruk.

16. Cylinder seals replaced the much older stamp seals, first attested in Syria (Tell Bouqras and other sites) around 6,500 B.C.E. These objects apparently cover a number of different functions and can also be interpreted as amulets (Porada 1993).

17. Zimansky (1993) makes the point that at least some of these objects are better classified as beads.

18. Michalowski (1990, 1993).

19. Englund (2006, 21).

20. Englund (2006, 22).

21. A similar extension of the capability of the bullae's representational system is found on the Nuzi bulla, whose cuneiform inscription specifies what kind of animals (male and female, adult or not) are to be overseen by the shepherd. These details are otherwise not expressed by the undifferentiated tokens inside this specific clay envelope.

22. Lieberman (1980, 352).

23. Abusch (1981).

24. According to Dittmann (1986), the seal impressions replace actual personal names, as found in later tablets.

25. Cf. Englund (1998, 56) for a tentative reconstruction of the Susa stratigraphy. Nothing certain can be said for the situation in Uruk.

26. Englund (2006, 16).

27. Monaco (2014).

28. MacGinnis et al. (2014).

29. See, for instance, Englund (1998, 51, Figure 13).

30. Schmandt-Besserat (1992, 132n38).

31. Schmandt-Besserat (1992, 154); Englund (1998, 56).

32. Cf. the text W 6245,c in Englund (1998, 52).

33. Library catalogs exist from the period of the Third Dynasty of Ur (2,150–2,000 B.C.E.), though they are rare.

34. *Logonumeric* is preferred here because in later cuneiform tablets signs represent actual words and not just vague ideas or concepts (see also Cooper 2004).

35. Finkel (1985, 187–89).

36. The other possible explanation is that these tablets represent an earlier stage of writing, but this is not provable because of the disturbed stratigraphic context. Also, this idea is bound to the old and outdated view of Uruk colonies in the North. The Uruk presence there is complicated (cf. Stein 2002). This seems to be yet another "gray zone" of writing, which should be added to the array of possible outcomes of cultural contact between Uruk and indigenous cultures, including the imitation and the shallow adoption of writing according to local standards (Nagar?), sophisticated grammatogenesis (proto-Elamite script in Susa, contemporary with Uruk III), and reluctant attitudes to accept writing (Lamberg-Karlovsky 2003, 63). In the latter case, the explanation for the missed dissemination of the writing technology probably lies in the fact that rural areas, where dry agriculture is largely possible, are less likely to necessitate writing, as there is no need for canalizations, which in turn imply workforce management, storage, and the transformation of surplus. All these practices seem to be the prime movers for the invention of writing in southern Mesopotamia.

37. The proportion of lexical to administrative texts varies through time. Less than 1 percent of Uruk IV tablets are lexical, but the figure rises to 20 percent for Uruk III material (cf. Englund 2006, 28).

38. Veldhuis (2006).

39. Mullins, Whitehouse, and Atkinson (2013, 147–48).

40. Lévi-Strauss (1955, 354–55).

41. "If my hypothesis is correct, it would oblige us to recognize the fact that the primary function of written communication is to facilitate slavery. The use of writing for disinterested purposes, for the sake of intellectual and aesthetic pleasure, is a secondary result, and more often than not it may even be turned into a means of reinforcing, justifying, or dissimulating the other (i.e. its primary function)."

42. Woods (2015).

43. Steinkeller (2003, 2004).

44. The initial steps in the development of writing systems in ancient Egypt, China, and Mesoamerica are not as well documented as in Mesopotamia. It is therefore more difficult to assess, for instance, whether there was a primary medium used to write those scripts (Postgate, Wang, and Wilkinson 1995). Space constraints prevent a more detailed description of later script phases for individual writing systems and the possible implications for the consequent development of the relative media.

45. Baynes (2004); Stauder (2010).

46. Bagley (2004); Boltz (1986); Bottéro (2004); Shaughnessy (2010).

47. Houston (2004, 287–88); Palka (2010).

48. Cf. preliminary remarks in Rowton (1977).

49. As far as Mesopotamia is concerned, privileged epigraphic sources for the study of the seminomadic component within the urban scenario are the archives of Mari (Tell Hariri), on the Middle Euphrates, dated to the early second millennium B.C.E. (cf. Charpin and Durand 1986; Durand 2004).

50. Porter (2009).

51. Englund (1995; 1998, 94–95).

52. One can describe the development of writing systems as a purely Darwinian process (cf. Lock and Gers 2012). Although the family tree of writing systems' typologies is a useful tool, it does not do justice to the fact that no system is "pure" (as already acknowledged by Gelb [1952]). For instance, a syllabic or logosyllabic script under circumstances such as the case of writing foreign names or loanwords may use syllable-signs that are meant to represent only the consonantal part (the last vowel remains silent). Within cuneiform, certain archives, such as the merchant letters of Old Assyrian entrepreneurs in Anatolia (1900 B.C.E.), are written mostly syllabically and with a limited repertoire of signs. Regardless of its simplicity, this system did not spread, likely for reasons of prestige associated with the old tradition and politics. A logoconsonantal script, such as Egyptian hieroglyphics, has the built-in capability of a consonantal alphabet but remains mostly unexploited. Conversely, written English is alphabetic but has a remarkable tendency to maintain historical spellings that only loosely represent their spoken counterparts at a phonemic level. As for scripts, it is true that scribal hand can be transmitted over generations, but the fitness of a script depends on both material (the availability of media, pupils, etc.) and immaterial factors (politics, culture, esthetic appeal, etc.). If dual-inheritance theories are reformulated to fit the script scenario, then they must take account of this complex and heterogeneous environment.

REFERENCES

Abusch, T. 1981. "Notes on a Pair of Matching Texts: A Shepherd's Bulla and an Owner's Receipt." In *In Honor of Ernest R. Lacheman on His Seventy-Fifth Birthday, April 29, 1981,* edited by M. A. Morrison and D. I. Owen, 1–9. Studies on the Civilization and Culture of Nuzi and the Hurrians 1. Winona Lake, Ind.: Eisenbrauns.

Amiet, P. 1966. "Il y a 5000 ans les elamites inventaient l'ecriture." *Areheologia* 12:16–23.

Bagley, R. W. 2004. "Anyang Writing and the Origin of the Chinese Writing System." In Houston 2004, 190–249.

Baynes, J. 2004. "The Earliest Egyptian Writing: Development, Context, Purpose." In Houston 2004, 150–89.

Biggs, R. D. 1967. "Semitic Names in the Fara Period." *Or NS* 36:55–66.

Boltz, W. G. 1986. "Early Chinese Writing." *World Archaeology* 17:420–36.

Bottéro, F. 2004. "Writing on Shell and Bone in Shang China." In Houston 2004, 250–61.

Buccellati, G. 2013. *Alle origini della politica: La formazione e la crescita dello Stato in Siro-Mesopotamia.* Milano: Jaca.

Cancik-Kirschbaum, E. 2012. "Writing, Language, and Textuality: Conditions for the Transmission of Knowledge in the Ancient Near East." In *The Globalization of Knowledge in History,* edited by Jürgen Renn, 125–52. http://www .edition-open-access.de/studies/1/toc.html.

Cavigneaux, A. 1989. "L'écriture et la réflexion linguistique en Mésopotamie." In *Histoire des idées linguistiques, tome 1: La naissance des métalangages en Orient et Occident,* edited by S. Auroux, 99–118. Liège, Belgium: Mardaga.

Charpin, D., and D. Durand. 1986. "Fils de Sim'al: Les origines tribales des rois de Mari." *Revue D'Assyriologie* 80:141–83.

Christy, T. C. 1989. "Humboldt on the Semiotic of Writing." In Rauch and Carr 1989, 339–46.

Civil, M. 1973. "The Sumerian Writing System: Some Problems." *Orientalia* 42:21–34.

Cooper, J. 2004. "Babylonian Beginnings—The Origin of the Cuneiform Writing System in Comparative Perspective." In Houston 2004, 71–99.

Damerow, P. 1998. "Prehistory and Cognitive Development." In *Piaget, Evolution, and Development,* edited by J. Langer and M. Killen, 247–70. Mahwah, N.J.: Lawrence Erlbaum.

Damerow, P. 2006. "The Origins of Writing as a Problem of Historical Epistemology." *Cuneiform Digital Library Journal* 1:1–10.

Damerow, P. 2012. "The Origins of Writing and Arithmetic." In *The Globalization of Knowledge in History,* edited by Jürgen Renn, 153–74. http://www.edi tion-open-access.de/studies/1/toc.html.

Daniels, P. T. 1992. "The Syllabic Origin of Writing and the Segmental Origin of the Alphabet." In *The Linguistics of Literacy,* edited by P. Downing, S. D. Lima, and M. Noonan, 83–110. Typological Studies in Language 21. Amsterdam: John Benjamins.

Daniels, P. T. 1996a. "The Invention of Writing." In Daniels and Bright 1996, 579–86.

Daniels, P. T. 1996b. "The Study of Writing Systems." In Daniels and Bright 1996, 3–17.

Daniels, P. T., and W. Bright, eds. 1996. *The World's Writing Systems.* Oxford: Oxford University Press.

Dittmann, R. 1986. "Seals, Sealings, and Tablets." In *Ǧamdat Naṣr: Period of Regional Style?* (Beihefte zum Tübinger Atlas des Vorderen Orients, reihe B 62), edited by U. Finkbeiner, and W. Röllig, 332–66. Wiesbaden, Germany: Dr. Ludwig Reichert.

Durand, J.-M. 2004. "Peuplement et sociétés a l'époque amorrite (I) Les clans bensim'alites." In *Nomades et sedentaires dans le Proche-Orient ancien: Compte rendu de la XLVIe Rencontre Assyriologique Internationale (Paris, 10–13 juillet 2000),* edited by C. Nicolle, 111–98. Amurru 3. Paris: Editions Recherches sur les Civilisations.

Englund, R. K. 1995. "Late Uruk Period Cattle and Dairy Products: Evidence from Proto-cuneiform Sources." *Bulletin of Sumerian Agriculture* 8:33–48.

Englund, R. K. 1998. "Texts from the Late Uruk Period." In *Mesopotamien: Späturuk-Zeit und Frühdynastische Zeit,* edited by J. Bauer, R. K. Englund, and M. Krebernik, 15–233. Orbis Biblicus et Orientalis 160/1. Göttingen, Germany: Vandenhoeck and Ruprecht.

Englund, R. K. 2004. "Proto-Cuneiform Account-Books and Journals." In *Creating Economic Order: Record-Keeping, Standardization and the Development of Accounting in the Ancient Near East,* edited by M. Hudson and C. Wunsch, 23–46. Bethesda, Md.: CDL Press.

Englund, R. K. 2006. "An Examination of the 'Textual' Witnesses to Late Uruk World Systems." In *A Collection of Papers on Ancient Civilizations of Western Asia, Asia Minor and North Africa,* edited by Yushu Gong and Yiyi Chen, 1–38. Oriental Studies Special Issue. Beijing.

Finkbeiner, U. 1991. *Uruk, Kampagne 35–37, 1982–1984: Die archäologische Ober˘äschenuntersuchung (Survey).* AUWE 4. Mainz: von Zabern.

Finkel, I. 1985. "Inscriptions from Tell Brak 1984." *Iraq* 47:187–201.

Gelb, I. 1952. *A Study of Writing: The Foundations of Grammatology.* Chicago: University of Chicago Press.

Glassner, J.-J. 2003. *The Invention of Cuneiform: Writing in Sumer.* Baltimore: Johns Hopkins University Press. Translation of *Ecrire à Sumer: L'invention du cunéiforme.* Paris: n.p, 2000.

Gnanadesikan, A. 2009. *The Writing Revolution: Cuneiform to the Internet.* Chichester, UK: Wiley-Blackwell.

Houston, S. D., ed. 2004. *The First Writing: Script Invention as History and Process.* Cambridge: Cambridge University Press.

Krebernik, M. 1998. "Die Texte aus Fāra und Tell Abū Ṣālābīḫ." In *Mesopotamien. Späturuk-Zeit und Frühdynastische Zeit,* edited by J. Bauer, R. Englund, and M. Krebernik, 237–427. Orbis Biblicus et Orientalis 160/1. Fribourg, Switzerland: Fribourg University Press.

Lamberg-Karlovsky, C. C. 2003. "To Write or Not to Write." In *Culture through Objects: Ancient Near Eastern Studies in Honour of P. R. S. Moorey,* edited by T. F. Potts, M. Roaf, and D. L. Stein, 59–75. Oxford: Griffith Institute.

Lévi-Strauss, C. 1955. *Tristes tropiques.* Paris: Plon.

Lieberman, S. J. 1980. "Of Clay Pebbles, Hollow Clay Balls, and Writing: A Sumerian View." *American Journal of Archaeology* 84:339–58.

Liverani, M. 2006. *Uruk The First City.* London: Equinox. (Original Italian edition 1998.)

Lock, A., and M. Gers. 2012. "The Cultural Evolution of Written Language and Its Effects." In *Writing: A Mosaic of New Perspectives,* edited by E. Grigorenko, E. Mambrino, and D. Preiss, 11–35. New York: Psychology Press.

MacGinnis, J., M. W. Monroe, D. Wicke, and T. Matney. 2014. "Artefacts of Cognition: The Use of Clay Tokens in a Neo-Assyrian Provincial Administration." *Cambridge Archaeological Journal* 24:289–306.

Michalowski, P. 1990. "Early Mesopotamian Communicative Systems: Art, Literature, and Writing." In *Investigating Artistic Environments in the Ancient Near East: Essays Presented at a Symposium Held in April 1988 at the Arthur M. Sackler Gallery,* edited by A. C. Gunter, 53–69. Washington, D.C.: Smithsonian Institution.

Michalowski, P. 1993. "Tokenism." *American Anthropologist* 95:996–99.

Michalowski, P. 1994. "Writing and Literacy in Early States: A Mesopotamian Perspective." In *Literacy: Interdisciplinary Conversations,* edited by D. Keller-Cohen, 49–70. Cresskil, N.J.: Hampton Press.

Michalowski, P. 1996. "Mesopotamian Cuneiform: Origin." In Daniels and Bright 1996, 33–36.

Michalowski, P. 2008. "Sumerian." In *The Ancient Languages of Mesopotamia, Egypt, and Aksum,* edited by R. Woodard, 6–46. Cambridge: Cambridge University Press.

Monaco, S. 2014. *Archaic Bullae and Tablets in the Cornell University Collections.* Cornell University Studies in Assyriology and Sumerology (CUSAS) 21. Bethesda: CDL Press.

Mullins, D. A., H. Whitehouse, and Q. D. Atkinson. 2013. "The Role of Writing and Recordkeeping in the Cultural Evolution of Human Cooperation." *Journal of Economic Behavior & Organization* 90: 141–51.

Nissen, H. 1986. "The Archaic Texts from Uruk." *World Archaeology* 17:317–34.

Nissen, H. 1988. *The Early History of the Ancient Near East, 9000–2000 B.C.* Chicago: University of Chicago Press.

Nissen, H., P. Damerow, and R. Englund. 1993. *Archaic Bookkeeping: Writing and Techniques of Economic Administration in the Ancient Near East.* Chicago: University of Chicago Press. (Original German edition 1990.)

Palka, J. 2010. "The Development of Maya Writing." In Woods, Teeter, and Emberling 2010, 225–29.

Porada, E. 1993. "Why Cylinder Seals? Engraved Cylindrical Seal Stones of the Ancient Near East, Fourth to First Millennium B.C." *Art Bulletin* 75:563–82.

Porter, A. 2009. "Beyond Dimorphism: Ideologies and Materialities of Kinship as Time-Space Distanciation." In *Nomads, Tribes, and the State in the Ancient Near East: Cross-Disciplinary Perspectives,* edited by J. Szuchman, 201–25. Oriental Institute Seminars 5. Chicago: Oriental Institute of the University of Chicago.

Postgate, N., T. Wang, and T. Wilkinson. 1995. "The Evidence for Early Writing: Utilitarian or Ceremonial?" *Antiquity* 69:459–80.

Pournelle, J. R. 2013. "Physical Geography." In *The Sumerian World,* edited by H. Crawford, 13–32. London: Routledge.

Rauch, I., and J. Carr, eds. 1989. *The Semiotic Bridge.* Approaches to Semiotics 86. New York: Mouton de Gruyter.

Rowton, M. B. 1977. "Dimorphic Structure and the Parasocial Element." *Journal of Near Eastern Studies* 36:181–98.

Schmandt-Besserat, D. 1992. *Before Writing. Volume I: From Counting to Cuneiform.* Austin: University of Texas Press.

Schmandt-Besserat, D. 1995. "Record Keeping Before Writing." In Vol. 4 of *Civilizations of the Ancient Near East,* edited by Jack M. Sasson, 2097–106. New York: Charles Scribner's Sons.

Schmandt-Besserat, D. 2010. "The Token System of the Ancient Near East: Its Role in Counting, Writing, the Economy and Cognition." In *The Archaeology of Measurement, Comprehending Heaven, Earth and Time in Ancient Societies,* edited by C. Renfrew, 27–34. Cambridge: Cambridge University Press.

Schmandt-Besserat, D. 2012. "Tokens as Precursors of Writing." In *Writing: A Mosaic of New Perspectives,* edited by E. Grigorenko, E. Mambrino, and D. Preiss, 3–10. New York: Psychology Press.

Shaughnessy, E. L. 2010. "The Beginning of Writing in China." In Woods, Teeter, and Emberling 2010, 215–24. Oriental Institute Museum Publications 32. Chicago: Oriental Institute of the University of Chicago.

Stauder, A. 2010. "The Earliest Egyptian Writing." In Woods, Teeter, and Emberling 2010, 137–47.

Stein, G. 2002. "Colonies without Colonialism: A Trade Diaspora Model of 4th Millennium BC Mesopotamian Enclaves in Anatolia." In *The Archaeology of Colonialism,* edited by C. Lyons and J. Papadopoulos, 26–64. Los Angeles: J. Paul Getty Museum.

Steinkeller, P. 2003. "Archival Practices in Babylonia in the Third Millennium." In *Ancient Archives and Archival Traditions: Concepts of Record-Keeping in the Ancient World,* edited by M. Brosius, 37–58. Oxford: Oxford University Press.

Steinkeller, P. 2004. "The Function of Written Documentation in the Administrative Praxis of Early Babylonia." In *Creating Economic Order: Record-Keeping, Standardization, and the Development of Accounting in the Ancient Near East,* edited by M. Hudson and C. Wunsch, 65–88. Bethesda: CDL Press.

Stordeur, D. 2010. "Domestication of Plants and Animals, Domestication of Symbols?" In *The Development of Pre-state Communities in the Ancient Near East: Studies in Honour of Edgar Peltenburg,* edited by D. Bolger and L. C. Maguire, 123–30. Oxford: Oxbow Books.

Veldhuis, N. 2006. "How Did They Learn Cuneiform? Tribute/Word List C as an Elementary Exercise." In *Approaches to Sumerian Literature: Studies in Honour of Stip (H. L. J. Vanstiphout),* edited by P. Michalowski and N. Veldhuis, 181–200. Cuneiform Monographs 35. Leiden: Brill.

Veldhuis, N. 2012. "Cuneiform: Changes and Developments." In *The Shape of Script: How and Why Writing Systems Change,* edited by S. D. Houston, 3–24. Santa Fe: School for Advanced Research Press.

Verhoeven, M. 2010. "Social Complexity and Archaeology: A Contextual Approach." In *The Development of Pre-state Communities in the Ancient Near*

East: Studies in Honour of Edgar Peltenburg, edited by D. Bolger and L. C. Maguire, 11–21. Oxford: Oxbow Books.

Wimsatt, W. 2014. "Entrenchment and Scaffolding: An Architecture for a Theory of Cultural Change." In *Developing Scaffolds in Evolution, Culture, and Cognition,* edited by L. Caporael, J. R. Griesemer, and W. Wimsatt, 77–105. Cambridge, Mass.: MIT Press.

Woods, C. 2006. "Bilingualism, Scribal Learning, and the Death of Sumerian." In *Margins of Writing, Origins of Culture,* edited by S. L. Sanders, 91–120. Chicago: Oriental Institute of the University of Chicago.

Woods, C. 2009. "New Light on the Sumerian Language." *Canadian Society for Mesopotamian Studies Journal* 4:77–85.

Woods, C. 2010. "The Earliest Mesopotamian Writing." In Woods, Teeter, and Emberling 2010, 33–84.

Woods, C. 2015. "Contingency Tables and Economic Forecasting in the Earliest Texts from Mesopotamia." In *Texts and Contexts: The Circulation and Transmission of Cuneiform Texts in Social Space,* edited by P. Delnero, J. Lauinger, 121–42. Studies in Ancient Near Eastern Records 9. Boston: De Gruyter.

Woods, C., E. Teeter, and G. Emberling, eds. 2010. *Visible Language: Inventions of Writing in the Ancient Middle East and Beyond.* Oriental Institute Museum Publications 32. Chicago: Oriental Institute of the University of Chicago.

Zimansky, P. 1993. "Review of Schmandt-Besserat D. 1992, before Writing." *Journal of Field Archaeology* 20:513–17.

11

CULTURAL SCAFFOLDING AND TECHNOLOGICAL CHANGE
A Preliminary Framework

JOSEPH D. MARTIN

T ECHNOLOGY HELPS US to do new things or to do old things in new ways. This, at least, is our common understanding and continual hope. Technologies, however, only become useful when guided by human means to human ends, and they therefore do not add to our arsenal of abilities in an unproblematic, straightforward manner. Rather, they must confront a complex and preexisting set of biological traits and cultural practices before their potentialities and consequences are clear. My goal here is to sketch an account of how technologies interact with the innate and socially supported human capacities to learn and develop, using cultural scaffolding as an interpretive tool.

To realize that goal, I should first sketch some key terminology. *Developmental scaffolding* refers to the structures that support growth in developmental systems. It can be fruitfully applied to many types of systems, at scales from the microscopic to the institutional (Caporael, Griesemer, and Wimsatt 2014). *Cultural scaffolding* is a type of developmental scaffolding that describes the self-perpetuating patterns of systematic behavior—what Linnda Caporael (2014) calls repeated assemblies—organized to confer valuable skills or competencies to individuals or groups. I rely here on earlier formulations of the concept by William Wimsatt (2014) and Wimsatt and Griesemer (2007), who point out that cultural scaffolding, as a conceptual tool, can fruitfully describe a diverse array of systems. It can capture phenomena from the caregiver interactions that help children acquire language, to the repetitive training that allows a tennis player to hit a consistent backhand, to the rehearsal schedule that prepares an orchestra to perform a concert as a cohesive unit. This is, by design, a tool for a messy world, wrought broad and flexible to capture a wide variety of cultural practices.

Identifying the limits of what should qualify as technology, and what should not, is a similarly imprecise exercise. Historian of technology Robert Friedel, for example, notes this difficulty while provisionally defining technology as "the knowledge and instruments that humans use to accomplish the purposes of life" (Friedel 2007, 1). This definition would include many cultural-scaffolding processes. Military drilling, so critical to compelling soldiers to function as a cohesive unit, might be easily classified as both a scaffolding process and a technology. Friedel points out that it is easier to say what is *not* technology than to say what is. In that spirit, I exclude organized behavioral interactions among human beings from the technological realm. Some technologies, of course, might lack material form; although many technological ideas result in a material manifestation, like a better mousetrap or an improved transistor design, others, such as software programs, might not. For the purposes of this analysis, though, I will not consider to be technology those ideas that are enacted through human interactions—such as political and institutional organizations, games, and other social practices.

By characterizing these concepts, I do not aim to draw bright lines between them but to map out two realms, which are overlapping and interacting but nevertheless discernable. The location and the substance of their interaction is the subject of the following discussion, which describes the ways in which technological change influences cultural scaffolding and suggests how this understanding can be used to guide technology policy. I propose three ways of describing the interaction between technology and cultural scaffolding. First, technology can displace existing scaffolds. Second, it might combine with existing scaffolds and assume a role in scaffolding practices themselves. Third, it can catalyze the assembly and growth of new scaffolding structures. These effects are not exclusive of one another, but for the sake of clarity, I treat them separately. After outlining these interactions, I present a matter of current policy interest—the use of digital information technology in the K–12 classroom—to demonstrate how the scaffolding perspective offers policy utility.

TECHNOLOGY'S EFFECTS ON CULTURAL SCAFFOLDING

This section describes three primary ways cultural scaffolding can change in the face of technological change: displacement, combination, and catalysis. These are related processes and will often occur simultaneously. As a result, their boundaries are not sharply defined. Which one we perceive to dominate in any specific example will depend upon our frame of reference and interpretive goals.[1]

To offer a simple example, the widespread use of digital technologies, and therefore keyboards, to generate text has led to a reduction in the amount of time young people spend writing by hand, both in school and on their own. Recent research has uncovered evidence that this comes at a cost, not just to handwriting ability but to basic fine motor skills as well (Sülzenbrück et al. 2011). In this sense, digital technology is in the process of displacing scaffolding processes—formal handwriting instruction and the informal practice that accompanies it—that promote basic fine motor skills. The proliferation of computing technologies nevertheless also demands new scaffolding processes to confer new skills, such as touch typing, and also makes itself a part of existing scaffolding processes, for instance, by serving as a delivery mechanism for books and other content used in educational contexts.

This example illustrates that technologies are apt to have multiple, simultaneous influences on preexisting scaffolding structures. But for the purposes of generating a framework with practical utility, these three types of effects may be fruitfully considered independently. The distinctions drawn here are not designed to carry deep ontological implications but rather to describe features of technology-scaffolding interactions that are self-similar enough to be taken as discrete for the purposes of drawing attention to features of technological change that might otherwise escape our attention.

Displacement

Scaffolding displacement is the process by which new technologies encourage the cessation of capacity-promoting cultural-scaffolding activities. The following criteria describe the conditions under which it can occur:

1. Cultural-scaffolding processes support the acquisition of capabilities and competencies on the part of individuals and organizations.
2. New or improved technologies often replicate capabilities that individuals possess autonomously after benefiting from the existing scaffolding processes.
3. If a technology replicates a capability that is supported by an extant scaffolding process and the technology becomes prevalent, then the cultural activities that compose the scaffolding structure, which confers

autonomous agency, can be halted or altered. In such cases, the technology becomes the predominant way in which the capability is exercised.

4. If 1–3 occur, capabilities that individuals could once exercise without immediate technological assistance become difficult or impossible to accomplish without the aid of the technology that displaced the capability-conferring scaffold. By this mechanism, technologies become increasingly necessary and can thereby become entrenched (Martin 2015a, 5–6).

We often think of technology as expanding our range of capabilities, and it often does. But just as often, it encourages us to do things we were already able to do in new, technologically dependent ways. The history and philosophy of technology offer plentiful examples of technology displacing scaffolding processes. Langdon Winner's classic essay "Do Artifacts Have Politics?" describes the introduction of pneumatic molding machines in the 1880s into Cyrus McCormick II's Chicago plant, where the iconic McCormick reaper was manufactured. The machines were expensive to buy, install, and operate but nevertheless produced an inferior product than did the skilled casters who were deskilled by the machines. So why install them at all?

Citing Robert Ozanne's study of labor-management relations at the same factory (Ozanne 1967), Winner points out that the molding machines *settled an issue;* by replacing the skilled labor of hand molding with unskilled machine operator jobs, McCormick could undermine union action and tilt the balance of factory relations in favor of management (Winner 1986, 24–25). The machines, we can say, displaced a scaffolding process. In labor contexts, workers become skilled through years, sometimes decades, of apprenticeship and experience. The depth of their investment in scaffolding processes gives them social capital in the workplace because their expertise is resource-intensive to replicate, and so they are difficult to replace. The introduction of a machine that does the same work, even if that work is substandard, renders slow and demanding scaffolding processes redundant, with consequences for the social dynamics of the workplace. The consequence, once the scaffolds that supported skilled labor erode, is increased dependence on the new technology and further entrenchment of the technological system of which it is a part. Indeed, the process by which technology replaces skilled labor can be understood as a cultural analog of what Wimsatt (1986) calls *generative entrenchment* in molecular genetics, in which a gene is more resistant to evolutionary perturbation the greater the diversity of its developmental consequences. The mechanization of labor, which integrates machines more richly into the processes of production, causes a manufacturing system, in the absence of the expertise capable of replicating machine work, to become less flexible.

The example above is one instance of a more general phenomenon, one not limited to labor contexts. Scaffolding can also be displaced in the conduct of quotidian affairs, such as the example of the decline of handwriting and the consequences for fine motor skills. Relying on technology to accomplish tasks that we are able to complete independently—perhaps with the commitment of a bit more time and the exertion of a bit more cognitive effort—can inhibit acquisition of the capabilities in question. Scaffolding displacement, in short, occurs when new technologies discourage individuals or groups from partaking in particular kinds of scaffolding processes that would otherwise have allowed them to develop capabilities independent of the new technologies.

I begin with displacement because it is the subtlest aspect of the interface between technology and cultural scaffolding. First-generation users of new technologies typically will have acquired any capability the technology replicates with the benefit of existing scaffolding structures. Their comfort with these structures will sometimes generate resistance to adopting the new technology. But even if they do so, they will retain at least some of the skill they derived from older scaffolding processes, even in the event that their facility decays through disuse. As a result, displacement effects will appear most markedly in the generation that grows up using the new technology, and so lacks the incentive or the opportunity to participate in the scaffolding processes the technology displaced. This is a crucial policy consideration, which I will discuss in more detail below, but first let us consider the other ways in which technology and cultural scaffolding interact.[2]

Combination

Technologies are often themselves components in scaffolding processes. We require scaffolding to learn how to use existing technologies effectively, and technologies can help us attain capabilities that we can then exercise more or less independently, without immediate external aid. When new technologies appear, they can therefore change, supplement, or replace parts of existing scaffolding processes so that the relevant skill is scaffolded by different means. That is, new technologies encounter existing cultural practices with which they interact, and that interaction can result in their incorporation into those practices. The examples below describe three ways in which technological developments can combine with existing scaffolding practices.

Technological change has had a marked effect on the way sports are played. In ice hockey, for instance, skate construction has changed radically over the past fifty years. Hockey skates from the 1970s and earlier were predominantly of soft leather construction, much like leather boots, with little ankle support. They demanded considerable ankle strength on the part of the skater, favored long strides, and were not conducive to tight turns or quick stops and starts, which required tremendous strength and coordination from skaters attempting to keep their weight centered over the blade. Newer skates are much stiffer and more like ski boots, making it easier for skaters to center their weight. This does not make strong skating any less essential to success in the game, but it has changed what it means to be a strong skater. In a game played with hard-shell skates, quick stops and starts and hard, tight turns are more essential than long, graceful strides. The way young players are taught to skate has changed correspondingly, responding to new expectations created by the shift in skate construction. It would be problematic, though, to claim that modern skaters are any more or less proficient than their earlier counterparts. The technology, in this example, has produced a *qualitative change,* without enabling novel new possibilities.

In a different context, we see a raft of changes introduced by digital indexing and cataloging services in libraries. The library catalog is as old as the library itself, and it scaffolds the effective use of library resources—not merely helping researchers find materials but helping them think about how those materials relate to one another. The advent of computing technologies made it natural to migrate older physical systems, such as card catalogs, to digital format. Simple iterations of digital catalogs offer little beyond standard browsing and searching functions that were previously available with physical resources. The functions available through digital databases, though, have expanded substantially in recent decades, and in this case replacing one scaffold with another has opened up new possibilities. These capabilities are exerting subtle but substantial forces on the way scholars approach the research process (Martin 2015b). It is now possible to juxtapose a wide array of sources, which might be physically housed in geographically disparate locations, in a very short span of time, allowing researchers to make comparisons that would have been theoretically possible but practically infeasible using older analog catalogs. In this instance, the technological changes in existing scaffolding offer *novel capabilities.*

Combination, however, can also have a deleterious effect on how skills are scaffolded, especially when the technology in question offers advantages in other areas. Take audio recordings, which are now frequently used, usually in the form of software programs, to help language learners develop vocabulary and correct pronunciation and that combine with existing scaffolds for language instruction. Before recording technologies became widespread, language learners would need access to a native speaker to develop any reasonable grasp of pronunciation. With the advent of recordings, examples of correct pronunciation could be more widely distributed and easily accessed. This is a technological change that has little influence over the *way* the skill is scaffolded. It might, however, have consequences for how *effectively* it is scaffolded, considering that interactions with recordings are one-directional and so are less flexible than interactions with a native speaker, which would allow the learner to ask, for example, for additional examples or for immediate feedback. Although they can be distributed widely and thus may generate more occasions to hear a native speaker, recordings do not permit the learner to interact directly with a native speaker, who can take questions, provide context, and identify errors that might not be obvious to an untrained ear.

The case of language instruction shows how combination, in addition to generating qualitative differences in the way skills are exercised and enabling new capabilities, can also produce *interference* with skill conferral. Research on the efficacy of language instruction software has indeed indicated that, although software offers the learner greater convenience and autonomy, more resource-intensive teaching practices are better at conferring competency (Nielson 2011). This case of combination interfering with a scaffolding process is similar to scaffolding displacement, as outlined above. It differs from displacement, however, because the technology aims to scaffold the same skill as the preexisting cultural practice—that is, the end goal of the process is to enable the learner to speak on his or her own, rather than to make the technology the principle means by which people communicate in a second language (such as in the case of speech-to-speech translation apps). In the case of combination producing a reduction in scaffolding efficacy, we instead see the technology sacrifice some features of the learning process, in this case flexibility and adaptability, in favor of others, such as convenience and accessibility.

Our greatest hopes for new technology often turn on combination, assuming that we will be able to exercise or acquire existing skills in new, better ways. As these examples show, that can indeed be one effect of integrating new technologies into scaffolding processes. Offering new affordances is just one of the ways combination can have an effect on existing scaffolding, however. It might, as in the case of changes in hockey skates, enact a qualitative difference without offering anything strictly new; novice skaters still learn to skate but with an emphasis on a different style and differential emphasis on various fundamental skills. Combination might also, as in the case of language instruction software, reduce the efficacy of skill conferral or performance in favor of emphasizing some other value, such as convenience or access.

How technology combines with scaffolds depends as much on its implementation as its function. Language software is capable of supplementing language instruction by providing additional practice and consistent repetition, but if it is used to the exclusion of scaffolded interactions with experienced speakers, then its rigidity limits its utility. Here, we begin to see how this framework can be useful in a policy context. If the goal of new technologies is to confer existing competencies in a better way and to promote the generation of new capabilities, then policy-making around new technology should attend to how it combines with existing scaffolding structures in order to ensure that it does not compromise the elements of those structures that make them effective.

Catalysis

Scaffolding catalysis occurs when new technologies require the assembly of new scaffolds. Consider the following poem, "First Snowfall in St. Paul," by Katrina Vandenberg:

> *This morning in the untouched lots*
> *of Target, St. Agnes, and Lake*
> *Phalen, girls all over the city*
> *in the first snowfall*
> *of their sixteenth year are being asked*
> *by brothers, fathers—my cousin*
> *Warren—to drive too fast then lock*
> *their brakes, to teach them how to right*
> *themselves. The whine of the wheels, the jerk*
> *when they catch—from Sears to Como Park*
> *to Harding High, the smoke*

that bellows from their lungs,
the silver sets of jagged
keys, the spray of snow,
the driver's seat, the encouraging Go

From *Atlas* by Katrina Vandenberg (Minneapolis: Milkweed Editions, 2004). Copyright 2004 by Katrina Vandenberg. Reprinted with permission from Milkweed Editions. milkweed.org.

Vandenberg describes a ritual that will be familiar to anyone who came of age in the snowbelt of the United States, where the subtle skill of winter driving is essential. It is one example of the many informal scaffolding processes that have grown up around the automobile. New technologies, even while they displace and combine with existing scaffolding structures, can also prompt the growth of new ones. Rarely are the full potentials of a new technology, or the best practices for using it, self-evident. As a result, they require new and sometimes elaborate scaffolding to prepare new users.

The automobile necessitated both informal scaffolding processes, such as those described in the poem above, and formalized scaffolding practices, such as drivers' education programs, certification exams, and road tests. The car is part of a large technological system—entrenched by the highway and roadway infrastructure, urban planning decisions that assume its presence, and cultural traditions and expectations—with which it is all but essential for full participation in modern American society. Vandenberg's poem channels the idea that technology's success also depends on constructive interaction with local cultural practices—such as those that develop in response to local environmental constraints.

A notable feature of catalysis is that the scaffolding necessary to navigate new technological landscapes does not appear spontaneously for all groups who might benefit from it. Uncertainty about when and how that scaffolding matures can be a source of considerable friction as new technologies proliferate. Keeping, for the nonce, with the autovehicular theme, consider the number of traffic fatalities per year since 1900. Figure 11.1 shows how motor vehicle fatalities rose dramatically beginning in the 1910s, as automobiles became widespread. Many of these deaths would have been due to structural challenges: cars had to operate alongside horse-drawn carriages, streetcars and trollies, and pedestrians unaccustomed to heeding fast-moving vehicles. But it is also critical to note that the know-how needed to operate a

Figure 11.1. Fatalities caused by motor vehicles in the United States, 1900–2010. The x-axis is the year; the y-axis measures deaths as a proportion of the total population. Data from the National Highway Traffic Safety Administration and the Federal Highway Administration; courtesy of Wikimedia Commons.

motor vehicle, and operate it safely, took some time to penetrate the population. The scaffolding necessary to support safe motor vehicle operation took some time to catalyze.

It was not until 1932 that Amos Neyhart, an engineer associated with the Pennsylvania State College, instituted the first high school driver's education program in the United States. Neyhart's program, along with parallel industry initiative and government investigations to raise awareness of traffic safety, responded to a rash of traffic deaths beginning to be understood as an urgent public health threat (Damon 1958).

Driver education programs proliferated through the mid-1930s. Many of them attempted to teach driver safety as an element of good citizenship (Packer 2008). In his introduction to *Man and the Motor Car,* the widely adopted driver safety manual first published in 1936, Albert W. Whitney lamented: "We have shown little fighting spirit in the face of the hazard that the automobile has created,—perhaps because we have not been willing to discipline ourselves, perhaps because we have felt the pleasure and convenience that it has brought us was something that we could not have except at a price" (Whitney 1938, xi). The recipe for lowering that price, according to Whitney, was safer drivers, who would be made so by education programs in America's high schools.

The rise of driver's education programs in the mid- to late 1930s corresponds neatly with the leveling off in the rate of traffic fatalities in the same period, as seen in Figure 11.1. That is not, of course, the only factor to consider. Infrastructure was improving on the strength of New Deal programs, cars themselves were getting safer, and the population was becoming more aware of the dangers automobiles posed. The outbreak of World War II introduced additional complications. Gasoline rationing, the cessation of civilian automobile production, and other changes would have suppressed traffic mortality. Similarly, the introduction of a national speed limit in 1974 likely played some role in the reduction at that time. It is therefore difficult to disentangle the effect of education programs from other factors conspiring to suppress traffic mortality throughout the mid-twentieth century. Nevertheless, the realization that many drivers were ill-prepared to operate cars safely, and the concerted efforts to build the scaffolding structures that would allow them to do so, are an illustrative case of scaffolding catalysis. Contemporaries perceived poor driver preparation to be a contributing factor to high mortality rates, and they responded by developing formal systems of scaffolding to address the problem.

Catalysis shows us that the challenges posed by technology sometimes require more than technological solutions (see also Weinberg 1966). The automobile did not come prepackaged with the scaffolding necessary for people to use it safely and effectively, and the local, informal scaffolding procedures that would allow more experienced operators to confer their expertise to novices proved insufficient to stem the traffic fatalities that had become an epidemic by the early 1930s. The more rapidly new technologies are adopted, the more likely the scaffolding required to use them safely or effectively will lag behind. As of this writing, automobile safety is again a matter of widespread public and policy interest, this time focusing on the issue of self-driving cars. Should they come to fruition, autonomous vehicles would indeed offer a technological response to the problem of traffic fatalities and would displace the elaborate scaffolding that now prepares drivers to operate motor vehicles safely—perhaps even to the satisfaction of all but the most committed gearheads. But this solution, if it comes at all, will not be feasible until well over a century after the problem first arose. The intervening years required the catalysis of scaffolding structures, both formal and informal, to ease the integration of one of the most ubiquitous pieces of modern technology into American life.

CULTURAL SCAFFOLDING FOR EDUCATION POLICY

The framework outlined above can provide a practical guide for describing several key features of the interface between scaffolding and technology, especially for thinking about the policy challenges new technologies pose. Considering how to implement new technologies often emphasizes their potential—that is, how they will allow us do things better, faster, or more easily. But responsible implementation requires understanding technological change at a higher resolution, and that finer-grained perspective is something scaffolding language can offer. Managing technological change requires taking into account displacement effects and catalysis requirements, for example, alongside the potential efficiencies and new affordances combination can sometimes provide. This section illustrates how a lack of attention to the full extent of these factors has, in one case, undercut the stated policy goals for the implementation of a new technology.

When examining the consequences of technological change and evaluating policy responses, the considerations sketched above should be taken in conjunction: What scaffolding does the technology displace, how does it combine with existing scaffolding, and what new scaffolding might it require? Using the example of classroom-based digital information technology, it is possible to sketch how such an assessment can anticipate the challenges posed when deploying new technology in order to enact education policy goals. Bringing technology into the classroom has been the subject of a number of initiatives in the United States recently, both on the local and state levels. This section considers how one of those efforts has fared with respect to its stated aims and suggests how the framework developed in this paper can be used to assess it.

Recent efforts in the state of Maine to populate K–12 classrooms with laptops are notable for their scope. The Maine Learning Technology Initiative (MLTI), which began in 1999, was formed to implement a $50 million endowment. The MLTI aimed "to ensure a basic level of access to technology, the Internet and training and learning opportunities for all Maine public schools, students and teachers" (State of Maine 2001, 39). The rhetoric around the MLTI gave special emphasis to vocational skills. Maine's governor, Paul

LePage, praised the program by saying, "It is important that our students are using technology that they will see and use in the workplace" (quoted in Woodard 2013). MLTI rhetoric also developed the argument that giving every student a laptop would empower students to find the information they want. The task force reported: "Students learn largely by working on projects that connect with their own interests—their own visions of a place where they want to be, a thing they want to make or a subject they want to explore" (State of Maine 2001, 8). The MLTI, then, focused first on giving all students a baseline proficiency with digital technologies—a proficiency that potential employers expect—and, second, on allowing students to more easily manipulate digital technologies in a self-directed way and thereby to access information faster and more efficiently.

Close attention to the manner in which the MLTI was implemented, however, shows that the program fell short of these goals in some important respects. Karen Kusiak, in a doctoral dissertation based on extensive in-classroom observations at schools implementing the MLTI program, concluded that laptops were frequently introduced to Maine classrooms with little consideration to how they would integrate with the existing curriculum: "Students might unintentionally be directed to engage in classroom activities that do little to promote their skills and competencies. . . . Laptop use provides tremendous support for students to engage with high school curriculum and to benefit from instruction, however the underlying goals of the curriculum must be examined to be sure the use of laptops is for laudable purposes" (Kusiak 2011, 13, 254). Kusiak's observations showed that although having laptops in the classroom did help students acquire and maintain basic computer skills, their role in supporting the traditional curriculum had not always been thought out, and their presence therefore sometimes distracted from curricular goals and disrupted the learning environment. The introduction of laptops did proceed more smoothly in some cases because those individual teachers and school administrators took it upon themselves to ensure that the devices worked within the existing curriculum.

Ambivalence about information technology's role in developing the skills necessary to manipulate information is mirrored elsewhere in the media studies literature. Sonia Livingstone's research on how children use the Internet challenges the preconception that those who grow up in an environment rich with digital technologies—"digital natives," in Marc Prensky's (2001) terms—are naturally facile with them: "Watching children click links quickly or juggle multiple windows does not, necessarily, confirm that they are engaging with online resources wisely or, even, as they themselves may have hoped" (Livingstone 2009b, 5). Similar conclusions emerged from the Ethnographic Research in Illinois Academic Libraries (ERIAL) project, which observed student research habits at a number of Illinois universities. Andrew Asher and Lynda Duke (2012) concluded from observing student researchers at Illinois Wesleyan University that when considered in terms of their abilities to locate relevant resources using library catalogs and databases, "the seeming simplicity of tools like Google belies a complex and iterative process that requires the integration of numerous analytical and technical steps as well as knowledge and experience on the part of the user" (71). The consequences, Asher and Duke noted, were that students were less adept at understanding how information is organized, at evaluating sources successfully, and at figuring out how to access library resources.

These examples suggest that early efforts to exploit the potential of digital information technology to confer both new proficiency with computing technology and old research capabilities to students have been overly sanguine. Understanding these programs within the framework developed in this chapter can clarify why. The principle aim of the MLTI was catalysis. Maine's students, the initiative presupposed, were not exposed to enough computing technology in their existing educational and home environments to ensure that they graduated into college or the workforce with a baseline level of computer skills. This goal was laudable. The problem of differential access to the scaffolding necessary to successfully use new technologies is the source of considerable social justice challenges. The rise of digital information technologies, like the rise of the automobile, requires a new set of structures and processes to scaffold their effective use. Access to informal scaffolds supporting computer skills can differ according to geography, race, gender, and socioeconomic status, and formalizing previously informal scaffolds within the context of the public school system is one way to address this inequality.[3]

What the scaffolding perspective makes clear in this case, however, is that mere access to a new technology is not enough to catalyze the scaffolding required for its most effective use. Just like access to automobiles was insufficient for the skills necessary to operate them safely to penetrate a population, putting laptops in K–12 classrooms is insufficient to support competency with the variety of skills they support. This is doubly true in the absence of careful consideration of the way laptops influence the scaffolding processes already in place within the classroom. Kusiak's observations showed that the MLTI did not, in its inception, account for the variety of ways in which technologies interact with cultural scaffolding. Efforts to introduce laptops into Maine classrooms often equated access with catalysis and paid little heed to how the new technology would combine with the existing formal scaffolding of the school curriculum. In these cases, laptops combined with existing curricular scaffolding in a way that sometimes interfered with the goals of that existing scaffolding, rather than aiding them.

Kusiak's comparison of two Maine high schools that adopted laptops for English class instruction shows that their success depended principally on how well the curriculum was structured, rather than on the presence or absence of the technology itself. When care was not taken to ensure that writing tasks integrated with discussions of source material, laptops, although perhaps helping students gain experience with computing technology, led students to incorporate into their assignments commercial messaging and content that was orthogonal, if not counterproductive, to course goals (Kusiak 2011, 230–32). By focusing on catalyzing skill with computing technology, the MLTI sometimes failed to heed combination effects that could interfere with other curricular goals.

Displacement is also an evident consequence of the ubiquity of information technology. The ERIAL study's observation that, although they are proficient at manipulating digital interfaces, digital natives often lack basic skills associated with manipulating the information those interfaces organize, suggests that student are not participating in the types of activities that might have conferred those skills. The same technology that helps students access *content* about a new subject can also discourage developing familiarity with the *process* that supports competency finding, assessing, and organizing information about that same subject. ERIAL research suggested that the sense of ease conferred by digital tools could prevent students from engaging in behaviors that might have helped them learn: "Although the majority of . . . students struggled with finding the correct database to use, their search terms, locating a known item, and/or technical problems, not one student sought assistance from a librarian during an observed search" (Asher and Duke 2012, 83). These observations support the rationale that underwrites the decision to restrict calculator use in early math classes. Understanding the mechanics of arithmetic is critical for developing mathematical proficiency, even when a calculator might allow one to do sums more quickly.

From a policy standpoint, then, responsible implementation of digital information technology in K–12 classrooms requires attention to displacement, combination, and catalysis. Such attention suggests asking some straightforward but often neglected questions: What practices does the technology discourage that are worth preserving? How do existing practices need to shift to accommodate the new technology? What new practices need to be encouraged to ensure that the new technology is being used effectively?

Thinking with scaffolding also suggests a route to some preliminary answers. Observing that digital tools can disincentivize students from participating in the processes that allow them to think clearly and flexibly about locating resources indicates that we cannot assume that digital natives are natively proficient with digital research tools and that we should teach them accordingly. The scaffolding that can help students manipulate digital tools effectively needs to be catalyzed. Furthermore, the ways in which digital sources can be marshaled in support of a research project share much in common with the ways in which analog sources can be assembled for the same purpose. For students to learn how to conduct research, digital tools need to combine constructively with foundational training in identifying sources and assessing the reliability of evidence. In a similar vein, if and when we introduce laptops into classrooms, we should think about the balance between combination and catalysis: To what extent is our goal to development proficiency with new tools and to what extent are we using those tools to address existing curricular goals?

Scaffolding provides a vocabulary for describing the complex array of ways technology influences the capabilities we value and want to encourage. In so doing, it offers a potent and necessary antidote to the rhetoric of innovation and progress that often accompanies the introduction of new technologies (Russell and Vinsel 2016), rhetoric that focuses attention disproportionately on the novel capabilities new technologies allow, without attending to the possibility of displacement or recognizing the hard work of combination and catalysis they will require.

TECHNOLOGY AND cultural scaffolding interact in complex and multifaceted ways, a fact that should be no surprise given the diversity that exists within each. I nevertheless contend that we can characterize their interaction with enough clarity and resolution to ground reasoned and effective policy planning. I have outlined three modes of interaction between technology and cultural scaffolding and have argued that a lack of attention to one or more of these is likely to be the culprit when new technologies produce unanticipated effects or fail to meet the high expectations we often set for them.

Because discussions of technology often focus on what it allows us to do, we might pay less attention to the things it stops us from doing. Attending to these things highlights the ways in which technology can displace cultural scaffolds. When a new technology becomes an important part of navigating the world, the practices that technology causes us to cease might be critical components of scaffolding processes and worth the upkeep for their own sake. New technologies therefore present the challenge of identifying the capacities they might threaten by displacing scaffolds and concocting ways to preserve them, either by maintaining those scaffolds or by scaffolding the relevant capacities in different ways.

Ensuring that new technologies combine with existing scaffolds in a way that serves our desired ends demands a great deal of spadework that, as seen in the case of the MLTI, is easily neglected amid an enthusiastic, wide-scope embrace of new technology. When new technologies are called upon to accomplish existing tasks, it will not always be obvious how they can best be used to do so. The MLTI shows that conscious attention to how laptops in the classroom could be used to complement and supplement established curricular tasks is necessary to ensure a successful rollout that realizes the potential of the technology while guarding against its pitfalls.

One reason we might neglect combination—a reason, for example, we might give short shrift to questions about how we need to think through K–12 curricula to accommodate the changes to classroom practice that come with laptops—is that it is often easier to recognize that new technologies require the catalysis of new scaffolding to support competency with new technologies. Catalysis cannot be assumed to be spontaneous. It is not implied by the logic of the technology itself, and effective management of the systems that will support competency with new technologies alongside their safe and effective implementation is essential to responsible technology policy.

The assumption that new technologies will make their own way in the world and present to us their own optimal modes of use is seductive. If, however, we are committed to the idea that technologies can only be useful to the extent to which they modify, enhance, or expand human capabilities, then we must be sensitive to the processes that allow them to do so. I have offered one account of those processes with the hope that making them explicit is the first step to developing useful conceptual tools for navigating a world defined in many ways by technological change.

NOTES

Discussions with the participants at "Beyond the Meme: Articulating Dynamic Structures in Cultural Evolution" at the University of Minnesota helped bring this piece into focus. I am further indebted to Alan Love and Bill Wimsatt for their incisive comments and careful editorial work.

1. Note also that these all describe one-way effects, from technology to cultural scaffolding. This is not to imply that reciprocal effects are not possible, or even uncommon; the shapes new technologies take depend integrally on how people already participate in scaffolding processes, and they might shift and adapt in response to feedback from scaffolding processes. I limit my focus to the effects of technology on cultural scaffolding with the goal of establishing a clear, preliminary framework for discussing this interaction.

2. Displacement, it is worth noting, need not always be complete. In instances where scaffolding is partially displaced, it might be more appropriate to refer to it as *suppression*.

3. I thank Malik Horton for bringing this point to my attention.

REFERENCES

Asher, Andrew D., and Lynda M. Duke. 2012. "Searching for Answers: Student Research Behavior at Illinois Wesleyan University." In *College Libraries and Student Culture: What We Now Know,* edited by Lynda M. Duke and Andrew D. Asher, 71–86. Chicago: American Library Association.

Caporael, Linnda R. 2014. "Evolution, Groups, and Scaffolded Minds." In *Developing Scaffolds in Evolution, Cognition, and Culture,* edited by Linnda R. Caporael, James R. Griesemer, and William C. Wimsatt, 57–76. Cambridge, Mass.: MIT Press.

Caporael, Linnda R., James R. Griesemer, and William C. Wimsatt. 2014. "Developing Scaffolds: An Introduction." In *Developing Scaffolds in Evolution, Cognition, and Culture,* edited by Linnda R. Caporael, James R. Griesemer, and William C. Wimsatt, 1–20. Cambridge, Mass.: MIT Press.

Damon, Norman. 1958. "The Action Program for Highway Safety." *Annals of the American Academy of Political and Social Science* 320 (1): 15–26.

Friedel, Robert D. 2007. *A Culture of Improvement: Technology and the Western Millennium.* Cambridge, Mass.: MIT Press.

Kusiak, Karen. 2011. "Students and Laptops: Identity Construction in One-to-One Classrooms." PhD diss. University of Maine.

Livingstone, Sonia. 2009a. *Children and the Internet.* Cambridge: Polity Press.

Livingstone, Sonia. 2009b. "Enabling Media Literacy for 'Digital Natives'—A Contradiction in Terms?" In *'Digital Natives': A Myth? A Report of the Panel Held at the London School of Economics and Political Science, on 24th November 2009,* edited by Ranjana Das and Charlie Beckett, 4–6. London: POLIS, London School of Economics and Political Science.

Martin, Joseph D. 2015a. "Evaluating Hidden Costs of Technological Change: Scaffolding, Agency, and Entrenchment." *Techné: Research in Philosophy and Technology* 19:3–32.

Martin, Joseph D. 2015b. "New Straw for the Old Broom." *Studies in the History and Philosophy of Science Part A* 54:138–43.

Nielson, Katherine B. 2011. "Self-Study with Language Learning Software in the Workplace: What Happens?" *Language Learning and Technology* 15:110–29.

Ozanne, Robert W. 1967. *A Century of Labor-Management Relations at McCormick and International Harvester.* Madison, Wis.: University of Wisconsin Press.

Packer, Jeremy. 2008. *Mobility without Mayhem: Safety, Cars, and Citizenship.* Durham, N.C.: Duke University Press.

Prensky, Marc. 2001. "Digital Natives, Digital Immigrants." *On the Horizon* 9 (5): 1–6.

Russell, Andrew, and Lee Vinsel. 2016. "Hail the Maintainers." *Aeon.* https://aeon.co/essays/innovation-is-overvalued-maintenance-often-matters-more.

State of Maine. 2001. "Teaching and Learning for Tomorrow: A Learning Technology Plan for Maine's Future." *Final Report of the Task Force on the Maine Learning Technology Endowment.* 119th leg. 2nd sess.

Sülzenbrück, Sandra, Mathias Hegele, Gerhard Rinkenauer, and Herbert Heuer. 2011. "The Death of Handwriting: Secondary Effects of Frequent Computer Use on Basic Motor Skills." *Journal of Motor Behavior* 43:247–51.

Weinberg, Alvin M. 1966. "Can Technology Replace Social Engineering?" *Bulletin of the Atomic Scientists* 22 (10): 4–8.

Whitney, Albert W. 1938. *Man and the Motor Car.* 12th printing. Detroit: Board of Education.

Wimsatt, William C. 1986. "Developmental Constraints, Generative Entrenchment, and the Innate-Acquired Distinction." In *Integrating Scientific*

Disciplines: Case Studies from the Life Sciences, edited by William Bechtel, 185–208. Dortrecht, The Netherlands: Martinus Nijhoff.

Wimsatt, William C. 2014. "Entrenchment and Scaffolding: An Architecture for a Theory of Cultural Change." In *Developing Scaffolds in Evolution, Cognition, and Culture,* edited by L. Caporael, J. Griesemer, and W. Wimsatt, 77–106. Cambridge, Mass.: MIT Press.

Wimsatt, William C., and James R. Griesemer. 2007. "Reproducing Entrenchments to Scaffold Culture: The Central Role of Development in Cultural Evolution." In *Integrating Evolution and Development: From Theory to Practice,* edited by R. Sansome and R. Brandon, 228–323. Cambridge, Mass.: MIT Press.

Winner, Langdon. 1986. "Do Artifacts Have Politics?" In *The Whale and the Reactor: A Search for Limits in an Age of High Technology,* 19–39. Chicago: University of Chicago Press.

Woodard, Colin. 2013. "Maine Laptop Choice Leaves Schools Relieved, Confused." *Portland Press Herald,* April 30. http://www.pressherald.com/2013/04/30/state-laptop-choice-leaves-schools-relieved-confused_2013-04-30/.

12

THE EVOLUTION OF THE SOCIAL SELF
Multidimensionality of Social Identity Solves
the Coordination Problems of a Society

PAUL E. SMALDINO

WE EACH CONTAIN MULTITUDES

My mother grew up in a largely Jewish neighborhood in Brooklyn and resided in the New York metropolitan area until just a few years ago, when my parents retired to Colorado. Since the move, my mother talks much more often about being culturally Jewish and actively seeks out interactions with fellow Jews. She prepares traditional Jewish dishes such as kugel and matzo ball soup with increasing frequency, occasionally refers to an idiosyncrasy as "a Jewish thing," and has hung in her foyer a poster featuring an Asian boy holding a sandwich with the caption "You don't have to be Jewish to love Levy's real Jewish Rye." She cherishes a coffee mug that features the quotation: "I never think about being Jewish until I leave New York."

The coffee mug makes sense. The New York metropolitan area is home to the largest Jewish population outside of Israel. In Brooklyn, New York's most populous borough, 23 percent of its 2.6 million residents are culturally Jewish (Cohen, Ukeles, and Miller 2012). These numbers are especially impressive given that Jews comprise only about 2.1 percent of the total U.S. population. No matter where you go in the United States, when you leave New York, there are fewer Jews. This is important because American Jews have many cultural traits in common by virtue of being Jewish and American, irrespective of whether they are found in New York, Los Angeles, Denver, or Atlanta (Whitfield 1999). The difference is that in New York being Jewish is both so common and so pervasive in the larger culture of the city that *being* Jewish is not, in itself, a particularly useful signal of an individual's norms and perspectives, and so fades into the background of many Jews' identity palettes. It may be more informative to identify as a psychotherapist, or a Buddhist, or a Libertarian; these identities appear in smaller numbers, and so by announcing oneself as such, one can more effectively find others with similar values.

In suburban Colorado, on the other hand, Jews may want to be more proactive in seeking each other out by signaling their Jewishness. This is both because being Jewish is now an informative signal (in the information theoretic sense that it is surprising or unusual) and because the associated norms and perspectives diverge more noticeably from those of the general population. Most people are worthless if you're looking for a decent knish.

My intention in this chapter is neither to talk about my mother nor about Judaism.[1] Rather, I want to talk about a facet of human existence that has been largely underplayed in discussions of cultural evolution: social identity (though see Moffett 2013). This is puzzling because social identities often serve as cultural demarcations and, as I will argue, help humans to solve a crucial coordination problem that would otherwise impede the large-scale cooperation that, some say, defines our species (Bowles and Gintis 2011).

The discussion above highlights several important features of social identity. First, social identity is important. Humans place an immense value on clearly identifying to others *who they are* and *to which groups they belong.* Second, social identity is context dependent. Who I am, and how I express that to you, depends on where I am, who you are, and who else is around. A corollary of this is that social identity is multidimensional. Each of us contains multitudes. We are all many things, and we are different things in different contexts, with different people, in different times and places. These shifting identities help us to act and respond appropriately, both to identify ourselves to the right individuals and to differentiate ourselves from the crowd.

These facts have obvious implications for organizational psychology and the social sciences. Less obvious, perhaps, are their implications for human cultural evolution. In this chapter, I want to talk about how the complex nature of human social identity helps solve a key problem in the evolution of human societies: cooperative group formation. Following that, I will discuss how the role of social identity in facilitating cooperation has changed as human societies themselves have changed.

COOPERATION, COORDINATION, AND GROUP-LEVEL TRAITS

When compared with other social mammals, human cooperation is astounding. Much has been made of the extreme propensity human beings possess for altruism and other forms of cooperation with their fellow humans. As an evocative example, Sarah Hrdy (2009) has pointed out that the ability of three hundred or more strangers to sit calmly in an airplane for a transoceanic flight—replete with crying babies, snoring neighbors, and ever-shrinking seat sizes—is a marvel in the animal kingdom. Three hundred chimpanzees similarly locked in a metal cabin for eight hours would rip each other to pieces.

This predilection for prosociality is not adequately explained by the mechanisms traditionally employed to explain cooperation in nonhuman species—namely, inclusive fitness and reciprocity. For example, humans in contemporary industrialized societies often cooperate with unrelated strangers in one-shot interactions. Explaining this type of large-scale cooperation probably requires consideration of how cultural transmission (and forms of cultural inertia such as niche inheritance and technological lock-in) interacts with a developmental psychology predisposed to social learning, conformity, and empathy to create a species that has come to dominate the global ecosystem through its ability to cooperate with relative strangers instead of attack them (Laland, Odling-Smee, and Feldman 2000; Arthur 2007; Chudek and Henrich 2011; Tomasello et al. 2012; Smaldino 2014; Wimsatt 2014; Richerson et al. 2016).

Most theory on the evolution of cooperation has treated it as an individual's propensity for prosocially helping another, even if that entails a cost on the part of the helper. In other words, cooperation is an individual-level trait. This characterization is unsurprising. In general, theories of both biological and cultural evolution have generally focused on the evolution of individual-level traits—physical properties and behaviors that are heritable through genetic or cultural transmission. Such traits are generally presumed to be the property of a particular organism, and it is through selective survival and reproduction that evolution hones the trait-environment fit of a species. Yet traits need not only describe properties of individual organisms. As a classic example, the cellular slime mold *Dictyostelium discoideum* forms a slug-like proto-organism when resources are scarce, enabling a group of otherwise free-living amoebae to move to higher ground and for a select few to disperse to more nutrient-rich territory (Savill and Hogeweg 1997). The structure and behavior of this slug is crucial to the life cycle of *D. discoideum,* yet it is not accurate to describe these features as a trait of any individual amoeba. Instead, they are emergent group-level traits.

Groups of humans also exhibit many emergent group-level traits (Smaldino 2014). These groups may often be ephemeral, with a group coming together for an activity and disbanding. A major difference between emergent group-level traits in humans and in other organisms is that, in the case of humans, the process of group formation for any specific trait is only minimally controlled by genetics (even in other species, principles of self-organization and environmental feedback likely play a large role).[2] Compared with other species in which there is widespread division of labor—the ants and termites, for example—humans are more morphologically uniform and yet much more behaviorally diverse. This is easy to see if one considers the enormous variety in the nature and behavior of groups of humans working together in organized, coordinated, and often differentiated roles. A cappella choirs, sailing crews, hunting parties, soccer teams, drum circles, policy institutions, farming collectives, winter harvest festivals, urban infrastructure, software development teams, film crews, pickup basketball, military service, commerce. There are myriad ways in which people can work together (for an excellent review of teamwork in humans and other species, see Anderson and Franks [2003]).

Human cooperation often involves groups of individuals working together in a coordinated fashion toward common or mutually beneficial goals. How humans coordinate to *form* cooperative groups, which often involve the emergence of group-level traits, is a major problem for the development of theories of cultural evolution (Smaldino 2014). In order to proceed, it will help to discuss the general problems associated with cooperation.

The Cooperation Problem

The problem of cooperation is often stated in the language of evolutionary game theory: How can individuals with cooperative strategies invade and continuously outperform free riders? In other words, cooperating is risky. If you help your partner but she doesn't help you, you are a sucker as well as an evolutionary dead end. So how can cooperation evolve so that cooperators aren't suckers?

Decades of research have been put into this question. The overly simplistic but largely correct answer is that most of it has to do with positive assortment. As long as there is some mechanism that allows cooperative individuals to interact preferentially with each other, they can outperform free riders who can't reap the benefits of synergy, while avoiding being played for a sucker. There are a bunch of mechanisms that allow this to happen. Interacting preferentially with kin is a good one. Cooperation can evolve and stabilize through inclusive fitness when closely related individuals interact with one another, through either proximity or some sort of recognition mechanism. Critically, these mechanisms work even if the individuals are *not* closely related, as long as they each share cooperative traits that they can pass on either genetically or culturally (Hamilton 1964; McElreath and Boyd 2007; Gintis 2014). One way this kind of assortment can occur is through limited dispersal—when offspring live their lives near the location in which they were born (Koella 2000; Mitteldorf and Wilson 2000; Kümmerli et al. 2009; Smaldino and Schank 2012). Another way to stabilize cooperation is to make it costly to do otherwise. Partner selection and explicit punishment are among the ways to get this done, and in humans explicit institutions have arisen to do just this (Richerson and Henrich 2012; Ostrom 2014).

Yet another mechanism is to signal with group markers or tags, which can aid assortment by signaling whether an individual is in your group and so is likely to cooperate again either with you or someone you know (Axelrod, Hammond, and Grafen 2004; Hammond and Axelrod 2006; Cohen and Haun 2013). This last mechanism speaks to our earlier discussion of social identity, which I will argue functions as a sort of multidimensional, context-dependent marker for assortment. But, if cooperators can effectively signal to each other with simple tags, why might such a complex mechanism as context-dependent, multidimensional social identities be necessary for assortment? The reason is that finding other cooperators is only part of the problem associated with effective cooperation.

The Hermione Dilemma

The central problem of cooperation is usually framed in terms of how cooperators can invade and outperform free riders. This cooperation problem is largely solved, even if some but-fors and nitty-gritties remain to be worked out. In the case of humans, people are often cooperative. We are the cooperative species, after all (Bowles and Gintis 2011). Problem solved. However, we are still left with the *other* problem of cooperation: how to best generate a benefit between two or more cooperators (Calcott 2008; Smaldino 2014). Often, the question for an individual is not how to find *someone* who will cooperate but how to find the *best* person to cooperate with (Nöe and Hammerstein 1994; Tooby and Cosmides 1996; Barclay and Willer 2007).

For illustrative purposes, consider the characters in J. K. Rowling's popular *Harry Potter* fantasies.[3] Clever and resourceful Hermione wants to fight the Dark Lord Voldemort, and luckily she has a bevy of helpful would-be heroes just waiting to assist her! On her left is the one and only Harry Potter: holder of the most telling of scars, Harry is brave, talented, and buoyed by throngs of admirers and supporters. On her right is bumbling Neville Longbottom: kindhearted but clumsy, socially isolated, and possibly a bit dim. Which of these two should she choose to join her in her quest to rid the wizarding world of evil? The problem here is categorically *not* how to pick the cooperator instead of the free rider. Instead, the difficulty is to choose the *best* cooperator, given the task at hand and Hermione's extant personality and skill set. Hermione is doing more than choosing a cooperator. She is choosing a collaborator: someone with whom she will have at least partially aligned goals and with whom she will coordinate to generate synergistic benefits. To make her choice, Hermione is aided by the overt and tacit signals sent by Harry and Neville, advertising their vices and virtues.

How to choose whom to cooperate with is a general problem that humans face all the time. From among the pool of potential partners who might be willing to cooperate, an individual must find a partner or team with whom interests are aligned, norms of behavior and communication are shared, and skills and experience are either common or complementary, depending on the task. To form successful collaborative partnerships or teams, individuals have to find the right people and make themselves desirable to them.

THE ROLE OF SOCIAL IDENTITY IN COLLABORATIVE GROUP FORMATION

Among cooperative individuals, there are myriad ways in which they might cooperate. This is often discussed as a problem of coordination. Assuming two or more individuals have the psychological machinery for shared attention and joint behaviors (Tomasello et al. 2005; Gallotti and Frith 2013; Heyes 2013), it is beneficial for them to maximize the degree to which they can harmoniously coordinate their efforts to generate the most productive synergistic outcome. If they share goals, vocabulary, and behavioral norms, coordination may go deeper and more smoothly, generating a larger benefit compared with individuals who cooperate out of obligation or necessity but must struggle to find common ground (McElreath, Boyd, and Richerson 2003; Calcott 2008). Thus, individuals must find a way to assort not only according to their cooperative tendencies but according to their norms and values. It is proposed here that social identity facilitates this kind of assortment.

For the purpose of this discussion, I follow the social psychologist Kay Deaux (1993) in allowing for a fairly broad definition of social identity: social identities are those roles or membership categories that a person claims as representative. These can include groups such as "Asian Americans" or roles such as "mother." This definition is by and large aligned with the sociological concept of the reference group and is also consistent with how identity is discussed in sociocultural anthropology. In his well-known chapter on ethnic groups and boundaries, Barth (1969) writes:

> It makes no difference how dissimilar members may be in their overt behavior—if they say they are A, in contrast to another cognate category B, they are willing to be treated and let their own behavior be interpreted and judged as A's and not B's; in other words, they declare their allegiance to the shared culture of A's. (15).

A key point here is that identity is not just something that is felt internally, as is the view from psychoanalytic theories concerned with the "struggle for identity" (e.g., Erikson 1968) as well as social psychological theories concerned with self-conceptualization (e.g., Brewer 1991; Hogg 2000). Self-concept is an interesting and surely important factor in explaining human behavior, but it is neither highly relevant to the present discussion of cooperative assortment nor easily measured in any sort of experimental paradigm. Instead, I am concerned with social identity as something that is actively and outwardly expressed.

Social Identity as a Signal

The expression of a social identity might take the form of an overt declaration ("I love socialism!"), covert signals such as encrypted jokes referencing shared experiences (Flamson and Bryant 2013), or markers such as clothing or vocabulary. Because of the high dimensionality of social identity, however, an individual cannot and should not express every facet of his or her identity. Rather, a subset gets expressed depending on context. But which subset? This question has been investigated by social psychologists who fall broadly into two camps. Both camps focus on the need to distinguish one-self from others but differ on the reasons for doing so and from whom one should differentiate oneself.

Theories related to distinctiveness or uniqueness focus on carving out a niche for oneself and thereby differentiating oneself from similar others (Snyder and Fromkin 1980; Brewer 1991; Vignoles 2011). In particular, *optimal distinctiveness theory* (Brewer 1991) posits that individuals adapt their self-concept to balance opposing needs for assimilation and differentiation. This adaptation is presumed to be based on the relative distinctiveness of the various components of their overall social identity in the current social landscape. For example, if I am a Socialist Muslim, I might identify more strongly as a Socialist when Socialists are rare and Muslims common, and as a Muslim in the opposite case. Though unable to test for internal self-concept, experiments have shown that Western college students do alter their expression of social identity based on the relative distinctiveness of those components in at least some settings (Pickett, Silver, and Brewer 2002).

Optimal distinctiveness theorists sometimes adopt an adaptationist rationale for their posited innate psychological desire for belonging to groups of relatively moderate size. For example, Leonardelli, Pickett, and Brewer (2010) suggest that such preferences allowed hominins to optimize the size of their cooperative groups, reaping the benefits of scale while avoiding the free rider problems found in large collectives. Without considering if the genetic evolution of such preferences is even feasible (see Gould 1991), we can first ask whether such individual preferences would, in fact, give rise to "optimally" sized groups that maximize the benefits to their constituents. Mathematical modeling suggests that this is unlikely. Smaldino et al. (2012) modeled a simple scenario in which all agents had group identities, had identical preferences for a moderate relative group size, and switched groups when another group had a preferable size. They showed that this scenario led to assortment into overly large groups in which no one's preferences were satisfied, except in the case where rigid network structures were imposed. In other words, preferences for relative distinctiveness did not result in group sizes that reflected those preferences. Moreover, group size is likely to be determined by the specifics of the task at hand and the resources available to group members, rather than by the aggregate preferences of its members. If there are benefits to group membership, then the interests of those who want to join a group may be opposed to the interests of those already in the group, who would be better off keeping them out (Smith 1985; Giraldeau and Caraco

2000; Smaldino and Lubell 2011, 2014). In this case, group size will equilibrate to the point where the benefit lost to group members by adding a member is equal to the cost of barring a new member from entering (Smith 1985; Giraldeau and Caraco 2000). In addition, the optimal size for cooperative groups will likely be task dependent. As a result, a passive mechanism for determining group size—such as a general preference for joining groups of a particular size—will be insufficient to facilitate optimal assortment in most cases. It is therefore quite unlikely that strategies of social identity expression have evolved to optimize group size for cooperative endeavors. Instead, it seems more likely that the expression of social identity is geared toward assortment into groups in which the constellation of social identities satisfies the *group-level* needs for coordination and division of labor.

This line of reasoning does not invalidate the experimental findings of the optimal distinctiveness theorists, nor does it suggest that individuals do not strive to differentiate themselves from similar others. The most obvious benefit to differentiating oneself from the crowd is that it allows one to more easily find collaborators. But once this has been achieved, another mechanism is required to facilitate further assortment into groups.

Another camp of social identity theorists, the *identity signaling theorists* (Berger and Heath 2008), suggests that the expression of social identity functions largely to differentiate oneself from those who are *different,* in order to ensure that others understand who they are and do not mistake them for those with opposed norms or values. In other words, people understand when there is a chance they may be mistaken for a member of another group and take active precautions against this. For example, Stanford students in a typically "jocky" dorm were sold a one-dollar "Livestrong" bracelet, as was another dorm across campus as a control (to test the effect of boredom). A week later, bracelets were also sold to members of a neighboring "dorky" academic dorm, and these students tended to interact heavily with members of the first dorm in classes, dining halls, and so on. After another week, 32 percent of the jocks but only 6 percent of the control dorm members had stopped wearing the bracelets (Berger and Heath 2008, Study 2). If the expression of social identity is to make sure others know who you are, then one should abandon a signal when it is not reliable.

Both camps of social identity expression get something right. Social identity helps an individual to stick out from the crowd *and* find similar others, the latter being achieved both through the disassociation from dissimilar others discussed by the identity signaling theorists as well as through direct assortment for similarity, or homophily (McPherson, Smith-Lovin, and Cook 2001). In other words, *the expression of social identity functions as a signal to facilitate assortment for successful coordination.*

SOCIETAL STRUCTURE AND THE MULTIDIMENSIONALITY OF SOCIAL IDENTITY

In his 1961 novel *Mother Night*, Kurt Vonnegut tells his readers, "We are what we pretend to be, so we must be careful about what we pretend to be." Therein lies an important lesson about the difference between self-concept and self-expression, and a reminder that we are judged by our actions, not our thoughts. However, as a psychological theory, Vonnegut's analysis is lacking. We may be what we pretend to be, but we pretend to be *lots of things.*

Each of us has multiple identities. In-group biases are well documented in intergroup interactions, but we should recall that we all belong to multiple in-groups. We identify and are identified by family, friends, work, gender, politics, race, the sports teams we support, and the music we listen to (and the corresponding t-shirts we wear). Which of these identities is most salient is dependent on the context. Our lives are multifaceted, and different aspects of our social identity are expressed and utilized in different social and behavioral contexts (Long 1958; Deaux 1993; Putnam 2000; Roccas and Brewer 2002; Ashmore, Deaux, and McLaughlin-Volpe 2004). There are, of course, individual differences in the ways in which the multidimensionality of social identity is conceptualized and expressed (Roccas and Brewer 2002), but this aspect of that multidimensionality is not what concerns me here. Let us simply assume that humans express social identities in a manner that accounts for their multidimensional and context-dependent natures. The point I want to make is that, because humans have to cooperate in many different contexts, *the multidimensionality of social identity is important for successful coordination.*

Different Societies Imply Different Roles for Social Identity

This is the point at which we finally encounter the topic of cultural evolution, because the structure of society will determine the contexts for cooperation and therefore impinge on the multidimensional nature of social identities.

Different societal structures create different strategic opportunities and necessities for interaction. The sociologist Miller McPherson notes, for example, that as societies transition from small-scale foraging structures to large-scale agricultural structures, "the activation of entirely new dimensions such as education, occupational prestige, and other distinctions come into play" (McPherson 2004, 266). Of course, certain identities are specific to certain cultures, corresponding to particular idiosyncrasies of history, religion, or climate. These differences will affect the specific ways in which social identities are expressed. However, there may be certain regularities found in the expression of social identity across cultures. These regularities include the diversity of social identities in a population, as well as the manners in which social identities are expressed to facilitate coordination.

I will simplistically talk about human societies as varying on a continuum of "complexity." I want to be careful to note that I mean to make no judgments or appraisals related to the worth or quality of a society in referring to one as more or less complex. In particular, the complexity, depth, or intelligence of the *individuals* in those societies is completely orthogonal to this discussion. Rather, let the complexity of a society be a gestalt measure encompassing the size of its population and the diversity of the specialized social and economic roles held by its members. By this measure, a foraging society of a few thousand would be relatively simple, and a modern international community would be maximally complex. Other, more precise measures are obviously desirable, but this is somewhere to start.

Social Identity Expression in "Simple" Societies

For most of human history, people lived in relatively small groups. Studies of modern hunter–gatherers show that although extended societies can number in the thousands, most of an individual's time is spent with small foraging groups numbering between thirty to one hundred adults (Hamilton et al. 2007; cf. Caporael 2014). Most discussion of identity as a facilitator of cooperation and coordination in both contemporary and prehistorical foraging societies has focused on the role of overt ethnic markers or tags (such as language, accent, or clothing) for distinguishing between cultural groups (Barth 1969; Cohen and Haun 2013; Hammond and Axelrod 2006; McElreath et al. 2003; Moffett 2013). In such societies, only outsiders had to be identified by a tag, as in-group members could be known directly, through either personal experience or reputation (Apicella et al. 2012; Hamilton et al. 2007).

In a small, perfectly egalitarian society—which may, some have claimed, describe the conditions under which typical ancestral Pleistocene humans lived (Boehm 1999; though see Smith et al. 2010)—the diversity of roles should be quite low. Individuals would be identified by the persistence of role diversification based largely on sex, age, or skill. The expression of social identity in such a society should therefore be minimal. Although individuals vary in their skills, experiences, and outlooks, these differences tend to be known by everyone in the community in very small-scale societies.

This is not to say that, in small-scale societies, assortment based on shared norms and values is necessarily a trivial problem. Individual differences abound even in small groups, and individuals in all societies develop friendships with their preferred interaction partners, solving the problem of assortment by identifying specific individuals to cooperate with (Hruschka 2010). Fostering deep friendships takes time, and finding potential friends is still a challenge. Flamson and Bryant (2013) have raised the interesting proposition that within small communities, jokes and other forms of humor serve as encrypted signals that allow similarly minded individuals to preferentially assort without alienating dissimilar in-group members, with whom they must still occasionally cooperate. Such a strategy solves the problem of coordination without a need for overt identity expression.

Social Identity Expression in "Complex" Societies

With the introduction of hierarchy and social classes, social identity within the community can become concerned with largely prescribed roles and with facilitating the proper behavior between two or more actors in consideration of their positions. For example, in the Indian caste system, individuals from different castes may cooperate in the domain of farming, but not intermarry (Waring 2012). Here, social identity can facilitate smoother assortment for coordination, as individuals can be placed into categories of potential or forbidden partners based on their belonging to a particular class, saving the individuals the trouble of getting to know every other individual in depth to make such an assessment.

After the rise of agriculture, societies became larger, more complex, and more entangled with other societies via trade networks (Johnson and Earle 2000; Moffett 2013; Richerson and Boyd 1999; Gowdy and Krall 2016). In a population that is both large and has a high degree of diversity in social roles, individuals must often interact cooperatively with relative strangers in a large variety of contexts. In complex societies, therefore, mechanisms for establishing trust, compatible skills, and common norms and values become increasingly important to the formation of cooperative groups.

Complex societies pose two new problems for human cooperation. First, as human societies grew larger, members of cooperative groups would increasingly have to interact with individuals whom they had not previously encountered or otherwise knew little about, making finding partners for cooperation and coordination increasingly difficult. Second, as the diversity of roles within a society became greater, individuals would increasingly have to modify the expression of their social identities to relate to others in a larger variety of contexts. To solve these problems, individuals in complex societies must make rapid, accurate decisions regarding both whether to cooperate with a potential partner and how to do so.[4] Individuals looking to join a group as well or let new members into their groups facilitate this process through the expression and evaluation of social identities.

To be clear, I am not suggesting that individuals' social identities need be more or less rich in different societies. Individuals have complex and well-developed identities in all known societies. Rather, I argue that in more complex societies, the landscape of possible identities is more heterogeneous, and the multidimensionality of social identity is employed more directly as a coordination device. Thus, the advent of social identities in modern complex societies, such as national or regional identities; religious affiliations; or various fan communities for sport teams, film, or music, may be indicative of a cultural evolved solution set to the problem of assortment for cooperation and coordination in an expanding world.

A hint at how this kind of psychological transformation could occur through cultural evolutionary processes is suggested by the results of a recent study by Isabel Scott and her colleagues (2014). They looked at male and female perceptions of facial characteristics in potential sexual partners across twelve populations that included complex, urbanized societies as well as smaller-scale pastoral and foraging societies. They found that highly dimorphic preferences—square jawlines in men and softer, rounder faces in women—were more prevalent in urban, large-scale societies. Individuals in smaller-scale societies, in contrast, did not rely on such signals. They also found that the degree to which more masculinized faces were perceived as more aggressive was strongly correlated with the percent of the population that was urbanized—that is, living in a large, complex social environment. An interesting aspect of this finding is that masculinity *is* a reasonably strong predictor of aggression, in part because it correlates with circulating testosterone levels (Pound, Penton-Voak, and Surridge 2009). It seems quite possible that the stereotype of square-jawed men as more aggressive never developed in smaller-scale societies because there was no need for it—when aggressive men can be known individually and by reputation, such a stereotype would equate to discarding a strong signal for a weaker one. In large-scale societies, however, where rapid evaluation of strangers is paramount, such stereotypes may become useful heuristics.

Hogwarts Revisited

Let us return briefly to the problems of partner selection facing Hermione. In the earlier example, she was able to use firsthand knowledge of Harry and Neville in order to contrast them as potential cooperation partners. All three students are members of Gryffindor House and have regular contact, including extensive interactions on their very first day of school. This example reinforces the point that in complex societies, social organization takes many forms, and social identity is hardly required for *all* partner selection problems. However, consider a scenario in which Hermione, being of above-average intelligence and skill, is allowed to skip a grade. She finds herself in mixed classes with unfamiliar, older students from both Gryffindor and Ravenclaw. For partnered activities, she might use membership in Gryffindor as a first-pass signal to reduce the set of potential partners to those she knows share her values of bravery. This would be an especially good strategy if the activity required bravery, such as taming a wild hippogriff. On the other hand, suppose the activity was something requiring exceptional cleverness, such as devising an empirical test for how magical ability is transmitted from parents to offspring. In this case, Hermione might favor her identity as a clever person over her identity as a Gryffindor, and choose to partner with a Ravenclaw. The complex societal structure of this world facilitates many identities and many permutations for assortment.

COGNITIVE TRADE-OFFS BETWEEN DEPTH AND BREADTH OF SOCIAL IDENTITY

The ways individuals in the West conceptualize and express social identity are often written about as if they represent universal features of human nature. Indeed, proposed adaptationist explanations for these social identity strategies (e.g., Leonardelli, Pickett, and Brewer 2010) assume that our Pleistocene ancestors thought about other people in a manner similar to how we now think about ourselves and others. Such assumptions are dubious. Instead, the psychological nature of social identity is highly constrained by the structure of one's cultural milieu. Thus, as the structures of cultural societies have changed, so too have the ways in which humans have conceived of and expressed social identity. Indeed, the relative recency of the transition to agriculture and the emergence of complex societies suggest that many strategies related to the expression of social identity likely arose through cultural rather than genetic evolution. It would hardly be surprising if a rethinking of the psychology of social identity with an eye toward cultural evolution is necessary, given the extent to which psychological research has been skewed by a focus on minds belonging to individuals raised in Western, industrialized societies (Henrich, Heine, and Norenzayan 2010).

A few anthropologists have noted to me in private that individuals in "simpler" societies appear to have, if anything, a richer conception of identity. By this, I believe these anthropologists mean that, for people in these societies, the differences between the individuals they know are more pronounced, and the number of labels they can put on any individual is higher. If true, it strikes me as possible that there exists a cognitive trade-off in the depth and breadth of social identities an individual might use to identify himself or herself and others. As the diversity of contexts for identifying people increases, the depth at which any one person can be identified would decrease. This is surely one of the more speculative elements in this chapter, but it may nevertheless have legs.

To see how this might work, we can remind ourselves that the phenomenological aspects of the human mind emerge from the activity of the meat brains residing in our skulls, and that these brains are made of neurons. We can therefore draw insights by analogy to artificial neural networks. Consider a network with a fixed number of nodes, tasked with pattern discovery. In developing this example, I assumed the relatively simple architecture of feature discovery through competitive learning developed by Rumelhart and Zipser (1985), but any number of other architectures would suffice. The purpose of this type of network is to classify a large set of stimuli into discrete categories, such that the number of categories is not predetermined but has an upper limit of M. Here, the network is tasked with categorizing an individual's interaction partners based on a suite of expressible traits (their identities). If an individual's interaction partners are all drawn from the same group, then the network should discover systematic differences that exist between clusters of individuals within that group. On the other hand, if the set of interaction partners is drawn from across many groups, such that individuals within each group have correlated traits, then the network is likely to cluster individuals by group and, by extension, treat all individuals from each group as identical.

This is obviously an oversimplification of a complex aspect of human cognition, but it helps to explain how there might be a trade-off between depth and breadth in how we categorize those we interact with. Depth is sacrificed for breadth when one must coordinate with individuals with whom one shares less cultural and developmental overlap. Such a trade-off might be investigated empirically both through computer simulation with artificial neural networks as well as through direct cross-cultural comparison of how individuals in differently structured societies categorize themselves and others. More generally, this discussion points to the fact that social identity serves a social purpose, and that purpose is dependent on the strategic needs, opportunities, and affordances of the individuals in a given society.

Using Social Identity to Talk about Others

An alternative perspective on the trade-offs between simple and complex societies stems from the role of social identity as a way of communicating, not only between potential partners, but also with third parties. This idea is complementary rather than oppositional to the idea on cognitive trade-offs just presented. In small societies in which everyone knows one another, discussions about individuals who are not present can be had using direct references to those individuals, and the need to discuss personality or behavioral properties of strangers may be minimal. Complex societies, on the other hand, necessitate the existence of norms for describing people one knows in one social context (e.g., work colleagues) to people one knows in another social context (e.g., college friends). Indeed, people require common ground in order to describe anything, including other people, to their conversation partners (Clark and Brennan 1991). Social identities can serve as scaffolds for learning about new people, providing schemata for their potential behaviors and personalities. Moreover, the wider the variety of people we encounter, the more categories we will need to discuss them all. Requiring more categories, in turn, might lead to a shallower description of any specific person, as descriptions rely more on broad categories and less on detailed behavioral analysis.

In any society, individuals face problems of assortment, both to find cooperators while avoiding free riders or bullies, and to maximize the benefits of coordination with like-minded partners. I have proposed that (1) social identity helps to facilitate assortment for successful coordination, (2) the structure of social identity, and the extent to which it is used for assortment, is tied to the structure of the society individuals find themselves in, and (3) the multidimensionality of social identity evolved culturally to facilitate cooperation with different individuals serving different needs in different contexts.

Solving the problems of cooperation necessary for the synergistic underpinnings of human culture requires assortment and coordination (Tomasello et al. 2012). As societies grew larger and more complex, social identity enabled people to solve the problem of assortment as other solutions—such as kin recognition, reciprocity, or monitoring—increasingly failed. Using social identity as a tool for assortment would have piggybacked on preexisting psychological structures related to identifying one's place within a group, which evolved in the context of simpler societies. The *cultural* evolution of highly multidimensional social identity profiles in response to the changing demands of complex societies might, in turn, explain a potential trade-off between depth and breadth in social identity.

After groups assort, there exist well-known feedback processes in which group members grow closer together, are more strongly identified, and are increasingly better at coordinated activities (Sherif 1988; Theiner and O'Connor 2010; Gallotti and Frith 2013). More generally, social identities may be shaped through the course of group membership, such that group membership acts as a scaffold toward role development. Abrams (2014) has proposed that as social organization persists and individuals take on different social roles, the intrinsic organizational structure may encourage "cohesive cognitive subnetworks," which in turn will cause individuals who take on similar roles in different groups to become increasingly similar. In other words, the environmental effect of participating in a group-level endeavor may lead to a number of cognitive and behavioral similarities among individuals occupying similar social roles, above and beyond those necessary for performing those roles.

The details of any particular culture will constrain the options of its members in many ways (Smaldino and Richerson 2012), including options related to social identity. For example, Michele Gelfand and her colleagues (2011) characterized thirty-three national cultures on a spectrum between "tight" and "loose." Tight cultures are defined by strong norms and a low tolerance of deviant behavior, with the reverse true of loose cultures. In tight cultures, social identity fluidity should be lower than in looser cultures, independent of the society's "complexity." In general, the psychology of social identity and the evolution of human cultural complexity are complex and complicated topics, and it is unlikely that any single hypothesis or line of reasoning will be able to explain either of them, as Sterelny (2012) argues more generally for explanations of human social complexity. My goal in this chapter is simply to provide a new perspective on the overlooked connection between these two areas of research.

Cultures are shaped not only by how individuals use social identities to assort with others, but also by the specific norms and goals associated with those identities. I have largely ignored this distinction, and in particular I have ignored those social identities that have shaped human cultural evolution perhaps more than any others: religious identities. These are often accompanied by heavily enforced institutions that promote social cohesion and group-adaptive behaviors, and have likely been critical in the emergence of large-scale, hierarchical societies (Wilson 2002; Norenzayan 2013; see also Watts et al. 2015). This omission is certainly not a reflection of a lack of importance but is instead a tactic to focus on the multidimensionality of social identity and its role in facilitating coordination.

The thesis put forth here is necessarily somewhat imprecise, dealing with the interaction of many complex topics that are not always so well defined, perhaps none more so than "social identity." I view this chapter as a first attempt to organize some thoughts on the relationship between social identity, cooperative coordination, and the evolution of human social complexity. In the future, formal mathematical and computational models will be useful to constrain the problems discussed here more precisely. Such models will hone these arguments and guide empirical research to look for particular patterns in the data, much as models of social learning and cumulative culture inspired subsequent laboratory research on the relationship between group size and the maintenance of complex technologies (Derex et al. 2013; Muthukrishna et al. 2013; Kempe and Mesoudi 2014).

Cultural traits are inherently social traits. Thus, in order to think about how cultural ideas, technologies, and institutions spread and evolve, we need to think about the emergence and evolution of traits that are properly described at the level of groups (Smaldino 2014). A part of this picture is the recognition that humans identify themselves in terms of others. Understanding the role of social identity in cultural evolution will help us move beyond the meme, beyond the focus on individual traits and simplified models from population genetics or epidemiology, and toward models of cultural evolution that capture the essential "we"-ness of human beings.

NOTES

I am grateful to Bill Wimsatt and Alan Love for organizing and inviting me to the "Beyond the Meme" workshop at the University of Minnesota, where the ideas in this chapter were originally presented and fleshed out and many stimulating conversations had. For helpful comments and discussion on the topics presented in this chapter, I also thank Bob Bettinger, Monique Borgerhoff Mulder, Marilynn Brewer, John Bunce, Jimmy Calanchini, Tom Flamson, Michelle Kline, Alan Love, Richard McElreath, Nicole Naar, Lesley Newson, Emily Newton, Karthik Panchanathan, Pete Richerson, Gil Tostevin, and Bill Wimsatt.

1. Though I could hardly be more Jewish by opening with a discussion of my mother.

2. For example, a tapered body shape and random movement in a confined space is sufficient to produce the huddling behavior observed in rat pups *(Rattus norvegicus),* which is thought to be critical in thermoregulation and energy conservation (May et al. 2006).

3. I hope the die-hard Potter fans will forgive me for this noncanonical interpretation.

4. These problems would be easier to overcome in smaller, less diverse groups. However, the advantages of size and efficiency likely outweighed their costs as some complex societies emerged and began to compete with other cultural groups. For discussions of this transition, see Turchin and Gavrilets (2009) and Richerson et al. (2016).

REFERENCES

Abrams, M. 2014. "Maintenance of Cultural Diversity: Social Roles, Social Networks, and Cognitive Networks." *Behavioral and Brain Sciences* 37 (3): 254–55.

Anderson, C., and N. R. Franks. 2003. "Teamwork in Animals, Robots, and Humans." *Advances in the Study of Behavior* 33:1–48.

Apicella, C. L., F. W. Marlowe, J. H. Fowler, and N. A. Christakis. 2012. "Social Networks and Cooperation in Hunter-Gatherers." *Nature* 481:497–501.

Arthur, W. B. 2007. "The Structure of Invention." *Research Policy* 36:274–87.

Ashmore, R. D., K. Deaux, and T. McLaughlin-Volpe. 2004. "An Organizing Framework for Collective Identity: Articulation and Significance of Multi-dimensionality." *Psychological Bulletin* 130:80–114.

Axelrod, R., R. A. Hammond, and A. Grafen. 2004. "Altruism via Kin-Selection Strategies That Rely on Arbitrary Tags with Which They Coevolve." *Evolution* 58:1833–38.

Barclay, P., and R. Willer. 2007. "Partner Choice Creates Competitive Altruism in Humans." *Proceedings of the Royal Society B* 274:749–53.

Barth, F. 1969. "Introduction." In *Ethnic Groups and Boundaries,* edited by F. Barth, 9–38. New York: Little, Brown.

Berger, J., and C. Heath. 2008. "Who Drives Divergence? Identity Signaling, Out-group Dissimilarity, and the Abandonment of Cultural Tastes." *Journal of Personality and Social Psychology* 95:593–607.

Boehm, C. 1999. *Hierarchy in the Forest: The Evolution of Egalitarian Behavior.* Cambridge, Mass.: Harvard University Press.

Bowles, S., and H. Gintis. 2011. *A Cooperative Species: Human Reciprocity and Its Evolution.* Princeton, N.J.: Princeton University Press.

Brewer, M. B. 1991. "The Social Self: On Being the Same and Different at the Same Time." *Personality and Social Psychology Bulletin* 17:475–82.

Calcott, B. 2008. "The Other Cooperation Problem: Generating Benefit." *Biology and Philosophy* 23:179–203.

Caporael, L. R. 2014. "Evolution, Groups, and Scaffolded Minds." In *Developing Scaffolds in Evolution, Culture, and Cognition,* edited by L. R. Caporael, J. R. Griesemer, and W. C. Wimsatt. Cambridge, Mass.: MIT Press.

Chudek, M., and J. Henrich. 2011. "Culture-Gene Coevolution, Norm-Psychology, and the Emergence of Human Prosociality." *Trends in Cognitive Sciences* 15:218–26.

Clark, H. H., and S. E. Brennan. 1991. "Grounding in Communication." In *Perspectives on Socially Shared Cognition,* edited by L. B. Resnick, J. M. Levine, and S. D. Teasley, 127–49. New York: American Psychological Association.

Cohen, E., and D. Haun. 2013. "The Development of Tag-Based Cooperation via a Socially Acquired Trait." *Evolution and Human Behavior* 34:230–35.

Cohen, S. M., J. B. Ukeles, and R. Miller. 2012. *Jewish Community Study of New York: 2011.* New York: UJA Federation of New York.

Deaux, K. 1993. "Reconstructing Social Identity." *Personality and Social Psychology Bulletin* 19:4–12.

Derex, M., M-P. Beugin, B. Godelle, and M. Raymond. 2013. "Experimental Evidence for the Influence of Group Size on Cultural Complexity." *Nature* 503:389–91.

Erikson, E. H. 1968. *Identity: Youth and Crisis.* New York: W. W. Norton.

Flamson, T. J., and G. A. Bryant. 2013. "Signals of Humor: Encryption and Laughter in Social Interaction." In *Developments in Linguistic Humour Theory,* edited by M. Dynel, 49–73. Amsterdam: John Benjamins.

Gallotti, C., and C. D. Frith. 2013. "Social Cognition in the We-Mode." *Trends in Cognitive Sciences* 17:160–65.

Gelfand, M. J., J. L. Raver, L. Nishii, L. M. Leslie, J. Lun, and B. C. Lim et al. 2011. "Differences between Tight and Loose Cultures: A 33-Nation Study." *Science* 332:1100–04.

Gintis, H. 2014. "Inclusive Fitness and the Sociobiology of the Genome." *Biology and Philosophy* 29:477–515.

Giraldeau, L. A., and T. Caraco. 2000. *Social Foraging Theory.* Princeton, N.J.: Princeton University Press.

Gould, S. J. 1991. "Exaptation: A Crucial Tool for an Evolutionary Psychology." *Journal of Social Issues* 47:43–65.

Gowdy, J., and L. Krall. 2016. "The Economic Origins of Ultrasociality." *Behavioral and Brain Sciences* 39:e92.

Hamilton, M. J., B. T. Milne, R. S. Walker, O. Burger, and J. H. Brown. 2007. "The Complex Structure of Hunter-Gatherer Social Networks." *Proceedings of the Royal Society B* 274:2195–202.

Hamilton, W. D. 1964. "The Genetical Evolution of Social Behaviour." *Journal of Theoretical Biology* 7:1–16.

Hammond, R. A., and R. Axelrod. 2006. "Evolution of Contingent Altruism When Cooperation Is Expensive." *Theoretical Population Biology* 69:333–38.

Henrich, J., S. J. Heine, and A. Norenzayan. 2010. "The Weirdest People in the World?" *Behavioral and Brain Sciences* 33:61–135.

Heyes, C. M. 2013. "What Can Imitation Do for Cooperation?" In *Cooperation and Its Evolution,* edited by K. Sterelny, R. Joyce, B. Calcott, and B. Fraser. Cambridge, Mass.: MIT Press.

Hogg, M. A. 2000. "Subjective Uncertainty Reduction through Self-Categorization: A Motivational Theory of Social Identity Processes." *European Review of Social Psychology* 11:223–55.

Hrdy, S. B. 2009. *Mothers and Others: The Evolutionary Origins of Mutual Understanding.* Cambridge, Mass.: Harvard University Press.

Hruschka, D. J. 2010. *Friendship: Development, Ecology, and Evolution of a Relationship.* Berkeley: University of California Press.

Johnson, A. W., and T. Earle. 2000. *The Evolution of Human Societies.* Stanford, Calif.: Stanford University Press.

Kempe, M., and A. Mesoudi. 2014. "An Experimental Demonstration of the Effect of Group Size on Cultural Accumulation." *Evolution and Human Behavior* 35:285–90.

Koella, J. C. 2000. "The Spatial Spread of Altruism versus the Evolutionary Response of Egoists." *Proceedings of the Royal Society B: Biological Sciences* 267:1979–85.

Kümmerli, R., A. Gardner, S. A. West, and A. S. Griffin. 2009. "Limited Dispersal, Budding Dispersal, and Cooperation: An Experimental Study." *Evolution* 63:939–49.

Laland, K. N., J. Odling-Smee, and M. W. Feldman. 2000. "Niche Construction, Biological Evolution, and Cultural Change." *Behavioral and Brain Sciences* 23:131–75.

Leonardelli, G. J., C. L. Pickett, and M. B. Brewer. 2010. "Optimal Distinctiveness Theory: A Framework for Social Identity, Social Cognition, and Intergroup Relations." *Advances in Experimental Social Psychology* 43: 63–113.

Long, N. E. 1958. "The Local Community as an Ecology of Games." *American Journal of Sociology* 64:251–61.

May, C. J., J. C. Schank, S. Joshi, J. Tran, R. J. Taylor, and I. Scott. 2006. "Rat Pups and Random Robots Generate Similar Self-Organized and Intentional Behavior." *Complexity* 12 (1): 53–66.

McElreath, R., R. Boyd, and P. J. Richerson. 2003. "Shared Norms and the Evolution of Ethnic Markers." *Current Anthropology* 44:122–130.

McElreath, R., and R. Boyd. 2007. *Mathematical Models of Social Evolution.* Chicago: University of Chicago Press.

McPherson, M. 2004. "A Blau Space Primer: Prolegomenon to an Ecology of Affiliation." *Industrial and Corporate Change* 13:263–80.

McPherson, M., L. Smith-Lovin, and J. M. Cook. 2001. "Birds of a Feather: Homophily in Social Networks." *Annual Review of Sociology* 27:415–44.

Mitteldorf, J., and D. S. Wilson. 2000. "Population Viscosity and the Evolution of Altruism." *Journal of Theoretical Biology* 204:481–96.

Moffett, M. W. 2013. "Human Identity and the Evolution of Societies." *Human Nature* 24:219–67.

Muthukrishna, M., B. W. Shulman, V. Vasilescu, and J. Henrich. 2013. "Sociality Influences Cultural Complexity." *Proceedings of the Royal Society B* 281:201 32511.

Noë, R., and P. Hammerstein. 1994. "Biological Markets: Supply and Demand Determine the Effect of Partner Choice in Cooperation, Mutualism, and Mating." *Behavioral Ecology and Sociobiology* 35:1–11.

Norenzayan, A. 2013. *Big Gods: How Religion Transformed Cooperation and Conflict.* Princeton, N.J.: Princeton University Press.

Ostrom, E. 2014. "Do Institutions for Collective Action Evolve?" *Journal of Bioeconomics* 16:3–30.

Pickett, C. L., M. D. Silver, and M. B. Brewer. 2002. "The Impact of Assimilation and Differentiation Needs on Perceived Group Importance and Judgments of Ingroup Size." *Personality and Social Psychology Bulletin* 28:546–58.

Pound, N., I. S. Penton-Voak, and A. K. Surridge. 2009. "Testosterone Responses to Competition in Men Are Related to Facial Masculinity." *Proceedings of the Royal Society B* 276:153–59.

Putnam, R. D. 2000. *Bowling Alone: The Collapse and Revival of American Community.* New York: Simon and Schuster.

Richerson, P. J., R. Baldini, A. Bell, K. Demps, K. Frost, V. Hillis, and S. Mathew et al. 2016. "Cultural Group Selection Plays an Essential Role in Explaining Human Cooperation: A Sketch of the Evidence." *Behavioral and Brain Sciences* 39:e30.

Richerson, P. J., and R. Boyd. 1999. "Complex Societies: The Evolution of a Crude Superorganism." *Human Nature* 10:253–89.

Richerson, P. J., and J. Henrich. 2012. "Tribal Social Instincts and the Cultural Evolution of Institutions to Solve Collective Action Problems." *Cliodynamics* 3:38–80.

Roccas, S., and M. B. Brewer. 2002. "Social Identity Complexity." *Personality and Social Psychology Review* 6:88–106.

Rumelhart, D. E., and D. Zipser. 1985. "Feature Discovery by Competitive Learning." *Cognitive Science* 9:75–112.

Savill, N. J., and P. Hogeweg. 1997. "Modelling Morphogenesis: From Single Cells to Crawling Slugs." *Journal of Theoretical Biology* 184:229–35.

Scott, I. M., A. P. Clark, S. C. Josephs, A. H. Boyette, I. C. Cuthill, and R. L. Fried et al. 2014. "Human Preferences for Sexually Dimorphic Faces May Be Evolutionarily Novel." *Proceedings of the National Academy of Sciences* 111:14 388–93.

Sherif, M. 1988. *The Robber's Cave Experiment: Intergroup Conflict and Cooperation*. Middletown, Conn.: Wesleyan University Press.

Smaldino, P. E. 2014. "The Cultural Evolution of Emergence Group-Level Traits." *Behavioral and Brain Sciences* 37 (3): 243–95.

Smaldino, P. E., and M. Lubell. 2011. "An Institutional Mechanism for Assortment in an Ecology of Games." *PLOS ONE* 6 (8): e23019.

Smaldino, P. E., and M. Lubell. 2014. "Institutions and Cooperation in an Ecology of Games." *Artificial Life* 20:207–21.

Smaldino, P. E., C. L. Pickett, J. W. Sherman, and J. C. Schank. 2012. "An Agent-Based Model of Social Identity Dynamics." *Journal of Artificial Societies and Social Simulation* 15 (4): 7.

Smaldino, P. E., and P. J. Richerson. 2012. "The Origins of Options." *Frontiers in Neuroscience* 6:50.

Smaldino, P. E., and J. C. Schank. 2012. "Movement Patterns, Social Dynamics, and the Evolution of Cooperation." *Theoretical Population Biology* 82:48–58.

Smith, E. A. 1985. "Inuit Foraging Groups: Some Simple Models Incorporating Conflicts of Interest, Relatedness, and Central-Place Sharing." *Ethology and Sociobiology* 6:27–47.

Smith, E. A., K. Hill, F. Marlowe, D. Nolin, P. Wiessner, M. Gurven, S. Bowles, M. Borgerhoff Mulder, T. Hertz, and A. Bell. 2010. "Wealth Transmission and Inequality among Hunter-Gatherers." *Current Anthropology* 51:19–34.

Snyder, C. R., and H. L. Fromkin. 1980. *Uniqueness: The Human Pursuit of Difference*. New York: Plenum.

Sterelny, K. 2012. *The Evolved Apprentice: How Evolution Made Humans Unique*. Cambridge, Mass.: MIT Press.

Theiner, G., and T. O'Connor. 2010. "The Emergence of Group Cognition." In *Emergence in Science and Philosophy,* edited by A. Corradini and T. O'Connor, 78–117. New York: Routledge.

Tomasello, M., M. Carpenter, J. Call, T. Behne, and H. Moll. 2005. "Understanding and Sharing Intentions: The Origins of Cultural Cognition." *Behavioral and Brain Sciences* 28:675–35.

Tomasello, M., A. P. Melis, C. Tennie, E. Wyman, and E. Herrmann. 2012. "Two Key Steps in the Evolution of Human Cooperation: The Interdependence Hypothesis." *Current Anthropology* 53:673–92.

Tooby, J., and L. Cosmides. 1996. "Friendship and the Banker's Paradox: Other Pathways to the Evolution of Adaptations for Altruism." *Proceedings of the British Academy* 88:119–43.

Turchin, P., and S. Gavrilets. 2009. "Evolution of Complex Hierarchical Societies." *Social Evolution and History* 8 (2): 167–98.

Vignoles, V. L. 2011. "Identity Motives." In *Handbook of Identity Theory and Research,* edited by K. Luycke, S. J. Schwartz, and V. L. Vignoles, 403–32. New York: Springer.

Waring, T. M. 2012. "Cooperation Dynamics in a Multi-Ethnic Society: A Case Study from Tamil Nadu." *Current Anthropology* 53 (5): 642–49.

Watts, J., S. J. Greenhill, Q. D. Atkinson, T. E. Currie, J. Bulbulia, and R. D. Gray. 2015. "Broad Supernatural Punishment but Not Moralizing High Gods Precede the Evolution of Political Complexity in Austronesia." *Proceedings of the Royal Society B* 282:20142556.

Whitfield, S. J. 1999. *In Search of American Jewish Culture.* Hanover, N.H.: Brandeis University Press.

Wilson, D. S. 2002. *Darwin's Cathedral: Evolution, Religion, and the Nature of Society.* Chicago: University of Chicago Press.

Wimsatt, W. C. 2014. "Entrenchment and Scaffolding: An Architecture for a Theory of Cultural Change." In *Developing Scaffolds in Evolution, Culture, and Cognition,* edited by L. R. Caporael, J. R. Griesemer, and W. C. Wimsatt. Cambridge, Mass.: MIT Press.

13

WICKED SYSTEMS AND THE FRICTION
BETWEEN "OLD THEORY AND NEW DATA"
IN BIOLOGY, SOCIAL SCIENCE, AND
ARCHAEOLOGY

CLAES ANDERSSON, ANTON TÖRNBERG,
AND PETTER TÖRNBERG

SOME OF THE most challenging and important problems facing us both scientifically and as citizens emanate from large-scale complex adaptive systems, such as societies and ecosystems. Diverse examples include social exclusion, credit crises, environmental unsustainability, biological evolution (with its myriad subproblems), and the evolution of hominin culture. These problems are not only important and complex. They are also interlinked in bewildering ways, are difficult to even define or delimit, and appear to have become increasingly pressing lately. We seem to see and feel the shortcomings of our understanding more acutely now than we used to only a few decades ago.

The reasons for the increasing saliency of these sorts of problems are manifold, and they are expressed differently in different fields, but we believe that three important and interlinked factors can be identified: (1) theories that used to be thought of as safe ground are being undermined by (2) an explosion of new data, while at the same time, (3) the development of complexity science has provided models and concepts for expressing and detecting these types of problems. For societal problems, an additional point is germane: the societal system itself is becoming more and more complex (i.e., more interconnected, less predictable, and less stable, as well as more and more energy- and material-intensive, with a stronger effect on ecology and climate as a result).

We believe a major transformation is occurring that spans several disciplines. This transformation is unfolding at different paces and along somewhat different trajectories in different disciplines, which reflects differences in theoretical and empirical backgrounds, in what constitutes central questions and aims. It is generating a growing substrate of semicongruent critiques and new ideas, but an understanding of what is wrong with "old theory" is, overall, more developed than an account of what would work better. The situation presents a need, and indeed an opportunity, for conceptual tools that act to align and direct this substrate of critiques and new ideas on an abstract level so that it can reach across disciplinary boundaries.

This chapter aims to contribute some elements of such a toolbox. We begin with a review and analysis of recent empirical and theoretical trajectories, in and across three important areas where this transformation is having strong effects: evolutionary biology, social science, and archaeology. Our exploration leads to the introduction of two interrelated tools: a metarepresentational diagram and a new class of *wicked systems*. We probe the value of these tools by addressing questions such as: What is "wickedness"? What approaches have been applied successfully to understanding it? What can we do to make further headway?

TRANSFORMATIONS IN AND ACROSS DISCIPLINES

Evolutionary biology is undergoing a dramatic transformation driven by strong empirical advances that have occurred over the past two decades. These advances have been interpreted as revealing a mechanistic basis for evolution (Wagner, Chiu, and Laubichler 2000; Laubichler and Maienschein 2013; Laubichler and Renn 2015)—from molecular to ecological scales—at a level of detail that was hardly imaginable only twenty years ago. In this emerging picture of evolution, age-old disciplinary boundaries break down, and trusted models are being undermined, both in terms of the predictions they make and the assumptions that underpin them (e.g., Erwin 2008; Odling-Smee et al. 2013; Laland 2014). Yet this new picture is not yet unified. It emerges—tantalizing in outline but still somewhat out of focus—from a range of perspectives on the evolutionary process, including the complex mapping between genotype and phenotype and the structuring of phenotypic spaces, the role of organization and history in evolution, multiple channels of inheritance, selection on multiple levels, and macroevolutionary patterns.

These new perspectives are embodied as an evolving system of diverse and interacting theoretical elements, such as evolutionary developmental biology (e.g., Arthur 2011), niche construction theory and ecological inheritance (e.g., Odling-Smee, Laland, and Feldman 2003), ecosystems engineering (e.g., Jones, Lawton, and Shachak 1996), ecoevolutionary dynamics (e.g., Pelletier, Garant, and Hendry 2009; Loreau 2010), facilitated variation theory (e.g., Bruno, Stachowicz, and Bertness 2003; Gerhart and Kirschner 2007), developmental systems theory (e.g., Oyama, Griffiths, and Gray 2001), generative entrenchment (e.g., Wimsatt 1986, 2001), developmental innovation (e.g., Erwin and Krakauer 2004), and "public goods" theories of evolutionary transitions (e.g., Erwin and Valentine 2013; Erwin 2015). The basis for this new system of theories derives from empirical fields enabled by technological advances, such as comparative genomics and developmental genetics (e.g., O'Brien et al., 1999). The lesson is that biological evolution is a much broader and more complex problem than we previously imagined.

Laubichler and Maienschein (2013) identify two alternative narratives about the history and future of evolutionary theory. The first, and most widespread, is that we are seeing a completion of the Modern Synthesis—that is, an "extended synthesis" (e.g., Pigliucci and Müller 2010) where this new mechanistic understanding is being accommodated in a cumulative fashion and within current theoretical frameworks. The second is that there is a more fundamental challenge to how evolutionary dynamics is understood in the Modern Synthesis. Laubichler and Maienschein (2013) argue that we are seeing the emergence of a new "causal mechanistic" evolutionary biology that has its roots in old complementary approaches to evolutionary biology (Laubichler and Maienschein 2007).

As the detailed mechanistic basis of biological evolution is comprehended better and better, evolutionary biology is increasingly being forced to view its subject area similar to the way that qualitative social sciences and the humanities have always viewed their subject areas. In this view, biological evolution is composed of hierarchical and historical complex systems in which contingent details matter greatly, problems and subsystems are potentially impossible to delimit, and important interconnections exist across levels of organization. This recent confluence between biology and social science appears to be "spontaneous"—that is, there is no suggestion that theoretical exchanges across this academic divide drove these developments. The emerging view of biology reveals features that are deeply congruent with corresponding features of societal systems and, as a consequence, has directed theory, problem formulation, and debates in similar directions.

This new way of viewing biological evolution enlists historical and interpretative approaches, akin to the narrative case studies used in qualitative social science (e.g., Ragin 2009; Byrne and Callaghan 2014), and is evident in the novel theoretical trajectories listed above. Specific, contingent processes and histories frequently must be described with diagrams and narratives in order to capture their causal structure. It is not a matter of abandoning formal modeling; rather, it is the limitations of various formalisms that have become more acutely felt. The emerging empirical picture of evolution forces biologists to go outside the bounds that formal modeling imposes on inquiry.

The social sciences have a long tradition of qualitative theorizing about the detailed causal structure of society. This has nurtured an internal animosity and fragmentation between qualitative and formal quantitative approaches. The social sciences are also under increasing pressure to deliver in terms of policy. This pressure emanates from the empirical developments of information and communication technology, as well as from new demands in a rapidly changing reality (e.g., Beddoe et al. 2009; Zalasiewicz et al. 2011; Steffen et al. 2015). Not least, a mounting scale and frequency of societal and environmental crises—a "metacrisis" (Lane and van der Leeuw 2011)—has made the limits of our understanding and our control over society and the global environment simultaneously more obvious and threatening.

In the wake of this metacrisis, which was neither predicted nor hindered by our current understanding, there is a widespread sentiment that we must broaden the range of factors we think affect the direction of society: from the primacy of economic values to an inclusion of societal and environmental values; from a reductionist view to a more holistic and inclusive view. Although there is no consensus about what this really means or entails, most now agree that society is highly complex, and we must attend to its complexity much more explicitly.

All of these developments have changed the landscape for policy, which represents a normative dimension that biology and archaeology largely lack (see, e.g., Byrne 2005; Scoones et al. 2007; Leach, Scoones, and Stirling 2010; also, reflecting this fragmentation, Ball 2012; Helbing 2013). The question of how to predict and optimize the future is yielding to an acceptance of the futility of such aims and an embrace of other goals, such as resilience and sustainability. On the one hand, this raises serious questions about the efficacy of many standard policy tools, most of which were designed under different assumptions about how societal systems work (most notably, neoclassical economics). Indeed, this challenges even our basic intuitions about how societies evolve. But, on the other hand, it also has opened up the promise of entirely new types of analytical tools, based on ideas about how we can dynamically steer and scaffold society by engaging more directly with its causal mechanics. Possibilities include more bottom-up approaches like the management and design of social networks of actors (e.g., Lane and van der Leeuw 2011), the historical study of sociotechnical transitions (e.g., Geels 2002), or the management of innovation "pathways" (e.g., Leach, Scoones, and Stirling 2007, 2010; Loorbach 2010; Wise et al. 2014).

Archaeology and paleoanthropology provide an interesting third case, not least because biological and societal evolution fuse in a common coevolutionary history. These fields have been hit by an empirical revolution that is at least as dramatic as the one described for evolutionary biology. Again, new and improved laboratory techniques constituted a major driving force; for example, in biomolecular analysis (e.g., Brown and Brown 2013), use-wear analysis (e.g., Lerner et al. 2007), palynology (e.g., Holt and Bennett 2014), dating techniques (e.g., Aitken 2014), and methodology (Tostevin 2012). Details about the lives of ancient hominins that were unimaginable until lately are now coming to light (e.g., Kristiansen 2014); the mechanistic basis of cultural evolution is being revealed, and similar to biology, there is considerable friction between "old theory and new data."

Two dominant "old theory" approaches to understanding Paleolithic cultural evolution can be characterized in a schematic fashion:

1. A cognitive/physiological approach (CPA) that emphasizes cognition, as both an enabler of and constraint on culture. The CPA is pervasive but is rarely championed explicitly; Richard Klein typically serves as its embodiment (and lightning rod) in the literature (e.g., Klein and Edgar 2002). Its logic dictates that periods of cultural stability must express what is maximally attainable at certain levels of cognitive capability because strong selection would rapidly exhaust cognitive potential to produce adaptive artifacts and strategies. Transitions, consequently, would be the result of genetic novelty that confer new and distinct behavioral "packages."

2. An ecological/economical approach (EEA) largely based on behavioral ecology and economic constraints like time consumption and energy costs (inter alia). The EEA is often contrasted to the CPA approach, which focuses more on artifacts and hominin taxonomy. The EEA emphasizes geographically and temporally varying environmental selection pressures as the prime mover of change in the past (e.g., Foley and Gamble 2009). In this approach, cognition is seen more as a contributor to variability than as a set of fixed capacities.

Neither of these approaches is sufficient on its own, and controversy typically concerns the relative importance of the features they emphasize. Moreover, both share a common Modern Synthesis model of adaptation where the fitness of physiological and cultural expressions, relative to an external environment, is the sole provider of evolutionary direction and where evolution is a process of constrained optimization.

As we move into the Holocene with sedentary farming communities, the discourse shifts weight from biology to sociology and anthropology; the EEA remains, but the CPA falls to the side as cognitive evolution loses its centrality.

This is natural since the more recent empirical record presents archaeologists with considerably more detail than earlier periods. But, despite this different theoretical emphasis, the prescriptions of older universal models in those traditions are not that different from those of the CPA and neither is the friction caused by emerging empirical patterns. For example, the idea of transition as a sudden appearance of a new "package" has been equally important in Neolithic research (e.g., Çilingiroglu 2005; Barker 2006) as it has been for Paleolithic research (e.g., McBrearty and Brooks 2000; Belfer-Cohen and Hovers 2010).

Formal Darwinian approaches to cultural evolution were introduced and developed starting in the early 1980s (e.g., dual-inheritance theory and evolutionary archaeology; see Cavalli-Sforza and Feldman 1981; Boyd and Richerson 1985; Mesoudi 2011). These go outside of the mainstream that the CPA and EEA describe, but they largely fit into the same pattern as they seek to identify a unified theoretical basis from a fundamental principle of organization (e.g., *population thinking*). These approaches adapt models from population genetics and rational choice theory, with population dynamics coming out as so fundamental and dominant that other factors become secondary or peripheral.

The new wealth of detailed data is being interpreted in terms of a more gradualist pattern (e.g., McBrearty and Brooks 2000; McBrearty 2007; Maher, Richter, and Stock 2012), often characterized as more complex, messy, intermittent, and in need of attention to detail (e.g., Barker 2006; Hovers and Kuhn 2006; Habgood and Franklin 2008; Belfer-Cohen and Goring-Morris 2011; Hovers and Belfer-Cohen 2013). There is a search for new theoretical traction (e.g., Hauser 2012; Zeder and Smith 2009; Zeder 2014; Stiner et al. 2014; Stiner and Kuhn 2016) as older *prime mover* and *single origins* theories are undermined.

Overall, the weight of evidence tells us that ecological, evolutionary, and societal systems do not work as previously assumed for the sake of methodological expediency. In and across these disciplinary fields, new empirical knowledge undermines old theory in three major ways: (1) it buttresses old complaints about poor predictions, explanations, and policy advice from these traditional approaches, (2) it refutes central assumptions, many with an axiomatic status, that underpin old theory, and (3) old theory is frequently an obstacle to making sense of these new data. The reason is that we are not just dealing with more data but new types of data, which older theory was designed specifically to ignore since they could not be accessed with confidence. As a consequence, old theory is frequently criticized for not being extendable in the required directions, necessitating more radical theoretical innovation.

A wealth of new theoretical elements emerges in this friction between old theory and new data, but they are at present not strongly aligned, not even within the fields. The "search for new theoretical traction" is still very much unfolding; obtaining it is the challenge and opportunity that lies before us.

A WICKED THEORETICAL CRISIS

What we see in this emerging picture is the outline of a class of systems that exhibit a deep similarity and encompass both societies and ecosystems. This similarity provides a common platform from which to search for new theory. The outline revolves around features known to be methodologically problematic, the key elements of this new empirical picture (e.g., complexity, lacking clear levels of organization, heterogeneous structure, etc.). The proposed deep similarity is expressed as similar sets of problems, theoretical responses, debates, models, and concepts; it crystallizes more and more clearly as we explore and discover more about these systems. Call this class of systems *wicked systems* because they are distinct from and yet related to complex systems, which would otherwise be a natural label.

The label *wicked systems* accents a potentially deep connection (whose exact nature remains to be worked out) between this class of systems and what have been called *wicked problems* in social science. The term *wicked problems* was first coined in management research by Horst Rittel (briefly introduced by West Churchman [1967]) to characterize a class of problems that did not fit into the mold of formal systems theoretical models that were being applied widely and with considerable confidence at the time. Most large-scale societal problems fall naturally into the category of wicked problems: starvation, climate change, geopolitical conflicts, social disenfranchisement, and so on. These problems resist definitional characterization, and the efficacy of proposed solutions is called into question frequently, not only with regard to feasibility and adequacy but also with respect to the risk of creating cascades of unforeseen problems that may be worse than the initial problem (see also Leach, Scoones, and Stirling 2007; Scoones et al. 2007). With wicked problems, we either tame them by creating "an aura of good feeling and consensus" or by "carving off a piece of the problem and finding a rational and feasible solution to this piece" (West Churchman 1967). This also describes the problems we discussed initially across all three fields, encapsulating the troubles we face when applying old theory to new data. By considering "wickedness" as a system quality, we can generalize to speak of *wicked dynamics, wicked phenomena,* and *wicked systems.* This allows us to refer to these crises and transformations in a unified way and articulate the growing realization that the sciences must face wickedness more directly on its own terms. But how? Will our old weapons and battle plans work? Can they be incrementally changed and combined to meet the challenge? What sort of understanding, prediction, or control can we expect, realistically? To begin answering such questions, we need to better characterize wickedness as a system quality. One place to start is with a review of approaches that have been applied historically to evaluate how they have succeeded and failed.

HOW DO WE DEAL WITH WICKEDNESS?

A battery of approaches have been used in the past to deal with wicked systems. These approaches fall into four broad categories: narrative theory, analytical models, systems theory, and complexity science. Narrative-based theory—basically disciplined or systematic thinking and communication—is very old, whereas the latter three formal approaches are newer additions to the toolbox that we apply to understand the world (Figure 13.1). The confidence we place in these to understand wicked systems is typically buttressed by a strong track record of success in understanding other, less unruly, systems (often in the physical sciences). They encourage us to see that we only need bring these unrulier systems under the umbrella of "proper science" and away from the interminable talk, hairsplitting, and subjective opinion that is seen as inherent to narrative approaches. But we are still waiting for the breakthrough, and, to varying degrees, there are signs of stagnation in all three approaches.

Where does this leave us with regard to wicked systems? Although complexity is a crucial concept and complexity science arouses the most enthusiasm as a problem-solving strategy, the latter appears to have hit considerable resistance in the face of these systems. Although concepts like path-dependency, attractors, tipping points, and chaos have transformed ideas about causality in society and biology, these highly general lessons have proven hard to operationalize for wicked systems. Complexity science appears to offer a perpetual promissory note, but by examining its lack of achievements, we can glean insights into how these systems work.

Complexity, as invoked so far, is not well defined; that is precisely our point of entry for the remainder of our exploration. Setting aside the vast and fragmented literature that attempts to define complexity, both because it is unnecessary and likely a misguided project, we will approach complexity ostensively by assuming that "complexity is what complexity science does (well)."

COMPLEX, COMPLICATED, AND WICKED

Complexity scientists often distinguish between complexity and complicatedness (or dynamical versus structural complexity; see, e.g., Erdi 2008). These two system qualities are often contrasted for the purpose of explaining what complexity science focuses on: complexity is associated with bottom-up self-organization, such as the behavior of a school of fish or a crowd, whereas complicatedness is associated with top-down organization, such as in engineering. Though not a formal definition, it helps to illuminate the practice of complexity science, which deals with complexity, not complicatedness, even though the latter can be seen as a subset of the former.

The history of complexity science helps to illuminate how this practice and (largely tacit) meaning of the term *complexity* emerged. The Santa Fe Institute (SFI) acted as a powerful uniting and aligning force in what today is referred to as *complexity science*. Founded in 1984 by a group of highly influential scientists, many of whom were active at the nearby Los Alamos National Laboratory, the SFI was the first dedicated research center for complexity science. Because of the founders, it was tightly linked to the origins of scientific computing and dynamic systems theory (see, e.g., Galison 1997). Although many important ideas about complexity predate SFI, such as are found in qualitative social science and systems theory (see, e.g., Sawyer 2005; Vasileiadou and Safarzyska 2010), it remains the case that the SFI came to define a mainstream of complexity science and thereby also, in practice, the concept of complexity as understood by scientists, policymakers, and the public.

The SFI was created as a multidisciplinary center. Although it remains highly multidisciplinary, it is not as methodologically diversified. The primary methodology that was (and still is) pursued at the SFI is formal and quantitative, much closer to natural science and quantitative social science.[1] Computer simulation is at the heart of this methodology, which puts into motion the entities and interaction rules of dynamic systems. This extremely flexible methodology makes it possible to study and visualize dynamics that were previously inaccessible to the human mind—aided or unaided. Above all, it makes possible a systematic inquiry into emergent properties in dynamic systems. This capability provided a powerful impetus to the formation of complexity science worldwide with the SFI as its central hub.

The typical model in this tradition has a microlevel of abstract agents or nodes existing in a predefined environment. Complexity scientists study and probe the patterns that arise on an emergent macrolevel from the dynamic interaction between these agents or nodes. This is what complexity science does well. Individual traditions and scientists may be more or less strongly aligned with it, but anyone claiming to work with "complex systems" must relate to this methodology in one way or another. Thus, complexity is a concept whose meaning is constructed mainly by the complexity science community working with it.

Making this typical construal of complexity explicit helps to reveal the limitations of its applicability and delimit the class of systems that are amenable to analysis using it.[2] In short, although complexity and complicatedness are linked in numerous ways, they present radically different sets of methodological and theoretical challenges.

Are societal systems and ecosystems complex or complicated? On the one hand, they are undeniably complicated, with multilevel organization and a bewildering array of qualitatively different and interacting entities. Systems theories seize upon what appears to be an irreducible complicatedness of societal systems. Yet society is also a complex system in the bottom-up self-organization sense (e.g., Sawyer 2005; Castellani and Hafferty 2009; Ball 2012). One can even argue that much of its complicated structure arises from bottom-up rather than top-down processes. The story is similar for ecosystems. There is no reason why systems cannot be both complicated and complex at the same time; our two wicked systems appear to be excellent examples of this type of system.

	General	Evolutionary biology	Archaeology	Social science and humanities
Narrative theory	Narrative theorizing is sometimes referred to as conceptual, qualitative, or interpretative. Employs language and cognition.	Darwin's "long argument"; presynthesis evolutionary biology is largely narrative based. Heterodox twentieth-century traditions (e.g., Gould 2002; Lewontin 2000). Philosophy of biology.	Postprocessual archaeology (e.g., Hodder 1982). Narrative is overall in wide use as a way of providing cohesive explanations across systems, space, and time.	Historical case studies (e.g., Ragin 2009). Very widespread but seen as a second-rate approach in quantitative social science. Divides the fields.
Analytical models	Analysis in terms of variables and symbolic operations. Reductionist in the sense of reducing degrees of freedom in models but otherwise applicable regardless of scale and level of organization.	Modern Synthesis evolutionary biology is strongly based on analytical models. Evolutionary game theory (e.g., Axelrod and Hamilton 1981) and several other bodies of biological theory.	Dual-inheritance theory based on models from population genetics (e.g., Boyd and Richerson 1985), human behavioral ecology (e.g., Bird and Connell 2006). Statistics.	Neoclassical economics. Rational choice theory. Game theory. Statistics. Defines quantitative social science.

System theories	Most prevalently, cybernetics and general systems theory. Holistic view, focus on information and control. Lasting legacy but declined as disciplines in their own rights. Generally, "systems thinking" (e.g., Weinberg 2001) is very widespread.	Not influential in Modern Synthesis theory. In developmental approaches explicitly in developmental systems theory (e.g., Waddington, Gottlieb; see Griffiths and Tabery 2013 for review).	Important especially in the 1960s and 1970s; e.g., Clarke (1968); Flannery (1969). See, e.g., Kohler (2012) for a review.	Very widespread across quantitative and qualitative social science.
Complexity science	A toolbox of approaches that emerged with cheap computing from the 1980s. Includes, e.g., cellular automata (e.g., Wolfram 1994), agent-based modeling (e.g., Gilbert 2008), and complex networks (e.g., Newman 2003).	Basic dynamic evolutionary phenomena, such as cooperation (e.g., Lindgren 1992). Genetic and other networks (e.g., Clauset, Moore, and Newman 2008). Artificial Life (e.g., Bedau et al. 2000).	Agent-based models, e.g., Anasazi (Dean et al. 2000). Extending Dual-Inheritance Theory (e.g., Shennan 2009). Simulation is increasingly common.	Two traditions: In qualitative areas extending from systems theories (e.g., Byrne and Callaghan 2014). In quantitative areas, simulation, networks, etc. (e.g., Epstein 2007).

Figure 13.1. Examples and illustrations of how the four major approaches have been employed in the three areas of study. The lists are not intended to be exhaustive but merely to provide an overview and point to representative examples.

Figure 13.2 uses these two dimensions as axes to map out a space of possibilities (see Andersson, Törnberg, and Törnberg [2014b] for a discussion focused on societal systems). We thereby obtain a separation between types of systems that we otherwise tend to conflate as "highly complex" and our wicked systems cluster where both dimensions are emphasized (i.e., complex *and* complicated). Surprisingly, the possibility of systematically exploring the consequences of systems exhibiting both complexity and complicatedness has not been pursued explicitly.[3] Complexity science may be aware that complexity and complicatedness are distinct qualities, but complicatedness in complex systems is not seen as a fundamental problem. That they are complex is fundamentally important; extending mainstream complexity science to deal with them is seen as challenging but, essentially, gradual and cumulative work. However, wicked systems are not a type of complex system but rather fall within a system where complexity and complicatedness are both present. This combination—wickedness—is not something that complexity science, systems approaches, analytical models, or combinations thereof address very well.

UNDERSTANDING THE COEVOLUTION OF METHODS, PROBLEMS, AND SYSTEMS

Why is it so difficult to extend mainstream complexity science to wicked systems? This is a question about the relations between methods, problems, and systems. The answer boils down to why formal approaches, in general, will be incapable of dealing comprehensively with these systems. In Figure 13.3, the four basic approaches—narrative theory, analytical models, systems theory, and complexity science—are mapped onto the complexity-complicatedness plane. Narrative approaches are not married to assumptions of low complexity and complicatedness, but they quickly run into problems when complexity or complicatedness become too prevalent. Thus, narrative approaches fall roughly in the middle of the diagram. The introduction of analytical models over the past few centuries, such as Newtonian physics, neoclassical economics, and Modern Synthesis evolutionary biology, achieve analytical power by abstracting away from the richness of real-world systems (much of which was inaccessible empirically). These successes, especially in the natural sciences, dovetail with the Platonic idea that nature is at the bottom, governed by simple and elegant laws that the sciences uncover. In the mid-twentieth century, the "complicated flank" was occupied

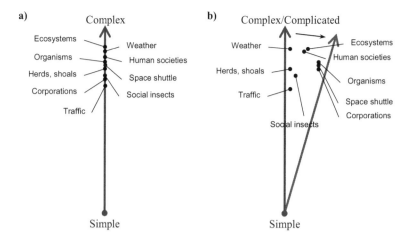

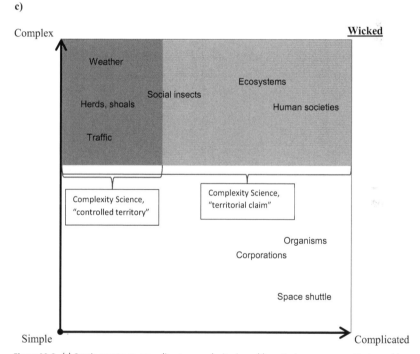

Figure 13.2. (a) Sorting systems according to complexity is problematic in many ways. Most would probably agree that the examples listed are in some sense complex systems, but it remains unclear what we mean by complexity. There is a lingering feeling that we are comparing apples and oranges. (b) Differentiating between complexity and complicatedness is commonplace, but here we are using this differentiation in a novel way: to open up a space where we hope that our examples will be better spaced and cluster in a more interesting way, yielding something like the diagram (c), in which we indicate also the region in which complexity science has been successful contra the region corresponding to systems of high complexity in general. There is considerable room for argument about where the examples are placed, how they should extend across the diagram, and what exceptions may exist. It is a strength of the diagram that it can serve as a basis for such discussions.

increasingly with the development of general systems theory and cybernetics. This was motivated in part by a need to match the macroscopic organization of systems but also by the conviction that elegant laws resided in holistic systems. As cheap computing became widely available in the 1980s, complexity science entered the scene and covered the "complex flank." The search for universal laws focused on emergent patterns in dynamic systems.

Formal approaches are unable to address many of the problems that wicked systems present us with (see above, section 2). Instead, they selectively address subproblems that happen to fall in their domains or transplant problems from near the wicked corner to the corners of their methodological preference (Figure 13.4). In the former case, important but limited "snapshots" of the system in question can be obtained, typically with a taste of revealing laws of great generality. Although these often reveal important major principles, we are faced with the problem of how to combine the snapshots (see also Wimsatt 1975). In the latter case, we may get spurious results because strong assumptions mean that the benefit of using formal methods of analysis does not warrant the price in realism.

What Figures 13.2 and 13.3 primarily accent is that there is a theoretical lacuna for wicked systems. This theoretical lacuna does not emerge clearly unless we systematically make a separation between complexity and complicatedness. But our diagram also emphasizes further questions: What is "wickedness"? Why is it so analytically recalcitrant?

THE GENESIS OF WICKED SYSTEMS

Complexity and complicatedness can be seen as mutually reinforcing in our two principal examples of wicked systems: societies and ecosystems. Self-organization generates, changes, and maintains macrostructure. Macrostructure, in turn, scaffolds and creates a multitude of arenas for self-organization. How do wicked systems originate? How do complexity and complicatedness become fused into wickedness? How are wicked systems maintained? What sets them apart from systems where either complexity or complicatedness dominates?

Complicated systems, such as machines and organisms, have distinct life cycles and tend to be adapted to specific functions in the context of an external environment. They have an initial phase of assembly or development, during which they are shielded from the rigors that face the completed system during a subsequent use phase. Automobiles, for example, get

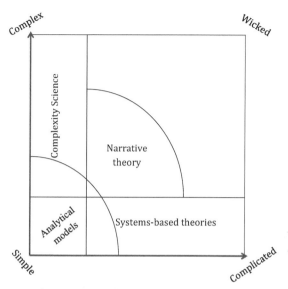

Figure 13.3. The types of problems that different approaches are competent at dealing with, as seen on a plane described by a complexity and a complicatedness axis.

assembled according to strict plans, and they are used as automobiles only once fully assembled. Organisms develop under some use-phase requirements, the most basic being they must remain alive throughout development, and juveniles of many species (not least, humans) gradually become exposed to the rigors of adult life, while metamorphosing species undergo one or more transitions between different ecologically adapted forms. What we describe is more salient in K-strategists; r-strategists invest in large numbers rather than robustness in, and shielding of, the offspring. In both cases, investments are needed to minimize the effects of conflicts between functional and developmental requirements. The life spans of complicated systems are usually sufficiently short, so the environments they are adapted to can be assumed to change very little. They are fundamentally not designed to be very flexibility and their flexibility mainly resides between rather than within life cycles, such as in design processes, variation and selection regimes, and so on. The strong and fundamental nature of this constraint on evolutionary adaptation is evident from the high investment that organs provide even limited intragenerational adaptability represent—for example, the adaptive immune system and brains that provide a capacity for learning or even culture.

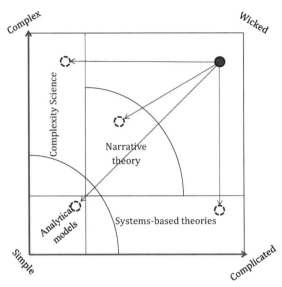

Figure 13.4. Wicked problems can be treated as complex, simple, or complicated problems in order to apply formal machineries. Although human cognition (represented as *narrative*) appears adapted to dealing with wickedness, likely as a result of coevolving with human cultural systems, insurmountable problems arise for systems as wicked as ecosystems and modern societies.

Wicked systems are complicated, but they are neither designed nor assembled. They have contingent histories and lack a separation between assembly and use phases. However, they also are complex, though they do not self-assemble from simple (e.g., regular or random) initial states, which is how complexity science models complex systems. Change, adaptation, and mitigation in these systems typically constitute what is meant by *wicked problems* (see above, section 6).

Something that seems to characterize adapted complex and complicated systems is the lack of incentive for parts to benefit at the expense of the whole. Along with external shielding during an assembly phase, this is an important factor that lends flexibility and precision to their processes of formation: they may be designed, assembled, and changed "in peace" so as to fulfill an overall functionality. When interactions are not necessarily symbiotic, such as when humans are part of social systems, one finds distinct mechanisms for eliminating such interactions and for enforcing an overall alignment of the parts (e.g., various forms of detection and policing, such as by immune systems, legal systems, intelligence agencies, the military, etc.). The relation between somatic and germ line cells (e.g., Michod and Nedelcu 2003) represents the most potent example because it signifies how even the potential of parts to rebel against the whole is eliminated.

Wicked systems are open arenas for complex interactions between complicated systems, where these interact freely (complexity), creating and dissolving structure (complicatedness) across levels of organization and constantly changing the environment for the interacting components (again, complexity). In these arenas, we find a range of types of interactions. Sandén and Hillman (2011) describe this heterogeneity of interactions in society using terminology borrowed from ecology. These stem from the set of possible outcomes for agents in pair-wise interaction: favorable (+), neutral (0), and unfavorable (–). For three or more agents, there are symbiotic (++), competitive (––), neutral (00), parasitic (+–), commensal (+0), and amensal (–0) interactions. The presence of a full set of ecological interactions is central to the quality of wickedness because the organization of such systems is constantly in upheaval—it never settles into something that is easy to understand, adapt, or design.

WICKED SYSTEMS AS POORLY DECOMPOSABLE SYSTEMS

It is not hard to imagine why systems exhibiting this range of interactions would be challenging to understand formally, but can we understand formally why they cannot be understood formally? This could offer a more detailed map of the exact methodological problems with these systems. Such an understanding might inform us about what approaches are likely to work in particular contexts and could serve to make it easier for formalists and nonformalists to collaborate. One way of understanding why wicked systems are so recalcitrant that is both formal and intuitive and accessible to formalists and nonformalists alike is in terms of not exhibiting near-decomposability (Simon 1996). Near-decomposability turns out to be necessary for almost all formal theorizing.

Simon (1996) introduced the concept of *near-decomposability* to explain in a clear and systematic way what conditions need to be fulfilled for a system to be studied in a formal and controlled manner. In order to study a system in isolation, its dynamics cannot be disturbed significantly by outside influences. We should be able to identify an internal environment where the dynamics under scrutiny take place and an external environment that is assumed to be stable, or variable only in highly regular ways. The boundary between the internal and external environment is referred to as the *interface* (Figure 13.5). What we study with a model is the internal environment.

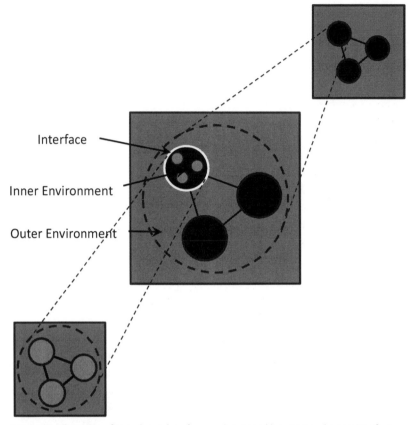

Figure 13.5. Illustration of Simon's ontology for near-decomposable systems, where an interface subsumes all components and dynamics that happen within objects (in their inner environment). The object can then be treated as a cohesive whole, interacting via interfaces with other objects against the constant background of an outer environment.

Hierarchical system organization is also important: our internal environment constitutes the external environment of the objects that populate it. Objects are dealt with only in the form of interfaces, such as via their interactions with other objects in the studied system. The beauty of all of this is that it makes the world manageable: we declare our system autonomous from external disturbance and hide any complexity or complicatedness at lower levels of the hierarchical organization.

We study this internal environment on a timescale that is long enough for our objects' interfaces to be meaningful and for important dynamics to have time to occur and short enough for our assumptions about the interfaces to remain valid.[4] The greater the separation of scales between the internal and the external environment, the greater the difference in size and speed of the dynamics on these two levels, giving more interesting things time to happen. For example, models of particle physics can be formulated in this way because those systems exhibit clear scale separation. Engineered systems are designed to fit this description (Simon 1996; see Figure 13.6).

In many important cases, we can make assumptions of near-decomposability for wicked systems and thereby bring powerful scientific approaches to bear on them. Certain subsystems, such as crowd behavior, protein folding, or the ceteris paribus fate of a new trait in a population, can fit this description and become amenable to complexity science modeling. The dynamics of cars and people play out over much shorter timescales than urban systems, roads, and traffic regulation change. Such phenomena are often ephemeral, which bounds the problem even further. For example, at night the traffic jam dissipates and leaves no traces that affect tomorrow's traffic. Similar features obtain for abstractly conceived phenomena that depend on persistent features, such as network dynamics, geography, basic resource constraints, or strategic dilemmas.

What about evolutionary and ecological phenomena more generally? For example, what about sociotechnical transitions, evolutionary radiation events, or other wicked problems? Wicked systems are open systems in which many and diverse types of processes coexist, coevolve, and have an impact on each other across overlapping timescales and levels of organization. They involve discontinuous, qualitative change as well as cascade effects (e.g., Lane 2011), whereby change strongly and rapidly feeds back on the conditions for further change. Such systems are difficult to contain in a suitable timescale for transitions to be studied against the background of an unchanging

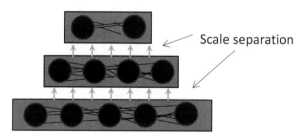

Figure 13.6. An illustration of Simon's ontology for near-decomposability, nested hierarchically into levels with clear scale separations. This facilitates focusing on one level of organization at a time, subjecting microlevels and macrolevels to strong simplifying assumptions.

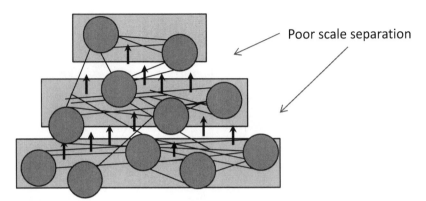

Figure 13.7. In wicked systems, levels of organization break down, with interactions going across scales, not least due to a continual formation, change, and dissolution of objects (see section 7).

external environment. The fundamental problem in this context for complexity science, or any approach that relies on these ontological assumptions, is that there is no way of partitioning wicked systems into distinct and persistent levels of organization (Figure 13.7).

WE BEGAN by reviewing recent developments in evolutionary biology, archaeology, and social science, pointing out interconnections and similarities in terms of their problems, debates, new ideas, and new data. A historical dearth of data—both in terms of volume and types of data—made it easier to base elegant theoretical models upon strong assumptions about complexities that were unknown. Considerations of methodological expediency were thereby not resisted by empirical knowledge about the causal and material bases. Strong general bodies of theory, emphasizing the broad strokes, emerged in this environment, as did disciplines dedicated to their study. But merit is not the only thing that conserves these theories and disciplines and the boundaries that we have erected between them: they have been subject to generative entrenchment and hardening. This is what we refer to as *old theory*. *New data* is much more multifaceted, and voluminous bodies of empirical knowledge that call into question many of those generatively entrenched basic simplifying assumptions—about human behavior, culture, development, and so on—are now causing friction against the hardened pillars of old theory.

The cautions that we would like to make are, first, that the conclusion that old theory can accommodate old data is a convenient one and should be scrutinized critically: it may be so, but there are structural reasons why we may shun the effort of going for more fundamental reevaluations. Another reason is that generative entrenchment partly consists in, precisely, a priming of our imagination as we innovatively search for new solutions: it has accustomed us to think about dynamics and causation, what legitimate problems look like, what tools are deemed secure, and so on. This type of constraining facilitation of innovation has many names across these fields, such as *facilitated variation* in evo-devo (Gerhart and Kirschner 2007), *design spaces* in innovation theory (Stankiewicz 2000), and *developmental canalization* in behavioral biology. The second caution is that we live in an "innovation society," in times that value novelty and change highly (Lane 2016). There is the opposite risk of overreaction and of equating "old" with "bad." What is called for, we believe, is a delicate and thoughtful reevaluation of the old in the light of the new.

The emerging empirical pictures in these fields are frequently referred to as a *complex*. To get a better idea about these issues we factored *complexity* into two component qualities: complexity and complicatedness—corresponding to dynamic and structural complexity, respectively. This made it possible to map four different approaches (narrative theory, analytical models, systems theory, and complexity science) onto distinct system classes that are simple, complex but not complicated, complicated but not complex, or both complex and complicated (i.e., wicked). This mapping displayed interactions between methods, systems, and problems, while leaving a conspicuous lacuna near the "wicked corner."

We argued that the described friction may, moreover, be understood as a much broader encounter with a distinct system quality: *wickedness*. The term originated, indeed, precisely in the interaction between strictly formulated systems theory and the societal problems to which they were applied (Rittel and Webber 1973); the friction in this case did not have to wait for advanced empirical instruments to emerge.

Next, we turned to answering how wicked systems work and where they come from, aiming to differentiate them from, yet keeping them in relation to, complex and complicated systems. Wicked systems were deemed best described as arenas for complex interaction between complicated systems. They remain "enclosed" and in recurrent interaction, but the absence of effective top-down alignment means that any structure that emerges will constantly be in flux, with constant qualitative change flowing up and down a poorly scale-separated hierarchy. As a consequence, they conform poorly to the key requirements needed for formal modeling to be effective, such as near-decomposability (Simon 1996).

Even though wicked systems are distinct from complex and complicated systems, they are typically addressed as if they belonged to one of these classes. Grappling with wicked systems *as distinct types of systems* is a promising direction for future research. Any resulting understanding could contribute to a more systematic metatheoretical discussion, foster new transdisciplinary connections, and improve our capability to use new empirical assets and tackle pressing problems.

NOTES

1. In the SFI "mission and vision" statement (http://santafe.edu/about /mission-and-vision/), the commitment to quantitative approaches is explicit: "SFI combines expertise in quantitative theory and model building with a community and infrastructure able to support cutting-edge, distributed and team-based science."

2. See also Byrne's (2005) concept of "simple complexity" and Morin's (2007) concept of "restricted complexity," which align with the complexity science mainstream; Byrne and Callaghan (2014) discuss the dominance of this mainstream.

3. At least, not recently, and with the benefit of more developed sciences of complexity at hand. See Wimsatt (1975) for an early and interesting analysis that goes a long way in this direction (also reprinted in Wimsatt [2007]).

4. For instance, a human can make decisions (a typical interface feature) over a timescale of minutes but hardly on a timescale of milliseconds.

REFERENCES

Aitken, M. J. 2014. *Science-Based Dating in Archaeology.* 3rd ed. New York: Routledge.

Andersson, C., A. Törnberg, and P. Törnberg. 2014a. "An Evolutionary Developmental Approach to Cultural Evolution." *Current Anthropology* 55, no. 2 (April): 154–74.

Andersson, C., A. Törnberg, and P. Törnberg. 2014b. "Societal Systems: Complex or Worse?" *Futures* 63: 145–17.

Arthur, W. 2011. *Evolution: A Developmental Approach.* Chicester, UK: Wiley-Blackwell.

Ball, P. 2012. *Why Society Is a Complex Matter: Meeting Twenty-First Century Challenges with a New Kind of Science.* Berlin: Springer Verlag.

Barker, G. 2006. *The Agricultural Revolution in Prehistory: Why Did Foragers Become Farmers?* Oxford: Oxford University Press.

Beddoe, R., R. Costanza, J. Farley, E. Garza, J. Kent, I. Kubiszewski, and J. Woodward et al. 2009. "Overcoming Systemic Roadblocks to Sustainability: The Evolutionary Redesign of Worldviews, Institutions, and Technologies." *Proceedings of the National Academy of Sciences of the United States of America* 106 (8): 2483–89.

Belfer-Cohen, A., and N. Goring-Morris. 2011. "Becoming Farmers: The Inside Story." *Current Anthropology* 52 (S4): S209–20.

Belfer-Cohen, A., and E. Hovers. 2010. "Modernity, Enhanced Working Memory, and the Middle to Upper Paleolithic Record in the Levant." *Current Anthropology* 51 (Supp. 1): 167–75.

Boyd, R., and P. J. Richerson. 1985. *Culture and the Evolutionary Process.* Chicago: University of Chicago Press.

Brown, K. A., and T. A. Brown. 2013. "Biomolecular Archaeology." *Annual Review of Anthropology* 42:159–74.

Bruno, J. F., J. J. Stachowicz, and M. D. Bertness. 2003. "Inclusion of Facilitation into Ecological Theory." *Trends in Ecology and Evolution* 18, no. 3 (March): 119–25.

Byrne, D. 2005. "Complexity, Configurations, and Cases." *Theory, Culture, and Society* 22, no. 5 (October): 95–111.

Byrne, D., and G. Callaghan. 2014. *Complexity Theory and the Social Sciences: The State of the Art.* Abingdon, UK: Routledge.

Castellani, B., and F. W. Hafferty. 2009. *Sociology and Complexity Science: A New Field of Inquiry.* Berlin: Springer Verlag.

Cavalli-Sforza, L. L., and M. W. Feldman. 1981. *Cultural Transmission and Evolution.* Palo Alto, Calif.: Stanford University Press.

Çilingiroglu, C. 2005. "The Concept of Neolithic Package: Considering Its Meaning and Applicability." *Documenta Praehistorica* 2:1–13.

Erdi, P. 2008. *Complexity Explained.* Berlin: Springer Verlag.

Erwin, D. H. 2008. "Macroevolution of Ecosystem Engineering, Niche Construction and Diversity." *Trends in Ecology and Evolution* 23, no. 6 (June): 304–10.

Erwin, D. H. 2015. "A Public Goods Approach to Major Evolutionary Innovations." *Geobiology* 13 (4): 308–15.

Erwin, D. H., and D. C. Krakauer. 2004. "Insights into Innovation." *Science* 304:1117–19.

Erwin, D. H., and J. W. Valentine. 2013. *The Cambrian Explosion.* Greenwood Village, Col.: Roberts.

Foley, R., and C. Gamble. 2009. "The Ecology of Social Transitions in Human Evolution." *Philosophical Transactions of the Royal Society B: Biological Sciences* 364, no. 1533 (November): 3267–79.

Galison, P. 1997. *Image and Logic: A Material Culture of Microphysics.* Chicago: University of Chicago Press.

Geels, F. W. 2002. "Technological Transitions as Evolutionary Reconfiguration Processes: A Multi-level Perspective and a Case-Study." *Research Policy* 31, no. 8–9 (December): 1257–74.

Gerhart, J., and M. Kirschner. 2007. "The Theory of Facilitated Variation." *Proceedings of the National Academy of Sciences of the United States of America* 104 (May): 8582–89.

Habgood, P. J., and N. R. Franklin. 2008. "The Revolution That Didn't Arrive: A Review of Pleistocene Sahul." *Journal of Human Evolution* 55, no. 2 (August): 187–222.

Hauser, M. W. 2012. "Messy Data, Ordered Questions." *American Anthropologist* 114, no. 2 (June): 184–95.

Helbing, D. 2013. "Globally Networked Risks and How to Respond." *Nature* 497, no. 7447 (May): 51–59.

Holt, K., and K. Bennett. 2014. "Principles and Methods for Automated Palynology." *New Phytologist* 1996:735–42.

Hovers, E., and A. Belfer-Cohen. 2013. "On Variability and Complexity." *Current Anthropology* 54, no. S8 (December): S337–57.

Hovers, E., and S. L. Kuhn. 2006. *Transitions before the Transition.* Berlin: Springer Verlag.

Jones, C., J. Lawton, and M. Shachak. 1996. "Organisms as Ecosystem Engineers." In Vol. 59, *Ecosystem Management,* edited by F. B. Samson and F. L. Knopf, chap. 14, 373–86. New York: Springer.

Klein, R. G., and B. Edgar. 2002. *The Dawn of Human Culture.* New York: Wiley and Sons.

Kristiansen, K. 2014. "Towards a New Paradigm? The Third Science Revolution and Its Possible Consequences in Archaeology." *Current Swedish Archaeology* 22:11–13.

Laland, K. N. 2014. "On Evolutionary Causes and Evolutionary Processes." *Behavioural Processes* 117:97–104.

Lane, D. A. 2011. "Complexity and Innovation Dynamics." In *Handbook on the Economic Complexity of Technological Change,* edited by C. Antonelli, chap. 2, 63–80. Cheltenham, UK: Edward Elgar.

Lane, D. A. 2016. "Innovation Cascades: Artefacts, Organization, and Attributions." *Philosophical Transactions of the Royal Society B: Biological Sciences* 371 (1690): 20150194.

Lane, D., and S. van der Leeuw. 2011. "Innovation, Sustainability, and ICT." *Procedia Computer Science* 7:83–87.

Laubichler, M. D., and J. Maienschein. 2007. *From Embryology to Evo-Devo: A History of Developmental Evolution.* Cambridge, Mass.: MIT Press.

Laubichler, M. D., and J. Maienschein. 2013. "Developmental Evolution." In *The Cambridge Encyclopedia of Darwin and Evolutionary Thought,* edited by M. Ruse, chap. 46, 375–82. Cambridge: Cambridge University Press.

Laubichler, M. D., and J. Renn. 2015. "Extended Evolution: A Conceptual Framework for Integrating Regulatory Networks and Niche Construction." *Journal of Experimental Zoology Part B: Molecular and Developmental Evolution* 324B:565–77.

Leach, M., I. Scoones, and A. Stirling. 2007. *Pathways to Sustainability: An Overview of the STEPS Centre Approach.* STEPS Approach Paper, Brighton: STEPS Centre.

Leach, M., I. Scoones, and A. Stirling. 2010. "Governing Epidemics in an Age of Complexity: Narratives, Politics and Pathways to Sustainability." *Global Environmental Change* 20, no. 3 (August): 369–77.

Lerner, H., X. Du, A. Costopoulos, and M. Ostoja-Starzewski. 2007. "Lithic Raw Material Physical Properties and Use-Wear Accrual." *Journal of Archaeological Science* 34, no. 5 (May): 711–22.

Loorbach, D. 2010. "Transition Management for Sustainable Development: A Prescriptive, Complexity-Based Governance Framework." *Governance* 23, no. 1 (January): 161–83.

Loreau, M. 2010. *From Populations to Ecosystems: Theoretical Foundations for a New Ecological Synthesis.* Princeton, N.J.: Princeton University Press.

Maher, L. A., T. Richter, and J. T. Stock. 2012. "The Pre-Natufian Epipaleolithic: Long-Term Behavioral Trends in the Levant." *Evolutionary Anthropology* 21, no. 2 (March): 69–81.

McBrearty, S. 2007. "Down with the Revolution." In *The Human Revolution Revisited,* edited by P. A. Mellars, K. Boyle, O. Bar-Yosef, and C. B. Stringer, 133–51. Cambridge: McDonald Institute for Archaeological Research.

McBrearty, S., and A. S. Brooks. 2000. "The Revolution That Wasn't: A New Interpretation of the Origin of Modern Human Behavior." *Journal of Human Evolution* 39:453–563.

Mesoudi, A. 2011. *Cultural Evolution*. Chicago: Chicago University Press.

Michod, R. E., and A. M. Nedelcu. 2003. "On the Reorganization of Fitness during Evolutionary Transitions in Individuality." *Integrative and Comparative Biology* 43 (1): 64–73.

Morin, E. 2007. "Restricted Complexity, General Complexity." In *Worldviews, Science and Us: Philosophy and Complexity*, edited by C. Gershenson, D. Aerts, and B. Edmonds, 5–29. Singapore: World Scientific.

O'Brien, S. J., M. Menotti-Raymond, W. J. Murphy, W. G. Nash, J. Wienberg, R. Stanyon, N. G. Copeland, N. A. Jenkins, J. E. Womack, and J. A. Marshall Graves. 1999. "The Promise of Comparative Genomics in Mammals." *Science* 286, no. 5439 (October): 458–81.

Odling-Smee, F. J., D. H. Erwin, E. P. Palovacs, M. W. Feldman, and K. N. Laland. 2013. "Niche Construction Theory: A Practical Guide for Ecologists." *Quarterly Review of Biology* 88 (1): 3–28.

Odling-Smee, F. J., K. N. Laland, and M. W. Feldman. 2003. *Niche Construction: The Neglected Process in Evolution*. Monographs in Population Biology 37. Princeton, N.J.: Princeton University Press.

Oyama, S., P. E. Griffiths, and R. D. Gray. 2001. *Cycles of Contingency*. Cambridge, Mass.: MIT Press.

Pelletier, F., D. Garant, and A. P. Hendry. 2009. "Eco-evolutionary Dynamics." *Philosophical Transactions of the Royal Society B: Biological Sciences* 364, no. 1523 (June): 1483–89.

Pigliucci, M., and G. B. Müller. 2010. *Evolution, the Extended Synthesis*. Cambridge, Mass.: MIT Press.

Ragin, C. C. 2009. "Reflections on Casing and Case Oriented Research." In *The Sage Handbook of Case Based Methods*, edited by D. Byrne and C. C. Ragin, 522–34. London: Sage.

Rittel, H., and M. Webber. 1973. "Dilemmas in a General Theory of Planning." *Policy Sciences* 4:155–69.

Sandén, B. A., and K. M. Hillman. 2011. "A Framework for Analysis of Multimode Interaction among Technologies with Examples from the History of Alternative Transport Fuels in Sweden." *Research Policy* 40, no. 3 (April): 403–14.

Sawyer, R. K. 2005. *Social Emergence: Societies as Complex Systems*. Cambridge: Cambridge University Press.

Scoones, I., M. Leach, A. Smith, and S. Stagl. 2007. *Dynamic Systems and the Challenge of Sustainability*. STEPS Working Paper 1, Brighton: STEPS Centre.

Shott, M., and G. Tostevin. 2015. "Diversity Under the Bipolar Umbrella." *Lithic Technology* 40 (4): 377–84.

Simon, H. A. 1996. *The Sciences of the Artificial.* 3rd ed. Cambridge, Mass.: MIT Press.

Stankiewicz, R. 2000. "The Concept of 'Design Space.'" In *Technological Innovation as an Evolutionary Process,* edited by J. Ziman, 234–47. Cambridge: Cambridge University Press.

Steffen, W., W. Broadgate, L. Deutsch, O. Gaffney, and C. Ludwig. (2015). "The Trajectory of the Anthropocene: The Great Acceleration." *Anthropocene Review* 2 (1): 81–98.

Stiner, M. C., H. Buitenhuis, G. Duru, S. L. Kuhn, S. M. Mentzer, N. D. Munro, N. Pöllath, J. Quade, G. Tsartsidou, and M. Ozbaaran. 2014. "A Forager-Herder Trade-Off, from Broad-Spectrum Hunting to Sheep Management at Akl Höyük, Turkey." *Proceedings of the National Academy of Sciences of the United States of America* 111, no. 23 (June): 8404–9.

Stiner, M. C., and S. L. Kuhn. 2016. "Are We Missing the 'Sweet Spot' between Optimality Theory and Niche Construction Theory in Archaeology?" *Journal of Anthropological Archaeology* 44 (December): 177–84.

Tostevin, G. B. 2012. *Seeing Lithics.* Oxford: Oxbow Books.

Vasileiadou, E., and K. Safarzyska. 2010. "Transitions: Taking Complexity Seriously." *Futures* 42, no. 10 (December): 1176–86.

Wagner, G., C. Chiu, and M. Laubichler. 2000. "Developmental Evolution as a Mechanistic Science: The Inference from Developmental Mechanisms to Evolutionary Processes." *American Zoologist* 831:819–31.

West Churchman, C. 1967. "Guest Editorial." *Management Science* 14 (4): 141–42.

Wimsatt, W. C. 1975. "Complexity and Organization." In *Topics in the Philosophy of Biology.* Dordrecht, The Netherlands: Reidel, 174–93.

Wimsatt, W. C. 1986. "Developmental Constraints, Generative Entrenchment, and the Innate-Acquired Distinction." In *Integrating Scientific Disciplines,* edited by W. Bechtel, 185–208. Berlin: Springer.

Wimsatt, W. C. 2001. "Generative Entrenchment and the Developmental Systems Approach to Evolutionary Process." In *Cycles of Contingency,* edited by S. Oyama and P. E. Griffiths, chap. 17, 219–38. Cambridge, Mass.: MIT Press.

Wimsatt, W. C. 2007. *Re-engineering Philosophy for Limited Beings: Piece-Wise Approximations to Reality.* Cambridge, Mass.: Harvard University Press.

Wise, R. M., I. Fazey, M. Stafford Smith, S. E. Park, H. C. Eakin, E. R. M. Archer Van Garderen, and B. Campbell. 2014. "Reconceptualising Adaptation to Climate Change as Part of Pathways of Change and Response." *Global Environmental Change* 28:325–36.

Zalasiewicz, J., M. Williams, A. Haywood, and M. Ellis. 2011. "The Anthropocene: A New Epoch of Geological Time?" *Philosophical Transactions of the Royal Society A: Mathematical, Physical, and Engineering Sciences* 369 (1938): 835–41.

Zeder, M. 2014. "Alternative to Faith-Based Science." *Proceedings of the National Academy of Sciences of the United States of America* 111 (28): 2827.

Zeder, M., and B. Smith. 2009. "A Conversation on Agricultural Origins." *Current Anthropology* 50 (5): 681–91.

Marshall Abrams is associate professor in the Department of Philosophy at the University of Alabama at Birmingham. He has a PhD in philosophy from the University of Chicago and was a National Science Foundation–sponsored postdoctoral fellow at Duke University's Center for Philosophy of Biology. His research focuses on teasing out what is implied about natural processes by scientists' uses of models, on developing and applying new ideas about probability to biological and social sciences, and on using computer models and other methods to explore new roles for humanistic conceptions of culture in scientific research. He has written on mental representation and biological function and has collaborated on scientific research concerning natural selection on genes influencing obesity and diabetes.

Claes Andersson is associate professor and senior researcher in complex systems in the Department of Space, Earth, and Environment, Chalmers University of Technology, and external fellow at the European Center for Living Technology in Venice. He has a PhD in complex systems and has held research positions at Los Alamos National Laboratory and the University of Modena and Reggio Emilia. His research is focused on the long-term and large-scale evolution of societal systems, focusing on the deep human past and urban and regional dynamics, as well as on fundamental issues in complex systems.

Mark A. Bedau is professor of philosophy at Reed College and received his PhD in philosophy from the University of California, Berkeley. His work focuses on the philosophical implications of the emergence and open-ended evolution of life, mind, and technology, their social and ethical implications, and their study by high-throughput experimental methods as well as models and simulations. He has helped pioneer methods for measuring the evolutionary activity of evolving systems, for designing and creating complex biochemical systems with desired emergent properties, and for studying the evolution of technology by mining patent data. He has been editor in chief of the journal *Artificial Life* for more than a decade. His edited volumes include *Emergence* (with Paul Humphreys), *The Nature of Life* (with Carol Cleland), *Protocells: Bridging Nonliving and Living Matter* (with S. Rasmussen, L. Chen, D. Deamer, D. C. Krakauer, N. H. Packard, and P. F. Stadler), *The Ethics of Protocells: Moral and Social Implications of Creating Life in the Laboratory* (with Emily Parke), and *Living Technology: Five Questions* (with Pelle Guldborg Hansen, Emily Parke, and Steen Rasmussen).

James A. Evans is professor of sociology and member of the Committee on the Conceptual and Historical Studies of Science program at the University of Chicago. He directs the Knowledge Lab, a center focused on using large-scale publication data and machine learning to perform science studies research. He is current and founding faculty director of the Computational Social Science program at Chicago. His research focuses on the collective system of thinking and knowing, ranging from the distribution of attention and intuition, the origin of ideas, and shared habits of reasoning to processes of agreement (and dispute), the accumulation of certainty (and doubt), and the texture—novelty, ambiguity, topology—of collective understanding. He is especially interested in innovation—how new ideas and technologies emerge and evolve—and the role that social and technical institutions, including the Internet, artificial intelligence, markets, and collaborations, play in collective cognition and discovery. His work has been published in *Science, Social Studies of Science, Proceedings of the National Academy of Sciences, American Sociological Review, American Journal of Sociology,* and many other outlets.

Jacob G. Foster is assistant professor of sociology at the University of California, Los Angeles. He is a computational sociologist interested in the evolutionary dynamics of ideas, the social production of collective intelligence, and the mutual constitution of culture and cognition. His empirical work blends computational methods with qualitative insights from science studies to probe the strategies, dispositions, and social processes that shape the production and persistence of scientific ideas. He uses machine learning to mine the cultural meanings buried in text and computational methods from macroevolution to understand the dynamics of cultural populations. He also develops formal models of the structure and dynamics of ideas and institutions, with an emerging theoretical focus on the rich nexus of cognition, culture, and computation. After studying mathematical physics at Oxford as a Rhodes Scholar, he received his PhD in physics from the University of Calgary and was a postdoctoral scholar in the Department of Sociology at the University of Chicago. He is cofounder of the Metaknowledge Research Network, established with a generous grant from the John Templeton Foundation, and codirector of the Diverse Intelligences Summer Institute, established with a generous grant from the Templeton World Charity Foundation. His work has appeared in *American Sociological Review, Science, Proceedings of the National Academy of Sciences, Poetics, Sociological Science,* and *Social Networks.*

Michel Janssen holds a PhD in history and philosophy of science from the University of Pittsburgh and is professor of history of science at the University of Minnesota. His work mainly focuses on the genesis of relativity and quantum theory in the early decades of the twentieth century. Before coming to Minnesota in 2000, he was an editor at the Einstein Papers Project. More recently, he coedited *The Cambridge Companion to Einstein.*

Sabina Leonelli is a Turing Fellow and professor of philosophy and history of science at the University of Exeter, UK, where she codirects the Exeter Centre for the Study of the Life Sciences and leads the data studies research strand. Her research focuses on the philosophy of data-intensive science, especially the methods and assumptions involved in the production, dissemination, and use of big data for discovery; the ways in which the open science movement is redefining what counts as research and knowledge across different research environments; and the epistemic status of experimental organisms as models and data sources, particularly in plant science. She has published widely within the philosophy of science, as well as biology and science and technology studies, and is the author of *Data-Centric Biology: A Philosophical Study* (Lakatos Award 2018).

Alan C. Love holds a PhD in history and philosophy of science from the University of Pittsburgh and is professor of philosophy at the University of Minnesota. His research focuses on a variety of philosophical issues in biology, such as conceptual change, explanatory pluralism, the structure of evolutionary theory, reductionism, the nature of historical science, and interdisciplinary epistemology. Much of his work has concentrated on the concepts of innovation and novelty in evolutionary developmental biology (evo-devo), which has elucidated how the structure of problems serves to organize explanatory endeavors across disciplines. In particular, this illuminates how reductionist research programs in molecular biology articulate with inquiry at higher levels of organization in the life sciences to better understand biological complexity. More recently, he has explored similar issues in developmental biology to understand the intersection of genetics and physics in explaining complex biological phenomena.

Massimo Maiocchi is a research fellow in the Humanities Department of Ca' Foscari University of Venice. He is an expert in history of the Ancient Near East and Assyriology (Sumerian, Akkadian, Eblaite), with special regard to cuneiform texts from the fourth and third millennia B.C.E. He was a postdoctoral researcher and lecturer at the Department of Near Eastern Languages and Civilizations of the University of Chicago, where he expanded his interests to include grammatology. He published two monographs concerned with the edition of Old Akkadian cuneiform tablets from southern Mesopotamia (Cornell University Studies in Assyriology and Sumerology 13 and 19). His research focuses on how writing affected social and urban development in early Mesopotamia, approaching ancient sources through the prism of traditional philology, textual criticism, and Digital Humanities. He is associate editor of the Ebla Digital Archives project (http://ebda.cnr.it, under the direction of L. Milano), providing the online edition of the entire cuneiform corpus unearthed at Ebla (modern Tell Mardikh Syria). He participated in archaeological expeditions in Syria (Tell Mozan) and survey projects in Iraq.

Joseph D. Martin is assistant professor of history at Durham University. He holds a PhD in the history of science, technology, and medicine from the University of Minnesota; previously taught at Colby College, Michigan State University, and the University of Cambridge; and was a research fellow at the Consortium for History of Science Technology and Medicine in Philadelphia and the Centre for History and Philosophy of Science at the University of Leeds. His research interests span the history and philosophy of modern science and technology, with an emphasis on the United States during the Cold War. His book *Solid State Insurrection* examines the solid state and condensed matter physics community in the United States and exposes the importance of its distinctive disciplinary ideals and rich connections with technology for maintaining the prestige physics enjoyed throughout the Cold War era.

Salikoko S. Mufwene is the Frank J. McLoraine Distinguished Service Professor of Linguistics and the College, professor on the Committee on Evolutionary Biology, and professor on the Committee on the Conceptual and Historical Studies of Science at the University of Chicago. His current research is on language evolution from an ecological perspective. His books include *The Ecology of Language Evolution*; *Créoles, écologie sociale, evolution linguistique: Cours donnés au Collège de France durant l'automne 2003*; and *Language Evolution: Contact, Competition, and Change*. He has edited or coedited several books, including *Complexity in Language: Developmental and Evolutionary Perspectives* (with Christophe Coupé and François Pellegrino); *Iberian Imperialism and Language Evolution in Latin America; Colonisation, globalisation, vitalité du français* (coedited with Cécile B. Vigouroux); and *Globalization and Language Vitality: Perspectives from Africa* (with Cécile B. Vigouroux). He is the founding editor of *Cambridge Approaches to Language Contact* and was a fellow at the Institute for Advanced Study in Lyon.

Nancy J. Nersessian is Regents' Professor (Emerita), Georgia Institute of Technology, and research associate, Department of Psychology, Harvard University. Her research in philosophy of science, history of science, and cognitive science focuses on the creative research practices of scientists and engineers, especially how their modeling practices lead to fundamentally new ways of understanding the world. Her research on the bioengineering sciences seeks to understand the dynamic interplay of cognition and culture in pioneering research laboratories and how these laboratories foster and sustain creative and innovative practices. This research has been funded by the National Science Foundation, the National Endowment for the Humanities, and the Radcliffe Institute for Advanced Study. She is a fellow of the American Association for the Advancement of Science and the Cognitive Science Society and a Foreign Member of the Royal Netherlands Academy of Arts and Sciences. Her numerous publications include *Creating Scientific Concepts* (Patrick Suppes Prize in Philosophy of Science, 2011) and *Science as Psychology: Sense-Making and Identity in Science Practice* (with L. Osbeck, K. Malone, and W. Newstetter; William James Book Prize, 2012).

Paul E. Smaldino is assistant professor of cognitive and information sciences and a core faculty member of the quantitative and systems biology graduate group at the University of California, Merced. His research focuses on using mathematical and computational models to study social and evolutionary processes, including those related to cooperation, communication, and the behavior of scientific communities. His degrees and appointments have included residencies in departments of physics, psychology, medicine, anthropology, political science, and computer science.

Anton Törnberg is senior lecturer in sociology at Gothenburg University. His research interests lie on the boundary between sociology and complexity science, with a focus on social movements and collective action. His recent articles include "Combining Transition Studies and Social Movement Theory: Towards a New Research Agenda" in *Theory and Society*.

Petter Törnberg is assistant professor in political sociology at the University of Amsterdam. He has a PhD in complex systems and has held several international research positions as well as positions outside academia as a software developer and data analysis consultant. His research is highly interdisciplinary and tends to lie on the boundary between social science and complex systems, combining computational methods with qualitative approaches to contribute to sociological theory. This mixed-methods approach includes natural language processing, social network analysis, and computational simulations along with a more philosophical direction, including conceptual work on the epistemology of digital data and the complexity of social systems.

Gilbert B. Tostevin received his PhD in Paleolithic archaeology from Harvard University. He served as a visiting faculty member at Williams College (Massachusetts) from 1999 to 2001. Since 2001, he has been a faculty member in the Department of Anthropology at the University of Minnesota. He has conducted archaeological research at Middle and Upper Paleolithic sites in the Levant and Central Europe for more than twenty years. He is leading a new multidisciplinary excavation team investigating Neanderthal adaptations in Southeastern Europe at the rock shelter of Crvena Stijena "Red Rock" in Montenegro. His specialization is lithic analysis, focusing on its articulation with cultural transmission theory and behavioral archaeology. His more significant publications in this area include *Seeing Lithics: A Middle-Range Theory for Testing for Cultural Transmission in the Pleistocene*.

William C. Wimsatt studied engineering physics and philosophy at Cornell University and received his PhD in philosophy at the University of Pittsburgh. He is Peter B. Ritzma Professor in Philosophy, Evolutionary Biology at the University of Chicago Emeritus and was the Winton Chair of Liberal Arts at the University of Minnesota. He has written on functional organization, explanation, evolution, reductionism and reductionistic research strategies across the sciences, levels of organization and mechanistic explanation, units of selection, heuristics, emergence, mathematical modeling, robustness, satisficing, generative entrenchment, error-tolerant systems in biology and science, and methods and problems in studying complex systems. *Re-engineering Philosophy for Limited Beings* integrates many of these themes, some of which have been continued in the edited volumes *Characterizing the Robustness of Science* (with L. Soler, E. Trisio, and T. Nickles) and *Developing Scaffolds in Evolution, Culture, and Cognition* (with L. Caporael and J. Griesemer). He currently works primarily on cultural evolution and the role of generative entrenchment in evolutionary processes in biology, technology, and culture.

INDEX